Two Manuals Combined into One Book

Combat Risk Management

And

Joint Military Operations in Urban Terrain

Department of Defense

Pentagon
Publishing

FM 100-14

Field Manual
No. 100- 14

Headquarters
Department of the Army
Washington, DC, 23 April 1998

Risk Management

Contents

Preface

FM 100-14 applies across the wide range of Army operations. It explains the principles, procedures, and responsibilities to successfully apply the *risk management process* to conserve combat power and resources. The manual applies to both Army and civilian personnel during all Army activities, including joint, multinational, and interagency environments.

The manual is intended to help commanders,[1] their staffs, leaders,[2] and managers develop a framework to make risk management a routine part of planning, preparing, and executing operational missions and everyday tasks. This framework will allow soldiers to operate with maximum initiative, flexibility, and adaptability. Although the manual's prime focus is the operational Army, the principles of risk management apply to all Army activities.

Army operations—especially combat operations—are demanding and complex. They are inherently dangerous, including tough, realistic training. Managing risks related to such operations requires educated judgment and professional competence. The risk management process allows individuals to make informed, conscious decisions to accept risks at acceptable levels.

This manual is not a substitute for thought. Simply reading it will not make one adept in building protection around a mission.[3] Soldiers should compare the doctrine herein against their own experience and think about why, when, and how it applies to their situation and area of responsibility. If the doctrine herein is to be useful, it must become second nature.

The proponent of this manual is HQ TRADOC. Send comments and recommendations on DA Form 2028 directly to Commander, US Army Training and Doctrine Command, ATTN: ATBO-SO, Fort Monroe, VA 23651-5000.

Unless this publication states otherwise, masculine nouns and pronouns do not refer exclusively to men.

---

1 The term *commander* as used herein refers to personnel in a command position.

2 The term *leader* as used herein refers to commanders, personnel in the chain of command (team, squad, section, platoon leader), and staff members having personnel supervisory responsibility.

3 The term *mission* as used herein includes mission, operation, or task.

Introduction

Risk management is not an add-on feature to the decision-making process but rather a fully integrated element of planning and executing operations... Risk management helps us preserve combat power and retain the flexibility for bold and decisive action. Proper risk management is a combat multiplier that we can ill afford to squander.

General Dennis J. Reimer
Chief of Staff, Army
27 July 1995

The Army's fundamental purpose is to fight and win the nation's wars. For this purpose, the country gives the Army critical resources, including those most valuable—its sons and daughters. The Army uses its resources to generate overwhelming combat power to fight and win quickly, decisively, and with minimal losses. The Army's inherent responsibility to the nation is to protect and preserve its resources—a responsibility that resides at all levels. Risk management is an effective process for preserving resources. It is not an event. It is both an art and a science. Soldiers use it to identify tactical and accident risks, which they reduce by avoiding, controlling, or eliminating hazards.

The Army introduced the risk management process into training, the operational environments, and materiel acquisition in the late 1980s. Risk management was originally perceived as solely a safety officer function. However, by the early 1990s, the Army established a goal to integrate risk management into all Army processes and activities and into every individual's behavior, both on and off duty. Since the process was introduced, the personal involvement of commanders in preventing accidents—and their aggressive use of the process—have become driving factors in the steady downward trend in Army accidental losses.

Leaders must understand the importance of the process in conserving combat power and resources. Risk management, like reconnaissance and security, is an ongoing process that continues from mission to mission. Within the mission, leaders must know when the process begins and who has responsibility. It must be integral to the military decision. The process is an important means to enhance situational awareness.

Risk decisions are commanders' business. Such decisions are normally based on the next higher commander's guidance on how much risk he is willing to accept and delegate for the mission. Risk decisions should be made at the lowest possible level, except in extreme circumstances. Training operations, including those at combat training centers (CTCs), may be of such intensity that risk decision are retained at a higher level.

Both leaders and staffs manage risk. Staff members continuously look for hazards associated with their areas of expertise. They then recommend controls to reduce risks. Hazards and the resulting risks may vary as circumstances change and experience is gained. Leaders and individual soldiers become the assessors for ever-changing hazards such as those associated with environment (weather; visibility; contaminated air, water, and soil), equipment readiness, individual and unit experience, and fatigue. Leaders should advise the chain of command on risks and risk reduction measures.

Index

I

K

L

M

N

O

P

R

S

during training exercises. These losses are not caused by the enemy or an adversary. Factors include—

- An ever-changing operational environment.
- Effects of a fast-paced, high-operational tempo (OPTEMPO) and a high-personnel tempo (PERSTEMPO) on unit and human performance. Examples include leader or soldier error or failure to train or perform to standards.
- Equipment failure, support failure, and the effects of the physical environment.

PRINCIPLES

The basic principles that provide a framework for implementing the risk management process are—

- *Integrating risk management into mission planning, preparation, and execution.* Leaders and staffs continuously identify hazards and assess both accident and tactical risks. They then develop and coordinate control measures. They determine the level of residual risk for accident hazards in order to evaluate courses of action (COAs). They integrate control measures into staff estimates, operation plans (OPLANs), operation orders (OPORDs), and missions. Commanders assess the areas in which they might take tactical risks. They approve control measures that will reduce risks. Leaders ensure that all soldiers understand and properly execute risk controls. They continuously assess variable hazards and implement risk controls.
- *Making risk decisions at the appropriate level in the chain of command.* The commander should address risk guidance in his commander's guidance. He bases his risk guidance on established Army and other appropriate policies and on his higher commander's direction. He then gives guidance on how much risk he is willing to accept and delegate. Subordinates seek the higher commander's approval to accept risks that might imperil the next higher commander's intent.
- *Accepting no unnecessary risk.* Commanders compare and balance risks against mission expectations and accept risks only if the benefits outweigh the potential costs or losses. Commanders alone decide whether to accept the level of residual risk to accomplish the mission.

APPLICABILITY

Risk management applies to all situations and environments across the wide range of Army operations, activities, and processes. Risk management is useful in developing, fielding, and employing the total Army force. Figure 1-2 summarizes the key aspects of risk management.

DEVELOPMENT

Development concerns include force design, manpower allocation, training and training developments, and combat and materiel developments (equipment and weapons systems) and battle laboratories.

Risk management assists the commander or leader in—

- Conserving lives and resources and avoiding unnecessary risk.
- Making an informed decision to implement a COA.
- Identifying feasible and effective control measures where specific standards do not exist.
- Providing reasonable alternatives for mission accomplishment.

Risk management does not—

- Inhibit the commander's and leader's flexibility and initiative.
- Remove risk altogether, or support a zero defects mindset.
- Require a GO/NO-GO decision.
- Sanction or justify violating the law.
- Remove the necessity for standard drills, tactics, techniques, and procedures.

Figure 1-2. Key Aspects of Risk Management

Force Design

Concerns include risks introduced in trade-off decisions that involve the design and equipping of—

- Tables of organization and equipment (TOE).
- Modification tables of organization and equipment (MTOE).
- Tables of distribution and allowances (TDA) organizations.

Manpower Allocations

Concerns include shortfalls in manning that put unit readiness and full use of combat system capabilities at risk.

Training and Training Developments

Concerns include hazardous and critical training tasks and feasible risk reduction measures that provide leaders with the flexibility to safely conduct tough, realistic training.

Combat and Materiel Developments and Battle Laboratories

Concerns include providing a means to assist in making informed trade-off decisions such as—

- Balancing equipment form, fit, and function.
- Balancing the durability and cost of equipment and spare parts against their reliability, availability, and maintainability requirements.
- Determining the environmental impact.
- Determining whether to accept systems with less than the full capabilities prescribed in requirement documents and experimental procedures.

ARs 70-1 and 385-16 and MIL-STD-882 provide details on risk management application in the Army materiel acquisition process.

FIELDING

Fielding concerns include personnel assignments, sustainment and logistics, training, and base operations.

Personnel Assignments

Concerns include making informed decisions in assigning replacement personnel. For example, a risk is associated with assigning a multiple launch rocket system crewmember as a replacement for a tube artillery cannon crewmember.

Sustainment and Logistics

Concerns include enhancing one's ability to determine support requirements, the order in which they should be received, and the potential impact of logistics decisions on operations.

Training

Concerns include helping leaders determine the—

- Balance between training realism and unnecessary risks in training.
- Impact of training operations on the environment.
- Level of proficiency and experience of soldiers and leaders.

Base Operations

Concerns include prioritizing the execution of base operations functions to get the most benefit from available resources. Examples include allocating resources for pollution prevention, correcting safety and health hazards, and correcting violations of environmental protection regulations. FM 20-400 provides specific guidance on environmental protection in military operations.

EMPLOYMENT

Employment concerns include force protection and deployment, operations, and redeployment.

Force Protection

Concerns include developing a plan that identifies threats and their associated hazards and balancing resource restraints against the risk.

Deployment, Operations, and Redeployment

Concerns include—

- Analyzing the factors of mission, enemy, terrain, troops, and time available (METT-T) to determine both tactical and accident risks and appropriate risk reduction measures.
- Determining the correct units, equipment composition, and sequence.
- Identifying controls essential to safety and environmental protection.

Military Decision-Making Process	Risk Management Steps: Step 1 Identify Hazards	Step 2 Assess Hazards	Step 3 Develop Controls and Make Risk Decision	Step 4 Implement Controls	Step 5 Supervise and Evaluate
Mission Receipt	X				
Mission Analysis	X	X			
COA Development	X	X	X		
COA Analysis	X	X	X		
COA Comparison			X		
COA Approval			X		
Orders Production				X	
Rehearsal[1]	X	X	X	X	X
Execution and Assessment[1]	X	X	X	X	X

[1]All boxes are marked to emphasize the continued use of the risk management process throughout the mission

Figure 2-1. Risk Management Steps Correlated with Military Decision-Making Tasks

Risk decisions should be based upon awareness rather than mechanical habit. Leaders should act on a keen appreciation for the essential factors that make each situation unique instead of from conditioned response. Throughout the entire operational continuum, the commander must consider US Government civilians and contract support personnel in his risk management process. Hazards can exist, regardless of enemy or adversary actions, in areas with no direct enemy contact and in areas outside the enemy's or adversary's influence. The two types of risk that exist across the wide range of Army operations are *tactical risks* and *accident risks.*

- *Tactical risk* is risk concerned with hazards that exist because of the presence of either the enemy or an adversary. It applies to all levels of war and across the spectrum of operations.
- *Accident risk* includes all operational risk considerations other than tactical risk. It includes risks to the friendly force. It also includes risks posed to civilians by an operation, as well as an operations impact on the environment. It can include activities associated with hazards concerning friendly personnel, civilians, equipment readiness, and environmental conditions.

STEPS 1 AND 2

Steps 1 and 2 together comprise the risk assessment. In Step 1, individuals identify the hazards that may be encountered in executing a mission. In Step 2, they determine the direct impact of each hazard on the operation. The risk assessment provides for enhanced situational awareness. This awareness builds confidence and allows soldiers and units to take timely, efficient, and effective protective measures.

STEPS 3 THROUGH 5

Steps 3 through 5 are the essential follow-through actions to effectively manage risk. In these steps, leaders balance risk against costs—political, economic, environmental, and to combat power—and take appropriate actions to eliminate unnecessary risk. During execution, as well as during planning and preparation, leaders continuously assess the risk to the overall mission and to those involved in the task. Finally, leaders and individuals evaluate the effectiveness of controls and provide lessons learned so that others may benefit from the experience.

THE FIVE STEPS APPLIED

STEP 1. IDENTIFY HAZARDS

A *hazard* is an actual or potential condition where the following can occur due to exposure to the hazard:

- Injury, illness, or death of personnel.
- Damage to or loss of equipment and property.
- Mission degradation.

Hazards are sources of danger or risks due to enemy or adversary presence and other conditions not due to enemy or adversary capabilities. Hazards are found in all operational environments. Combat operations, stability operations, base support operations, and training present unique hazards for units involved in these kinds of missions. Hazards are identified during the first four steps of the military decision-making process: *mission receipt, mission analysis, COA development*, and *COA analysis*.

The ability of unit leaders and staffs to identify hazards is key. One reality of today's missions is that the aspect of a hazard can change rapidly. Things of little risk initially can quickly become major threats due to unforeseen natural or man-made events. Leaders should be aware of this possibility. Complacency to the fact that existing controls may not continue to control hazards in rapidly changing situations should be viewed as a hazard in itself.

The factors of METT-T provide a sound framework for identifying hazards when planning, preparing, and executing operations. When applying risk management to METT-T during mission analysis, leaders and staffs should look for hazards that affect both tactical and accident risks. They must identify all hazards that may present significant risks to the mission.

Mission

Leaders first analyze the assigned mission. They look at the type of mission to be accomplished and consider possible subsequent missions. Certain kinds of operations are inherently more dangerous than others. For example, a deliberate frontal attack, because of the associated movement, is more likely to expose a unit to losses than would a defense from prepared positions. Identifying missions that routinely present great risk is imperative. Leaders also look for hazards associated with complexity of the plan such as—

- A scheme of maneuver that is difficult to understand or too complex for accurate communications down to the lowest level.
- The impact of operating under a fragmentary order (FRAGO).

Enemy

Commanders look for enemy capabilities that pose significant hazards to the operation. For example, "What can the enemy do to defeat my operation?" Common shortfalls that can create hazards during operations against an enemy include failure to—

- Assess potential advantages to the enemy provided by the battlefield environment.
- Fully assess the enemy's capabilities.
- Understand enemy collection capabilities and friendly vulnerabilities to those capabilities.
- Accurately determine the enemy's probable COAs.
- Plan and coordinate active ground and aerial reconnaissance activities.
- Disseminate intelligence about the enemy to lower levels.
- Identifying terrorist threats and capabilities.

Intelligence plays a critical part in identifying hazards associated with tactical risk. Intelligence-preparation-of-the-battlefield (IPB) is a dynamic staff process that continually integrates new information and intelligence that ultimately becomes input to the commander's risk assessment process. Intelligence assists in identifying hazards during operations by—

- Identifying the opportunities and constraints the battlefield environment offers to threat and friendly forces.
- Thoroughly portraying threat capabilities and vulnerabilities.
- Collecting information on populations, governments, and infrastructures.

FMs 34-130 and 34-60, respectively, provide detailed information on IPB and on counterintelligence operations and multidiscipline counterintelligence analysis.

Terrain and Weather

In addition to those due to the enemy or adversaries, the most obvious hazards to military operations are due to terrain and weather. Terrain and weather affect the type of hazard encountered. When the enemy uses terrain to his advantage, the risk is clearly tactical. The aspects of terrain and weather may create situations where accident risks predominate. When looking at this from a purely mission perspective, familiarity of the unit with the terrain and its associated environment must be paramount. Basic issues include—

- How long the unit has operated in the environment and climate.
- Whether the terrain has been crossed before.

time can result in shortages of time for subordinate and adjacent units to accomplish their mission.

Civilians

The commander's legal responsibility is to consider hazards to, and safeguarding of, civilians in his area of operations. *Civilians* include nongovernmental organizations (NGOs), private voluntary organizations (PVOs), US Government civilians, foreign national civilians, the media, and dislocated civilians put at risk by military operations. The commander must consider hazards that can occur across the range of operations, such as—

- *In a wartime environment.* The commander must consider the hazard of collateral damage which may result in creating new adversaries.
- *In a peacetime environment.* The commander must consider the political attitudes and previous actions of civilians in identifying hazards to friendly forces and the populace itself.

Adversaries are hostile elements other than the enemy that may be encountered during any operation. They present additional hazards. They may be organized opposition or individuals that challenge authority. They may include such diverse elements as rioters, criminals, rogues, or gangs that might want to harass a peace enforcement mission.

STEP 2. ASSESS HAZARDS

Step 2 completes the risk assessment. Risk is the chance of hazard or bad consequences. This step examines each hazard in terms of probability and severity to determine the risk level of one or more hazardous incidents that can result from exposure to the hazard. This step is conducted during three steps of the military decision-making process—*mission analysis, COA development,* and *COA analysis.* This step is also conducted after controls are developed.

The incident must be credible in that it must have a reasonable expectation of happening. The end result is an estimate of risk from each hazard and an estimate of the overall risk to the mission caused by hazards that cannot be eliminated. Leaders must also assess the risk to civilians posed by the operation. They may need to assess the operations' impact on the environment. This step is conducted in three substeps.

Substep A

Leaders and staffs assess each hazard in relation to the *probability* of a hazardous incident. The probability levels estimated for each hazard may be based on the mission, COAs being developed and analyzed, or frequency of a similar event. Figure 2-2 provides a summary of the five degrees of probability. The letters in parentheses following each degree (A through E) provide a symbol for depicting probability. For example, the letter *A* represents *frequent* probability.

FREQUENT (A) Occurs very often, continuously experienced	
Single item	Occurs very often in service life. Expected to occur several times over duration of a specific mission or operation. Always occurs.
Fleet or inventory of items	Occurs continuously during a specific mission or operation, or over a service life.
Individual soldier	Occurs very often in career. Expected to occur several times during mission or operation. Always occurs.
All soldiers exposed	Occurs continuously during a specific mission or operation.
LIKELY (B) Occurs several times	
Single item	Occurs several times in service life. Expected to occur during a specific mission or operation.
Fleet or inventory of items	Occurs at a high rate, but experienced intermittently (regular intervals, generally often,).
Individual soldier	Occurs several times in career. Expected to occur during a specific mission or operation.
All soldiers exposed	Occurs at a high rate, but experienced intermittently.
OCCASIONAL (C) Occurs sporadically	
Single item	Occurs some time in service life. May occur about as often as not during a specific mission or operation.
Fleet or inventory of items	Occurs several times in service life.
Individual soldier	Occurs some time in career. May occur during a specific mission or operation, but not often.
All soldiers exposed	Occurs sporadically (irregularly, sparsely, or sometimes).

Figure 2-2. Hazard Probability

SELDOM (D) Remotely possible; could occur at some time	
Single item	Occurs in service life, but only remotely possible. Not expected to occur during a specific mission or operation.
Fleet or inventory of items	Occurs as isolated incidents. Possible to occur some time in service life, but rarely. Usually does not occur.
Individual soldier	Occurs as isolated incident during a career. Remotely possible, but not expected to occur during a specific mission or operation.
All soldiers exposed	Occurs rarely within exposed population as isolated incidents.
UNLIKELY (E) Can assume will not occur, but not impossible	
Single item	Occurrence not impossible, but can assume will almost never occur in service life. Can assume will not occur during a specific mission or operation.
Fleet or inventory of items	Occurs very rarely (almost never or improbable). Incidents may occur over service life.
Individual soldier	Occurrence not impossible, but may assume will not occur in career or during a specific mission or operation.
All soldiers exposed	Occurs very rarely, but not impossible.

Figure 2-2. Hazard Probability (continued)

Substep B

Substep B addresses the *severity* of each hazard. It is expressed in terms of—

- Degree of injury or illness.
- Loss of or damage to equipment or property.
- Environmental damage.
- Other mission-impairing factors such as lost combat power.

The degree of severity estimated for each hazard may be based on knowledge of the results of similar past events. Figure 2-3 provides a summary of the four degrees of hazard severity. The Roman numerals in parentheses following each degree (I through IV) provide a convenient symbol for depicting severity. For example, *I* represents the *catastrophic* degree of severity.

CATASTROPHIC (I)	Loss of ability to accomplish the mission or mission failure. Death or permanent total disability (accident risk). Loss of major or mission-critical system or equipment. Major property (facility) damage. Severe environmental damage. Mission-critical security failure. Unacceptable collateral damage.
CRITICAL (II)	Significantly (severely) degraded mission capability or unit readiness. Permanent partial disability, temporary total disability exceeding 3 months time (accident risk). Extensive (major) damage to equipment or systems. Significant damage to property or the environment. Security failure. Significant collateral damage.
MARGINAL (III)	Degraded mission capability or unit readiness. Minor damage to equipment or systems, property, or the environment. Lost day due to injury or illness not exceeding 3 months (accident risk). Minor damage to property or the environment.
NEGLIGIBLE (IV)	Little or no adverse impact on mission capability. First aid or minor medical treatment (accident risk). Slight equipment or system damage, but fully functional and serviceable. Little or no property or environmental damage.

Figure 2-3. Hazard Severity

Substep C

In this substep leaders and staffs expand what they understand about probable hazardous incidents into estimates of levels of risk for each identified hazard and an estimate of the overall risk for the operation. Estimating risk follows from examining the outcomes of Substeps A and B; that is, both the probability and severity of hazardous incidents. This substep is more art than science. Much depends on the use of historical lessons learned, intuitive analysis, ammunition caches will critically impact his combat power. He considers that his unit is deployed in excellent defensive positions. He has repelled two attacks that resulted in the destruction of an estimated 50 percent of the enemy's combat power. He receives information that the probability of the enemy attacking is *likely*, but that the probability of the enemy being reinforced and attacking in overwhelming force is remote *(seldom)*. The commander concludes that the risk of conducting a counterattack with limited ammunition is greater than the *moderate* risk of the enemy pushing him back.

L - Low: Expected losses have little or no impact on accomplishing the mission. The probability of *critical* loss is *unlikely* (IIE), while that of *marginal* loss is *seldom* (IIID) or *unlikely* (IIIE). The probability of a *negligible* loss is *likely* or *less* (IVB through (IVE).

Example: A mechanized task force (TF) conducting a movement to contact in a desert environment is overtaken by nightfall before reaching its limit of advance (LOA). The terrain along the axis of advance is flat and open. Visibility is about 800 meters under a clear sky illuminated by a full moon. Estimates put the enemy, which has been hastily withdrawing for the past three days, at approximately 30 percent strength. Contact has been light with no defensible terrain along the TF's axis. The TF commander considers all the factors. In addition, the TF is 100 percent operational in using night vision devices. The TF commander estimates that it is *unlikely* that his unit will incur losses of *critical* severity by being surprised by the enemy or lose *critical* combat power due to an accident. He estimates the risk to his force in continuing a nighttime movement is *low*.

Figure 2-5. Levels of Risk (continued)

STEP 3. DEVELOP CONTROLS AND MAKE RISK DECISIONS

> *Risk management is the recognition that decision making occurs under conditions of uncertainty. Decisions must remain consistent with the commander's stated intent and offer a good expectation of success. The risk-taking skill requires competency as a prerequisite.*
>
> FM 100-7, Decisive Force:
> *The Army in Theater Operations,* May 1995

Step 3 is accomplished in two substeps: develop controls and make risk decisions. This is done during the COA development, COA analysis, COA comparison, and COA approval of the military decision-making process.

Substep A - Develop Controls

After assessing each hazard, leaders develop one or more controls that either eliminate the hazard or reduce the risk (probability and/or severity) of a hazardous incident. When developing controls, they consider the reason for the hazard not just the hazard itself.

Types of Controls. Controls can take many forms, but fall into three basic categories—*educational controls, physical controls,* and *avoidance.*

- *Educational controls.* These controls are based on the knowledge and skills of the units and individuals. Effective control is implemented through individual and collective training that ensures performance to standard.
- *Physical controls.* These controls may take the form of barriers and guards or signs to warn individuals and units that a hazard exists. Additionally, special controller or oversight personnel responsible for locating specific hazards fall into this category.
- *Avoidance.* These controls are applied when leaders take positive action to prevent contact with an identified hazard.

Criteria for Controls. To be effective, each control developed must meet the following criteria:

- *Suitability.* It must remove the hazard or mitigate (reduce) the residual risk to an acceptable level.
- *Feasibility.* The unit must have the capability to implement the control.
- *Acceptability.* The benefit gained by implementing the control must justify the cost in resources and time. The assessment of acceptability is largely subjective. Figure 2-6 gives criteria for determining acceptability of controls for each identified hazard.

Support	Availability of adequate personnel, equipment, supplies, and facilities necessary to implement a suitable controls.
Standards	Guidance and procedures for implementing a control are clear, practical, and specific.
Training	Knowledge and skills are adequate to implement a control.
Leadership	Leaders are competent to implement a control.
Individual	Individual soldiers are sufficiently self-disciplined to implement a control.

Figure 2-6. Criteria for Determining Acceptability of Controls

Examples of ControlsExamples of controls include—

- Engineering or designing to eliminate or control hazards.
- Selecting a COA that avoids identified hazards.
- Limiting the number of people and the amount of time they are exposed to hazards, consistent with mission requirements.
- Selecting personnel with appropriate mental, emotional, and physical capabilities.
- Providing protective clothing, equipment, and safety and security devices.
- Providing such services as adequate sanitation facilities and water purification capabilities.
- Providing warning signs and signals.
- Scheduling vehicle and aircraft silhouette drills.
- Planning training, including rehearsals, rock drills, battle drills, and so forth.
- Programming communications links for key civilian organizations.
- Establishing battlefield controls such as areas of operations and boundaries, direct fire control measures, fire support coordination measures, rules of engagement, airspace control measures, bridge classification, traffic control, and so forth.
- Developing terrorist attack warning systems and response plans.

The key is to specify who, what, where, when, and how each control is to be used. For example—

- Planning and scheduling intensive threat and friendly vehicle identification refresher training for all antiarmor and air defense weapons crews before the mission reduces the probability of engaging a friendly vehicle or aircraft (fratricide).
- Programming installation of crashworthy passenger seats in the UH-60 Blackhawk, when mission circumstances do not indicate their removal, can reduce the severity of injuries in crashes.
- Requiring soldiers to wear flak vests and helmets during movement to contact, or when riding in vehicles in areas where enemy fire is likely, can reduce the probability and severity of a wound from small arms fire or fragments.

- Establishing strong continuity documents and planning overlap tours for key leaders facilitate smooth transitions during extended operations.

Residual RiskOnce the responsible leader develops and accepts controls, he determines the residual risk associated with each hazard and the overall residual risk for the mission.

- *Residual risk* is the risk remaining after controls have been selected for the hazard. Residual risk is valid (true) only if the controls for it are implemented. As controls for hazards are identified and selected, the hazards are reassessed as in Step 2 and the level of risk is then revised. This process is repeated until the level of residual risk is acceptable to the commander or leader or cannot be further reduced. See Figures A-3 through A-5.
- *Overall residual risk* of a mission must be determined when more than one hazard is identified. The residual risk for each of these hazards may have a different level, depending on the assessed probability and severity of the hazardous incident. Overall residual mission risk should be determined based on the incident having the greatest residual risk. Determining overall mission risk by averaging the risks of all hazards is not valid. If one hazard has high risk, the overall residual risk of the mission is high, no matter how many moderate or low risk hazards are present.

Substep B - Make Risk Decision

A key element of the risk decision is determining if the risk is justified. The commander must compare and balance the risk against mission expectations. He alone decides if controls are sufficient and acceptable and whether to accept the resulting residual risk. If he determines the risk level is too high, he directs the development of additional controls or alternate controls, or he modifies, changes, or rejects the COA.

Leaders can use the risk assessment matrix in Figure 2-4—in conjunction with their commanders' guidance—to communicate how much risk they are willing to delegate. For example, a commander may place constraints on his subordinates that restrict their freedom of action to accept risk in instances where the risk might imperil his intent, his higher commander's intent, or a critical capability of the unit.

- Capture and disseminate lessons learned so that others may benefit from the experience.
- Consider the effectiveness of the risk assessment in identifying and accurately assessing the probability and severity of hazards that resulted in mission degradation.
- Determine whether the level of residual risk of each hazard and of the overall mission were accurately estimated.
- Evaluate the effectiveness of each control in reducing or removing risk, including whether controls were effectively communicated, implemented and enforced.

Leaders and individuals determine why some controls were ineffective and what should be done when the hazard is encountered again. A control may be altered; the way it is implemented or supervised may be changed to make it effective; or a completely different control may be more effective. Leaders must energize the system to fix systemic problems that hinder combat effectiveness.

Figure 2-7 shows that the risk management process continues throughout a mission as well as from mission to mission. It is integral to the military decision-making process. Its application requires good judgment and intuitive analysis borne of confidence, experience, and situational awareness.

TOOLS AND PITFALLS

The appendix provides examples of risk management tools to help leaders assess identified hazards, develop controls, and make risk decisions. The tools should be tailored to suit particular situations and missions. The examples in Figures A-3 through A-5 are tools to manage risk at the tactical level. The example in Figure A-6 is a tool to manage risk at the operational level. Units may develop additional tools suitable for their needs.

Units train to a standard. They operate and train regardless of the degree of real or perceived difficulty. Risk reduction begins with commanders identifying their METLs. Commanders use the risk management process to assess the degree of risk related to each METL their unit must perform. From this assessment, risk reducing standard operating procedures evolve.

Figure A-7 provides an example of risk management considerations integrated into a mission training plan (MTP) task.

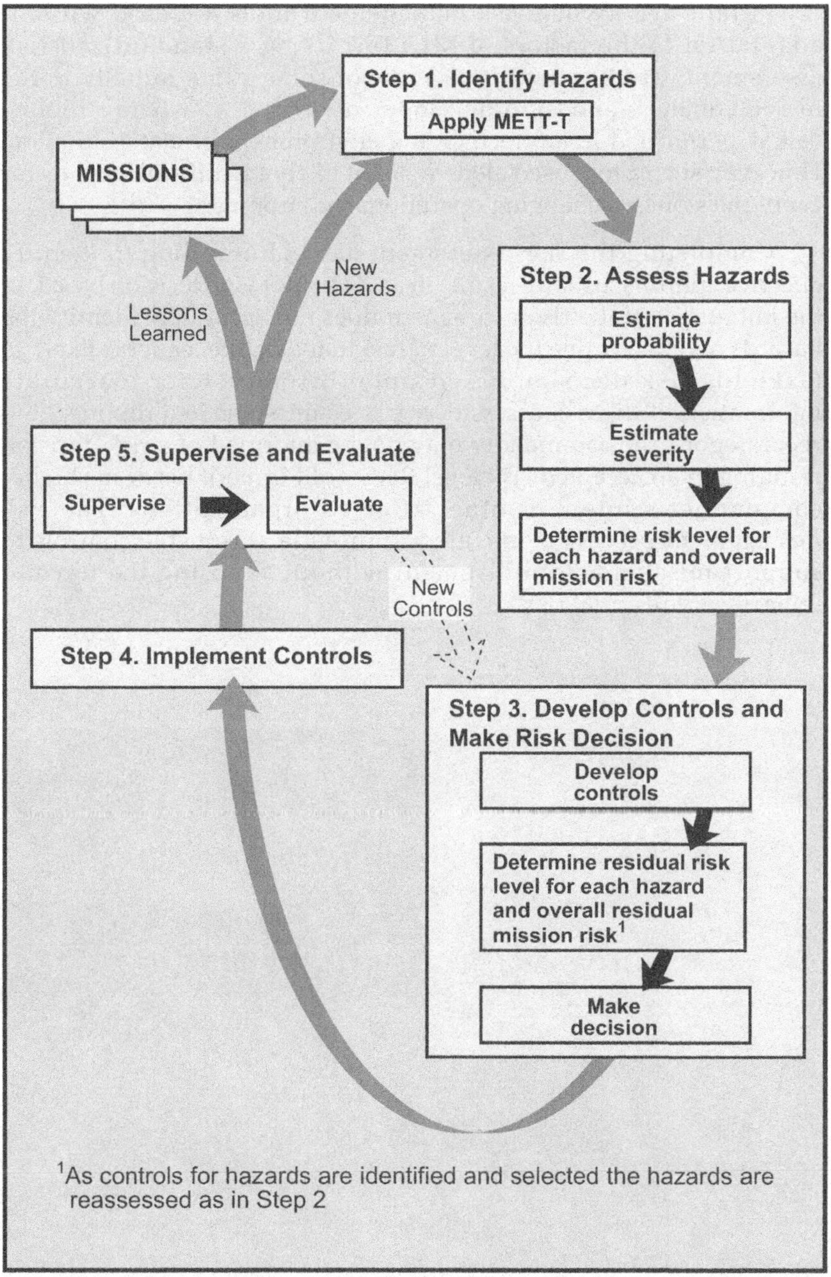

Figure 2-7. Continuous Application of Risk Management

Pitfalls arise when risk management tools are used without adaptation to the factors of METT-T. Using a standardized risk assessment card or checklist may be of some value initially in the mission analysis and COA development or in cases where a routine task is performed in an unchanging environment or static situation. However, such a tool used alone will not likely identify all hazards for every mission in a changing operational environment.

Completing the risk assessment alone, but failing to identify effective controls, usually results in a GO or NO-GO decision based on the initial risk. If the risk assessment does not accurately identify the hazards and determine the level of residual risk, the leader is likely to make his risk decision based upon incomplete or inaccurate information. If the risk assessment places missions in a routine, low-risk category, the commander may not be informed of a risk decision resulting in an accepted risk level that could imperil his or his higher commander's intent or other affected organizations. The risk management process is intended to provide reasonable controls to support mission accomplishment without exposing the force to unnecessary residual risk.

Chapter 3

Risk Management Implementation

It is imperative to develop twenty-first century leaders who trust their subordinates' abilities and judgment. Such leaders must be willing to underwrite their subordinates' honest errors and coach them on to excellence, without tolerating incompetence or laxity. We must recognize that Army leaders are not perfect, and that activity at the ragged edge of audacity sometimes leaves debris in its wake.

GEN William W. Hartzog
Commanding General
US Army Training and Doctrine Command

This chapter presents the moral and ethical implications of risk management. It outlines essential responsibilities and considerations necessary to effectively integrate and assess risk management within the Army. Leaders should tailor these essentials to develop specific how-to procedures suited to their circumstances, available resources, and mission.

MORAL AND ETHICAL IMPLICATIONS FOR LEADERS

To be successful, risk management must be underwritten by the chain of command. Leaders should not expect that all missions will be accomplished with zero defects—free from errors, flaws, or less-than-perfect performance. Demanding such rigid standards leads to oversupervision and paralysis; it produces timid leaders, afraid to make tough decisions in crisis and unwilling to take risks necessary for success in military operations. A zero defects mindset creates conditions that will lead inevitably, in the larger sense, to failure in battle and higher casualties. Leaders are morally bound to support a subordinate's decision to accept risks that are within his commander's intent and guidance, as he understands it.

Leaders accept that things may go wrong, even with the certain knowledge that a subordinate has done all within his power to prevent an incident. In such an event, the leader steps forward and

- Obtaining and providing to subordinates the necessary assets to control risk.
- Knowing their own limitations, their leaders' and soldiers' limitations, and their unit's capabilities.
- Preventing a zero-defects mindset from creeping into their command's culture.
- Allowing subordinates to make mistakes and learn from them.
- Demonstrating full confidence in subordinates' mastery of their trade and their ability to execute a chosen COA.
- Keeping subordinates informed; consulting with subordinate leaders before making a decision, if feasible.
- Listening to subordinates.

Commanders are responsible and accountable for their own actions and those of units under their charge. Commanders must weigh the repercussions of casualties, damage to the environment, and loss of equipment. They must also consider the level of public reaction to loss against national, strategic, operational, or tactical objectives. Commanders are also responsible for keeping soldiers from falling into complacency. During sustained operations in particular, complacency can creep in as a result of accepting the status quo. For example, a unit can be deployed for several months and nothing may happen. The unit then gets very satisfied with itself and its performance. It gets conditioned to its initial circumstances—being well-armed, well-equipped, well-disciplined, and well-trained. It thinks it has risks under control and does not need to change. Such complacency, and the associated loss of situational awareness, can result in leaders and soldiers taking a gamble instead of a prudent risk. A risk is the accepted result of an informed decision; a gamble is an uninformed bet or guess on a hopeful outcome. Leaders and soldiers must clearly understand the difference.

Command is often exercised in conditions of uncertainty and ambiguity, where violence, danger, fear, and friction abound, and under the ever present time constraints driven by OPTEMPO. Risk decisions are frequently required by and dependent on the immediate situation. Judgment is required; a formula, rule, or checklist, by itself, is not appropriate under such circumstances.

Avoiding the zero-risk mindset requires the exercise of positive leadership. The commander's approach to managing risk should be through empowering leaders by pushing risk decisions as far down the chain of command as feasible within the next higher commander's guidance. Commanders must insist that subordinates exercise their freedom of action to act decisively and aggressively to complete assigned missions and promote success of the larger force. Subordinates must consider hazards outside their assigned responsibilities that impact the mission. The result is to encourage coordination and communication—laterally and up and down the chain of command. This requires and encourages initiative, which demands well-trained, determined, disciplined soldiers.

Risk management is a two-way street. It is important that those involved in mission preparation and execution be fully aware of the amount of command involvement and actions necessary to control or remove hazards. The higher commander's guidance specifies the degree of damage or risk to subordinate units that he is willing to accept during the current operation. Subordinates ensure they understand and implement their commander's intent and guidance. If, during the planning process, the accident and/or tactical residual risk exceeds that which the higher commander is willing to accept, the subordinate informs his commander. He requests the resources necessary to mitigate the risk. If, during mission execution, the subordinate determines the risk is too great, he directs the development of additional or alternate controls or modifies or changes the COA. He should notify his next higher commander of his decision. Requiring subordinates to report to the higher commander when a risk decision point is reached during mission execution can result in paralysis.

The objective of managing risk is not to remove all risk, but to eliminate unnecessary risk. Commanders conduct tough, realistic training, knowing that they may put lives and property at risk in the course of military operations. Nothing is worth the cost of a life as the result of taking unnecessary risk. If an action will result in an unacceptable risk, measures should be taken to mitigate it. If the risk cannot be mitigated to an acceptable level, the action should not be executed. Circumstances may occur during mission execution when a decision to stop and defer execution of the operation should be made to avoid taking unwarranted risk. Such a situation will generally occur at the tactical level. For example, circumstances may determine if a trade-off between maintaining the momentum of the attack or risking fratricide or serious accidents is justified. For example, during the deployment of TF Eagle to Bosnia-Herzegovina, there was pressure to complete the Sava River bridge as quickly as possible. The commander assessed the situation and decided to call for a short rest period because his soldiers were tired and getting careless. Consequently, they completed the bridge ahead of schedule under the most difficult conditions imaginable and without injuries.

LEADERS

Many of the tasks identified for commanders apply to all leaders. Leaders' specific responsibilities in managing risk include—

- Establishing clear, feasible risk management policies and goals.
- Conducting detailed planning within time constraints; assessing each mission and task in terms of its risk; continuously reassessing risk as the mission and conditions change and experience is gained.
- Making informed risk decisions and establishing and clearly communicating risk guidance.
- Training the risk management process. Ensuring subordinates understand the who, what, where, when, how, and why of managing risk and how the process applies to their circumstances and assigned responsibilities.
- Examining how subordinates manage risk and how soldiers protect themselves.
- Supervising and evaluating the unit's execution of risk controls during the mission to correct areas needing improvement.
- Advising his chain of command on risks and risk-reduction measures.
- Providing subordinates with feedback on their performance and ways to improve.
- Assessing the effectiveness of their unit's risk management program.
- Capturing and disseminating lessons learned to ensure they are continued from mission to mission so that others may benefit from the experience.

STAFFS

The chief of staff or executive officer is responsible for supervising integration of risk management across the staff. He coordinates development of risk reduction controls with emphasis on deconflicting controls that affect multiple functional areas and adjacent units. The staff officer helps the commander eliminate unnecessary risks by—

- Analyzing his functional area and applying risk management during the military decision-making process.
- Identifying constraints in the higher commander's risk guidance.
- Including hazards and their risks in the mission analysis briefing.
- Including a risk assessment for the commander's estimate.
- Considering the risk assessment in the operations estimate.
- Including risks and recommending ways to reduce their impact in the staff estimate.
- Implementing risk controls by coordinating and integrating them into the appropriate paragraphs and graphics of the OPORD and into products such as SOPs and OPLANs.
- Establishing procedures and standards that are clear and practical.
- Determining the effectiveness of hazard/risk controls and continuously assessing their suitability, feasibility, and acceptability.
- Supervising, evaluating, and assessing the integration of risk management during an operation.
- Continuously identifying hazards, assessing initial and residual risks for each hazard, recommending control measures to reduce the risk to the force.
- Identifying and assessing hazards associated with complacency, especially during extended operations, and recommending appropriate actions to the commander.

At the operational level, staffs focus on hazards and their risks across the spectrum of protecting the force

— Maintaining situational awareness, especially of vulnerabilities.

— Avoiding unnecessary accident risk.

Risk management works best when incorporated into existing command training and operational planning cycles. They then act as risk management tools and provide for—

- Development of policy, goals, objectives, and priorities in the commander's quarterly training guidance.
- The commander's training assessment, by identifying hazards and risk controls.
- Systematic observation and assessment of the unit's risk management performance and feedback into the training management cycle and SOPs.

Leaders and soldiers must have the skills, knowledge, and attitude to effectively manage risks inherent in all operations. Effective training helps soldiers become proficient. It qualifies them technically and tactically, and as leaders, to accomplish the mission without unnecessary risk.

Doctrine developers must integrate risk management into planning for all Army processes, especially the training management cycle defined in FMs 25-100 and 25-101. Unit leaders and their staffs must continually assess and evaluate the integration of risk management into short-, near-, and long-term training plans. They must continually review METLs to ensure that training is supported by realistic risk management objectives. In the past, unprepared or improperly trained units paid a high price for veteran status. The Army must learn from past experiences if it is to avoid repeating such losses.

ASSESSMENT OF THE RISK MANAGEMENT PROCESS

To assess the risk management process is to determine a unit's current level of proficiency in implementing the process. The term *assessment,* as discussed here, differs from *evaluation* as used in Step 5 of the process. Evaluation is used to measure demonstrated ability to accomplish specified objectives within a discrete operation or exercise. Assessment, as used here, also differs from the aspect of assessing hazards for probability and severity, as addressed in Step 2 of the process.

How well risk is managed affects readiness. Leaders need to know the current status and effectiveness of their organization's risk management program. They self-assess their unit's effectiveness in managing risk in order to gain insight into areas for improvement and get feedback on subordinates' understanding and application of risk guidance. The assessment objectives are to determine how—

- Effectively risk management is embedded into planning and preparing for operations.
- Well risk management is understood by subordinate leaders and soldiers.
- Effectively risk management is used to execute operations.

Leaders assess the effectiveness of their units by reviewing how well hazards are identified and risk controls are—

- Specified in oral and written OPORDs, OPLANs, and SOPs.
- Communicated to lowest level of chain of command.
- Included in short-, near-, and long-term training plans.
- Implemented into training and activities on and off duty.
- Embedded into protect-the-force programs such as safety and health and antiterrorism.
- Part of after-action reviews and fed into lessons learned.

Risk management cannot be seen as a competitive program whereby a unit or leader is judged or compared in a competitive sense. Focus is strictly on both reduction of risk and risk behavior.

Take calculated risks. That is quite different from being rash.

General George S. Patton, Jr.

Appendix

Examples of Risk Management Application

The examples in this appendix are designed to help those charged with managing risk.

TRACKING TOOL

The work sheet instructions are in Figure A-1.

Work Sheet Instructions	
Block	
A – D	Self explanatory
E	Identify task relating to the mission or task in Block A
F	**Identify Hazards** – Identify hazards by reviewing METT-T factors for the mission or task. Additional factors include historical lessons learned, experience, judgment, equipment characteristics and warnings, and environmental considerations.
G	**Assess Hazards** – Assessment includes historical lessons learned, intuitive analyses, experience, judgment, equipment characteristics and warnings, and environmental considerations. Determine initial risk for each hazard by applying risk assessment matrix (Figure 2-4). Enter the risk level for each hazard.
H	**Develop Controls** – Develop one or more controls for each hazard that will either eliminate the hazard or reduce the risk (probability and/or severity) of a hazardous incident. Specify who, what, where, why, when, and how for each control. Enter controls.
I	**Determine Residual Risk** – Determine the residual risk for each hazard by applying the risk assessment matrix (Figure 2-4). Enter the residual risk level for each hazard.
J	**Implement Controls** – Decide how each control will be put into effect or communicated to the personnel who will make it happen (written or verbal instruction; tactical, safety, garrison SOPs, rehearsals). Enter controls.
K	**Determine Overall Mission/Task Risk** – Select the highest residual risk level and circle it. This becomes the overall mission or task risk level. The commander decides whether the controls are sufficient to accept the level of residual risk. If the risk is too great to continue the mission or task, the commander directs development of additional controls or modifies, changes, or rejects the COA.
	Supervise and Evaluate – This last step is not on the worksheet. Plan how each control will be monitored for implementation (continuous supervision, spot-checks) and reassess hazards as the situation changes. Determine if the controls worked and if they can be improved. Pass on lessons learned.

Figure A-1. Risk Management Work Sheet Instructions

The work sheet (Figure A-2) provides a starting point to logically track the process of hazards and risks. It can be used to document risk management steps taken during planning, preparation, and execution of training and combat missions and tasks.

A. Mission or Task:			B. Date/Time Group Begin: End:		C. Date Prepared:
D. Prepared By: (Rank, Last Name, Duty Position)					
E. Task	F. Identify Hazards	G. Assess Hazards	H. Develop Controls	I. Determine Residual Risk	J. Implement Controls ("How To")
K. Determine overall mission/task risk level after controls are implemented (circle one) LOW (L) MODERATE (M) HIGH (H) EXTREMELY HIGH (E)					

SAMPLE

Figure A-2. Sample Risk Management Work Sheet

A. Mission or Task: Redeploy unit to home station			B. Date/Time Group Begin: 010035R May XX End: 010600R May XX		C. Date Prepared: 29 April XX
D. Prepared By: (Rank, Last Name, Duty Position) MAJ Woolsey, S3					
E. Task	**F. Identify Hazards**	**G. Assess Hazards**	**H. Develop Controls**	**I. Determine Residual Risk**	**J. Implement Controls** ("How To")
Conduct convoy operations-movement from Grafenwohr to home base	Fatigue leading to poor judgment/ accident-causing errors	Moderate (M)	1. Ensure all drivers receive adequate rest the night before movement 2. Brief planned rest stops and actions to take if driver becomes too tired/ill or has doubts about alertness 3. Have NCOs check physical alertness of drivers prior to movement	Low (L)	Unit SOP, OPORD on Tactical Road March, STP 21-2-MQS (Tactical Road March)
	Fast moving traffic mix of cars and large trucks with trailers	Moderate (M)	1. Use extreme caution when passing 2. Continue to scan (be aware of) traffic 3. Brief all drivers and vehicle commanders on lane-changing procedures	Low (L)	German laws and unit SOP
	Accidents and breakdowns blocking road	Moderate (M)	1. Brief drivers on actions to take a. Pull as far off the road as possible b. Get out on passenger side c. Post triangle sign at least 100 meters behind vehicle d. Stay as far away from vehicle and traffic as possible e. Brief all personnel on maintenance plan/action if vehicle is disabled f. Brief all personnel on procedures for vehicle exits on the traffic side	Low (L)	Unit SOP, OPORD on Tactical Road March

Figure A-5. Example of Completed Risk Management Work Sheet for Battalion/Task Force

A. Mission or Task: Redeploy unit to home station (continued)		B. Date/Time Group Begin: 010035R Dec XX End: 010600R Dec XX		C. Date Prepared: 29 Nov XX	
D. Prepared By: (Rank, Last Name, Duty Position) MAJ Woolsey, S3					
E. Task	**F. Identify Hazards**	**G. Assess Hazards**	**H. Develop Controls**	**I. Determine Residual Risk**	**J. Implement Controls** ("How To")
Conduct convoy operations- movement from Grafenwohr to home base	Accidents and breakdowns blocking road (continued)	Moderate (M)	2. Leaders ensure that proper maintenance are done prior to move (PMCS), dispatching and licensing procedure. 3. Contact team follows convoy and repairs or removes disabled vehicles from highway.	Low (L)	
	Weather conditions– high winds, sleet, snow	High (H)	1. Reduce speed according to environmental conditions less stopping distance available. 2. Road conditions especially right after rains – hydroplaning – review procedures for skids or hydroplanes. 3. Maintain convoy interval to allow for braking and assist civilian traffic in passing.	Moderate (M)	Unit SOP and vehicle operations manual
	High winds– reduction visibility, sudden reduction due to passing traffic	Moderate (M)	1. Change lanes carefully; ensure clearance and avoid erratic moves 2. Be aware of winds and effects on vehicles. 3. Drive defensively.	Low (L)	German traffic laws
	20% downgrade at checkpoint 1	Moderate (M)	1. Use lower gears. 2. Reduce speed to 25 mph	Low (L)	Unit SOP and vehicle operations manual

Figure A-5. Example of Completed Risk Management Work Sheet for Battalion/Task Force (continued)

A. Mission or Task: Redeploy unit to home station (continued)

B. Date/Time Group **Begin:** 010035R Dec XX **End:** 010600R Dec XX

C. Date Prepared: 29 Nov XX

D. Prepared By: (Rank, Last Name, Duty Position) MAJ Woolsey, S3

E. Task	F. Identify Hazards	G. Assess Hazards	H. Develop Controls	I. Determine Residual Risk	J. Implement Controls ("How To")
Conduct convoy operations-movement from Grafenwohr to home base	Heavy traffic, traffic jams, and congestion	Moderate (M)	1. Use extreme caution when passing 2. Use headlights at all times 3. Brief convoy speeds and vehicle intervals 4. Brief actions to take if vehicle pulls in front a. Maintain clearance from vehicle b. Maintain braking distance 5. Stay in right lane in construction zones 6. Ensure AMVs 2 or more meters wide remain in right lane 7. Continuously scan surroundings (situational awareness)	Low (L)	Provide risk assessment to convoy leaders. Provide safety brief information to leaders and drivers

K. Determine overall mission/task risk level after controls are implemented (circle one)

LOW (L) MODERATE (M) [circled] HIGH (H) EXTREMELY HIGH (E)

Figure A-5. Example of Completed Risk Management Work Sheet for Battalion/Task Force (continued)

A. Mission or Task: Peace Enforcement			B. Date/Time Group Begin: 010035R May XX End:		C. Date Prepared: 29 April XX
D. Prepared By: (Rank, Last Name, Duty Position) G-3 Div/Corps/EAC					
E. Task	**F. Identify Hazards**	**G. Assess Hazards**	**H. Develop Controls**	**I. Determine Residual Risk**	**J. Implement Controls** ("How To")
Occupy area of operations	Assault on facilities	Moderate (M)	Identify and isolate combatants, terrorists, supporters Harden potential terrorist targets	Low (L)	Coordination with local law enforcement agencies FM 5-103, *Survivability* *General /Order*
	Ambush	Moderate (M)	Kevlar helmets and flak jackets required outside cantonment area Minimum of 4 vehicles for movement Two qualified drivers per vehicle	Low (L)	Rehearsals Immediate action drills General orders
	Mines	High (H)	Countermine plan Mine awareness training Predeployment training	Moderate (M)	FM 20-32, *Mine/Countermine Operations* Lessons learned

Figure A-6. Example of Completed Risk Management Work Sheet for Division/Corps/EAC Task Force

MISSION TRAINING PLAN

Figure A-7 shows risk management integrated into an MTP task. The example portrays a possible method for integrating the risk management process and products into MTPs. The text in bold italics indicates suggested areas in which to integrate risk management.

TASK: C^2 THE BATTALION (7-1-1901)(FM 7-20)

ITERATION: 1 2 3 4 5 (circle)

TRAINING STATUS: T P U (circle)

CONDITION: The brigade issues an OPORD or FRAGO

TASK STANDARD :

a. The battalion plan accomplishes the directed mission and specified tasks according to the brigade commander's concept and intent. The plan is received and understood by the leadership of the battalion, who makes the plan successful. It is coordinated with higher, adjacent, and supporting elements.

b. The plan is as fully prepared as time allows to initiate the mission at the directed time.

c. The battalion controls and synchronizes subordinate and supporting elements so that it accomplishes the mission and preserves the force. ***Include commander's risk guidance.***

d. The battalion keeps higher, adjacent, subordinate, supporting, and supported headquarters informed of essential information key to controlling the battle or making required decisions.

SUBTASKS AND STANDARDS: GO NO-GO

+1. Battalion leaders issue the warning order.

a. A complete warning order is issued within 15 minutes of receipt of the brigade order.

b. Warning order is received by all platoons within 45 minutes of issuance of battalion warning order.

+ Critical task.
* Leader task.

Figure A-7. Integration of Risk Management into MTP Task

*2. Battalion commander analyzes mission and gives initial guidance.

a. Guidance includes restated mission, which includes brigade commander's intent for battalion and identifies all specified and implied tasks.

b. Guidance is given within 30 minutes of receipt of order.

c. Guidance includes instructions on information requirements and initially required preparation actions (movement, resupply) to start. ***Guidance includes chain of command authorized to accept risk (extremely high, high, moderate, and low) affecting higher command's intent, resources, or adjacent units' missions.***

+3. Battalion accomplishes reconnaissance and other actions to gather needed information.

a. Reconnaissance actions begin to physically gain information on the enemy and terrain as early as possible.

b. Commander conducts a personal reconnaissance, when possible. If not, the commander conducts a detailed map reconnaissance.

c. Subordinate leaders perform a personal reconnaissance when possible. See subordinate company (ARTEP 7-10-MTP) and platoon (ARTEP 7-8-MTP) T&EOs.

d. Staff coordinates with subordinate, higher, supporting, and adjacent headquarters to gather information for planning. ***Staff analyzes mission to identify hazards.***

e. Staff provides operations, intelligence, and CSS estimates to include all critical METT-T factors and ***risk assessment considerations (identified hazards/risks).***

+4. Battalion commander develops and wargames courses of action and selects one.

Figure A-7. Integration of Risk Management into MTP Task (continued)

a. Tactically feasible courses of action (include CS and maneuver) are made and wargamed with the available staff (commander, S3, and FSO are best for quick planning sequences; XO, S2, engineer, S4, S3 (Safety), and ADA officer are best in more deliberate situations). ***Each course of action contains assessment of hazards, risk level, and control measures identified to lower or control the risk.***

b. Best COA is selected.

c. COA is wargamed and refined by the command and staff. The staff must understand the concept to produce a sound OPORD and rehearse. ***The staff continually assesses the situation to identify new hazards, assess the risk from each hazard, develop acceptable control measures, and reassess for residual risk to the mission. COA should pose minimum risk to soldiers, equipment, and mission accomplishment. Risk acceptance decisions for the COA are made at the appropriate level in the chain of command.***

*5. Staff develops an OPLAN and OPORD from the commander's guidance.

OPLAN and OPORD successfully accomplish the mission according to higher commander's intent. ***Hazards and risk control measures are included in the appropriate paragraphs and graphics.***

*6. Battalion commander and staff issue implementing FRAGOs (accident risk controls).

*7. Battalion commander issues the OPORD/FRAGO.

a. OPORD/FRAGO is issued IAW the one-third, two-thirds rule and makes full use of daylight time.

b. OPORD/FRAGO accomplishes all directed missions and tasks, complies with the brigade commander's intent, controls risk, and is doctrinally sound. It is based on evaluator judgment and on comparison of brigade and battalion OPORDs.

Figure A-7. Integration of Risk Management into MTP Task (continued)

c. All subordinate and supporting elements receive the OPORD/FRAGO.

d. OPORD/FRAGO contains task organization, mission, concept, ***accident risk controls,*** intent for maneuver, supporting fires, and obstacles; missions and tasks for each subordinate; fire support and CSS instructions; and coordinating instructions to synchronize the efforts of maneuver forces and CS.

e. If more time is available, the battalion commander issues a fully developed OPORD (although an initial FRAGO may be issued to allow subordinates to begin preparation and followed by a full OPORD.)

f. Order is given at a location that reduces travel time, allows observation of the zone/sector, and promotes OPSEC. (Depending upon the METT-T factors, observation of the zone/sector may not be possible.)

g. Battalion commander should perform brief-backs and war gaming, ***to include identifying hazards and control measures,*** immediately after the order to improve subordinate understanding and reaction.

h. Subordinate leaders and staff should perform lateral coordination before leaving the orders site.

+8. Commander and staff coordinate and refine the plan.

a. Time is well used to continue gathering information and to improve the plan (contingency plans, ***hazard identification and controls,*** fire plans) .

b. New information is disseminated and coordinated with higher, adjacent, and supporting headquarters to include—

- Changes or refinements in plan.
- Information on the enemy in the sector or zone.

Figure A-7. Integration of Risk Management into MTP Task (continued)

e. Changes that affect the battle are disseminated within five minutes.

+13. Subordinate commanders, leaders, and staff laterally coordinate actions during the battle.

All battle actions requiring coordination between elements are coordinated.

+*14. Battalion coordinates with adjacent and supporting headquarters.

All battle actions requiring coordination with other headquarters are laterally and promptly coordinated.

+*15. Battalion reports.

Battalion CPs submit all critical and required reports to brigade. They report events to adjacent and supporting elements that impact on them in time for those units to react. *They advise the chain of command, as appropriate, on risks and risk-reduction measures.*

Figure A-7. Integration of Risk Management into MTP Task (continued)

Glossary

AAR after-action review

ADA air defense artillery

ARFOR army forces headquarters

ARTEP Army Training and Evaluation Program

assessment an analytical process to determine an organization's current levels of proficiency on a specific objective (for example, a training objective or risk management implementation) (CJCSM 3500.03)

base operations support the provision of administrative and logistical services; includes supply operations, maintenance of materiel, personnel support, base services and administrative services rendered by or through activities of the supporting installation

BASOPS base operations support

C^2 command and control

COA course of action

CofS chief of staff

combat power the total means of destructive and/or disruptive force that a military unit or formation can apply against an opponent at a given time; a combination of the effects of maneuver, firepower, protection, and leadership

controls actions taken to eliminate hazards or reduce their risk

CP command post

CSS combat service support

CTC combat training center

danger	exposure or vulnerability to harm or risk; the balance between the chance or probability of a hazardous incident and the result of the hazardous incident
EAC	echelons above corps
evaluation	the process used to measure the demonstrated ability to accomplish specified objectives such as training within a discrete event or exercise (CJCSM 3500.03); measurement of the demonstrated ability of soldiers or units to perform a task and supporting skill and knowledge or learning objective against the established standard
exposure	the frequency and length of time personnel and equipment are subjected to a hazard
FM	field manual
FRAGO	fragmentary order
fratricide	the employment of friendly weapons and munitions with the intent to kill the enemy or destroy his equipment or facilities, which results in unforeseen and unintentional death or injury to friendly personnel
friction	the accumulation of chance errors, unexpected difficulties, enemy actions, and confusion of battle
FSO	fire support officer
G3	general staff operations section
hazard	any actual or potential condition that can cause injury, illness, or death of personnel, damage to or loss of equipment, property or mission degradation (FM 101-5); a condition or activity with potential to cause damage, loss or mission degradation (Joint Pub 1-02)
inherently dangerous	an activity or task containing a danger to life or limb that is a permanent and inseparable element of the activity
IPB	intelligence-preparation-of-the-battlefield
IR	infrared

LOA	limit of advance
LOGCAP	Logistics Civil Augmentation Program
METL	mission-essential task list
METT-T	mission, enemy, terrain, troops, and time available
MOS	military occupational specialty
MTP	mission training plan
NBC	nuclear, biological, chemical
NVD	night vision device
OCOKA	O - observation and fields of fire, C - cover and concealment, O - obstacles, K - key terrain and decisive terrain, A - avenues of approach
OPCON	operational control
operational tempo	the pace of an operation or operations; OPTEMPO includes all of the activities the unit is conducting; OPTEMPO can be a single activity or a series of operations
OPLAN	operations plan
OPORD	operations order
OPTEMPO	operational tempo
personnel tempo	unit work load level and number of deployed days per year
PERSTEMPO	personnel tempo
PIR	priority intelligence requirements
probability	the likelihood that a hazardous incident will occur
PVO	private voluntary organization
residual risk	the level of risk remaining after controls have been identified and selected for hazards that may result in loss of combat power
risk	chance of hazard or bad consequences; the probability of exposure to chance of injury or loss from a hazard; risk level is expressed in terms of hazard probability and severity (FM 101-5)

FIELD MANUAL
No. 3-06.11

HEADQUARTERS
DEPARTMENT OF THE ARMY
WASHINGTON, DC, 28 February 2002

COMBINED ARMS OPERATIONS IN URBAN TERRAIN

CONTENTS

DISTRIBUTION RESTRICTION: Approved for public release; distribution is unlimited.

*This publication supersedes FM 90-10-1, 12 May 1993, with Change 1, 3 Oct 95.

PREFACE

Worldwide urban growth and the shift of populations from rural to urban areas have affected Army operations. Urban areas will most probably constitute future battlefields. All major Army operations most likely include urban operations (UO) in the foreseeable future.

There is a high probability that the US Army may be engaged by threat forces that are intermingled with the civilian population. Units using the tactics, techniques, and procedures (TTP) outlined in this manual are bound by the specific rules of engagement (ROE) issued by their headquarters and the laws of land warfare.

This manual provides brigade and battalion commanders and staffs, company commanders, small-unit leaders, and individual Infantrymen with considerations and combined arms TTP for conducting full-spectrum urban operations (offense, defense, stability, and support). Some techniques for dealing with insurgents and terrorists or similar threats are included; however, the manuals which best address these issues are FM 7-98 and FM 90-8. This manual may also be used as a reference for other combat, combat support and combat service support commanders, leaders, and staffs that will be required to support combined arms urban operations.

The proponent of this publication is the US Army Infantry School. Send comments and recommendations to doctrine@benning.army.mil or on DA Form 2028 directly to Commandant, US Army Infantry School, ATTN: ATSH-ATD, Fort Benning, Georgia 31905-5410.

Unless this publication states otherwise, masculine nouns and pronouns do not refer exclusively to men.

CHAPTER 1

INTRODUCTION

"The rapid growth of the number and size of urban centers, especially in regions of political instability, increases the likelihood that US forces will be called upon to conduct MOUT."

Defense Science Board, October 1996

It is estimated that by the year 2010, seventy-five percent of the world's population will live in urban areas. Urban areas are expected to be the future battlefield and combat in urban areas cannot be avoided. This manual provides commanders, leaders, and staffs at brigade level and below with a discussion of the principles of urban operations and tactics, techniques, and procedures for fighting in urban areas.

Section I. GENERAL CONSIDERATIONS

Urban operations (UO) are not new to the US Army. Throughout its history the Army has fought an enemy on urban terrain. What **is** new is that urban areas and urban populations have grown significantly during the late twentieth century and have begun to exert a much greater influence on military operations. The worldwide shift from a rural to an urban society and the requirement to transition from combat to stability and support operations and vice-versa have affected the US Army's doctrine. The brigade will be the primary headquarters around which units will be task-organized to perform UO. Companies, platoons, and squads will seldom conduct UO independently, but will most probably conduct assigned missions as part of a battalion task force urban operation. This section provides the necessary background information that facilitates an understanding of how higher level commanders plan and conduct UO.

1-1. DEFINITIONS

Terms specific to UO are defined herein.

a. **Urban Operations.** UO are operations planned and conducted in an area of operations (AO) that includes one or more urban areas. An urban area consists of a topographical complex where man-made construction or high population density is the dominant feature. UO usually occur when—

- The assigned objective lays within an urban area and cannot be bypassed.
- The urban area is key (or decisive) in setting and or shaping the conditions for current or future operations.
- An urban area is between two natural obstacles and cannot be bypassed.
- The urban area is in the path of a general advance and cannot be surrounded or bypassed.
- Political or humanitarian concerns require the control of an urban area or necessitate operations within it.
- Defending from urban areas supports a more effective overall defense or cannot be avoided.

- Occupation, seizure, and control of the urban area will deny the threat control of the urban area and the ability to impose its influence on both friendly military forces and the local civilian population. Therefore, friendly forces can retain the initiative and dictate the conditions for future operations.

b. **METT-TC.** The tactics, techniques and procedures (TTP) the commander selects for each mission, whether in open or urban terrain, are always dependent upon the factors of mission, enemy, terrain, troops, and time available. Traditionally, the acronym "METT-T" has been used to help leaders remember this set of factors as they plan a mission. An effect of the increasing importance of urban areas is the addition of civil considerations (METT-TC).

c. **Urban Combat**. These offensive and defensive operations are the part of UO that include a high density of Infantry-specific tasks. Urban combat operations are conducted to defeat an enemy on urban terrain who may be intermingled with noncombatants. Because of this intermingling, and the necessity to limit collateral damage, the rules of engagement (ROE) and the restrictions placed on the use of combat power may be more restrictive than under other combat conditions.

d. **Categories of Urban Areas.** An urban area is a concentration of structures, facilities, and people that form the economic and cultural focus for the surrounding area. Operations are affected by all five categories of urban areas. Cities, metropolises, and megalopolises with associated urban sprawl cover hundreds of square kilometers. Brigades and below normally operate in these urban areas as part of a larger force. Extensive combat in these urban areas involves units of division level and above.

- *Villages (population of 3,000 inhabitants or less).* The brigade's area of operations (AO) may contain many villages. Battalions and companies bypass, move through, defend from, and attack objectives within villages as a normal part of brigade operations.
- *Towns (population of over 3,000 to 100,000 inhabitants and not part of a major urban complex).* Operations in such areas normally involve brigades or divisions. Brigades may bypass, move through, defend in, or attack enemy forces in towns as part of division operations.
- *City (population over 100,000 to 1 million inhabitants).*
- *Metropolis (population over 1 million to 10 million inhabitants).*
- *Megalopolis (population over 10 million inhabitants).*

e. **Conditions of Urban Operations**. Due to political and societal changes that have taken place in the late twentieth century, advances in technology, and the Army's growing role in maintaining regional stability, UO is conducted across the full spectrum of offense, defense, stability, and support. The full spectrum of UO will affect how units must plan and execute their assigned missions. The enemy's actions significantly affect the conditions of UO, which may transition from one condition to another rapidly. Units may be conducting operations under different conditions at two locations at the same time. The following definitions of the three general conditions of UO provide clarity, focus, and a mental framework for commanders and leaders conducting tactical planning for UO.

(1) ***Urban Operations Under Surgical Conditions.*** This condition is the least destructive and most tightly focused of all the conditions of UO. Operations conducted under surgical conditions include special-purpose raids, small precision strikes, or

small-scale personnel seizures or arrests, focused psychological or civil affairs operations, or recovery operations. They may closely resemble US police operations by special weapons and tactics (SWAT) teams. They may even involve cooperation between US forces and host nation police. Though conventional units may not be directly involved in the actual operation, they may support it by isolating the area or providing security or crowd control.

(2) ***Urban Operations Under Precision Conditions.*** Under precision conditions, either the threat is thoroughly mixed with noncombatants or political considerations require the use of combat power to be significantly more restrictive than UO under high-intensity conditions. Infantry units must routinely expect to operate under precision conditions, especially during stability and support operations.

(a) UO under precision conditions normally involve combat action, usually involving close combat. Some of this combat can be quite violent for short periods. It is marked, however, by the conscious acceptance by US forces of the need to focus and sometimes restrain the combat power used. The commander may bring overwhelming force to bear, but only on specific portions of the urban area occupied by the threat. He may choose different TTP in order to remain within the bounds of the more restrictive ROE. Tighter ROE demands strict accountability of individual and unit actions.

(b) When preparing for UO under precision conditions, commanders and leaders must realize that not only may the ROE change, but the TTP may change also. These changes require that soldiers be given time to train for the specific operation. For example, when clearing a room, units may modify the procedure of first throwing a grenade (fragmentation, concussion, stun) into the room before entering. This procedure may be done to lessen the possible casualties among noncombatants interspersed with the enemy. (See Chapter 3 for more information.)

(3) ***Urban Operations Under High-Intensity Conditions.*** These conditions include combat actions against a determined enemy occupying prepared positions or conducting planned attacks. UO under high-intensity conditions require the coordinated application of the full combat power of the joint combined arms team. Infantry units must be prepared at all times to conduct violent combat under conditions of high-intensity UO.

(a) An Infantry unit's mission is normally to seize, clear, or defend urban terrain, engaging and defeating the enemy by using whatever force is necessary. Although the changing world situation may have made high-intensity UO less likely, it represents the high end of the combat spectrum, and units must be trained for it.

(b) Urban combat under high-intensity conditions is the most stressful of all operations in urban areas and can be casualty-intensive for both sides. Even though the fully integrated firepower of the joint combined arms team is being used, commanders must still prevent unnecessary collateral damage and casualties among noncombatants.

f. **Stability Operations and Support Operations.** The Army has further categorized military operations other than war (MOOTW) as stability and support operations. Units conduct these operations, which are normally short of actual combat, to support national policy. Recent examples include famine relief operations in Mogadishu, Somalia; evacuation of noncombatants in Monrovia, Liberia; and peace enforcement in Bosnia.

(a) During a stability or support operation, units perform many activities not necessarily contained in their mission-essential task list (METL). Essentially, the unit accomplishes these activities through execution of tactical missions, such as security patrols, establishing roadblocks and checkpoints, base defense, and so forth.

(b) While stability and support operations can occur anywhere, they will most likely occur in an urban environment. These operations can resemble UO under precision conditions and can easily transition into combat operations. (Additional TTP and lesson plans are contained in Chapter 14 of TC 7-98-1, Stability and Support Training Support Package.)

g. **Confusion and Crossover Between Conditions.** As in Mogadishu, many types of operations may occur at the same time and certain types of operations can easily be transformed into others by enemy actions. The specific type of conditions may not have much meaning to the individual soldier, but the ROE must be understood and adhered to by all.

1-2. FULL SPECTRUM OPERATIONS/URBAN OPERATIONS CONCEPT

The UO are conducted within the operational framework of decisive, shaping, and sustaining operations (FM 3-0[FM 100-5]). Army units will conduct offensive, defensive, stability, and support (ODSS) operations within the operational framework shown in Figure 1-1. These operations comprise the spectrum of UO that a brigade must be prepared to conduct (Figure 1-2).

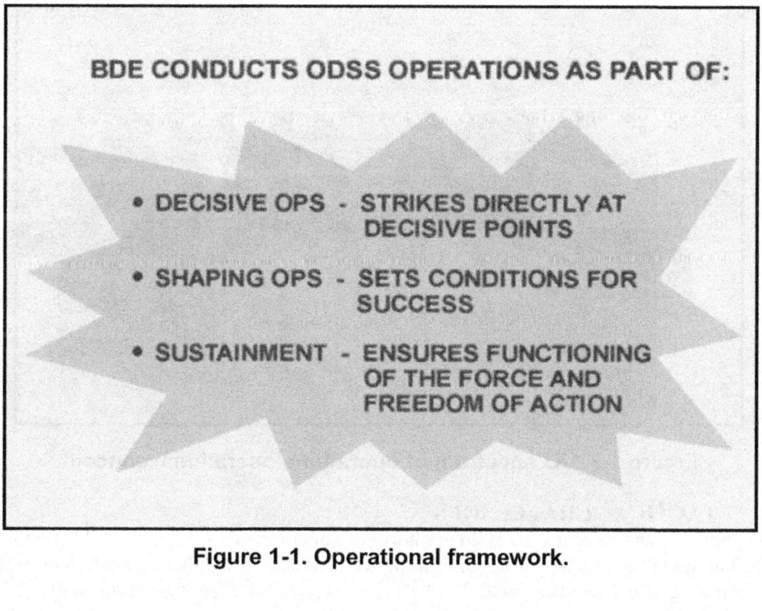

Figure 1-1. Operational framework.

a. Army operational commanders assigned to conduct UO—

- Continually *assess* the urban environment to determine effects on operations.
- Conduct *shaping* operations that emphasize isolation and set the conditions for decisive operations.
- *Dominate* through simultaneous and or sequential operations that establish and maintain preeminent military control over the enemy, geographical area, or population.
- Plan for and execute *transitions* between mission types and forces, and ultimately to the control of a non-Army agency.

b. Brigades must plan for and be prepared to conduct UO within the operational concept shown in Figure 1-2, which depicts the potential simultaneity of UO. Brigades must be prepared to transition from one type of ODSS operation to another. How brigades prepare for and execute ODSS operations will be determined by the factors of METT-TC. (Within mission considerations, the ROE will have significant importance.)

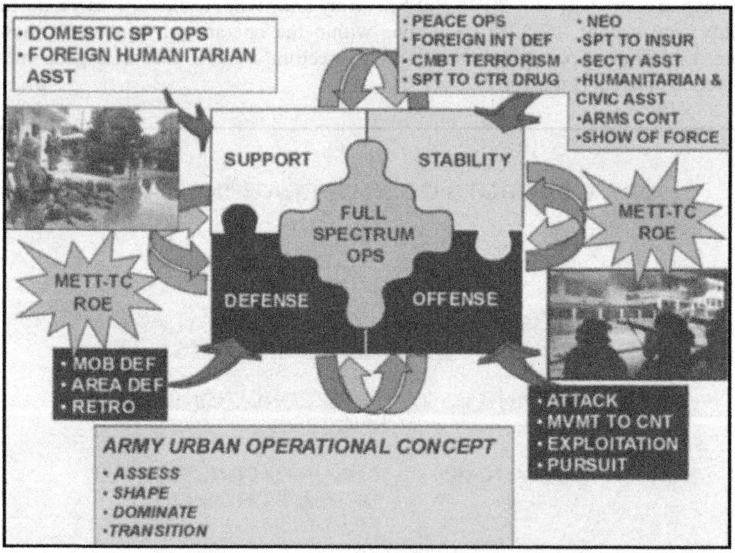

Figure 1-2. UO spectrum of operations/operational concept.

1-3. TACTICAL CHALLENGES

Companies, platoons and squads do not normally operate independently while conducting UO. The battalions to which they are assigned will face a number of challenges during the planning and execution of UO. The most likely challenges that units will face are discussed below.

a. **Contiguous and Noncontiguous Areas of Operations.** Brigades and battalions must be prepared to conduct ODSS operations in both contiguous and noncontiguous areas of operations (AO). They may be required to command and control subordinate units and elements over extended distances, which may include deploying subordinate battalions and companies individually in support of operations outside the brigade's immediate AO.

NOTE: Under the IBCT concept companies may operate independently.

(1) Contiguous operations are conducted in an AO that facilitates mutual support of combat, combat support (CS), and combat service support (CSS) elements. They have traditional linear features including identifiable, contiguous frontages and shared boundaries between forces. Contiguous operations are characterized by relatively close distances between subordinate units and elements.

(2) In noncontiguous operations, subordinate units may operate in isolated pockets, connected only through the integrating effects of an effective concept of operations. Noncontiguous operations place a premium on initiative, effective information operations, decentralized security operations, and innovative logistics measures. They make mutual support of combat, CS, and CSS elements complicated, or hinder it by extended distances between subordinate units and elements.

b. **Symmetrical and Asymmetrical Threats.** In addition to being required to face symmetrical threats, the brigade must be prepared to face threats of an asymmetrical nature.

(1) Symmetrical threats are generally "linear" in nature and include those threats that specifically confront the brigade's combat power and capabilities. Examples of symmetrical threats include conventional enemy forces conducting offensive or defensive operations against friendly forces.

(2) Asymmetrical threats are those that are specifically designed to avoid confrontation with the brigade's combat power and capabilities. These threats may use the civilian population and infrastructure to shield their capabilities from fires. Asymmetric threats may also attack the brigade and civilian population with weapons of mass destruction (WMD). Asymmetrical threats are most likely to be based in urban areas to take advantage of the density of civilian population and infrastructure. Examples of asymmetrical threats include terrorist attacks; EW, to include computer-based systems; criminal activity; guerilla warfare; and environmental attacks.

c. **Minimization of Collateral Damage and Noncombatant Casualties.** A condition that commanders and leaders will be required to confront during urban operations will be minimizing collateral damage and noncombatant casualties. This will have to be balanced with mission accomplishment and the requirement to provide force protection. Commanders must be aware of the ROE and be prepared to request modifications when the tactical situation requires them. Changes in ROE must be rapidly disseminated throughout the brigade. Commanders and leaders must ensure that changes to the ROE are clearly understood by all soldiers within the brigade.

d. **Quick Transition from Stability or Support Operations to Combat Operations and Back.** Commanders and leaders must ensure that contingencies are planned to transition quickly from stability and support to combat operations and vice-versa. For example, it may be tactically wise for commanders to plan a defensive contingency with on-order offensive missions for certain stability and support operations

Populations have dramatically increased in existing cities and urban sprawl has led to the increased number of slums and shantytowns within those urban areas. In many cases, this urbanization has occurred close to the seacoast, since the interior of many third world nations is undeveloped or uninhabitable.

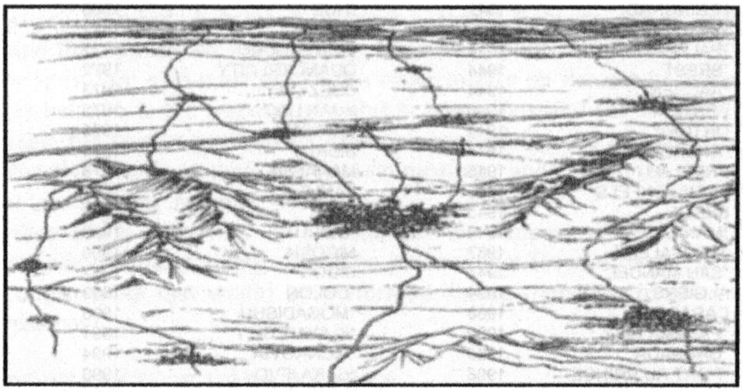

Figure 1-3. Urban areas blocking maneuver areas.

f. Extensive urbanization provides conditions that a threat force can exploit. Used with mobile forces on the adjacent terrain, conventional threat forces with antitank capabilities defending from urban areas can dominate avenues of approach, greatly improving the overall strength of the defense. Asymmetrical threats can use urban areas to offset US technological and firepower advantages.

g. Forces operating in such areas may have elements in open terrain, villages, towns, or small and large cities. Each of these areas calls for different tactics, task organization, fire support, and CSS.

1-5. FUNDAMENTALS OF URBAN OPERATIONS

The fundamentals described in this paragraph apply to UO regardless of the mission or geographical location. Some fundamentals may also apply to operations not conducted in an urban environment, but are particularly relevant in an environment dominated by manmade structures and a dense noncombatant population. Brigade and battalion commanders and staffs should use these fundamentals when planning UO.

a. **Perform Focused Information Operations and Aggressive Intelligence, Surveillance, Reconnaissance**. Information superiority efforts aimed at influencing non-Army sources of information are critical in UO. Because of the density of noncombatants and information sources, the media, the public, allies, coalition partners, neutral nations, and strategic leadership will likely scrutinize how units participate in UO. The proliferation of cell phones, Internet capability, and media outlets ensure close observation of unit activities. With information sources rapidly expanding, public information of Army operations will be available faster than the internal military information system (INFOSYS) can process it. Units can aggressively integrate information operations into every facet and at all levels of the operation to prevent negative impacts. Under media scrutiny, the actions of a single soldier may have significant strategic implications. The goal of information operations is to ensure that the information available to all interested parties, the public, the media, and other agencies, is accurate and placed in the proper context of the Army's mission. While many information operations will be planned at levels above the brigade, tactical units conducting UO may often be involved in the execution of information operations such as military deception, operations security (OPSEC), physical security, and psychological operations. Brigades and battalions must conduct aggressive intelligence, security, and reconnaissance operations that will allow them to properly apply the elements of assess, shape, dominate, and transition to specific UO.

b. **Conduct Close Combat.** Close combat is required in offensive and defensive UO. The capability must be present and visible in stability UO and may be required, by exception, in support UO. Close combat in any UO is resource intensive, requires properly trained and equipped forces, has the potential for high casualties, and can achieve decisive results when properly conducted. Units must always be prepared to conduct close combat as part of UO (Figure 1-4).

Figure 1-4. Soldiers conducting close combat in an urban area.

c. **Avoid the Attrition Approach.** Previous doctrine was inclined towards a systematic linear approach to urban combat. This approach placed an emphasis on standoff weapons and firepower. It can result in significant collateral damage, a lengthy operation, and be inconsistent with the political situation and strategic objectives. Enemy forces that defend urban areas often want units to adopt this approach because of the likely costs in resources. Commanders should only consider this tactical approach to urban combat only when the factors of METT-TC warrant its use.

d. **Control the Essential.** Many modern urban areas are too large to be completely occupied or even effectively controlled. Therefore, units must focus their efforts on controlling only the essentials to mission accomplishment. At a minimum, this requires control of key terrain (Figure 1-5). The definition of key terrain remains standard: terrain whose possession or control provides a marked advantage to one side or another. In the urban environment, functional, political, or social significance may be what makes terrain key. For example, a power station or a building may be key terrain. Units focus on control of the essential so they can concentrate combat power where it is needed and conserve it. This implies risk in those areas where units choose not to exercise control in order to be able to mass overwhelming power where it is needed.

Figure 1-5. Military airbase, an example of key terrain.

e. **Minimize Collateral Damage.** Units should use precision standoff fires, information operations, and nonlethal tactical systems to the greatest extent possible consistent with mission accomplishment. Operational commanders must develop unique ROE for each UO and provide necessary firepower restrictions. Information operations and nonlethal systems may compensate for some of these required restrictions, especially in stability operations and support operations. Moreover, commanders must consider the short and long term effects of firepower on the population, the infrastructure, and subsequent missions.

f. **Separate Combatants from Noncombatants.** Promptly separating noncombatants from combatants may make the operation more efficient and diminish some of the enemy's asymmetrical advantages. Separation of noncombatants may also reduce some of the restrictions on the use of firepower and enhance force protection. This important task becomes more difficult when the adversary is an unconventional force and can mix with the civil population.

g. **Restore Essential Services.** Tactical units may have to plan for the restoration of essential services that may fail to function upon their arrival or cease to function during an operation. Essential services include power, food, water, sewage, medical, and security. During planning for and the conduct of UO, the use of nonlethal and less destructive munitions and capabilities can help ensure that potentially vital infrastructure remains intact. Initially, Army forces may be the only force able to restore or provide essential services. However, units must transfer responsibility for providing essential services to other agencies, nongovernment organizations (NGOs), or the local government as quickly as possible.

h. **Preserve Critical Infrastructure.** Brigade and battalion commanders and staffs may have to analyze the urban area to identify critical infrastructure. Attempts to preserve the critical elements for post-combat sustainment operations, stability operations, support operations, or the health and well-being of the indigenous population may be required. Urban areas remain in the AO after combat operations have ceased and post-combat UO may be unavoidable. This requirement differs from simply avoiding collateral damage in that units may have to initiate actions to prevent adversaries from removing or destroying infrastructure that will be required in the future. In some cases, preserving critical infrastructure may be the assigned objective of the UO.

i. **Understand the Human Dimension.** Brigade and battalion commanders and staffs may have to carefully consider and manage the allegiance and morale of the civilian population that may decisively affect operations. The assessment of the urban environment must identify clearly and accurately the attitudes of the urban population toward units. Guidance to subordinates covering numerous subjects including ROE, force protection, logistics operations, and fraternization, is part of this assessment. Brigade and battalion commanders and staffs may also be required to consider the demographic variance in the attitudes of an urban population. Western cultural norms may not be appropriate if applied to a nonwestern urban population. Commanders and staffs must make their assessments based on a thorough understanding and appreciation of the local social and cultural norms of the population. Sound policies, discipline, and consideration will positively affect the attitudes of the population toward Army forces. Additionally, well-conceived information operations can also enhance the position of units relative to the urban population. Even during combat operations against a conventional enemy force, the sensitivity and awareness of units toward the civilian population will affect the post combat situation. The human dimension of the urban environment often has the most significance and greatest potential for affecting the outcome of UO.

j. **Transition Control.** UO of all types are resource intensive and thus commanders must plan to conclude UO expediently, yet consistent with successful mission accomplishment. The end state of all UO transfers control of the urban area to another agency or returns it to civilian control. This requires the successful completion of the Army force mission and a thorough transition plan. The transition plan may include returning control of the urban area to another agency a portion at a time as conditions permit. For brigades and below, transition may also include changing missions from combat operations to stability operations or support operations and vice–versa.

1-6. CHARACTERISTICS OF URBAN OPERATIONS

Many characteristics separate UO from other environments. US technological advantages are often not very useful during UO. Air power may not be of any assistance to an Infantry force fighting from buildings. An adept enemy will use the technique of "hugging" American forces to deny them use of their overwhelming firepower. The training and equipment for the fight against a mobile, armored threat may not necessarily be of much use in urban areas. Urban combat is primarily a small unit Infantry fight, requiring significant numbers of Infantry to accomplish the mission; however, combined arms must support the Infantry. Urban combat is characterized by moment-to-moment decisions by individual soldiers, which demonstrates the importance of ROE training. Commanders and leaders should facilitate this fight by anticipating what subordinates will need to accomplish the mission. Unit goals must be speed, precision, and minimization of soldiers in close combat with the enemy. The greatest threats might be snipers, grenade launchers, booby traps, and rocket-propelled grenades (RPGs). Soldiers can expect booby traps on doorways and windows and on entrances to underground passageways.

a. **Changing Conditions.** Platoons and squads may find themselves executing missions in changing conditions during UO. The change from stability and support operations to combat operations and vice-versa will often change conditions from high-intensity to precision or the opposite. METT-TC factors and the ROE determine this change. ROE changes are normally made at echelons much higher than company and battalion, but they normally require that units modify the way they fight in urban areas. Squads and platoons will be required to select different TTP based on the conditions they face. The ROE will ultimately determine these conditions for the Infantry platoon and squad.

b. **Small-Unit Battles.** Units fighting in urban areas often become isolated or feel like they are isolated, making combat a series of small-unit battles. Soldiers and squad or team leaders must have the initiative, skill, and courage to accomplish their missions while isolated from their parent units. A skilled, well-trained defender has tactical advantages over the attacker in this type of combat. The defender may occupy strong covered and concealed static positions and conduct three-tier ambushes, whereas the attacker must be exposed in order to advance. Greatly reduced line-of-sight ranges, built-in obstacles, and compartmented terrain may require the commitment of more troops for a given frontage. While the defense of an urban area can be conducted effectively with relatively small numbers of troops, the troop density required for an attack in urban areas may be greater than for an attack in open terrain. Individual soldiers must be trained and psychologically ready for this type of operation.

c. **Communications.** Urban operations require centralized planning and decentralized execution. Therefore, effective vertical and horizontal communications are critical. Leaders must trust their subordinates' initiative and skill, which can only occur through training. The state of a unit's training and cohesion are vital, decisive factors in the execution of operations in urban areas.

(1) Structures and a high concentration of electrical power lines normally degrade radio communications in urban areas. Many buildings are constructed so that radio waves will not pass through them. Frequently, units may not have enough radios to communicate with subordinate elements as they enter buildings and move through urban canyons and defiles.

(2) Visual signals may also be used but are often not effective because of the screening effects of buildings, walls, and so forth. Signals must be planned, widely disseminated, and understood by all assigned, attached, or OPCON units. Increased noise makes the effective use of sound signals difficult. Verbal signals may also reveal the unit's location and intent to the enemy.

(3) Messengers and wire can be used as other means of communication. Messengers are slow and susceptible to enemy fire when moving between buildings or crossing streets. Wire is the primary means of communication for controlling the defense of an urban area. It should be considered as an alternate means of communications during

offensive operations, if assets are available. However, wire communications can often be cut by falling debris, exploding munitions, and moving vehicles.

d. **Noncombatants.** Urban areas, by their very nature, are population centers. Noncombatants will be present and will affect both friendly and threat courses of action across the spectrum of UO. Besides the local inhabitants, refugees, governmental and NGOs, and the international media are likely to be present. For example, during the fighting in Grozny, 150,000 refugees were added to a prefight population of 450,000. There were 50,000 civilian casualties during the fight. Units must be prepared to deal with all categories of noncombatants

e. **High Expenditure of Ammunition.** Units conducting UO use large quantities of ammunition because of short ranges, limited visibility, briefly exposed targets, constant engagements, and requirements for suppression. AT4s, rifle and machine gun ammunition, 40-mm grenades, hand grenades, and explosives are high-usage items in this type of fighting. When possible, those items should be either stockpiled or brought forward on-call, so that they are easily available.

f. **Increased Casualties.** More casualties occur because of shattered glass, falling debris, rubble, ricochets, urban fires, and falls from heights. Difficulty in maintaining situational awareness also contributes to this problem because of increased risks of fratricide. Stress-related casualties and nonbattle injuries resulting from illnesses or environmental hazards, such as contaminated water, toxic industrial materials (TIM), and so forth, also increase the number of casualties.

g. **Limited Mounted Maneuver Space.** Buildings, street width, rubble, debris, and noncombatants all contribute to limited mounted maneuver space inside urban areas. Armored vehicles will rarely be able to operate inside an urban area without Infantry support.

h. **Three-Dimensional Terrain.** Friendly and threat forces will conduct operations in a three-dimensional battle space. Engagements can occur on the surface, above the surface, or below the surface of the urban area. Additionally, engagements can occur inside and outside of buildings. Multistory buildings will present the additional possibility of different floors within the same structure being controlled by either friendly or threat forces.

i. **Collateral Damage.** Depending on the nature of the operation and METT-TC factors, significant collateral damage may occur, especially under conditions of high-intensity UO. Commanders and leaders must ensure that ROE are disseminated and enforced.

j. **Reliance on Human Intelligence.** Until technological advancements provide more effective ways of gathering information, there is an increased need for human intelligence (HUMINT). Reconnaissance efforts of battalion and brigade assets can assist as well as the shaping operations of division or joint task force assets. Companies and below normally have to continue to rely on information provided to them from human sources.

k. **Need for Combined Arms.** While UO historically have consisted of a high density of Infantry-specific tasks, UO conducted purely by Infantry units have proven to be unsound. Properly tasked-organized combined arms teams consisting primarily of Infantry, engineers, and armor supported by other combat, CS, and CSS assets have proven to be more successful both in the offense and defense. The same concept is true for stability and support operations, when the main effort may not necessarily consist of combat units.

l. **Need to Isolate Critical Points.** During offensive operations, companies, platoons, and squads will be assaulting buildings and clearing rooms. More often, assets will not exist to isolate large portions of the urban area. Therefore, skillful use of direct and indirect fires, obscurants, and maneuver must occur to isolate key buildings or portions of buildings in order to secure footholds and clear.

m. **Snipers.** Historically, snipers have had increased utility in urban areas. They can provide long- and short-range precision fires and can be used effectively to assist company- and platoon-level isolation efforts. Snipers also have provided precision fires during stability operations. Along with engaging assigned targets, snipers are a valuable asset to the commander for providing observation along movement routes and suppressive fires during an assault.

n. **Support by Fire Positions.** Buildings, street width, rubble, debris, and noncombatants all dictate the positioning and fields of fire for crew-served and key weapons in urban areas.

1-7. URBAN BATTLE SPACE

Urban areas mainly consist of man-made features such as buildings that provide cover and concealment, limit fields of observation and fire, and block movement of forces, especially mechanized or armored forces. Thick-walled buildings provide ready-made, fortified positions. Thin-walled buildings may have fields of observation and fire that may prove important. Another important aspect is that urban areas complicate, confuse and degrade the commander's ability to identify and control his forces. All these factors will influence the urban battle space.

a. Commanders and leaders can enhance situational understanding by maintaining a clear understanding of their urban battle space (Figure 1-6, page 1-16). Urban battle space includes:

(1) ***Urban Airspace.*** Airspace provides a rapid avenue of approach into an urban area. While aviation assets are unaffected by obstacles such as rubble, vehicles, or constructed barriers, they must consider power lines, towers, sign poles, and billboards when flying. Task force reconnaissance elements can locate, identify, and report these obstacles to allow for improved flight planning.

(2) ***Supersurface (Tops of Buildings).*** The term "supersurface" refers only to the top, roof, or apex of a structure. These areas can provide cover and concealment, limit or enhance observation and fields of fire, and, depending on the situation, enhance, restrict, canalize, or block movement. Supersurface areas can also provide concealed positions for snipers, automatic weapons, light and medium antitank weapons, and man-portable air defense systems. In many cases, they enable top-down attacks against the weakest points of armored vehicles and unsuspecting aircraft.

(3) ***Intrasurface (Interior of Buildings).*** The intrasurface refers to the floors within the structural framework—the area from the surface level (ground) up to, but not including, the structure's permanent roof or apex. Intense combat engagements often occur in this intrasurface area, which is also known for its widely diverse and complex nature. The intrasurface of a building greatly limits what can be accomplished by reconnaissance and surveillance systems, but, at the same time, enhances cover and

concealment. Additionally, the intrasurface areas provide mobility corridors within and between structures at upper levels for both friendly and enemy forces. Intrasurface areas may also provide concealed locations for snipers, automatic weapons, light and medium antitank weapons, and man-portable air defense systems. In many cases, they enable top-down attacks against the weakest points of armored vehicles and unsuspecting aircraft.

(4) ***Surface (Ground, Street, and Water Level).*** Streets are usually avenues of approach. Streets and open areas provide a rapid approach for ground movement in urban terrain. Units moving along streets can be canalized by buildings and have little space for maneuver, while approaching across large open areas such as parks, athletic fields and parking areas. Streets also expose forces to observation and engagement by enemy elements. Obstacles on streets in towns are usually more effective than those on roads in open terrain since they are more difficult to bypass.

(5) ***Subsurface (Underwater and Subterranean).*** Common subsurface areas, which include subways, sewers, public utility systems, and cellars, can be used as avenues of movement for dismounted elements. Both attacker and defender can use subterranean routes to outflank or turn the opposition, or to conduct infiltration, ambushes, counterattacks, and sustaining operations. Subsurface systems in some urban areas are easily overlooked but can be important to the outcome of operations.

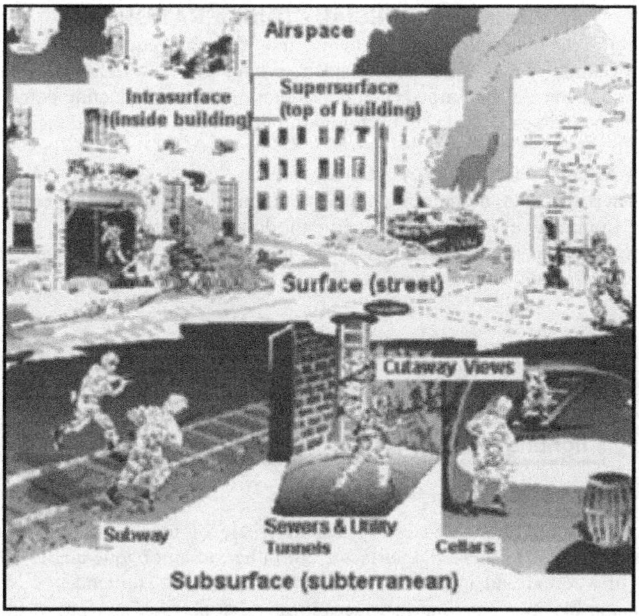

Figure 1-6. Urban battle space.

b. Commanders and leaders must be able to identify building types, construction materials, and building design and must understand the effectiveness and limitations of weapons against these factors. (See Chapters 3 and 8.) They must also understand that urban combat will require them to visualize a three-dimensional battle space. Commanders and leaders must be aware of how their urban battle space is changing as friendly and enemy forces and civilians move, and as weather and environmental conditions change. They can react to changes within the their battle space with the timely movement of assault, support, and breaching elements in the offense; repositioning of platoons and squads in the defense; and synchronization of CS and CSS assets. Other factors that impact battle space are:

- CASEVAC and resupply procedures.
- Handling EPWs and noncombatants.
- Rules of engagement. (See Appendix A.)
- Battlefield obscuration.
- Communications.
- Movement of vehicles. (How will the battle space affect movement and target engagement?)

Section II. SPECIAL CONSIDERATIONS

Employment of weapons, target engagements, munitions and equipment, noncombatants, disease prevention, stress, fratricide avoidance, situational awareness, the media, and unexploded ordnance all require special consideration when conducting combat in urban areas. (See Appendixes B, C, D, and E for additional special considerations.)

1-8. WEAPONS CONSIDERATIONS

The characteristics and nature of combat in urban areas affect the employment of weapons. Commanders and leaders at all levels must consider the following considerations in various combinations when choosing their weapons. (See Chapter 7 for detailed discussion of specific weapons.)

WARNING

Protecting personnel from backblast or fragmentation effects must be considered when fighting in urban areas.

a. Hard, smooth, flat surfaces are characteristic of urban targets. Rounds rarely impact perpendicular to these flat surfaces, but rather, at an oblique angle. This reduces the effect of a round and increases the threat of ricochets. The tendency of rounds to strike glancing blows against hard surfaces means that up to 25 percent of impact-fused explosive rounds may not detonate when fired into urban areas. Deflected rounds can easily ricochet or "rabbit" causing injury and death from strange angles. (A "rabbit" round is a round or fragment that strikes a surface at such a steep angle that it glances off and continues to travel parallel to that surface.)

b. Engagement ranges are close. Studies and historical analyses have shown that only 5 percent of all targets are more than 100 meters away. About 90 percent of all targets are located 50 meters or less from the identifying soldier. Few personnel targets will be visible beyond 50 meters and they usually occur at 35 meters or less. Engagement times are short. Enemy personnel present only fleeting targets.

c. Depression and elevation limits for some weapons create dead space. Tall buildings form deep canyons that are often safe from indirect fires. Some weapons can fire rounds to ricochet behind cover and inflict casualties. Target engagement from oblique angles, both horizontal and vertical, demands superior marksmanship skills.

d. Smoke from burning buildings, dust from explosions, shadows from tall buildings, and the lack of light penetrating inner rooms all combine to reduce visibility and increase a sense of isolation. Added to this is the masking of fires caused by rubble and manmade structures. Targets, even those at close range, tend to be indistinct.

e. Urban fighting can become confused mêlées with several small units attacking on converging axes. The risks from friendly fires, ricochets, and fratricide must be considered during the planning phase of operations, and control measures must be adjusted to lower these risks. Soldiers and leaders must maintain a sense of situational awareness.

f. The soldier and target may be inside or outside buildings, or they may both be inside the same or separate buildings. The enclosed nature of combat in urban areas means that all the weapon's effects including the muzzle blast and backblast, must be considered as well as the round's impact on the target.

g. Usually the man-made structure must be attacked before enemy personnel inside are attacked. Therefore, the decision to employ specific weapons and demolitions will often be based on their effects against masonry and concrete rather than against enemy personnel.

h. Modern engineering and design improvements mean that most large buildings constructed since World War II are resilient to the blast effects of bomb and artillery attack. Even though modern buildings may burn easily, they often retain their structural integrity and remain standing. Once high-rise buildings burn out, they may still have military utility and are almost impossible to damage further. A large structure can take 24 to 48 hours to burn out and become cool enough for soldiers to enter.

1-9. TARGET ENGAGEMENT

Most target engagements in urban areas are at ground level and above ground level. The following characteristics are considered when engaging targets.

a. **Ground Level.** At ground level the ranges of observation and fields of fire are reduced by structures as well as by the dust and smoke of battle. The density of urban terrain limits the ability of soldiers to employ their weapons out to the weapon's maximum effective range. Historically, engagements have been very close, often 100 meters or less. In urban areas, the ranges of observation and fields of fire are reduced by structures as well as by the dust and smoke of battle. As a result, urban engagements consist mostly of close, violent firefights. This situation requires an increase of precision/accurate small arms fire as well as strict fire control and proper identification of political opponents of US involvement may be hostile towards the US military presence but do not pose a threat to US forces.

b. **Noncombatants and Rules of Engagement.** All leaders and soldiers must understand the potential urban battlefield and the fact that they will most likely encounter noncombatants. If soldiers must deal with noncombatants, they should refer to their ROE. ROE should be very specific on treatment of each type of noncombatant [paragraph a(2)].

c. **Communication with Civilians.** Soldiers should learn basic commands and phrases in the language most common to their area of operations. When giving these commands or phrases, they should speak loudly and clearly at a normal rate and use gestures whenever possible. All soldiers should be given a basic language translation card. (See example in Table 1-2.)

ENGLISH WORD OR PHRASE	FOREIGN WORD OR PHRASE	PRONUNCIATION
HALT		
WHAT IS YOUR NAME?		
STAND UP		
WALK		
SIT DOWN		
YOU WILL BE SEARCHED NOW		
DON'T TALK		
YES		
NO		
NOT PERMITTED		
MEDICAL AID		
FOOD		
WATER		
USE THE LATRINE?		

Table 1-2. Example of basic language translation card.

d. **Cultural Issues.** Soldiers must be educated on the types of cultural issues that may offend the local inhabitants. For example, a gesture that may be innocent to Americans may deeply insult the inhabitants.

e. **Considerations for Handling Noncombatants.** Commanders and leaders should consider using CA, PSYOP, MPs, chaplains, and civil leaders and authorities, when available, if their mission involves handling noncombatants. Other considerations include the following:

(1) Carefully analyze the ROE concerning when deadly force can be used and what type of weapons may be employed (for example, using lethal as opposed to nonlethal weapons and capabilities).

(2) Do not assume that noncombatants will be predisposed for or against US troops. Always treat civilians with dignity and respect. Use force against civilians only in self-defense or, otherwise, in accordance with the ROE. Detain civilians only in accordance with command directives.

(3) When conducting offensive operations, plan to move any noncombatants away from firefights. Normally this task will be given to the support element after rooms and buildings have been secured. When available, PSYOP, CA, and MPs can assist with this task. A covered and concealed location away from the immediate combat area should be chosen. Noncombatants should be controlled and not permitted to enter the immediate combat area, unless they have been cleared to do so and will not compromise combat operations (for example, media personnel or governmental or NGO personnel that have a reason and authority to enter the combat area).

(4) When conducting defensive operations, plan to move noncombatants away from the immediate combat area. Companies and below are normally escort personnel to a designated location where they are turned over to civil authority, battalion, or higher control. In many cases, friendly or nonhostile civilians may be directed to a clearing point and allowed to go there without escort.

(5) Security is not normally provided for media or NGO personnel if they are permitted in the immediate combat area. Security requirements for civilians should be clarified at the mission briefing.

(6) Based on the factors of METT-TC, units may have to render some type of immediate humanitarian assistance (medical attention and feeding). Any plan that provides for the provision of medical care to the civilian population must be developed in conjunction with the staff judge advocate. If this type of assistance is necessary, clarify questions in the mission briefing. Battalion and brigade staffs can plan for and bring forward additional Class VIII and Class I, as appropriate.

f. **Determining the Status of Personnel.** Companies and below do not determine the status of individuals in the combat area. Any persons that are initially detained should be treated as PWs, and higher headquarters should be notified with a request for assistance in evacuating these individuals.

1-12. DISEASE PREVENTION

Many third world countries have poor sewage and refuse removal, as well as low-quality water supplies. Even cities with the most modern facilities revert quickly to substandard sanitation and have problems with disease when utilities are interrupted by natural catastrophes or urban combat. Exposure to disease can decimate a unit as quickly as combat. Normal field sanitation may be difficult due to lack of water and the inability to dig. Water utilities may not be working due to destruction of the water facilities, lack of power, or the lack of experienced personnel to run them. Even if the water is running, it cannot be expected to have been properly treated. Commanders and leaders must ensure that soldiers drink water only from designated sources or utilize water purification methods. Additionally, first sergeants must coordinate with the battalion S4 for waste removal and latrine facilities, if applicable. Medics must constantly monitor soldiers in an urban area for signs of disease or illness, and provide appropriate medical attention.

1-13. STRESS

The confusion, stress, and lethality of the modern battlefield place a burden on the individual soldier's endurance, courage, perseverance, and ability to perform in combat. Continuous close combat, intense pressure, high casualties, fleeting targets, and fire from a concealed enemy produce psychological strain and physical fatigue. Such stress requires consideration for the soldiers' and small-unit leaders' morale and the unit's esprit de corps. Rotating units that have been committed to heavy combat for long periods can reduce stress. Soldiers conducting combat operations must perform complex collective and individual tasks without adequate sleep and under stress. Commanders and leaders must be aware of what can cause stress, minimize those factors to the extent possible, and take every opportunity to rest soldiers. The following cause stress in combat and may be intensified in the urban environment.

a. **Anxiety.** The fear and anticipation of the unknown can have devastating effects on the mental and physical wellbeing of a person. Soldiers may experience the fear of death or being wounded or, because of the three-dimensional battlefield, the possibility of being engaged from all directions simultaneously. A lack of communication with others may cause a feeling of isolation and vulnerability.

b. **Intense Noise, Limited Visibility, and Low Light Levels**. Smoke, darkness, fog, rain, snow, ice, and glare make it hard to see. The extended wear of night vision goggles, protective masks, or laser protective lenses causes stress. Intense noise not only causes stress by itself, but it further isolates the soldier from human contact and interferes with situational awareness.

c. **Disrupted Wake/Sleep Cycle.** A soldier's performance suffers during normal sleeping hours due to the disruption of the normal schedule.

d. **Decision Making and Responsibility for Others.** Mental stress results from making vital decisions with little time and insufficient information. It is increased during times of great confusion and exposure to danger. Leaders are especially affected by the heavy load of responsibility they carry.

e. **Physical Fatigue and Illness.** Working the muscles faster than they can be supplied with oxygen and fuel can cause soldiers to function poorly without rest. Minor illnesses that do not completely disable the soldier add to his stress and hinder his ability to function at his full potential.

f. **Physical Discomfort.** Extreme cold, heat, wet, or thirst add greatly to the level of individual stress. Minor injuries or wounds can cause constant pain that, while not incapacitating to the soldier, add to his stress.

g. **Psychological Stress.** Commanders and leaders must remain alert for the signs of psychological stress. During the fighting in Grozny, 72 percent of the Russian soldiers demonstrated some kind of psychological disorder symptoms such as insomnia, lack of motivation, high anxiety, neuro-emotional stress, fatigue, and hypochondria. Brigade and battalion surgeons must be prepared for soldiers exhibiting signs of psychological stress.

1-14. FRATRICIDE AVOIDANCE

The overriding consideration in any tactical operation is accomplishing the mission. Commanders and leaders must consider fratricide in their planning process because of the decentralized nature of execution during UO. However, they must weigh the risk of fratricide against losses to enemy fire when considering a given course of action. Fratricide can be avoided by sound doctrine; proper selection and application of tactics, techniques, and procedures; detailed planning; disciplined execution; and rigorous, in-depth rehearsals.

a. **Doctrine**. Doctrine provides the basic framework for accomplishing the mission. Commanders and leaders must have a thorough understanding of US doctrine and, if operating with other services or nations, joint, combined, and host nation doctrine.

b. **Tactics, Techniques, and Procedures.** TTP provide a "how-to" that everyone understands. TTP are disseminated in doctrinal manuals and standing operating procedures (SOPs).

(1) ***Tactics.*** Tactics are the employment of units in combat or the ordered arrangement and maneuver of units in relation to each other and or the enemy to use their full potential. For example, a company employing support by fire elements from a secured foothold (intermediate objective) prior to conducting the assault on the objective.

(2) ***Techniques.*** Techniques are the general and detailed methods used by troops or commanders to perform assigned missions and functions. Specifically, techniques are the methods of using weapons and personnel. Techniques describe a method, but not the only method. An example is using precision room clearing techniques

(3) ***Procedures.*** Procedures are standard, detailed courses of action that describe how to accomplish a task. Examples might be using green colored squares to mark cleared rooms during an assault of a building, or marking each soldier with clear, identifiable markings that are IR visible as well.

c. **Planning.** A simple, flexible maneuver plan that is disseminated to the lowest level will aid in the prevention of fratricide. Plans should make the maximum use of SOPs and battle drills at the user level. They should incorporate adequate control measures and fire support planning and coordination to ensure the safety of friendly troops and allow changes after execution begins.

d. **Execution.** The execution of the plan must be monitored, especially with regard to the location of friendly troops and their relationship to friendly fires and the effects of those fires on the structural integrity of the building. For example, a fragmentation grenade used in a weakly constructed building may cause grenade fragments to pass through walls and injure friendly troops. Additionally, subordinate units must understand the importance of accurately reporting their position.

e. **Rehearsals and Training.** The most important factor in preventing fratricide is effective individual and collective training in the many tasks that support UO. Often the only combined arms training that will occur are the rehearsals with attached or OPCON assets such as engineers or armored vehicles.

1-15. SITUATIONAL AWARENESS

Situational awareness is the degree to which one is able to maintain a common operating picture of all aspects of the tactical situation. This picture includes an understanding of the friendly and enemy situation and the urban battle space. Since units will have to conduct operations in changing mission environments, it is imperative for commanders and leaders at all levels to achieve and maintain the best possible degree of situational awareness. Enhanced situational awareness will enhance lethality, survivability, and operational tempo.

a. **Urban Battle Space.** See paragraph 1-7 for detailed information.

b. **Questions.** To the company level leaders situational awareness means being able to answer certain questions:

- Where am I (in respect to the urban area or my assigned sector)?

of UXO becomes important to mission accomplishment. Again, the battalion and or brigade engineer should anticipate these requirements and plan the coordination with EOD teams for the detailed identification, removal, and or disposal of UXO, as applicable. Explosive ordnance disposal (EOD) teams, if available, should be used for the removal and or disposal of UXO. The battalion and or brigade engineer should plan to provide support to assist EOD teams with large-scale UXO operations.

b. **Company Team and Platoon Planning Considerations.** At the company level and below, the primary concern should be the identification and marking of UXO. Company teams and below should not attempt to move or destroy UXO. They may provide security to EOD teams and engineers during removal and or disposal operations. To assist in planning, the nine-line UXO spot report listed below should be used. (See FM 4-30.16.)

Line 1. Date-Time Group: DTG item was discovered.

Line 2. Reporting Activity: (Unit identification) and location (grid of UXO).

Line 3. Contact Method: Radio frequency, call sign, point of contact (POC), and telephone number.

Line 4. Type of Ordnance: Dropped, projected, placed, or thrown. If available, supply the subgroup. Give the number of items, if more than one.

Line 5. NBC Contamination: Be as specific as possible.

Line 6. Resources Threatened: Report any equipment, facilities, or other assets that are threatened.

Line 7. Impact on Mission: Provide a short description of the current tactical situation and how the presence of the UXO affects the status.

Line 8. Protective Measures: Describe any measures taken to protect personnel and equipment.

Line 9. Recommended Priority: Recommend a priority for response by EOD technicians or engineers.

Leaders at the company level and below must ensure that clear instructions are provided to the soldiers with reference to not handling any UXO. The UXO itself may not look particularly menacing and curiosity may lead soldiers to want to pick-up or move UXO. Soldiers must be briefed on the potential lethality of such actions. Proper reaction to UXO takes on increased significance during stability operations and support operations when soldiers are conducting missions in close proximity to noncombatants. (See Figures 1-8 through 1-11, pages 1-28 through 1-29.)

c. **Toxic Industrial Materials (TIM).** See Appendix F. Often UO are performed in areas that contain TIM. For example, damage or destruction of chemical or petroleum production facilities can produce extremely hazardous conditions and may actually prevent or seriously hamper mission accomplishment. Many of the UXO planning factors discussed in paragraphs a and b also apply to TIM. Battalion and or brigade chemical officers should coordinate with engineer officers and EOD battalion and or brigade elements to plan appropriate preventive or reactive measures in the event of a release of TIM. (See FM 4-30.16 for further information on UXO.)

Figure 1-8. Example of a threat ammunition supply point.

Figure 1-9. Unexploded rocket-propelled grenades (RPGs).

Figure 1-10. Unexploded foreign air delivered bomb.

Figure 1-11. Unexploded submunition on a street.

CHAPTER 2

URBAN ANALYSIS

Combat in urban areas requires thorough knowledge of the terrain and detailed intelligence preparation of the battlefield (IPB). To succeed in urban areas, commanders and leaders must know the nature of both the terrain and of the enemy they may face. They must analyze the effect the urban area has on both threat and friendly forces. The focus of the material presented in this chapter will be on those issues of urban analysis that commanders and leaders must consider before beginning the detailed planning process. (For a detailed explanation of urban IPB, see FM 34-130).

Section I. MODELS OF URBAN AREAS

Each model of an urban area has distinctive characteristics, but most resemble the generalized model shown in Figure 2-1.

Figure 2-1. Generalized model of an urban area.

2-1. GENERAL URBAN CHARACTERISTICS

Although different in many important ways, urban areas all over the world share many characteristics such as density of construction and population, street patterns, compartmentalization, affluent and poor sections, modernization, and presence of random and close orderly block are two common construction patterns that can be found within the city core and core periphery zones.

Figure 2-3. Core periphery.

(a) *Dense Random Construction* (Figure 2-4). This construction is a typical old inner city pattern with narrow winding streets radiating from a central area in an irregular manner. Buildings are closely located and frequently close to the edge of a roadway.

Figure 2-4. Dense random construction.

(b) *Close Orderly Block Construction* (Figure 2-5). Wider streets generally form rectangular patterns in this area. Buildings frequently form a continuous front along the blocks. Inner-block courtyards are common.

Figure 2-5. Close orderly block construction.

(3) ***Dispersed Residential Area*** (Figure 2-6). This type area is normally contiguous to close-orderly block areas in Europe. The pattern consists of row houses or single-family dwellings with yards, gardens, trees, and fences. Street patterns are normally rectangular or curving.

Figure 2-6. Dispersed residential area.

(4) ***High-Rise Area*** (Figure 2-7). Typical of modern construction in larger cities and towns, this area consists of multistoried apartments, separated open areas, and single-story buildings. Wide streets are laid out in rectangular patterns. These areas are often contiguous to industrial or transportation areas or interspersed with close-orderly block areas.

Figure 2-7. High-rise area.

(5) ***Industrial-Transportation Area*** (Figure 2-8). Industrial-transportation areas are generally located on or along major rail and highway routes in urban complexes. Older complexes may be located within dense, random construction or close-orderly block areas. New construction normally consists of low, flat-roofed factory and warehouse buildings. High-rise areas providing worker housing is normally located adjacent to these areas throughout the Orient. Identification of transportation facilities within these areas is critical because these facilities, especially rail facilities, pose significant obstacles to military movement.

Figure 2-8. Industrial-transportation area.

(6) ***Permanent or Fixed Fortifications*** (Figure 2-9). These include any of several different types and may be considered isolated forts, such as the Hue Citadel and the German fortifications that surrounded Metz, or as part of a fortified line (Siegfried and Maginot Lines). While most of these fortifications are found in Western Europe, many can be found in the Balkans, Middle East, Asia, Africa, and South America. Those in the United States are mostly of the coast defense type. Permanent fortifications can be made of earth, wood, rock, brick, concrete, steel-reinforced concrete, or any combination of the above. Some of the latest variants are built underground and employ heavy tank or warship armor, major caliber and other weapons, internal communications, service facilities, and NBC overpressure systems.

Figure 2-9. Permanent or fixed fortifications.

(7) ***Shantytowns.*** Shantytowns do not necessarily follow any of the above patterns and may be found in many different zones within urban areas. Many underdeveloped countries are composed of small towns and villages and very few large cities. Most of the structures in the small towns and villages may be constructed from materials ranging from cardboard to concrete block. Some countries in arid regions depend on adobe for construction. Even the larger cities can have shantytowns at the edge that consist of cardboard or tin shacks (Figure 2-10, page 2-8).

(a) These less structurally sound buildings have no common floor pattern and are more likely to have only one room. These types of substandard structures present a problem of weapons over-penetration. Weapons fired in one structure may penetrate the walls of one or more buildings. This penetration becomes a hazard for friendly forces as well as noncombatants. In order for buildings not to be structurally damaged or completely destroyed, reduce the explosive charges or do not use them. Fires are also more likely to develop and spread in shantytowns.

(b) Depending upon the type of operation, the temporary nature of the structures can mean that mobility can be either more or less restricted than other sections of an urban area. A unit with armored vehicles may easily knock down and traverse structures without affecting mobility at all. However, their destruction may cause unacceptable civilian casualties, in which case mobility becomes more restrictive as the narrow paths often do not accommodate vehicles. Regardless, commanders must carefully consider the effects of their operations in this area, to include vehicles and weapons, as the weak structures afford little protection increasing the risk of fratricide, civilian casualties, and large, rapidly spreading fires.

Figure 2-10. Shantytown construction.

b. **Street Patterns**. Knowledge of street patterns and widths gives commanders and leaders a good idea of whether or not mounted mobility corridors in different zones can permit wheeled or tracked vehicles and facilitate command and control. For example, a rectangular, radial, radial ring, or combined pattern facilitates movement and control better than irregular patterns. Common street patterns within the AI and AO are described in Figure 2-11.

Shape	Street Pattern	Effect
	Rectangular or Chessboard	Streets are grid-like, with parallel streets intersected by perpendicular streets.
	Rayed	Streets that fan out at various angles from a given focal point and through less than 360 degrees.
	Radial	Primary thoroughfares radiate out from a central point. These streets may be extended outward 360 degrees around the central point or within an arc from a point along a natural barrier, such as a coastline.
	Radial-Ring	Loops or rings are surrounded by successively larger ones. Usually found in conjunction with larger radial patterns. Radial rings incorporate the elements of both radial and ring/concentric designs.
	Contour Forming	Pronounced terrain relief influences construction of roadways along lines of elevation. Primary streets run parallel to contour lines, with intersecting roads connecting them.
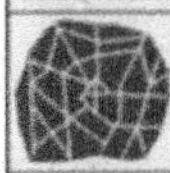	Irregular Pattern	Irregular street patterns have been specifically engineered without geometric patterns for aesthetic or functional reasons. An American subdivision with curving streets and cul-de-sacs is an example.
	Combined Pattern	Any combination of the above and is best demonstrated by the development of high rise and business districts in Medieval or pre-Medieval cities.
	Linear Pattern	A primary thoroughfare radiates down the center with buildings on either side. American strip malls and main shopping districts are patterned this way for ease and convenience.

Figure 2-11. Street patterns and effects.

2-4. SPECIAL TERRAIN CONSIDERATIONS

Several special considerations have implications in a terrain analysis and must be considered when developing the tactical plan for combat. Special terrain products must be developed to include specialized overlays, maps, and plans augmented by vertical or hand-held imagery. The depiction of obstacles, avenues of approach, key terrain, observation and fields of fire, and cover and concealment must focus on the terrain analysis. While much of this information is provided by the battalion S2 (Figure 2-12, page 2-10), the company commanders must analyze the terrain that affects the execution of their mission, and provide the results of the analysis to platoon leaders. This mission is accomplished in the same manner as other operations, using the OCOKA factors. Figure 2-13, on page 2-11, depicts a sample terrain analysis based on a company's specific objective area. The amount and type of information included in this analysis is time dependent.

	City Core	Core Periphery	Dispersed Residential Area	High – Rise Area	Industrial Transportation Area	Permanent or Fixed Fortifications
Observation & Fields of Fire **O**	-Bldgs 15-20 stories. -Short lines of sight reduced ranges of weapon systems. -Lighting may hinder NVGs	-Bldgs 5-10 stories. -Increased fields of view. -Longer lines of sight. -More distinct building types.	-Lighting may hinder NVGs. -Building normally 1-2 stories.	-Multi-story apartment and office -Building(10-50 stories) -Mid-long lines of sight.	-Potential obscuration due to fire. -Many open areas.	-Defender has advantage. -Pre-positioned observation posts.
Cover and Concealment **C**	-Mix of commercial buildings., and construction types. -Best cover available.	-Close grouping or attached housing. -Variety of construction types. -Potential limits cover.	-Lighter construction buildings.	-Large open areas adjacent to high-rise building framed heavy, light clad con-struction.	-Working materials and cargo offer multiple position. -Framed, light clad buildings, some brick.	-Normally limited for the attacker and reinforced to the defender.
Obstacles **O**	-Greater number of civilian vehicles.	-Greater number of civilian vehicles.	-Possible power lines, poles, and cell phone towers as obstacles for a/c. -Increased number of civilians.	-Access to stairwells may be limited.	-Potential secondary explosions due to flammable and explosive hazards and chemical hazards.	-Limited access due to modification of existing structures. -Obstacles supported by fire.
Key Terrain **K** **NOTE: Key terrain can be found in all urban zones**	-Radio -Subway -Telephone exchange -Television -Utilities -Intersection -Govt Center	-Water systems -Bridges -Stadiums -Sewer systems -Hospitals -Parks	-Major road network. -Schools -Churches -Municipal buildings (City Halls). -Community centers, -Helipads	-Banking center	-Chemical production plant. -Railroad control center. -Trucking center.	-None in AO. -One in AI.
Avenues of Approach **A**	-Average street width is 26 meters. -Increased vehicular movements. -Streets form rectangular patterns. -Best mounted movement.	-Narrow and or winding streets. -Connected buildings offer routes (rooftops). -Subterranean areas.	-Average street width is 14 meters. -Building 6-8m back from woods. -Street pat-terns rec-tangular or curved.	-Tall build-ings limited air avenues of approach.	-Normally lo-cated along major road and highway routes. -Flat roofs and open areas offer possible LZs.	-Canalized and covered by fire.

Figure 2-12. Sample S2's urban terrain analysis matrix.

NOTE: The figure above is only one example of an urban terrain analysis matrix; information may be provided in different forms. The S2 includes as much information as time permits. The battalion's AO and AI may or may not include all the zones described in paragraph 2-3a.

Observation & Fields of Fire O	Cover & Concealment C	Obstacles O	Key Terrain K	Avenues of Approach A
- Bldgs 1-3 stories. - Short lines of sight. . LOS from Bldg 1 to Bldg 3 and Bldg 1 to Bldg 2. - ATGM use restricted. - Observation & fields of fire improve on north side of objective. - Good tank field of fire from parking lot on east side of objective.	- Bldg 1: Mass, reinforced concrete. - Bldg 2: Mass, triple brick. - Bldg 3: Framed, light clad block. - Friendly and threat smoke.	- Noncombatants - Burned out vehicles. - Rubble in front of Bldg 2. - Probable booby traps in doors and windows.	- Bldg 3 (Co Obj) - Sewer system. - Bldg 1 (Foothold)	- Average street width 15-25 meters. - Mounted approaches from all cardinal directions. - Sewer system between Bldgs 2 & 3.

Figure 2-13. Company urban terrain analysis matrix.

a. Military maps, normally the basic tactical terrain analysis tool, do not provide sufficient detail for a terrain analysis in urban areas. Leaders' reconnaissance and aerial imagery become much more important in urban terrain. Subterranean features (sewer systems, subway systems, and underground water systems), elevated railways, mass transit routes, fuel and gas supply and storage facilities, electric power stations and emergency systems, and mass communications facilities (radio, telephone) are often not depicted on military maps.

(1) Sewer and subway systems provide covered infiltration and small-unit approach routes. Elevated railways and mass transit routes provide mobility between city sectors, and point to locations where obstacles might be expected. Utility facilities are key targets for insurgents and terrorists, and their destruction can hinder the capabilities of a defending force. Utility facilities may become key terrain for certain missions or may have to be safeguarded and protected under ROE.

(2) Due to growth, urban areas are constantly adding new structures and demolishing existing ones. Therefore, any map of an urban area, including city maps or plans published by the city, state, or national government, may be inaccurate and or obsolete.

b. The nature of combat can radically alter the terrain in an urban area in a short period. Incidental or intentional demolition of structures can change the topography of an area and destroy reference points, create obstacles to mobility, and provide additional defensive positions for defenders.

c. Certain public buildings must be identified during the terrain analysis. Hospitals, clinics, and surgical facilities are critical because the laws of war prohibit their attack when they are being used solely for medical support. The locations of civil defense, air raid shelters, and food supplies are critical in dealing with civilian affairs. The same is true during insurgency, guerrilla, or terrorist actions. Again, these facilities may become key terrain for certain missions or may have to be safeguarded and protected under ROE.

d. Stadiums, parks, sports fields, and school playgrounds are of high interest during operations in urban areas. They provide civilian holding areas, interrogation centers, insurgent segregation areas, and prisoner of war holding facilities. These open areas also provide landing zones (LZ) and pick up (PZ) zones. They provide logistical support areas and offer air resupply possibilities because they are often centrally located within a city or city district.

e. Construction sites and commercial operations, such as lumberyards, brickyards, steelyards, and railroad maintenance yards, serve as primary sources of obstacle and barrier construction materials when rubble is not present or is insufficient. They can also provide engineers with materials to strengthen existing rubble obstacles or with materials for antitank hedgehogs or crib-type roadblocks.

f. Roads, rivers, streams, and bridges provide high-speed avenues of approach. They also provide supporting engineer units locations to analyze demolition targets and to estimate requirements for explosives. When conducting navigation, do not rely on street signs as accurate navigation aids, and do not expect all bridges to be passable.

g. Public baths, swimming facilities, car washes, and cisterns are useful in providing bathing facilities. They also provide an alternate water source when public utilities break down or for use as decontamination facilities.

h. A close liaison and working relationship should be developed with local government officials and military forces. In addition to information on items of special interest, they may provide information on the population, size, and density of the urban area; fire-fighting capabilities; the location of toxic industrial materials (TIM); police and security capabilities; civil evacuation plans; and key public buildings. They may also provide English translators, if needed.

2-5. SPECIAL WEATHER CONSIDERATIONS

Terrain and weather analyses are inseparable. Leaders should include the weather's effects on the urban terrain during terrain analysis. Leaders must consider the military aspects of weather when planning missions.

a. **Visibility**. Light data have special significance during urban operations. Night and periods of reduced visibility, to include fog, favor surprise, infiltration, detailed reconnaissance, attacks across open areas, seizure of defended strong points, and reduction of defended obstacles. However, the difficulties of night navigation in restrictive terrain, without reference points and in close proximity to the threat, forces reliance on simple maneuver plans with easily recognizable objectives. Many urban areas are located along canals or rivers, which often create a potential for fog in low-lying areas. Industrial and transportation areas are the most affected by fog due to their proximity to waterways. In heavy industrial areas, smog can also limit observation under all light conditions.

b. **Winds**. Wind chill is not as pronounced in urban areas. However, the configuration of streets, especially in close-orderly block and high-rise areas, can cause wind canalization. This factor increases the effects of the wind on streets that parallel the wind direction, while cross-streets remain relatively well protected. Because of these factors, swirling winds occur and the wind speed and direction may constantly change. This factor also affects the use of smoke for both friendly and threat forces. Downwind predictions for NBC and TIM are also difficult.

c. **Precipitation**. Rain or melting snow often flood basements and subterranean areas, such as subways. This statement is especially true when automatic pumping facilities that normally handle rising water levels are deprived of power. Rain also makes storm and other sewer systems hazardous or impassable. Chemical agents and other TIM are washed into underground systems by precipitation. As a result, these systems may contain toxic concentrations much higher than surface areas and become contaminated "hot spots." These effects become more pronounced as chemical agents or TIM are absorbed by brick or unsealed concrete sewer walls.

d. **Temperature and Humidity.** Air inversion layers are common over cities, especially cities located in low-lying "bowls" or in river valleys. Inversion layers trap dust, chemical agents, and other pollutants, reducing visibility and often creating a greenhouse effect, which causes a rise in ground and air temperature.

(1) The heating of buildings during the winter and the reflection and absorption of summer heat make urban areas warmer than surrounding open areas during both summer and winter. This difference can be as great as 10 to 20 degrees, and can add to the already high logistics requirements of urban combat. Summer heat, body armor and other restrictive combat equipment, combined with the very physical requirements of urban combat, can cause severe heat-related injuries.

(2) Changes in temperature as a result of air inversions can also affect thermal sights during crossover periods of warm to cold and vice-versa. This period should be identified as it may differ from urban area to urban area.

(3) Extremely cold temperatures and heavily constructed buildings may affect target identification for thermal sights. For example, thick walls may make combat vehicle identification difficult by distorting hotspots and increased use of heaters and warming fires may clutter thermal sights with numerous hotspots.

2-6. ANALYSIS OF OTHER CHARACTERISTICS

Because these aspects vary greatly, a comprehensive list cannot be provided. However, other urban characteristics include:

- Sources of potable water.
- Bulk fuel and transport systems.
- Canals and waterways.
- Communications systems.
- Rail networks, airfields, and other transportation systems.
- Industries.
- Power production facilities and public utilities.
- Chemical and nuclear power production facilities.

2-7. APERTURE ANALYSIS

During offensive operations, a key function that the company commander must perform is an aperture analysis of the buildings that he is responsible for attacking. This analysis enables him to determine the number of apertures (windows, doors, holes due to weapons effects) in the building. It also provides key information that he needs to know about the buildings in order to accomplish his mission, such as apertures to be suppressed and where possible points of entry and exit are. (A similar application can be applied in the defense to determine how the enemy would attack buildings that friendly units are defending.) A technique for conducting this analysis is shown in Figures 2-14 and 2-15 on page 2-14. Information can be modified to suit specific situations. This information is obtained from the S2's IPB; reconnaissance from higher headquarters; company reconnaissance efforts; HUMINT sources, such as local inhabitants; and any other information sources available. As much mission-essential information is analyzed within the time constraints possible. The analysis then assists in determining how the company attacks. While the actual plan may have to be modified during execution, this process can assist in formulating what type of weapons or explosives may be effective and also assist in determining courses of action for mission execution.

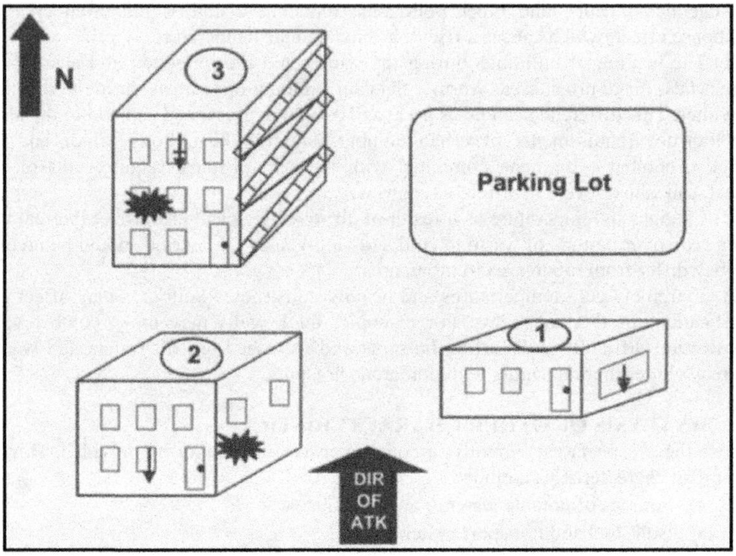

Figure 2-14. Aperture analysis (buildings comprising the company objective).

Bldg No	Construction Type	Floors	Rooms	Stair Wells	Basements Y/N/U	Apertures	Dir of Fire	Loc of Req Suppress (Entry/Exit)	Known Threat
1	Mass Brick	1	3	None	Unk	3 x S 1 x E	360-90 360-90	None Window E	Med MG
2	Mass Brick	2	6	1 Inside	Unk	6 x S 3 x E	270-90 270-360	Window 2, 1st Floor Entry thru Mouse Hole	Sqd (-) Light MG
3	Framed, Block	3	9	3 Inside/ Outside	Yes	9 X S	270-90	Window 2, 3rd Floor Entry thru Door	Sqd (-) Med MG

Figure 2-15. Completed aperture analysis for an attack.

2-8. QUESTIONS FOR COMMANDERS AND LEADERS

Commanders and leaders should be able to answer the following questions after they have completed their terrain and weather analyses.

- Where are the streets, alleys, "through-building" routes, subterranean passageways, that provide mounted and dismounted avenues of approach and mobility corridors within the company's AO?
- What are the number, types, and strength of buildings in the AO?
- What and where is the rubble that helps or hinders company movement?
- Which buildings present fire hazards to assault or support elements?
- Where are the building locations for support-by-fire positions (ability to withstand backblast or overpressure, ability to support vehicle weight)?
- How many kill zones (parking lots, streets, rooftops, wide boulevards) are in the AO?
- Which buildings, rooftops, intersections, or other surrounding terrain provides observation and fields of fire?
- What is the number of apertures for each building in the objective area, building composition, and likely weapons needed to suppress and breach?
- What are the current conditions of the objective area and the effects of preparatory fires?
- Where is a location for a reserve?
- Where are the counterattack routes?
- Where are the urban terrain features on which to place control measures?
- What are the locations for local medical treatment facilities?
- What are the effects on smoke and obscuration?
- Where are locations for company resupply operations?
- Where are the locations to procure barrier materials?
- Where are utilities and fuel sources?
- Where are phone systems and other potential communication systems?
- What are the effects of weather on men and equipment (visibility, temperature, precipitation, humidity, survivability, and mobility)?
- Where are the locations of noncombatants and what is their disposition to friendly and enemy forces (hostile, friendly, neutral)?

Section III. URBAN BUILDING ANALYSIS

This part of the analysis is very important for commanders, leaders, and staffs. Commanders and leaders must be capable of identifying the types of buildings that are in their company sectors, objective areas, and areas of influence. They must also understand the effects of weapons that are used against those buildings. The capability of identifying building types and understanding weapons effects enables commanders to give clear instructions to their subordinates concerning mission execution. It also assists the platoon and the squad leaders in choosing the appropriate weapons or explosives to accomplish their respective missions. (See Chapter 7 for more information.)

2-9. TYPES OF MASS-CONSTRUCTION BUILDINGS

Mass-construction buildings (Figure 2-16) are those in which the outside walls support the weight of the building and its contents. Additional support, especially in wide buildings, comes from using load-bearing interior walls, strongpoints (called pilasters) on the exterior walls, cast-iron interior columns, and arches or braces over the windows and doors.

a. **Modern Mass-Construction Buildings.** Modern types of mass-construction buildings are wall and slab structures, such as many modern apartments and hotels, and tilt-up structures commonly used for industry or storage. Mass-construction buildings are built in many ways:

- The walls can be built in place using brick, block, or poured-in-place concrete.
- The walls can be prefabricated and "tilt-up" or reinforced-concrete panels.
- The walls can be prefabricated and assembled like boxes.

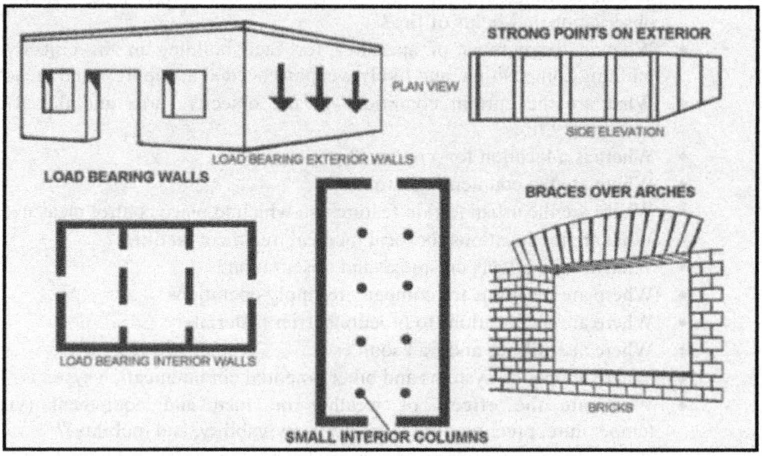

Figure 2-16. Mass-construction buildings.

b. **Brick Buildings.** Brick buildings are the most common mass-construction buildings. In Europe, brick buildings are commonly covered with a stucco veneer so that bricks do not show (Figure 2-17). One of the most common uses of brick buildings is the small store. These buildings are found in all urban areas but are most common in the dense random construction and close-orderly block areas (Figure 2-18).

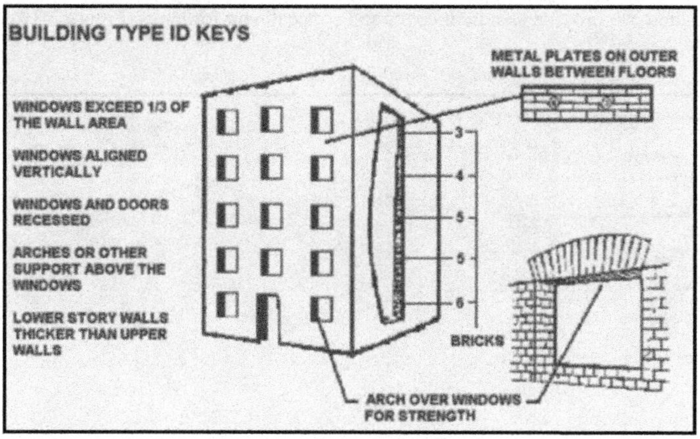

Figure 2-17. Brick buildings.

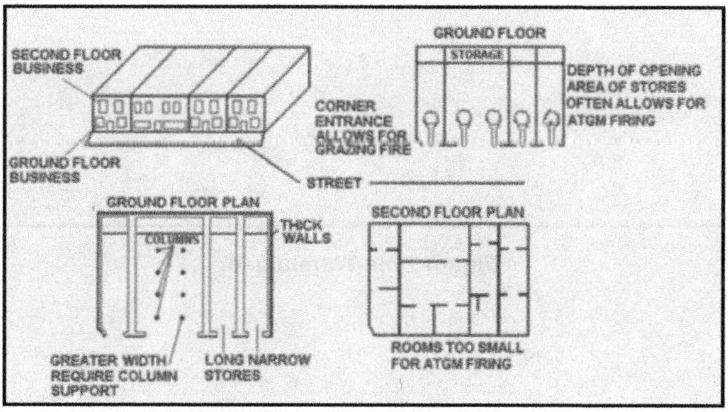

Figure 2-18. Brick structures.

c. **Warehouse.** Another common mass-construction building in the industrial-transportation zone is the warehouse. It is built of poured-in-place concrete reinforced with steel bars or of prefabricated "tilt-up" walls. The walls of warehouses provide good cover, although the roof is vulnerable. The warehouses' large open bays permit ATGM firing and, because they are normally found in outlying areas, often afford adequate fields of fire. These buildings are built on slabs, which can normally support the weight of vehicles and can provide excellent cover and concealment for tanks (Figure 2-19).

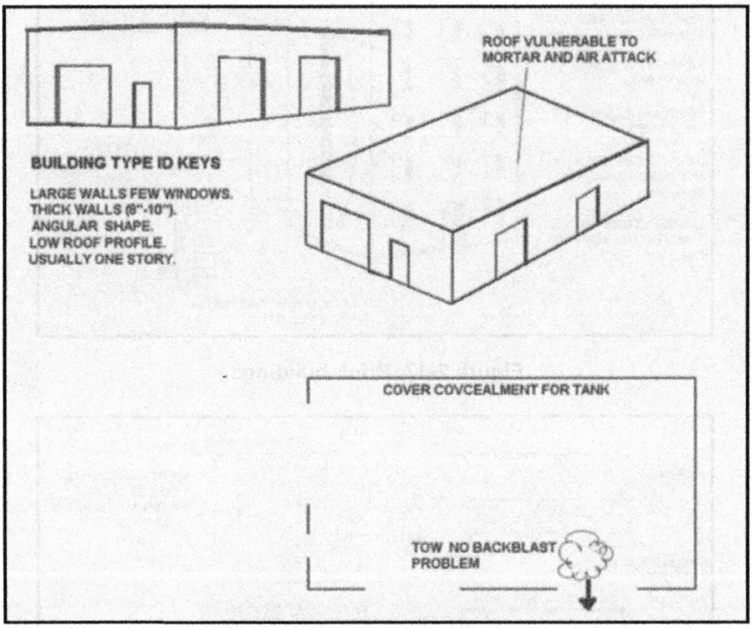

Figure 2-19. Warehouse.

d. **Box-Wall Type.** Another mass-construction building is the box-wall principle type. It is made from prefabricated concrete panels, which are made of 6- to 8-inch-thick reinforced concrete. The outside wall is often glass. The box-wall principle building provides good cover, except at the glass wall. The rooms are normally too small for ATGMs to be fired. A good circulation pattern exists from room to room and from floor to floor. These buildings are commonly used as hotels or apartments and are located in high-rise areas (Figure 2-20).

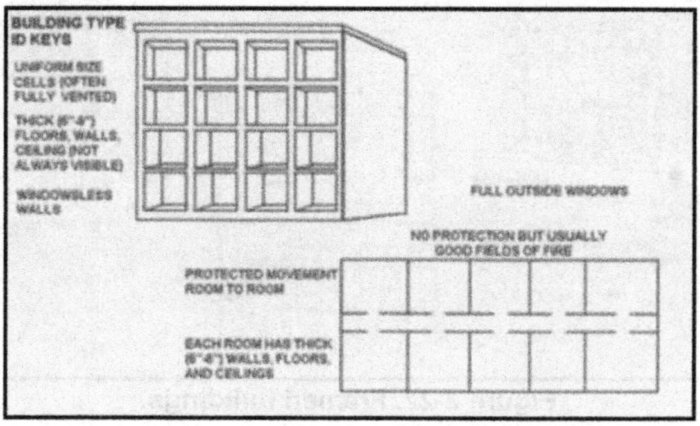

Figure 2-20. Box-wall principle building.

e. **Public Gathering Places.** Public gathering places (churches, theaters) are mass-construction buildings with large, open interiors. The walls provide good cover, but the roof does not. The interior walls are not load bearing and are normally easy to breach or remove. These buildings have adequate interior space for firing ATGMs. They are often located next to parks or other open areas and, therefore, have fields of fire long enough for ATGMs. Public gathering places are most common in the dispersed residential and high-rise areas (Figure 2-21).

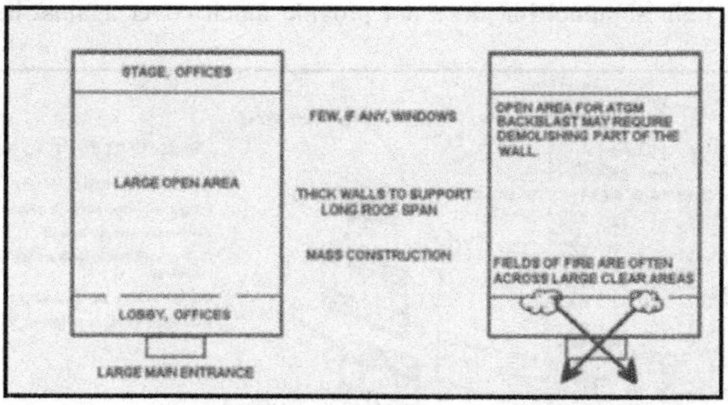

Figure 2-21. Public gathering places.

2-10. TYPES OF FRAMED BUILDINGS

Framed buildings are supported by a skeleton of columns and beams and are usually taller than frameless buildings (Figure 2-22). The exterior walls are not load bearing and are referred to as either heavy clad or light clad. Another type of framed building often found in cities is the garage, which has no cladding.

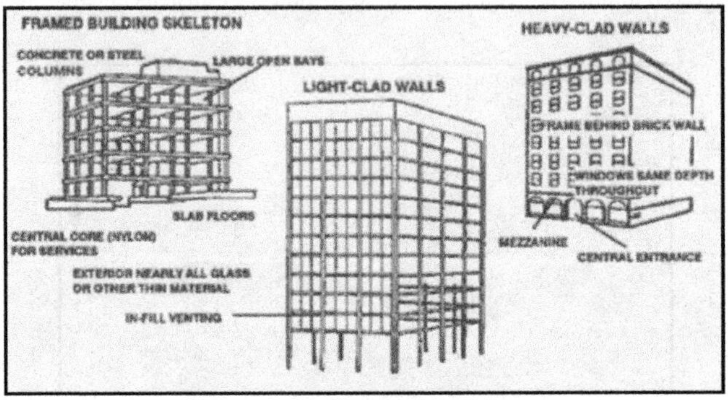

Figure 2-22. Framed buildings.

a. **Heavy-Clad Framed Buildings.** Heavy-clad buildings were common when framed buildings were first introduced. Their walls are made of brick and block that are sometimes almost as thick as frameless brick walls, although not as protective. Heavy-clad framed buildings are found in the city core or core periphery. They can be recognized by a classic style or architecture in which each building is designed with three sections: the pediment, shaft, and capital. Unlike the brick building, the walls are the same thickness on all floors, and the windows are set at the same depth throughout. Often the frame members (the columns) can be seen, especially at the ground floor. The cover provided from the cladding, consisting of layers of terra cotta blocks, brick, and stone veneer, is not as good as cover from the walls of brick buildings. It protects against small-arms fire and light shrapnel but does not provide much cover against heavy weapons (Figure 2-23).

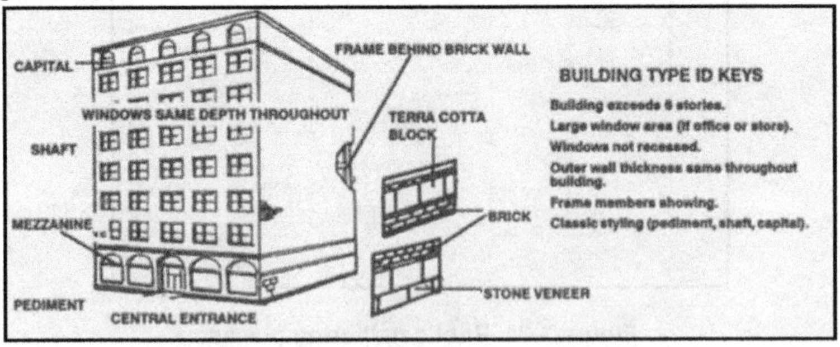

Figure 2-23. Heavy-clad framed building.

(1) The floor plans of these buildings depend upon their functions. Office buildings normally have small offices surrounding an interior hall. These offices have the same dimensions as the distance between columns (some large offices are as large as two times the distance between columns). These rooms are too small to permit firing of ATGMs but do provide some cover for snipers or machine gunners (Figure 2-24).

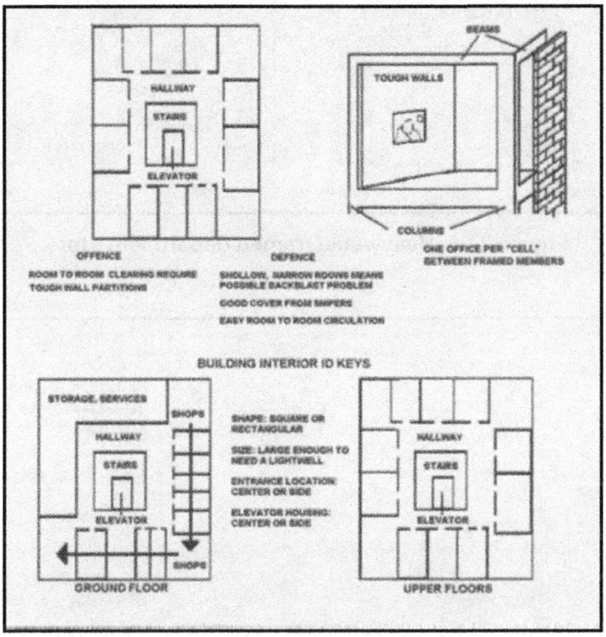

Figure 2-24. Floor plan of heavy-clad framed office.

(2) Department stores normally have large, open interiors (Figure 2-25, page 2-22). Such areas permit firing ATGMs (if there are adequate fields of fire). Often a mezzanine level with a large backblast area permits firing down onto tanks. Steel fire doors, which are activated by heat, often exist between sections of the store. Once closed, they are difficult to breach or force open, but they effectively divide the store into sections (Figure 2-26, page 2-22).

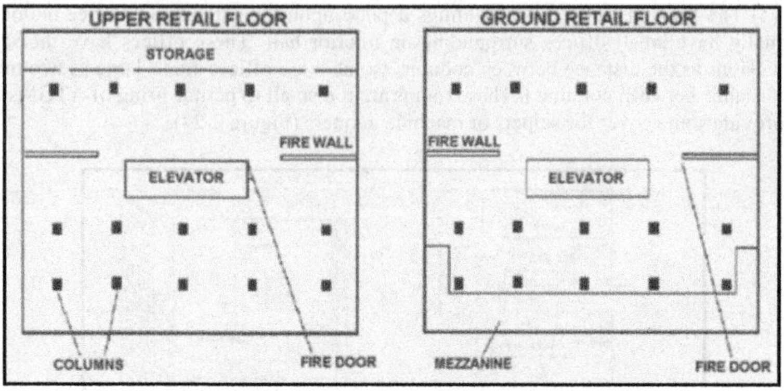

Figure 2-25. Heavy-clad framed department store.

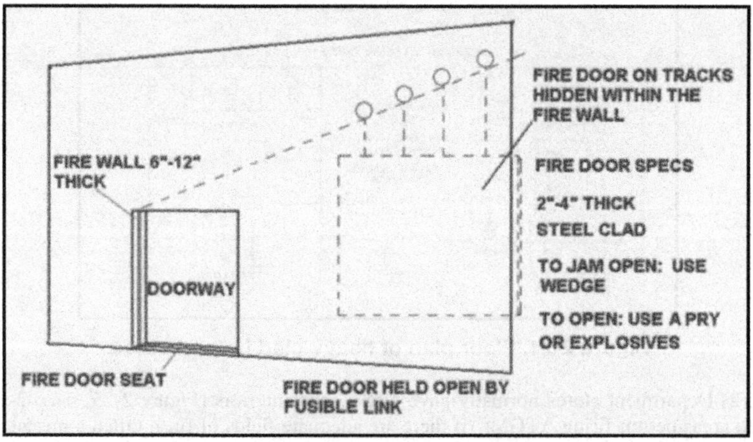

Figure 2-26. Fire wall and fire door.

(3) Another type of heavy-clad framed building is the high-rise factory (Figure 2-27). Such buildings are normally easily recognized because the concrete beams and columns are visible from the outside. They are usually located in older industrial areas. The large windows and open interior favor the use of ATGMs. Because the floors are often made to support heavy machinery, this building provides good overhead cover.

BUILDING TYPE ID KEYS

- Windows as full as possible (for light).
- Columns visible through windows.
- Open bay interiors.
- Loading docks, large doors on ground floor.
- Ventilation devices on roof.

Figure 2-27. High-rise factory.

b. **Light-Clad Buildings.** Light-clad buildings are more modern and may be constructed mostly of glass (Figure 2-28). Most framed buildings built since World War II are light-clad buildings. They are found in both core and outlying high-rise regions. Their walls consist of a thin layer of brick, lightweight concrete, or glass. Such materials provide minimal protection against any weapon. However, the floors of the buildings are much heavier and provide moderate overhead cover. The rooms in light-clad framed buildings are much bigger than those in heavy-clad. This feature, along with the fact that the buildings usually stand detached from other buildings, favors the employment of ATGMs. The interior partitions are thin, light, and easy to breach (Figure 2-29, page 2-30).

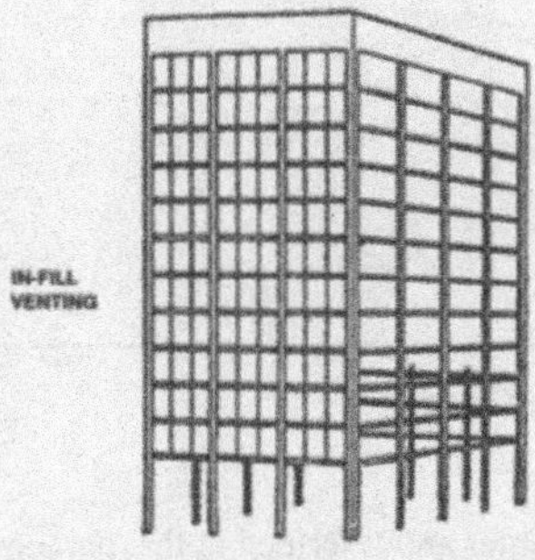

BUILDING TYPE ID KEYS

- Outer "skin" appears thin.
- Frame visible on ground floor through windows, at sides, rear.
- High proportion of windows.
- Building separated from others.
- Building exceeds four stories.

Figure 2-28. Light-clad framed building.

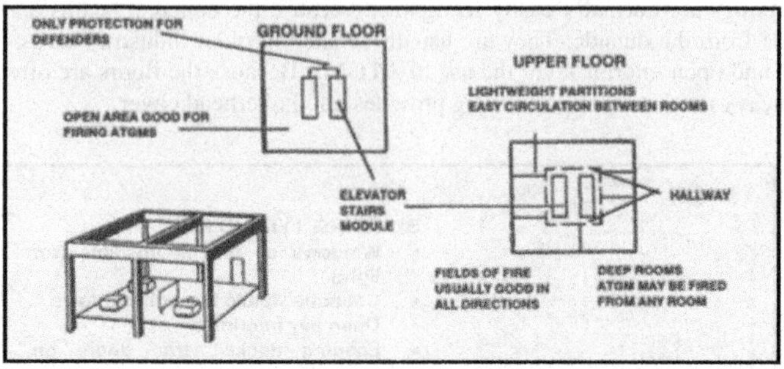

Figure 2-29. Light-clad framed room.

c. **Garage.** The garage (Figure 2-30) is one of the few buildings in an urban area in which all floors support vehicles. It provides a means to elevate vehicle-mounted TOWs, and the open interiors permit firing of ATGMs. Garages are normally high enough to provide a 360-degree field of fire for antiaircraft weapons. For example, a soldier equipped with a Stinger could hide under the top floor of the garage, come out to engage an aircraft, and then take cover back inside.

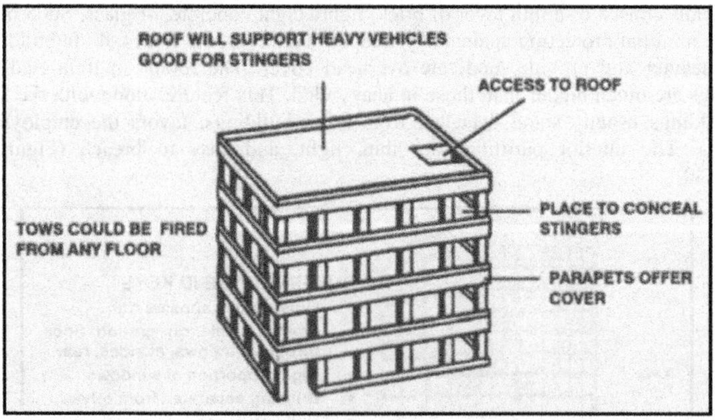

Figure 2-30. Garage.

2-11. FLOOR PLANS

The more common floor plans that units encounter are described in this paragraph.

a. **Brick Buildings.** The floor plans in brick buildings are usually different on ground floor levels than on upper levels (Figure 2-31).

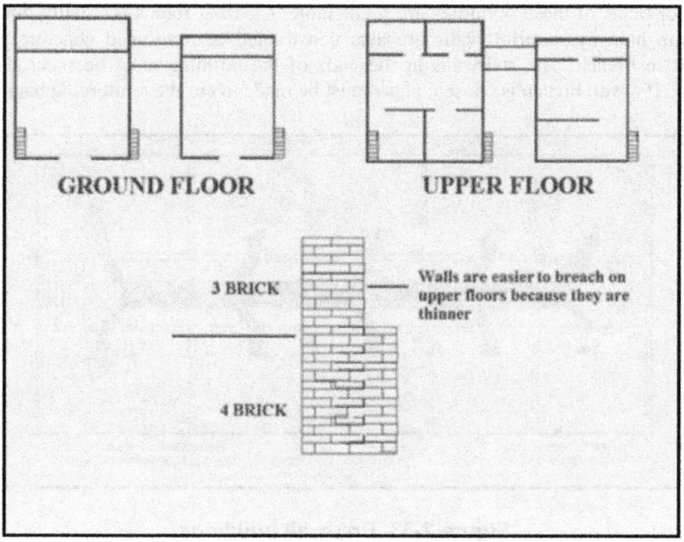

Figure 2-31. Brick building floor plans.

b. **Brick Houses.** Brick houses have similar floor plans on each floor (Figure 2-32); therefore, ground floors are cleared the same way as upper floors. The same recommendations for breaching and clearing brick buildings apply except that breaching deeply slanted roofs may be difficult.

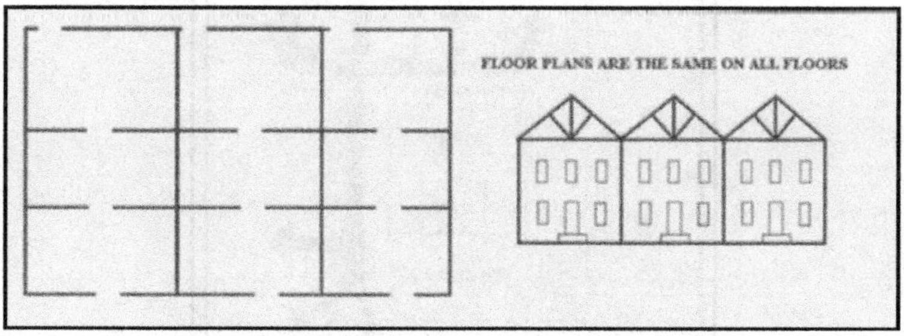

Figure 2-32. Brick house similar floor plans.

c. **Box-Wall Buildings.** Box-wall buildings often have reinforced concrete walls (Figure 2-33, page 2-26), which are difficult to breach due to the reinforcing bars. Therefore, the best way to enter is to breach through a door or one of the side windows.

The floor plans of these buildings are predictable. Clearing rooms is usually done from one main hallway. Interior walls are also constructed of reinforced concrete and are difficult to breach. The stairways at the ends of the building must be secured during clearing. If a wall breach is chosen, plans must be made to cut the reinforcing bars.

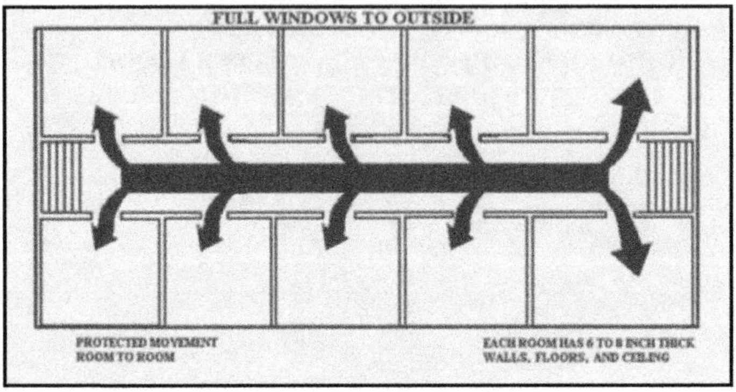

Figure 2-33. Box-wall buildings.

d. **Heavy Clad Framed Buildings.** Heavy-clad framed buildings are relatively easy to breach with explosive or ballistic breaching techniques. Their floor plans are oriented around a stairway or elevator, which must be secured during clearing (Figure 2-34). The interior walls of these buildings can be breached, although they may require use of explosives to be effective.

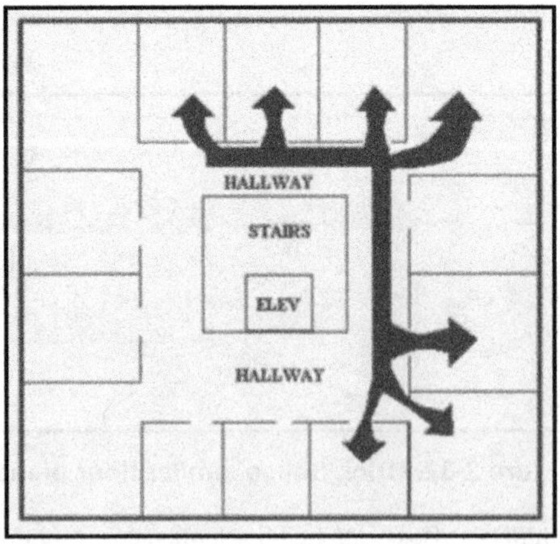

Figure 2-34. Heavy-clad framed building.

e. **Light-Clad Framed Buildings.** On light-clad framed buildings (Figure 2-35), the clearing tasks are usually the same: secure the central stairway and clear in a circular pattern. Walls are easier to breach since they are usually thinner.

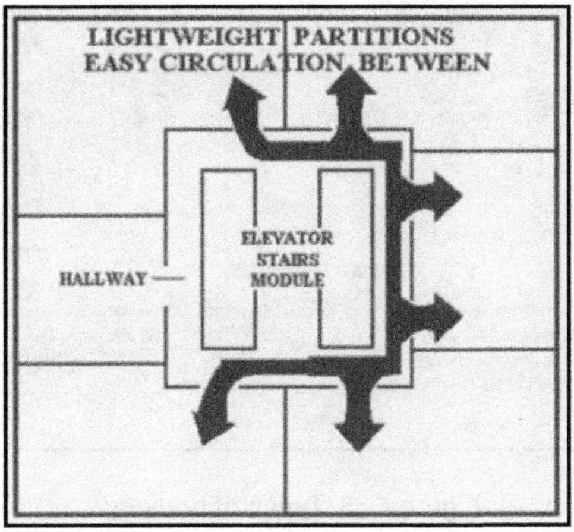

Figure 2-35. Light-clad framed building.

2-12. RESIDENTIAL AREAS

Residential houses in the western world are located in and around cities and in rural areas. Residential houses in cities are normally mass-construction brick buildings. Rural residential buildings in the continental US, South America, and Southeast Asia are commonly made of wood. In continental Europe, Southwest Asia, and Sub-Saharan Africa where wood is extremely scarce, rural buildings are normally constructed of concrete blocks (Figure 2-36, page 2-28). Another common type of building structure in cities with European influences is called the Hof-style apartment building (Figure 2-37, page 2-28). In the Mideast and tropical regions, the most common housing is the enclosed courtyard. Houses are added one to another with little regard to the street pattern. The result is a crooked, narrow maze, which is harder to move through or fire in than dense European areas (Figure 2-38, page 2-29).

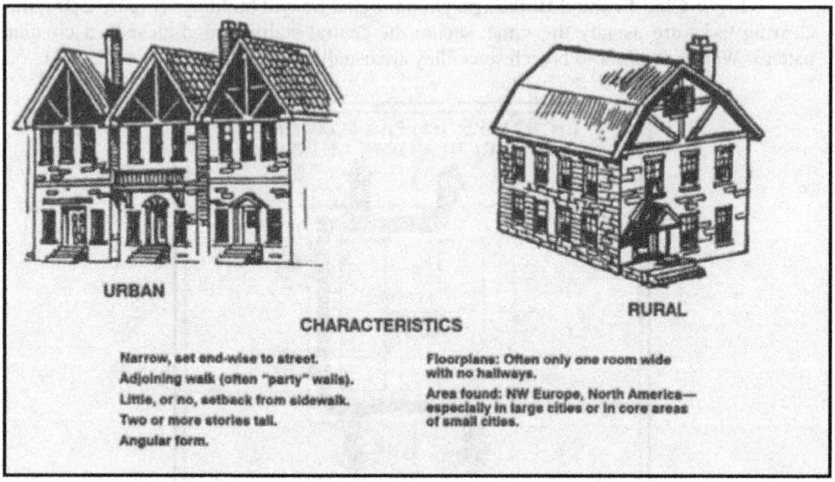

Figure 2-36. Types of housing.

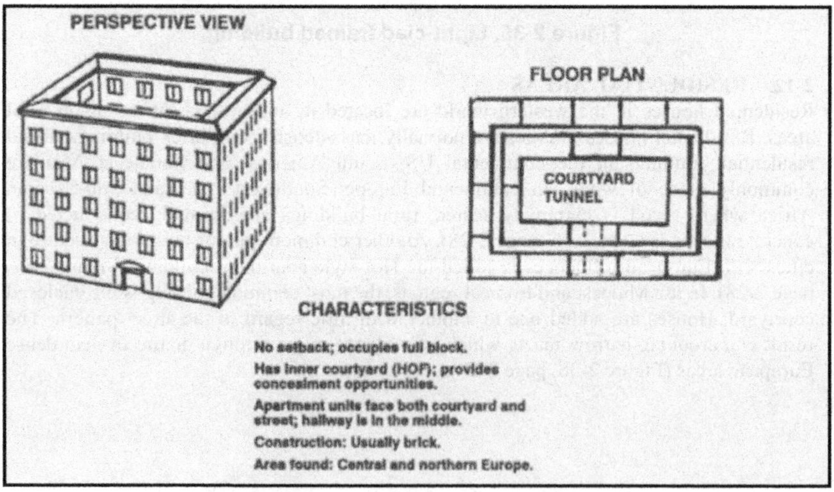

Figure 2-37. Hof-style apartment building.

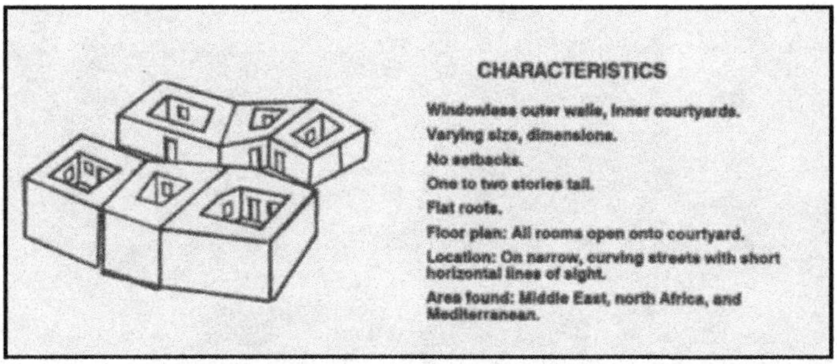

Figure 2-38. Enclosed courtyard.

2-13. CHARACTERISTICS OF BUILDINGS

Certain characteristics of both mass-construction and framed buildings can be helpful in analyzing a urban area. Leaders can use Table 2-1 to determine the height and wall thickness of buildings in relation to the type of weapon they plan to use. (See Chapter 7 for weapons effects.)

TYPE OF CONSTRUCTION	BUILDING MATERIAL	HEIGHT (STORIES)	AVERAGE WALL THICKNESS (inches)
Mass	Stone	1 to 10	30
Mass	Brick	1 to 3	8 to 12
Mass	Brick	3 to 6	12 to 24
Mass	Concrete Block	1 to 5	8
Mass	Concrete Wall and Slab	1 to 10	8 to 15
Mass	Concrete "Tilt-ups"	1 to 3	7
Framed	Wood	1 to 5	6 to 8
Framed	Steel (Heavy Cladding)	3 to 50	12
Framed	Concrete/Steel (Light Cladding)	3 to 100	1 to 3

Table 2-1. Types of construction.

2-14. DISTRIBUTION OF BUILDING TYPES

Certain types of buildings dominate certain parts of a city, which establishes patterns within a city. Analysis of the distribution and nature of these patterns has a direct bearing on military planning and weapon selection (Figure 2-39, page 2-30).

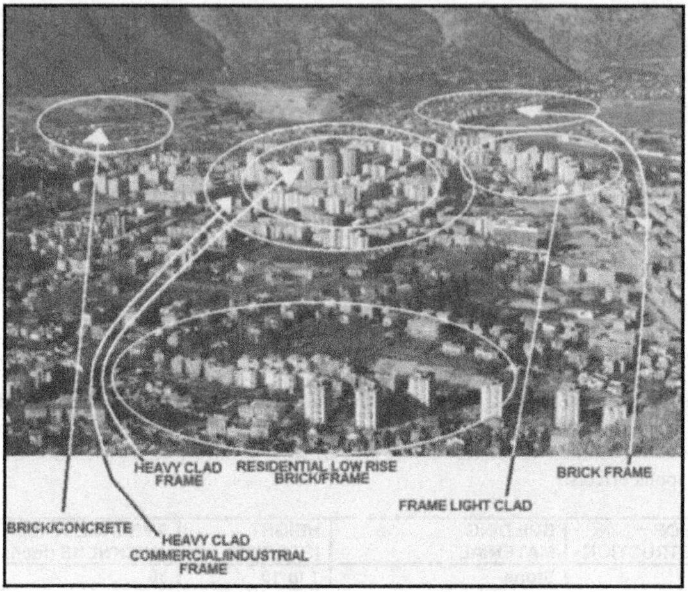

Figure 2-39. Distribution of building types.

a. **Mass Construction Buildings.** Mass-construction buildings are the most common structures in urban areas, forming about two-thirds of all building types. Brick structures account for nearly 60 percent of all buildings, especially in Europe.

b. **Framed Multistory Buildings.** Steel and concrete framed multistory buildings have an importance far beyond their one-third contribution to total ground floor area. They occupy core areas—a city's most valuable land—where, as centers of economic and political power, they have a high potential military significance.

c. **Open Space.** Open space accounts for about 15 percent of an average city's area. Many open spaces are grass-covered and are used for parks, athletic fields, and golf courses; some are broad, paved areas. The largest open spaces are associated with suburban housing developments where large tracts of land are recreation areas.

d. **Streets.** Streets serving areas consisting of mostly one type of building normally have a common pattern. In downtown areas, for example, high land values result in narrow streets.

(1) Street widths are grouped into three major classes:

- 7 to 15 meters, located in medieval sections of European cities.
- 15 to 25 meters, located in newer planned sections of most cities.
- 25 to 50 meters, located along broad boulevards or set far apart on large parcels of land.

(2) When a street is narrow, observing or firing into windows of a building across the street can be difficult because an observer is forced to look along the building rather than

into windows. When the street is wider, the observer has a better chance to look and fire into the window openings (Figure 2-40).

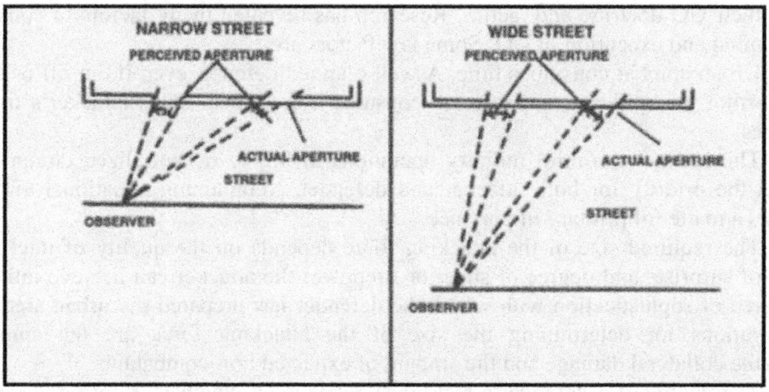

Figure 2-40. Line-of-sight distances and angles of obliquity.

Section IV. URBAN THREAT EVALUATION

Threat evaluation for urban combat uses a three-step process: developing a threat database, determining threat capabilities, and developing a doctrinal template file as threat evaluation for open terrain. The brigade and battalion S2 perform this function. Company commanders must use the information provided to determine the best method of mission accomplishment. (This process functions better against a conventional threat, such as a motorized rifle battalion. Units must also be prepared to fight unconventional threats. The process is essentially the same, however real-time or near real-time information must be used concerning the size of threat forces, arms and equipment, locations, and probable courses of action.) Due to the unique aspects of UO, certain operational factors and future threat capabilities must be recognized. These factors must be considered before preparing the required templates during threat integration of the IPB process.

2-15. OPERATIONAL FACTORS

The doctrinal concept of full spectrum operations and recent international developments presuppose the increasing chance of conflict or military intervention with regional and national threats. These conflicts may be combat operations against conventional forces of one or more Third World nations, combat operations against insurgent or asymmetrical threats, or military intervention in the form of stability and support operations.

a. **Conventional Threats and Planning Factors.** Most nations with conventional forces emphasize managing combined arms operations in urban areas. Among the conventional force structures, the poorer the nation, the less likely it is to field, maneuver, and sustain forces beyond logistic centers. Also, the extreme environment in some regions restricts operations beyond urban centers. Urban structural characteristics are shaped by social, cultural, and economic factors. These elements are the prime reasons that UO doctrine and tactics differ between nations. Coupled with the restrictive nature of urban combat, the differences in tactics may be superficial. More than any other factor, the advent of high technology, precision weapons has enabled nations to modify and update their UO doctrine and tactics. Research has revealed many factors to consider in the planning and execution of UO. Some key factors are:

(1) Urban combat consumes time. A well-planned defense, even if cut off or lacking in air, armor, or artillery weapons, can consume a great deal of an attacker's time and resources.

(2) The ability to control military operations in highly decentralized circumstances remains the priority for both attacker and defender. Maintaining situational awareness becomes a matter of primary importance.

(3) The required size of the attacking force depends on the quality of intelligence, degree of surprise, and degree of superior firepower the attacker can achieve rather than the degree of sophistication with which the defender has prepared the urban area. Other considerations for determining the size of the attacking force are the amount of acceptable collateral damage and the amount of expected non-combatants.

(4) The degree of a defender's resistance depends on whether or not he is separated from the local population, is wholly or partly cut off from external support, or has effective or ineffective communication systems. The local population's attitude toward the defender also affects his degree of resistance.

(5) The belief that armor has no role in city fighting is wrong. Tanks, Infantry fighting vehicles (IFVs), and armored personnel carriers (APCs) have proven vital to the attacker inside the city as long as they are protected by Infantry.

(6) If the attacker is subject to any constraints, such as restrictive ROE, the defender has a good chance of prolonging the battle or forcing the attacker into a mistake, thus raising the cost for the attacker.

(7) The defender has three tactical options: defense in depth, key sector defense, and mobile defense. Defense in depth suggests an outer and inner defense combination; key sector defense means strongpoint defense of vital positions, mainly those controlling major avenues of approach; and mobile defense is based on counterattack capability. These are not mutually exclusive options.

(8) Urban areas facilitate night exfiltration and movement by small groups.

(9) Preventing the enemy's reentry of cleared buildings is a significant challenge to both the attacker and defender. The requirement to secure cleared areas causes commanders and leaders to dedicate forces specifically to this task.

(10) Mortars may be used more heavily than artillery in urban areas due to their immediate response and high-angle fire capabilities.

(11) The employment of snipers and squad Marksmen in urban combat can prove to be extremely effective for both the attacker and defender.

(12) Ammunition consumption is probably five to ten times greater in an urban environment.

b. **Stability and Support Planning Factors.** The brigade and battalion S2s conduct the IPB process; commanders and leaders must use the information provided to determine the best method to accomplish the mission. Company teams often conduct certain stability and support operations, such as support to insurgency, and counterterrorist operations with special operations forces (SOF). For example, company teams may conduct supporting missions such as isolating the objective or providing security for the SOF. The following considerations assist in evaluating the threat during stability and support operations.

(1) Population status overlays are prepared for the urban area, showing potential neighborhoods or districts where a hostile population could be encountered. Overlays are also prepared showing insurgent or terrorist safe houses, headquarters, known operating areas, contact points, and weapons supply sources. These overlays must include buildings that are known, or could become, safe houses; and explosives, ammunition, and or weapons storage sites.

(2) Urban operations are three-dimensional. Subterranean routes are of primary concern when considering insurgent and terrorist avenues of approach and lines of communications. Sewers, subways, tunnels, cisterns, and basements provide mobility, concealment, cover, and storage sites for insurgents and terrorists.

(3) Elevated railways, pedestrian overpasses, rooftops, fire escapes, balconies, and access ladders provide mobility and concealment, and can serve as fighting or sniper positions.

(4) Although doctrinal templates are not developed for urban stability and support operations, pattern analysis may reveal how the threat operates, and what its primary targets are. Once the threat's method of operation is determined, insurgent situation maps can be developed. These maps should pinpoint likely targets based on the previous attack history. Electrical power generation and transmission facilities, gas production and holding facilities, water and sewer pumping and treatment plants, telephone exchanges and facilities, and radio and television stations should be considered as primary threat targets when developing these maps

(5) If the threat has become intermingled with the population, a greater degree of control is required for operations. While detection is more difficult, threat forces operating without uniforms share some common characteristics with insurgents or terrorists. ROE become more important to units in this situation.

(6) With operations of this type, intelligence and the careful application of nonlethal and lethal force, play dominant roles. Known members of the armed forces, their auxiliaries, and the underground must be identified and arrested and or removed from the populace. Use of minimum force and adherence to the ROE are critical.

(7) The local population's support to the threat may be either forced or given willingly. In either case, an effort must be made to separate the threat from the local population base. A population can be forced into giving support by a combination of terrorism (either by coercion or intimidation) and harassment. Commanders and leaders must be observant and sensitive to the local population's concerns before expecting the population to help friendly forces.

(8) The threat's logistical support occurs in smaller packages. The threat must rely on the local population to support the distribution of logistics; therefore, the identification and seizure or destruction of threat logistics (less medical materiel) bases are more difficult. To stop threat resupply operations, friendly forces would have to stop all movement within the urban area, which is practical. Therefore, intelligence and operational priorities should be the identification and destruction, neutralization or control of the threat's logistical bases.

(9) Commanders and leaders must be aware of the political and psychological impact of their actions if force is used. Although the local population may be neutral or demonstrate lukewarm support for friendly forces, excessive use of nonlethal or lethal force may cause the local civilians to support the threat. Of special concern is the media (newspapers, television, magazines, and so forth). Due to the large numbers of journalists and amateur and or professional photographers in urban areas, any negative images of friendly forces are probably publicized. Negative publicity could have an adverse effect on both US public opinion and interests. Conversely, positive publicity can greatly enhance friendly operations and morale. This publicity can also sway the local population away from the threat. Therefore, all media members are accompanied. Commanders and leaders must ensure that the ROE are enforced and that noncombatants are treated with dignity.

(10) Ground maneuver units have historically been used as a part of the effort to separate the threat from the local civilian populace. Some units may be detailed to provide civil services such as law enforcement patrols, trash pick-up, and or the restoration and maintenance of power, telephone, and water services. Commanders and staffs should anticipate these missions during transition to stability and support operations.

2-16. THREAT

The types of threats that units may face in an urban area and how they are to fight/operate are discussed in this paragraph.

a. **Conventional Forces.** Many Third World countries have adopted techniques of urban combat from either the United States or the Commonwealth of Independent States (some of the former USSR republics). Therefore, the future threat may consider the motorized or mechanized rifle battalion the most effective unit for urban combat because of its inherent mobility, armor protection, and ability to quickly adapt buildings and other structures for defense.

(1) These types of threat defenses are organized into echelons to provide greater depth and reserves.

(2) Company strongpoints are prepared for perimeter defense and form the basis for the battalion defensive position.

(3) The reserve is located in a separate strongpoint.

(4) Ambush locations are established in the gaps of the strongpoints, and dummy strongpoints are constructed to deceive the attacker.

(5) Positions for securing and defending the entrances to and exits from underground structures and routes are established.

(6) Security positions are prepared forward of first echelon defensive positions.

(7) A motorized or mechanized rifle company may defend several buildings with mutually supporting fires or use a single large building as part of a larger defensive system.

(8) Each platoon defends one or two buildings or defends one or two floors of a single building.

b. **Other Threats.** Different threats in many underdeveloped countries are neither conventional threats nor can they be described as an insurgency.

(1) Some forces may be semi-skilled light Infantry, as might be found in several African countries that have recently undergone civil wars such as the Democratic Republic of the Congo (formerly Zaire). These troops are normally poorly trained, poorly equipped, and poorly motivated. Using the case of the Congo as an example, 40 percent of the populace comes from the cities, which are the locations of most of the recent conflicts. These forces should not be underestimated. They know the terrain and may have the support of the population.

(2) Other types of forces may include personal armies of local warlords, such as those found in Somalia.

(3) Another type of force may include semi-organized groups of thugs, loosely organized under the control of an individual, such as the "Dignity Battalions" in Panama or the situation encountered in Haiti.

(4) Another source of organized force may be the police force or gendarme. In some countries, the gendarme may be a more effective fighting force than the army.

(5) In some instances, the threat may consist of organized criminals or paramilitary forces appearing on the battlefield. They may be actively hostile threats or act as support for other forces.

c. **Unconventional Forces.** These threats consist of urban insurgents and terrorist groups. (See FM 7-98 for detailed descriptions.)

d. **Threat Doctrine.** No single threat doctrine can be identified as the one the US military faces during UO. This does not mean that potential opponents do not have clearly identifiable doctrinal principles or that they have not developed specific TTP for urban combat based on those principles. Commanders and leaders at all levels must strive to identify the TTP an opponent uses and to anticipate his actions.

(1) Many of the world's largest modern armies are devoting an increasing amount of their training to UO. These armies realize that urban operations are unavoidable in future conflicts. Some, but not all, of these armies base their tactics on those developed and used by the former Soviet Union. Others use Western doctrinal principles to guide their development of specific TTP.

(2) The threat US forces may face during combat in urban areas is not limited to just large modern armies, however. Many regional powers and third world countries have developed philosophies that call for urban operations to be the centerpiece of their military policy. They continue to develop doctrinal principles and TTP that may be loosely based on a Western or former Soviet Union model, but are often unique to their specific geography, society, culture, and military capability.

(3) Insurgent groups and warring factions in many places around the world have discovered that the urban area offers many tactical advantages, especially against a modern army such as the US These groups may have some formal military training, or they may develop unique urban combat TTP based on specific characteristics of their beliefs or the personalities of their leaders. (Additional information on operations in this type of environment can be found in FM 7-98 and FM 90-8.)

(4) One threat in urban areas that has become increasingly common, regardless of the opponent, is the sniper. Worldwide, there has been an increase in the amount of military weapons designed to deliver long-range precision fire. Some of these weapons are now very powerful and accurate. The sniper has shown that he can be a dangerous opponent,

one that can sometimes obtain results far out of proportion to the number of personnel committed to sniping. (See Chapter 6 for more information.)

e. **Detailed Description of Threat Tactics.** The increasing availability of sophisticated technology has created unorthodox tactics that can be exploited by potential opponents. These tactics seek to counter the technological and numerical advantages of US joint systems and forces, and to exploit constraints placed on US forces due to cultural bias, media presence, ROE, and distance from the crisis location. Offsetting their inherent weaknesses, enemy forces seek an advantage in urban terrain to remain dispersed and decentralized, adapting their tactics to provide the best success in countering a US response. These threats range from units equipped with small arms, mortars, machine guns, antiarmor weapons, and mines to very capable mechanized and armor forces equipped with current generation equipment. Urban environments also provide many passive dangers such as disease from unsanitary conditions and psychological illnesses. While the active threats vary widely, many techniques are common to all. Figure 2-41 provides a set of tactics available to potential threats opposing US forces in urban areas.

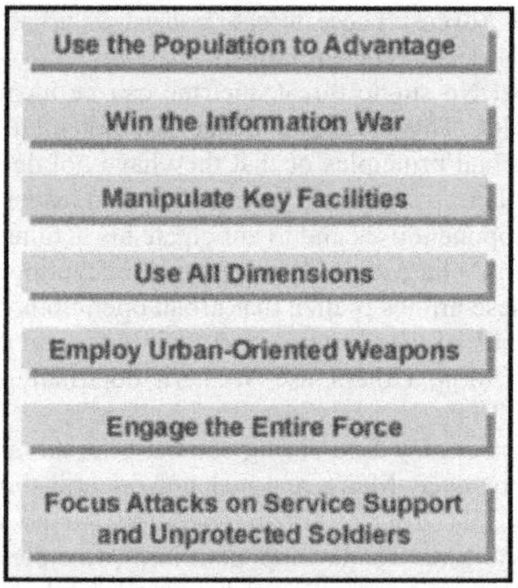

Figure 2-41. Urban threat tactics.

(1) ***Use the Population to an Advantage.*** The populace of a given urban area represents key terrain and the side that manages it best has a distinct advantage. Future urban battles may see large segments of the populace remain in place, as they did in Budapest, Hungary and Grozny, Chechnya. Units involved in urban stability and support operations certainly conduct missions in and among the residents of the area.

(a) Threat forces may use the population to provide camouflage, concealment, and deception for their operations. Guerilla and terrorist elements may look no different than any other members of the community. Even conventional and paramilitary troops may have a "civilian" look. Western military forces adopted the clean-shaven, close-cut hair standard at the end of the nineteenth century to combat disease and infection, but twenty-first century opponents might very well sport beards as well as civilian-looking clothing and other "nonmilitary" characteristics.

(b) The civil population may also provide cover for threat forces, enhancing their mobility close to friendly positions. Threat forces may take advantage of US moral responsibilities and attempt to make the civil population a burden on the Army's logistical and force protection resources. They may herd refugees into friendly controlled sectors, steal from US-paid local nationals, and hide among civilians during offensive operations.

(c) The civil population may also serve as an important intelligence source for the threat. Local hires serving among US soldiers, civilians with access to base camp perimeters, and refugees moving through friendly controlled sectors may be manipulated by threat forces to provide information on friendly dispositions, readiness, and intent. In addition, threat special purpose forces and hostile intelligence service assets may move among well-placed civilian groups.

(2) ***Win the Information War.*** Threat forces may try to win the information war as much as they may directly oppose friendly operations.

(a) Portable video cameras, Internet access, commercial radios, and cellular telephones are all tools that permit threat forces to tell their story. American "atrocities" may be staged and broadcast. Electronic mail may be transmitted to sympathetic groups to help undermine resolve. Internet web sites provide easy worldwide dissemination of threat propaganda and misinformation. Hackers may gain access to US sites to manipulate information to the threat's advantage.

(b) The threat may make skillful use of the news media. Insurgent campaigns, for example, need not be tactical military successes; they need only make the opposition's campaign appear unpalatable to gain domestic and world support. The media coverage of the Tet Offensive of 1968 affected the will of both the American people and their political leadership. Although the battle for Hue was a tactical victory for the US, the North Vietnamese clearly achieved strategic success by searing the American consciousness with the high costs of urban warfare.

(3) ***Manipulate Key Facilities.*** Threat forces may identify and quickly seize control of critical components of the urban area to help them shape the battle space to their own ends. Telephone exchanges provide simple and reliable communications that can be easily secured with off-the-shelf technologies. Sewage treatment plants and flood control machinery can be used to implement the use of weapons of mass destruction (WMD) or to make sections of the urban area uninhabitable. Media stations significantly improve the information operations position of the controlling force. Power generation and transmission sites provide means to control significant aspects of civilian society over a large area.

(4) ***Use the Three Dimensions of Urban Terrain.*** The threat thinks and operates throughout all dimensions of the urban environment. Upper floors and roofs provide urban threat forces excellent observation points and battle positions above the maximum elevation of many weapons. Shots from upper floors strike friendly armored vehicles in vulnerable points. Basements also provide firing points below many weapons' minimum depressions and strike at weaker armor. Sewers and subways provide covered and concealed access throughout the area of operations. Conventional lateral boundaries are

often not apply as threat forces control some stories of the same building while friendly forces control others.

(5) ***Employ Urban Oriented Weapons.*** Whether they are purpose-built or adapted, many weapons may have more utility in an urban environment while others may have significant disadvantages. Urban threat weapons are much like the nature of urbanization and the urban environment—inventive and varied. Small, man-portable weapons, along with improvised munitions, dominate the urban environment. Figure 2-42 lists examples of threat weapons favored in UO.

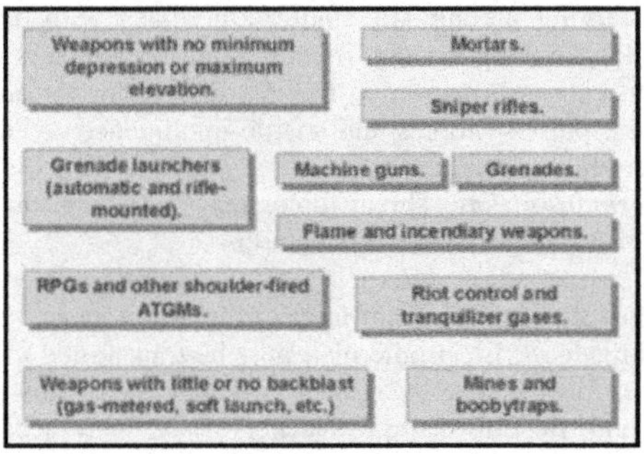

Figure 2-42. Favored threat weapons.

(6) ***Engage the Entire Enemy Force.*** Threat forces may "hug" units operating in an urban area to avoid the effects of high-firepower standoff weapon systems. They may also try to keep all or significant portions of the unit engaged in continuous operations to increase the susceptibility to stress-induced illnesses. UO, by their nature, produce an inordinate amount of combat stress casualties, and continuous operations exacerbate this problem. The threat may maintain a large reserve to lessen the effect on its own forces.

(7) ***Focus Attacks on Service Support and Unprotected Soldiers.*** Threat forces may prey on soldiers poorly trained in basic Infantry skills. Ambushes may focus on these soldiers while they are conducting resupply operations or moving in poorly guarded convoys. UO are characterized by the isolation of small groups and navigational challenges, and the threat may use the separation this creates to inflict maximum casualties even when there is no other direct military benefit from the action. As a result, ground maneuver units may find themselves providing security for logistical units during certain types of UO.

2-17. PROJECTED THREAT CAPABILITIES

The wealth of some nations are used to modernize their armed forces through the acquisition of new technologies. Projected future threat force capabilities are—

- New munitions such as fuel air explosives (FAE), enhanced blast, thermobaric, and other improved ballistic technologies.
- Systems with interchangeable warheads, some designed for urban combat.
- Precision-guided munitions.

- Robotics.
- Day or night target acquisition systems.
- Improved engineering abilities to breach or emplace obstacles.
- Soft-launch hand-held AT and flame weapons.
- Nonlethal incapacitating chemical or biological agents by conventional forces.
- Lethal chemical or biological agents used as an asymmetric threat.
- Improved self-protection (body armor).
- Improved communications.

2-18. MODERN URBAN BATTLE ANALYSIS

> *"What was needed was for us to act so that every house in which we had even one soldier became a fortress against the enemy. All would be well if every soldier fighting in a basement or under the stairs, knowing the general task facing the army, stood his ground alone and accomplished the task on his own. In street fighting, a soldier is on occasion his own general. He needed to be given correct guidance and so to speak, the trust of the generals."*
>
> Marshal Vasili I. Chuikov,
> Hero of the Soviet Union and Defender of Stalingrad
> Quoted in The Battle for Stalingrad

A historical analysis of recent experience in urban combat provides insight into the conduct of the urban fight. While urban combat remains primarily an infantry fight, the importance of maneuver warfare and combined arms in the urban environment cannot be neglected. The following information has been extracted from *Modern Experience in City Combat*, produced for the US Army Human Engineering Laboratory in 1987. It is based upon an analysis of urban combat from World War II, Korea, Vietnam, and the Middle East. (See Appendix H for additional information.)

a. **General.** An attack on urban terrain that is well planned and executed is successful. Such an attack does not have to be casualty or resource intensive. An attacker does not necessarily need a much higher force ratio in urban terrain than in other terrain. The force ratio does, however, have an impact on the duration of combat. Urban combat consumes time that the attacker may ill afford to spend. Urban battles may take two to three times longer than anticipated. Defense in an urban area does not appear to provide a significant advantage to the defender over a defense in other terrain. A well-planned defense in an urban environment can consume the attacker's time (even without combined arms forces or sophisticated weapons), which allows the defender to put other forces to better use or to prepare for other operations. Essentially, the attacker has a favorable situation over the defender during high-intensity combat. The more restrictive the ROE, the more difficult it is for the attacker.

b. **Intelligence.** Most recent urban battles lost by the attacker were lost because of a failure of intelligence. Some of these battles would never have taken place if the initial intelligence were clear.

c. **Surprise.** Surprise can be an important asset. It is a combat multiplier, but not necessarily a decisive one. Tactical surprise can preempt effective defensive preparation.

d. **Combined Arms.** As in other combat, the US Army fights the urban battle as a combined arms battle.

(1) ***Infantry.*** Urban combat is still Infantry intensive. Infantry units can be equipped and trained for urban operations. Infantrymen can negotiate types of urban terrain that no other combat arm can.

(2) ***Armored Vehicles.*** Urban combat is also an armor fight. Tank support of Infantry was a key element in many recent urban battles. Tanks act best as assault guns to reduce strongpoints. The use of armored vehicles has only been effective when they have been protected by Infantry. Lack of Infantry to protect armored forces leads to disaster on restricted urban terrain.

(3) ***Artillery.*** Artillery support can be significant to the outcome if the ROE allow its use. Delay fuzes allow penetration of buildings to cause the most casualties. Artillery rounds can be used with VT fuzes to clear rooftop observation and weapons positions with relatively little collateral damage. However, artillery to provide isolation of an objective and in the direct fire role to reduce strongpoints has been most effective in urban combat. In other than high-intensity conditions, artillery loses much of its ability because of the problems associated with collateral damage.

(4) ***Mortars.*** Mortars are the most effective indirect fire weapon in urban combat. The high angle of fire gives mortars flexibility to clear structures and place fires in streets where artillery cannot. Mortars suffer from the same limitations as artillery in conditions other than high-intensity.

(5) ***Antiaircraft Artillery.*** Antiaircraft artillery proved its worth in many urban battles because of its high rate of fire and destructive potential. Having phased out the Vulcan, the US Army does not have any remaining weapons of this type. Other multinational and coalition forces may have them, and they can be very useful in support.

(6) ***Aviation.*** Aviation assets are relatively ineffective by themselves in urban operations. They have had little impact on the defender's will or his ability to resist. Aviation assets can, however, help isolate an objective. AC-130 and helicopter gunships have much greater accuracy than other aviation assets and can reduce targets without much collateral damage to surrounding structures. These slow-moving assets are susceptible to small arms fire and shoulder-fired antiaircraft missiles.

e. **Time.** Urban battles can take two to three times longer than other battles and usually much longer than anticipated, which results in the expenditure of more logistics. Time may allow the defender to reorganize or redeploy forces elsewhere as he uses his forces in urban areas in economy-of-force missions.

f. **Isolation.** Where the attacker can isolate the defender, the outcome is certain. Even partial isolation provides the conditions for a favorable result for the attacker. Failure to isolate the defender significantly raises the cost of victory. The attacker must stem the uninterrupted flow of manpower, supplies, and weapons to the defender. No single factor appears to be as important as isolating the objective.

g. **Cost.** In the majority of urban battles, the cost to the attacker is high. Several factors affect the cost. Isolation of the urban area is a crucial element to the attacker; the higher the superiority of forces of the attacker, the lower the cost for the attacker. The operation must be carefully planned, and intelligence is critical. Finally, attacking forces must be trained in the TTP of urban combat. Understanding these considerations leads to lower costs. From the viewpoint of the defender, urban combat can be an economy-of-force operation or can cause attacking forces many casualties. If properly prepared, an urban defense can extract a significant cost from the attacker in men, resources, and time.

h. **Rules of Engagement.** The nature of modern urban combat may seriously restrict the use of weapons. Most recent urban battles have had one or both of the following restrictions imposed upon the attacker:

- To minimize civilian casualties and collateral destruction.
- To limit the use of specific ground or air weapons.

i. **Logistics.** Combat in urban environments may cause a dramatic increase in the demand for certain items and it certainly causes requests for items that have specific uses in urban terrain. High-intensity urban combat requires a continuous flow of ammunition and medical supplies. Commanders should consider the use of a *push* system to deliver supplies to units in urban terrain, based upon the type of operation they are conducting. This technique should include the push of an urban operations *kit* to units entering urban areas.

j. **Combat Health Support.** Although there is no difference in medical treatment (there is a greater reliance on self-aid/buddy aid and combat lifesavers), there are differences for evacuation in urban areas. Medical evacuation (MEDEVAC) in an urban environment presents many challenges in the location, acquisition, and evacuation of patients. Techniques may require modification to acquire and evacuate casualties from above, below, and at ground level. Further, during UO, the environment (rubble and debris) may dictate that evacuation be accomplished by litter carries rather than by vehicle or aircraft. Commanders should be prepared for evacuation from within buildings and for the possibility that air MEDEVAC may not be available due to the fragility of the aircraft and their susceptibility to small arms fire. Treatment facilities may have to be moved much farther forward than usual. Units need specific medical policies, directives, and SOPs for dealing with noncombatants.

CHAPTER 3

URBAN COMBAT SKILLS

Successful combat operations in urban areas depend on the proper employment of the rifle squad. Each member must be skilled in moving, entering buildings, clearing rooms, employing hand grenades, selecting and using fighting positions, navigating in urban areas, and camouflage.

Section I. MOVEMENT

Movement in urban areas is the first fundamental skill the soldier must master. Movement techniques must be practiced until they become habitual. To reduce exposure to enemy fire, the soldier avoids open areas, avoids silhouetting himself, and selects his next covered position before movement.

3-1. CROSSING OPEN AREAS

Open areas, such as streets, alleys, and parks, should be avoided. They are natural kill zones for enemy crew-served weapons or snipers. They can be crossed safely if the individual or small-unit leader applies certain fundamentals including using smoke from hand grenades or smoke pots to conceal movement. When employing smoke as an obscurant, keep in mind that thermal sighting systems can see through smoke. Also, when smoke has been thrown in an open area, the enemy may choose to engage with suppressive fires into the smoke cloud.

a. Before moving to another position, the soldier makes a visual reconnaissance, selects the position offering the best cover and concealment, and determines the route he takes to get to that position.

b. The soldier develops a plan for his own movement. He runs the shortest distance between buildings and moves along the far building to the next position, reducing the time he is exposed to enemy fire.

3-2. MOVEMENT PARALLEL TO BUILDINGS

Soldiers and small units may not always be able to use the inside of buildings as routes of advance and must move on the outside of the buildings (Figure 3-1, page 3-2). Smoke, suppressive fires, and cover and concealment should be used to hide movement. The soldier moves parallel to the side of the building (maintaining at least 12 inches of separation between himself and the wall to avoid *rabbit rounds*, ricochets and rubbing or bumping the wall), stays in the shadow, presents a low silhouette, and moves rapidly to his next position (Figure 3-2, page 3-2). If an enemy gunner inside the building fires on a soldier, he exposes himself to fire from other squad members providing overwatch. An enemy gunner farther down the street would have difficulty detecting and engaging the soldier.

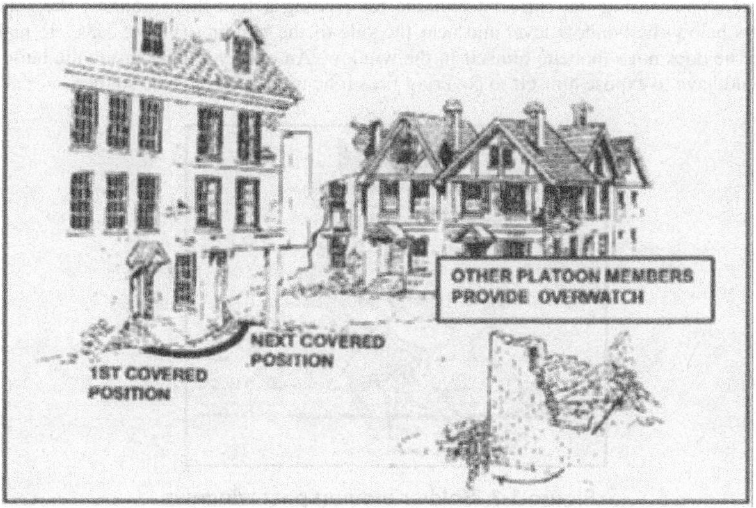

Figure 3-1. Selection of the next position.

Figure 3-2. Soldier moving outside building.

3-3. MOVEMENT PAST WINDOWS

Windows present another hazard to the soldier. The most common mistakes are exposing the head in a first-floor window and not being aware of basement windows.

a. When using the correct technique for passing a first-floor window, the soldier stays below the window level and near the side of the building (Figure 3-3). He makes sure he does not silhouette himself in the window. An enemy gunner inside the building would have to expose himself to covering fires if he tried to engage the soldier.

Figure 3-3. Soldier moving past windows.

b. The same techniques used in passing first-floor windows are used when passing basement windows. A soldier should not walk or run past a basement window, since he presents a good target to an enemy gunner inside the building. The soldier should stay close to the wall of the building and step or jump past the window without exposing his legs (Figure 3-4).

Figure 3-4. Soldier passing basement windows.

3-4. MOVEMENT AROUND CORNERS

The area around a corner must be observed before the soldier moves. The most common mistake a soldier makes at a corner is allowing his weapon to extend beyond the corner exposing his position (this mistake is known as *flagging* your weapon). He should show his head below the height an enemy soldier would expect to see it. The soldier lies flat on the ground and does not extend his weapon beyond the corner of the building. He wears his Kevlar helmet and only exposes his head (at ground level) enough to permit observation (Figure 3-5). Another corner clearing technique that is used when speed is required is the *pie-ing* method. This procedure is done by aiming the weapon beyond the corner into the direction of travel (without flagging) and side-stepping around the corner in a circular fashion with the muzzle as the pivot point (Figure 3-6).

Figure 3-5. Correct technique for looking around a corner.

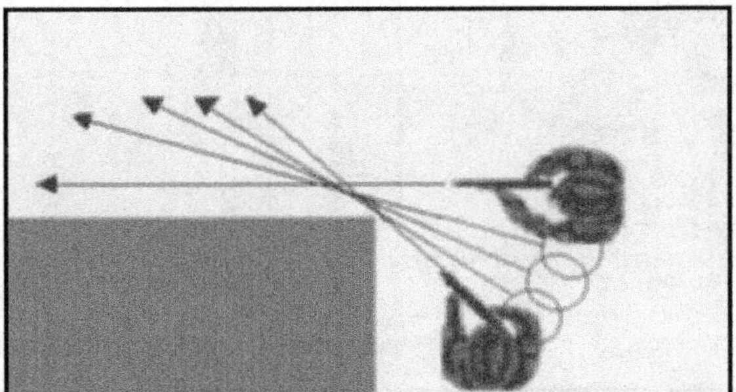

Figure 3-6. *Pie-ing* a corner.

3-5. CROSSING A WALL

Each soldier must learn the correct method of crossing a wall (Figure 3-7). After he has reconnoitered the other side, he rolls over the wall quickly, keeping a low silhouette. Speed of his move and a low silhouette deny the enemy a good target.

Figure 3-7. Soldier crossing a wall.

3-6. USE OF DOORWAYS

Doorways should not be used as entrances or exits since they are normally covered by enemy fire. If a soldier must use a doorway as an exit, he should move quickly to his next position, staying as low as possible to avoid silhouetting himself (Figure 3-8). Preselection of positions, speed, a low silhouette, and the use of covering fires must be emphasized in exiting doorways.

Figure 3-8. Soldier exiting a doorway.

3-7. MOVEMENT BETWEEN POSITIONS

When moving from position to position, each soldier must be careful not to mask his supporting fires. When he reaches his next position, he must be prepared to cover the movement of other members of his fire team or squad. He must use his new position effectively and fire his weapon from either shoulder depending on the position.

a. The most common errors a soldier makes when firing from a position are firing over the top of his cover and silhouetting himself against the building to his rear. Both provide the enemy an easy target. The correct technique for firing from a covered position is to fire around the side of the cover, which reduces exposure to the enemy (Figure 3-9).

b. Another common error is for a right-handed shooter to fire from the right shoulder around the left corner of a building. Firing left-handed around the left corner of a building takes advantage of the cover afforded by the building (Figure 3-10). Right-handed and left-handed soldiers should be trained to adapt cover and concealment to fit their manual orientation. Soldiers should be able to fire from the opposite shoulder.

Figure 3-9. Soldier firing from a covered position.

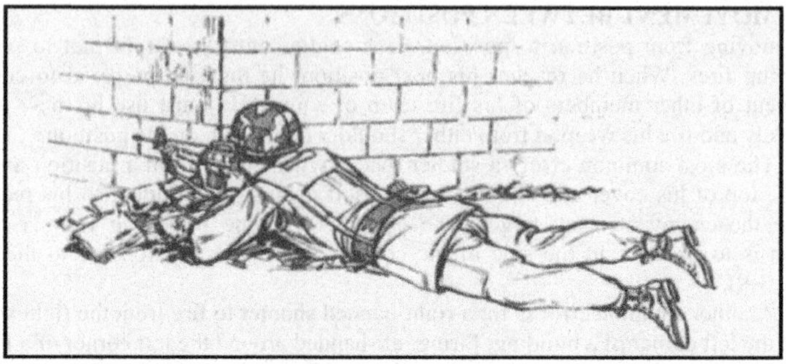

Figure 3-10. Firing left-handed around the corner of a building.

3-8. FIRE TEAM EMPLOYMENT

Moving as a fire team from building to building or between buildings presents a large target for enemy fire (Figure 3-11). When moving from the corner of one building to another, the fire team should move across the open area in a group. Moving from the side of one building to the side of another presents a similar problem and the technique of movement employed is the same. The fire team uses the building as cover. In moving to an adjacent building (Figure 3-12, page 3-8) team members should keep a distance of 3 to 5 meters between themselves and, using a planned signal, make an abrupt flanking movement (on line) across the open area to the next building.

Figure 3-11. Fire team movement.

Figure 3-12. Movement to adjacent building.

Section II. ENTRY TECHNIQUES

When entering buildings a soldier must minimize the time he is exposed. Before moving toward the building he must select the entry point. When moving to the entry point the soldier should use smoke to conceal his advance. He must avoid using windows and doors except as a last resort. He should consider the use of demolitions, tank rounds, and other means to make new entrances. If the situation permits he should precede his entry with a grenade, enter immediately after the grenade explodes, and be covered by one of his buddies.

3-9. UPPER BUILDING LEVELS

Although entering a building from any level other than the ground floor is difficult, clearing a building from the top down is the preferred method. Assaulting or defending a building is easier from an upper story. Gravity and the building's floor plan become assets when throwing hand grenades and moving from floor to floor.

a. An enemy who is forced to the top of a building may be cornered and fight desperately or escape over the roof. An enemy who is forced down to ground level may withdraw from the building, thus exposing himself to friendly fires from the outside.

b. Various means, such as ladders, drainpipes, vines, helicopters, or the roofs and windows of adjoining buildings, may be used to reach the top floor or roof of a building. One soldier can climb onto the shoulders of another and reach high enough to pull himself up.

c. Ladders offer the quickest method to access the upper levels of a building (Figure 3-13). Units deploying into an urban environment should be equipped with a lightweight, man-portable, collapsible ladder as referenced in the platoon urban operations kit.

Figure 3-13. Entering using portable ladder

(1) If portable ladders are not available, material to build ladders can be obtained through supply channels. Ladders can also be built with resources available throughout the urban area; for example, lumber can be taken from inside the walls of buildings (Figure 3-14).

Figure 3-14. Getting lumber from inside the walls.

(2) Although ladders do not permit access to the top of some buildings, they do offer security and safety through speed. Ladders can be used to conduct an exterior assault of an upper level if soldiers' exposure to enemy fire can be minimized.

3-10. USE OF GRAPPLING HOOK

The use of a grappling hook and rope to ascend into a building is not recommended. Experimentation and training has determined that using the grappling hook and rope to ascend is extremely difficult for the average soldier, and makes a unit more likely to fail their mission. Grappling hooks are still a viable tool for accomplishing the following tasks:

- Clearing concertina or other tangle wire.
- Clearing obstacles or barricades that may be booby trapped.
- Descending to lower floors.

3-11. SCALING OF WALLS

When required to scale a wall during exposure to enemy fire, all available concealment must be used. Smoke and diversionary measures improve the chances of success. When using smoke for concealment, soldiers must plan for wind direction. They should use suppressive fire, shouting, and distraction devices from other positions to divert the enemy's attention.

a. A soldier scaling an outside wall is vulnerable to enemy fire. Soldiers who are moving from building to building and climbing buildings should be covered by friendly fire. Properly positioned friendly weapons can suppress and eliminate enemy fire. The M203 grenade launcher is effective in suppressing or neutralizing the enemy from rooms inside buildings (Figure 3-15).

Figure 3-15. Employment of M203 grenade launcher for clearing enemy snipers.

b. If a soldier must scale a wall with a rope, he should avoid silhouetting himself in windows that are not cleared and avoid exposing himself to enemy fires from lower windows. He should climb with his weapon slung over the firing shoulder so he can bring it quickly to a firing position. If the ROE permits, the objective window and any lower level windows in the path of the climber should be engaged with grenades (hand or launcher) before the soldier begins his ascent.

c. The soldier enters the objective window with a low silhouette (Figure 3-16). Entry can be head first; however, the preferred method is to hook a leg over the window sill and enter sideways straddling the ledge.

Figure 3-16. Soldier entering the objective window.

3-12. RAPPELLING

Rappelling is an entry technique that soldiers can use to descend from the rooftop of a tall building into a window (Figure 3-17), or through a hole in the floor, in order to descend to the lower floor. (See TC 21-24 for more information on rappelling.)

Figure 3-17. Rappelling.

3-13. ENTRY AT LOWER LEVELS

Buildings should be cleared from the top down. However, entering a building at the top may be impossible. Entry at the bottom or lower level is common and may be the only course of action. When entering a building at lower levels, soldiers avoid entering through windows and doors since both can be easily booby trapped and are usually covered by enemy fire. (Specific lower-level entry techniques are shown in Figure 3-18 on pages 3-13 through 3-15. These techniques are used when soldiers can enter the building without receiving effective enemy fire.)

a. When entering at lower levels, demolitions, artillery, tank fire, antiarmor weapons fire, or similar means can be used to create a new entrance to avoid booby traps. This procedure is preferred if the ROE permit it. Quick entry is then required to take advantage of the effects of the blast and concussion.

b. When the only entry to a building is through a window or door, supporting fire is directed at that location to destroy or drive away enemy forces. The assaulting soldiers should not leave their covered positions before the support by fire element has accomplished this procedure.

c. Before entering, soldiers may throw a cooked off hand grenade into the new entrance to reinforce the effects of the original blast. The type grenade used, fragmentation, concussion, or stun, is based on METT-TC factors and the structural integrity of the building.

(1) When making a new entrance in a building, soldiers consider the effects of the blast on the building and on adjacent buildings. If there is the possibility of a fire in

adjacent building, soldiers coordinate with adjacent units and obtain permission before starting the operation.

(2) In wooden frame buildings, the blast may cause the building to collapse. In stone, brick, or cement buildings, supporting fires are aimed at the corner of the building or at weak points in the building construction.

NOTE: Armored vehicles can be positioned next to a building allowing soldiers to use the vehicle as a platform to enter a room or gain access to a roof.

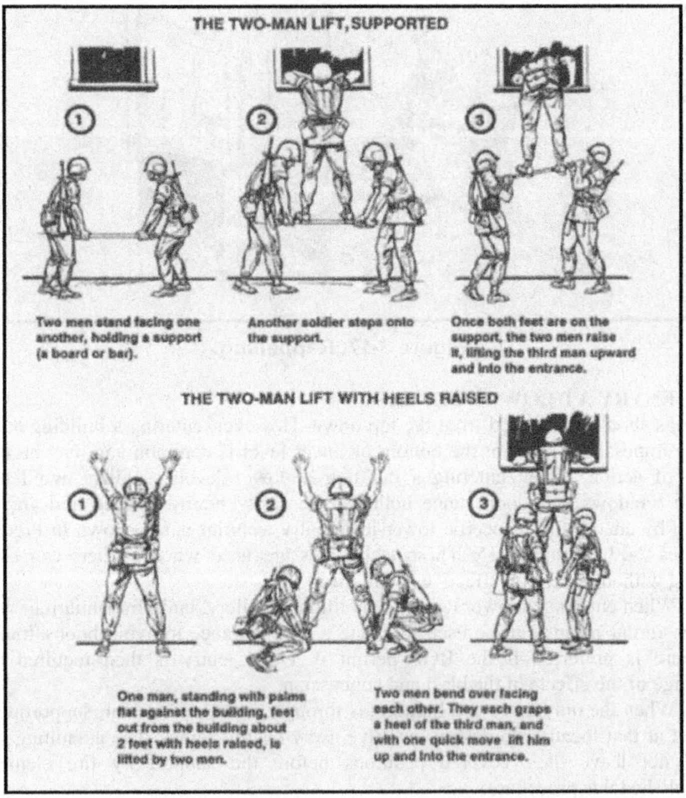

Figure 3-18. Lower-level entry technique.

Figure 3-18. Lower-level entry technique (continued).

Figure 3-18. Lower-level entry technique (continued).

3-14. USE OF HAND GRENADES

Combat in urban areas often requires extensive use of hand grenades. Unless the ROE prevent it, use grenades before assaulting defended areas, moving through breaches, or entering unsecured areas. Effective grenade use in urban areas may require throwing overhand or underhand, with both the left and right hand. Normally, the fragmentation grenade should be cooked off for two seconds to prevent the enemy from throwing them back.

a. Three types of hand grenades can be used when assaulting an urban objective: stun, concussion, and fragmentation. METT-TC factors and the type of construction materials used in the objective building influence the type of grenades that can be used.

(1) The M84 stun hand grenade is a *flash-bang* distraction device, which produces a brilliant flash and a loud bang to momentarily surprise and distract an enemy force (Figure 3-19, page 3-16). The M84 is often used under precision conditions and when the ROE demand use of a nonlethal grenade. The use of stun hand grenades under high intensity conditions is usually limited to situations where fragmentation and concussion grenades pose a risk to friendly troops or the structural integrity of the building.

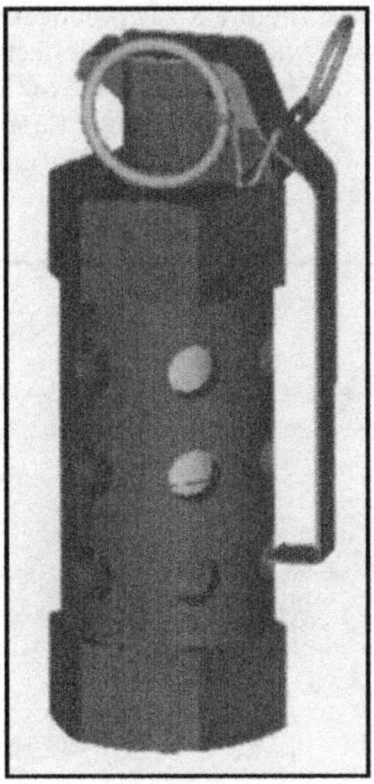

Figure 3-19. M84 stun hand grenade.

(2) The concussion grenade causes injury or death to persons in a room by blast overpressure and propelling debris within the room (Figure 3-20). While the concussion grenade does not discard a dangerous fragmentation from its body, the force of the explosion can create debris fallout that may penetrate thin walls.

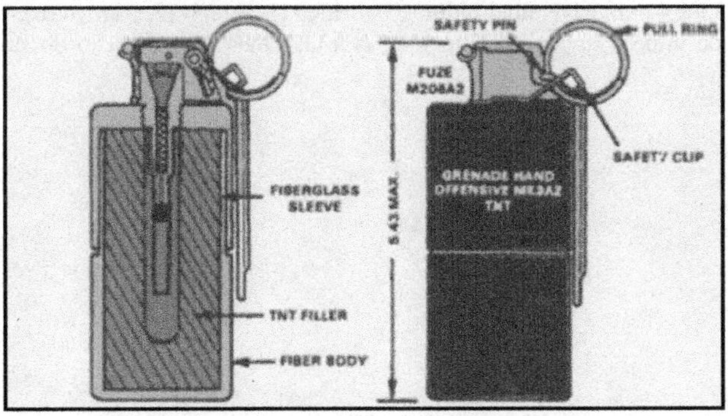

Figure 3-20. MK3A2 (concussion grenade).

(3) The fragmentation grenade (Figure 3-21) produces substantial overpressure when used inside buildings and, coupled with the shrapnel effects, can be extremely dangerous to friendly soldiers. If the walls of a building are made of thin material, such as Sheetrock or thin plywood, soldiers should either lie flat on the floor with their helmet towards the area of detonation, or move away from any wall that might be penetrated by grenade fragments.

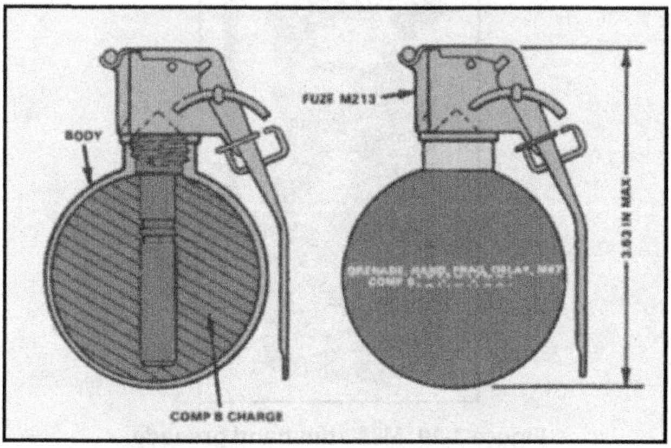

Figure 3-21. Fragmentation grenade.

b. Soldiers should engage upper-level openings with grenades (by hand or launcher) before entering to eliminate enemy that might be near the entrance.

(1) The M203 grenade launcher is the best method for putting a grenade in an upper-story window. The primary round of ammunition used for engaging an urban threat is the M433 high-explosive, dual-purpose cartridge (Figure 3-22, page 3-18). Throwing a hand grenade into an upper-story opening is a task that is difficult to do safely during combat.

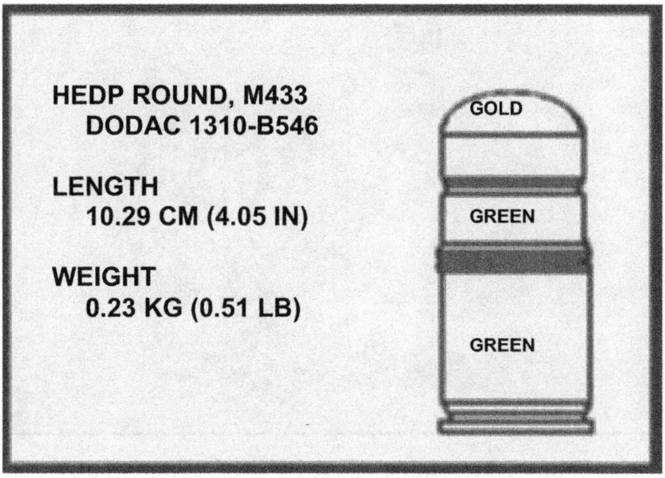

Figure 3-22. 40-mm, tube-launched, high-explosive, dual-purpose (HEDP) grenade.

(2) When a hand grenade must be thrown into an upper-story opening, the thrower should stand close to the building, using it for cover. This technique should only be employed when the window opening is free of glass or screen.

(3) The thrower should allow the grenade to cook off for at least two seconds, and then step out far enough to lob the grenade into the upper-story opening (Figure 3-23). He should keep his weapon in the nonthrowing hand, to be used if needed. The weapon should never be laid outside or inside the building. At the same time, everyone should have a planned area to move to for safety if the grenade does not go through the window but falls back to the ground.

(4) Once the grenade has been thrown into the opening and detonates, assaulting troops must move swiftly to enter the building.

Figure 3-23. Hand grenade thrown through window.

c. If soldiers must enter the building by the stairs, they must first look for booby traps, then engage the stairwell door with a grenade (by hand or launcher), let it detonate, and quickly move inside. They can then use the staircase for cover.

WARNINGS

1. **If stealth is not a factor, after throwing the grenade the soldier must immediately announce *frag out* to indicate that a grenade has been thrown. He then takes cover since the grenade may bounce back or be thrown back, or the enemy may fire at him.**

2. **When the M203 grenade launcher is used to deliver the grenade into a window or doorway, ensure proper standoff for arming the round. Also, the assaulting element should take cover around a corner or away from the target area.**

d. Breachholes and mouseholes are blown or cut through a wall so soldiers can enter a building. (See Chapters 4 and 7 for more information.) These are safer entrances than doors because doors can be easily booby trapped and should be avoided, unless explosive breaching is used against the door.

(1) A grenade should be thrown through the breach before entering. Use available cover, such as the lower corner of the building (Figure 3-24), for protection from fragments.

(2) Use stun and concussion grenades when engaging through thin walls.

Figure 3-24. Soldier entering through a mousehole.

e. When a door is the only means of entering a building, soldiers must beware of booby traps and fire from enemy soldiers within the room.

(1) Locked doors can be breached (forced open) using one of the four breaching methods: mechanical, ballistic, explosive, or thermal (see Chapter 8). If none of these methods are available, soldiers can resort to kicking the door open. This method is the least preferred since it is difficult and tiring to the soldier. It rarely works the first time, and gives any enemy soldiers in the room ample warning and time to shoot through the door. Once the door is breached, a grenade should precede the soldier's entry.

(2) When opening an unlocked door by hand, the assault team should be sure not to expose themselves to enemy fire through the door. The soldiers should stay close to one side of the doorway to minimize exposure in the open doorframe

(3) Once the door is open, a hand grenade should be tossed in. After the grenade explodes, soldiers enter and clear the room IAW the tactics, techniques, and procedures discussed in Section III.

f. Although buildings are best cleared from the top down, this procedure is not always possible. While clearing the bottom floor of a building, soldiers may encounter stairs, which must also be cleared. Once again, grenades play an important role.

(1) To climb stairs, first inspect for booby traps, then toss a grenade to the head of the stairs (Figure 3-25). Soldiers must use voice alerts when throwing grenades.

(2) Using the staircase for cover, soldiers throw the grenade underhand to reduce the risk of it bouncing back and rolling down the stairs.

(3) Once the first grenade has detonated, another grenade should be thrown over and behind the staircase banister and into the hallway, neutralizing any exposed enemy in the hallway.

(4) When the second hand grenade has detonated, soldiers proceed to clear the stairway in accordance with prescribed TTP.

NOTE: Large quantities of hand grenades are used when clearing buildings. A continuous supply must be available.

Figure 3-25. Soldier tossing grenade up stairway.

CAUTION

Throwing fragmentation grenades up a stairway has a high probability for the grenades to roll back down and cause fratricide. Soldiers should avoid clustering at the foot of the stairway and ensure that the structural integrity of the building permits the use of either a fragmentation or concussion grenade.

3-15. INDIVIDUAL WEAPONS CONTROL WHEN MOVING

As in all combat situations, the clearing team members must move tactically and safely. Individuals who are part of a clearing team must move in a standard manner, using practiced techniques known to all.

a. When moving, team members maintain *muzzle awareness* by holding their weapons with the muzzle pointed in the direction of travel. Soldiers keep the butt of the rifle in the pocket of their shoulder, with the muzzle slightly down to allow unobstructed vision. Soldiers keep both eyes open and swing the muzzle as they turn their head so the rifle is always aimed where the soldier is looking. This procedure allows to soldier to see what or who is entering their line of fire.

b. Team members avoid *flagging* (leading) with the weapon when working around windows, doors, corners, or areas where obstacles must be negotiated. Flagging the weapon gives advance warning to anyone looking in the soldier's direction, making it easier for an enemy to grab the weapon.

c. Team members should keep weapons on safe (selector switch on SAFE and index finger outside of trigger guard) until a hostile target is identified and engaged. After a team member clears his sector of all targets, he returns his weapon to the SAFE position.

d. If a soldier has a weapons malfunction during room clearing, he should immediately announce "gun down" and drop to one knee and conduct immediate action to reduce the malfunction. The other members of the team should engage targets in his sector. Once the weapon is operational, he should announce "gun up" and remain in the kneeling position until directed to stand-up by the team leader.

Section III. CLEARING

Infantry units often use close combat to enter and clear buildings and rooms. This section describes the TTP for clearing.

3-16. HIGH INTENSITY VERSUS PRECISION CLEARING TECHNIQUES

Precision clearing techniques do not replace other techniques currently being used to clear buildings and rooms during high-intensity combat. Specifically, they do not replace the clearing technique in which a fragmentation or concussion grenade is thrown into a room before the US forces enter. Precision room clearing techniques are used when the tactical situation calls for room-by-room clearing of a relatively intact building in which enemy combatants and noncombatants may be intermixed. They involve increased risk in order to clear a building methodically, rather than using overwhelming firepower to eliminate or neutralize all its inhabitants.

a. From a conceptual standpoint, standard high-intensity room clearing drills can be thought of as a deliberate attack. The task is to seize control of the room with the purpose being the neutralization of the enemy in the room. The fragmentation and or concussion grenades can be thought of as the preparatory fires used before the assault. As in a deliberate attack against any objective, the assaulting elements move into position using covered and concealed routes. The preparatory fires (fragmentation and or concussion grenades) are initiated when soldiers are as close to the objective as they can get without being injured by the fires. The assault element follows the preparatory fires onto the objective as closely as possible. A rapid, violent assault overwhelms and destroys the enemy force and seizes the objective.

b. Compared to the deliberate attack represented by high-intensity room clearing techniques, precision room clearing techniques are more conceptually like a reconnaissance in force or perhaps an infiltration attack. During a reconnaissance in force, the friendly unit seeks to determine the enemy's locations, dispositions, strength, and intentions. Once the enemy is located, the friendly force is fully prepared to engage and destroy it, especially if surprise is achieved. The friendly force retains the options of not employing preparatory fires (fragmentation and or concussion grenades) if they are not called for (the enemy is not in the room) or if they are inappropriate (there are noncombatants present also). The attacking unit may choose to create a diversion (use a stun grenade) to momentarily distract the defender while they enter and seize the objective.

c. The determination of which techniques to employ is up to the leader on the scene and is based on his analysis of the existing METT-TC conditions. The deliberate attack (high-intensity techniques), with its devastating suppressive and preparatory fires, neutralizes everyone in the room and is less dangerous to the assaulting troops. The reconnaissance in force (precision techniques) conserves ammunition, reduces damage, and minimizes the chance of noncombatant casualties. Unfortunately, even when well-executed, it is very stressful and hazardous for friendly troops.

d. Certain precision room clearing techniques, such as methods of squad and fire team movement, the various firing stances, weapon positioning, and reflexive shooting, are useful for all combat in confined areas. Other techniques, such as entering a room without first neutralizing known enemy occupants by fire or explosives, are appropriate in only some tactical situations.

e. Generally, if a room or building is occupied by an alerted enemy force that is determined to resist, and if most or all noncombatants are clear, overwhelming firepower should be employed to avoid friendly casualties. In such a situation, supporting fires, demolitions, and fragmentation grenades should be used to neutralize a space before friendly troops enter.

f. In some combat situations the use of heavy supporting fires and demolitions would cause unacceptable collateral damage or would unnecessarily slow the unit's movement. In other situations, often during stability and support operations, enemy combatants are so intermixed with noncombatants that US forces cannot, in good conscience, use all available supporting fires. Room-by-room clearing may be necessary. At such times, precision room clearing techniques are most appropriate.

3-17. PRINCIPLES OF PRECISION ROOM CLEARING

Battles that occur at close quarters, such as within a room or hallway, must be planned and executed with care. Units must train, practice, and rehearse precision room clearing techniques until each fire team and squad operates smoothly. Each unit member must understand the principles of precision room clearing: surprise, speed, and controlled violence of action.

a. **Surprise.** Surprise is the key to a successful assault at close quarters. The fire team or squad clearing the room must achieve surprise, if only for seconds, by deceiving, distracting, or startling the enemy. Sometimes stun grenades may be used to achieve surprise. These are more effective against a nonalert, poorly trained enemy than against alert, well-trained soldiers.

b. **Speed.** Speed provides a measure of security to the clearing unit. It allows soldiers to use the first few vital seconds provided by surprise to their maximum advantage. In precision room clearing, speed is not how fast you enter the room, rather it's how fast the threat is eliminated and the room is cleared.

c. **Controlled Violence of Action.** Controlled violence of action eliminates or neutralizes the enemy while giving him the least chance of inflicting friendly casualties. It is not limited to the application of firepower only, but also involves a soldier mind-set of complete domination. Each of the principles of precision room clearing has a synergistic relationship to the others. Controlled violence coupled with speed increases surprise. Hence, successful surprise allows increased speed.

3-18. FUNDAMENTALS OF PRECISION ROOM CLEARING

The ten fundamentals of precision room clearing address actions soldiers take while moving along confined corridors to the room to be cleared, while preparing to enter the room, during room entry and target engagement, and after contact. Team members—

- Move tactically and silently while securing the corridors to the room to be cleared.
- Carry only the minimum amount of equipment. (Rucksacks and loose items carried by soldiers tire them, slow their pace, and cause noise.)
- Arrive undetected at the entry to the room in the correct order of entrance, prepared to enter on a single command.
- Enter quickly and dominate the room. Move immediately to positions that allow complete control of the room and provide unobstructed fields of fire.
- Eliminate all enemy in the room by fast, accurate, and discriminating fires.
- Gain and maintain immediate control of the situation and all personnel in the room.
- Confirm whether enemy casualties are wounded or dead. Disarm, segregate, and treat the wounded. Search all enemy casualties.
- Perform a cursory search of the room. Determine if a detailed search is required.
- Evacuate all wounded and any friendly dead.
- Mark the room as cleared using a simple, clearly identifiable marking in accordance with the unit SOP.
- Maintain security and be prepared to react to more enemy contact at any moment. Do not neglect rear security.

3-19. COMPOSITION OF THE CLEARING TEAM

Precision room clearing techniques are designed to be executed by the standard four-man fire team. Because of the confined spaces typical of building- and room-clearing operations, units larger than squads quickly become unwieldy. When shortages of personnel demand it, room clearing can be conducted with two- or three-man teams, but four-man teams are preferred. Using fewer personnel greatly increases the combat strain and risks.

3-20. BREACHING

An integral part of precision room clearing is the ability to gain access quickly to the rooms to be cleared. Breaching techniques vary based on the type of construction encountered and the types of munitions available to the breaching element. Techniques range from simple mechanical breaching to complex, specialized demolitions.

a. A useful method of breaching is the *shotgun ballistic* breach for forced entry of standard doors. A 12-gauge shotgun loaded with buckshot or slugs can be used to breach most standard doors quickly. Number 9 shot works equally well with reduced collateral damage on the other side of the door. When done properly, the shotgun breach requires only a few seconds. The two standard techniques of shotgun breaching are the *doorknob breach* and the *hinge breach*. When attempting either technique, the gunner is announcing his presence by using the shotgun and is completely exposed to fire through the door. Therefore, exposure time must be minimized and the number 1 man must be ready to gain entry and return fire as soon as possible. While holding the stock of the shotgun in the pocket of his shoulder, the gunner places the muzzle tightly against the door, and aims down at a 45-degree angle.

NOTE: If the shotgun muzzle is not held tightly against the door, splatter may occur that could affect friendly troops. Also, buckshot and rifled slugs can overpenetrate doors and may kill or wound occupants in the room.

(1) For the doorknob breach, the aim point is a spot halfway between the doorknob and the frame, not at the doorknob itself. The gunner fires two quick shots in the same location, ensuring the second shot is aimed as carefully as the first. Weak locks may fly apart with the first shot, but the gunner should always fire twice. Some locks that appear to be blown apart have parts still connected that can delay entry. If the lock is not defeated by the second shot, the gunner repeats the procedure. Doors may not always open after firing. The gunner should be prepared to kick the door after firing to ensure opening of the entry point.

(2) The hinge breach technique is performed much the same as the doorknob breach, except the gunner aims at the hinges. He fires three shots per hinge—the first at the middle, then at the top and bottom (Figure 3-26, page 3-26). He fires all shots from less than an inch away from the hinge. Because the hinges are often hidden from view, the hinge breach is more difficult. Hinges are generally 8 to 10 inches from the top and bottom of the door; the center hinge is generally 36 inches from the top, centered on the door. Regardless of which technique the gunner uses, immediately after he fires, he kicks the door in or pulls it out. He then pulls the shotgun barrel sharply upward and quickly turns away from the doorway to signal that the breach point has been cleared. This rapid clearing of the doorway allows the following man in the fire team a clear shot at any enemy who may be blocking the immediate breach site.

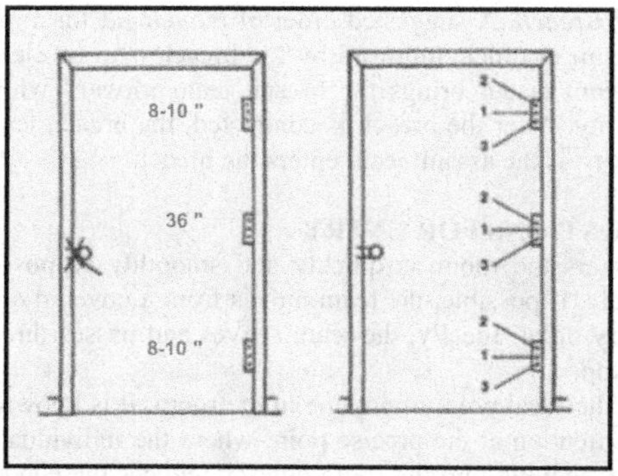

Figure 3-26. Aim points for shotgun breach of a standard door, doorknob target on left and hinge targets on right.

NOTE: The use of small arms (5.56-mm or 7.62-mm) as a ballistic breach on doorknobs and hinges is unsafe and should only be used as a last resort.

b. Demolitions are often needed to defeat more elaborate barriers or to produce a desired effect to aid the initial entry. (See Chapter 8 for a discussion of expedient demolitions for breaching common urban barriers.)

c. Mechanical breaching is planned as a backup to a ballistic or explosive breach. Mechanical breaching is an assumed capability within all units. Taking the time to defeat weak barriers, such as doors or windows, by means of crowbars, saws, sledgehammers, battering rams, axes, or other breaching tools is a decision that must be made based on the conditions of METT-TC.

d. Clearing team members must approach the breach point quickly, quietly, and in standard order. This approach preserves the element of surprise and allows for quick entry and domination of the room. The order of movement to the breach point is determined by the method of breach and intended actions at the breach point. The members of the fire team are assigned numbers 1 through 4, with the team leader normally designated number 2. If one member of the clearing team is armed with the SAW rather than an M16 rifle or carbine, he should be designated number 4.

(1) ***Ballistic (Shotgun) Breach.*** The order of movement for a shotgun breach has the gunner up front, followed by the number 1 man, number 2 man (team leader), and then the number 3 man. After the door is breached, the gunner moves to the rear of the lineup and assumes the position of the number 4 man.

(2) ***Explosive (Demolition) Breach.*** The order of movement for an explosive breach without engineer support is number 1, number 2 (team leader), number 3, and then number 4. The number 1 man provides security at the doorway. The number 2 man (team leader) carries the demolition charge and places it. The number 3 man provides security overhead, and the number 4 man provides rear security. After the demolition charge is placed, the team moves to covered positions and prepares to enter in the standard 1, 2, 3, 4 order. (Refer to Chapter 8 for information concerning minimum safe distances.)

(3) ***Mechanical Breach.*** A suggested order of movement for a mechanical breach is the initial assault team in order, followed by the breach man or element. At the breach point, the assault team leader brings the breach team forward while the assault team provides local security. After the breach is conducted, the breach team moves aside and provides local security as the assault team enters the breach.

3-21. CONSIDERATIONS FOR ENTRY

The entire team enters the room as quickly and smoothly as possible and clears the doorway immediately. If possible, the team moves from a covered or concealed position already in their entry order. Ideally, the team arrives and passes through the entry point without having to stop.

a. The door is the focal point of anyone in the room. It is known as the *fatal funnel*, because it focuses attention at the precise point where the individual team members are the most vulnerable. Moving into the room quickly reduces the chance anyone being hit by enemy fire directed at the doorway.

b. On the signal to go, the clearing team moves from covered or concealed positions through the door quickly and takes up positions inside the room that allow it to completely dominate the room and eliminate the threat. Team members stop movement only after they have cleared the door and reached their designated point of domination. The first man's position is deep into the near corner of the room. The depth of his movement is determined by the size of the room, any obstacles in the room, such as furniture, and by the number and location of enemy and noncombatants in the room.

c. To make precision room clearing techniques work, each member of the team must know his sector of fire and how his sector overlaps and links with the sectors of the other team members. Team members do not move to the point of domination and then engage their targets. They engage targets as they move to their designated point. However, engagements must not slow movement to their points of domination. Team members may shoot from as short a range as 1 to 2 inches. They engage the most immediate enemy threats first. Examples of immediate threats are enemy personnel who—

- Are armed and prepared to return fire immediately.
- Block movement to the position of domination.
- Are within arm's reach of a clearing team member.
- Are within 3 to 5 feet of the breach point.

d. Each clearing team member has a designated sector of fire unique to him initially and expands to overlap sectors of the other team members.

(1) The number 1 and number 2 men are initially concerned with the area directly to their front, then along the wall on either side of the door or entry point. This area is in their path of movement, and it is their primary sector of fire. Their alternate sector of fire is from the wall they are moving toward, back to the opposite far corner.

(2) The number 3 and number 4 men start at the center of the wall opposite their point of entry and clear to the left if moving toward the left, or to the right if moving toward the right. They stop short of their respective team member (either the number 1 man or the number 2 man).

e. The team members move toward their points of domination, engaging all targets in their sector. Team members must exercise fire control and discriminate between hostile and noncombatant room occupants. Shooting is done without stopping, using reflexive shooting techniques. Because the soldiers are moving and shooting at the same time, they must move using careful hurry. (Figure 3-31 in paragraph 3-23, page 3-32, shows all four team members at their points of domination and their overlapping sectors of fire.)

3-22. TECHNIQUES FOR ENTERING BUILDINGS AND CLEARING ROOMS

Battle Drill 6 is the standard technique used by the four-man fire team when they perform the task, Enter Building/Clear Room. However, ROE may not allow for, nor the enemy situation requires, such aggressive action on the part of the assaulting unit. Based on the aforementioned conditions, commanders may determine to use the following techniques when entering and clearing buildings and rooms.

a. **Situation.** Operating as part of a larger force (during daylight or darkness), the squad is tasked to participate in clearing a building. The platoon leader directs the squad to enter the building or to clear a room. An entry point breach has already been identified, or will be created before initiating the entry.

b. **Special Considerations.** Platoon and squad leaders must consider the task and purpose they have been given and the method they are to use to achieve the desired results.

(1) To seize or gain control of a building may not always require committing troops into the structure or closing with the enemy. The following steps describe effective techniques to be used when training soldiers to the toughest possible conditions. These techniques and procedures can be trained, rehearsed, and modified to a specific situation and mission. Before initiating this action the employment of all organic, crew-served, and supporting weapon systems should be directed onto the objective area in order to suppress and neutralize the threat, providing the mission and ROE permit.

(2) When conducting urban operations, soldiers must be equipped at all times with a night vision device or light source to illuminate the immediate area.

NOTE: The following discussion assumes that only the platoon's organic weapons are to support the infantry squad. Urban situations may require precise application of firepower. This situation is especially true of an urban environment where the enemy is mixed with noncombatants. Noncombatants may be found in the room, which can restrict the use of fires and reduce the combat power available to a squad leader. His squad may have to operate with *no fire* areas. Rules of engagement can prohibit the use of certain weapons until a specific hostile action takes place. All soldiers must be aware of the ROE. Leaders must include the specific use of weapons in their planning for precision operations in urban terrain.

c. **Required Actions.** Figures 3-27, 3-28, 3-29, and 3-30 (on pages 3-30 through 3-32) illustrate the required actions for performing this task.

(1) The squad leader designates the assault team and identifies the location of the entry point for them.

(2) The squad leader positions the follow-on assault team to provide overwatch and supporting fires for the initial assault team.

(3) Assault team members move as close to the entry point as possible, using available cover and concealment.

(a) If an explosive breach or a ballistic breach is to be performed by a supporting element, the assault team remains in a covered position until the breach is made. They may provide overwatch and fire support for the breaching element if necessary.

(b) All team members must signal one another that they are ready before the team moves to the entry point.

(c) Team members avoid the use of verbal signals, which may alert the enemy and remove the element of surprise.

(d) Assault team members must move quickly from the covered position to the entry point, minimizing the time they are exposed to enemy fire.

(4) The assault team enters through the breach. Unless a grenade is being thrown prior to entry, the team should avoid stopping outside the point of entry.

(a) The number 2 man may throw a grenade of some type (fragmentation, concussion, stun) into the room before entry.

(b) The use of grenades should be consistent with the ROE and building structure. The grenade should be cooked off before being thrown, if applicable to the type of grenade used.

(c) If stealth is not a factor, the thrower should sound off with a verbal indication that a grenade of some type is being thrown ("frag out," "concussion out," "stun out"). If stealth is a factor, only visual signals are given as the grenade is thrown.

CAUTION

If walls and floors are thin, fragments from fragmentation grenades and debris created by concussion grenades can injure soldiers outside the room. If the structure has been stressed by previous explosive engagements, the use of these grenades could cause it to collapse. Leaders must determine the effectiveness of these types of grenades compared to possibilities of harm to friendly troops.

(5) On the signal to go, or immediately after the grenade detonates, the assault team moves through the entry point (Figure 3-27, page 3-30) and quickly takes up positions inside the room that allow it to completely dominate the room and eliminate the threat (Figure 3-30). Unless restricted or impeded, team members stop movement only after they have cleared the door and reached their designated point of domination. In addition to dominating the room, all team members are responsible for identifying possible loopholes and mouseholes in the ceiling, walls and floor.

NOTE: Where enemy forces may be concentrated and the presence of noncombatants is highly unlikely, the assault team can precede their entry by throwing a fragmentation or concussion grenade (structure dependent) into the room, followed by bursts of automatic small-arms fire by the number one man as he enters.

(a) The first man (rifleman), enters the room and eliminates the immediate threat. He has the option of going left or right, normally moving along the path of least resistance to one of two corners. When using a doorway as the point of entry, the path of least resistance is determined initially based on the way the door opens; if the door opens inward he plans to move away from the hinges. If the door opens outward, he plans to move toward the hinged side. Upon entering, the size of the room, enemy situation, and furniture or other obstacles that hinder or channel movement become factors that influence the number1 man's direction of movement.

(b) The direction each man moves in should not be preplanned unless the exact room layout is known. Each man should go in a direction opposite the man in front of him (Figure 3-27). Every team member must know the sectors and duties of each position.

(c) As the first man goes through the entry point, he can usually see into the far corner of the room. He eliminates any immediate threat and continues to move along the wall if possible and to the first corner, where he assumes a position of domination facing into the room.

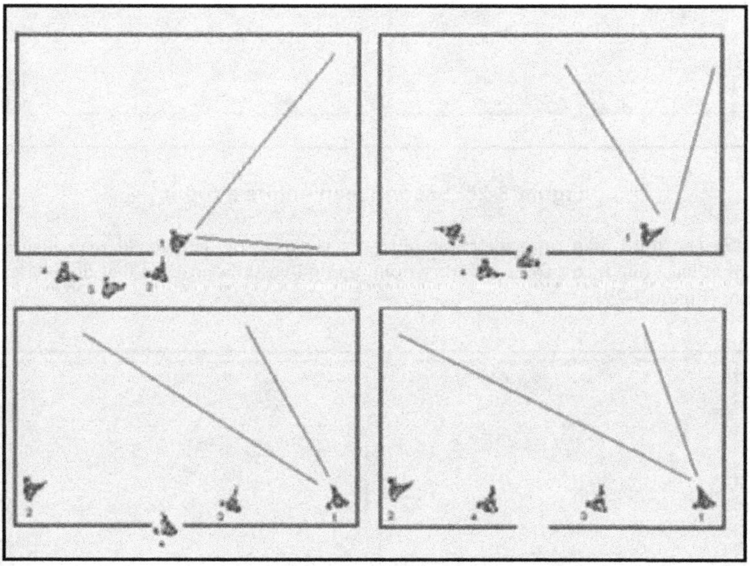

Figure 3-27. First man enters a room.

(6) The second man (team leader), entering almost simultaneously with the first, moves in the opposite direction, following the wall and staying out of the center

(Figure 3-28). The second man must clear the entry point, clear the immediate threat area, clear his corner, and move to a dominating position on his side of the room.

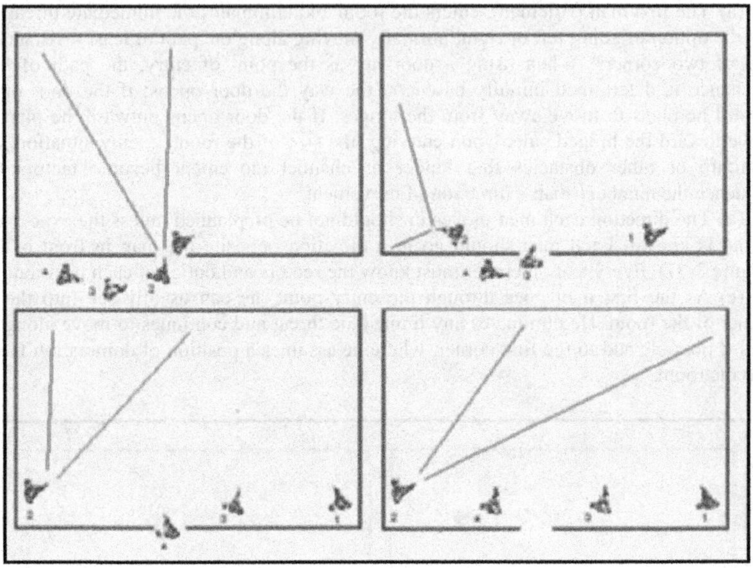

Figure 3-28. Second man enters a room.

(7) The third man (grenadier) simply goes opposite of the second man inside the room at least one meter from the entry point and moves to a position that dominates his sector (Figure 3-29).

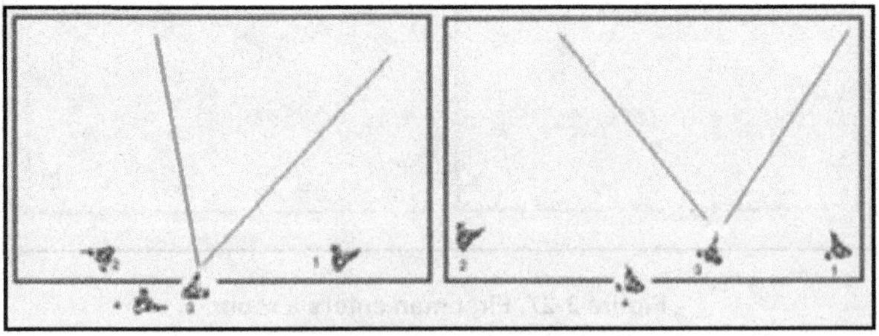

Figure 3-29. Third man enters a room.

(8) The fourth man (SAW gunner) moves opposite of the third man and moves to a position that dominates his sector (Figure 3-30).

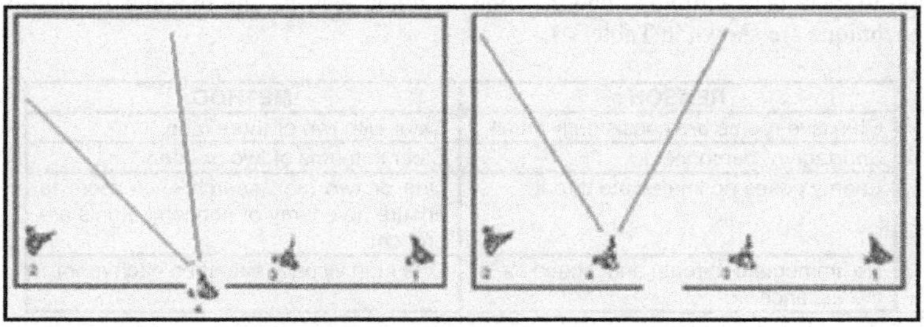

Figure 3-30. Fourth man in a room.

NOTE: If the path of least resistance takes the first man to the left, then all points of domination are the mirror image of those shown in the diagrams.

(9) Points of domination should not be in front of doors or windows so team members are not silhouetted to the outside of the room (Figure 3-31). No movement should mask the fire of any of the other team members.

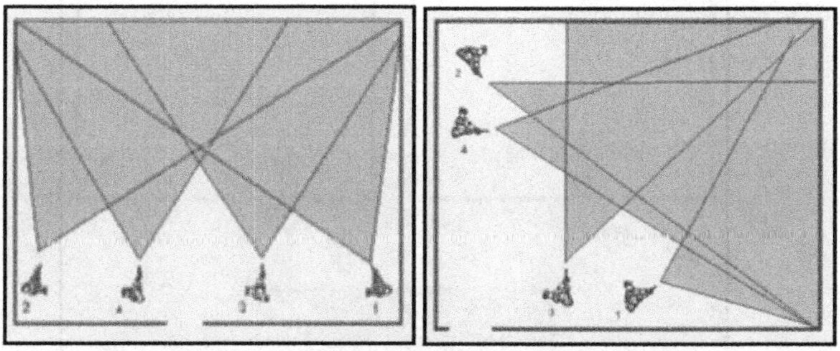

(Center door) **(Corner door)**

Figure 3-31. Points of domination and sectors of fire.

(10) On order, any member of the assault team may move deeper into the room overwatched by the other team members.

(11) Once the room is cleared, the team leader signals to the squad leader that the room has been cleared.

(12) The squad leader marks the room (IAW unit SOP). The squad leader determines whether or not his squad can continue to clear through the building.

(13) The squad reorganizes as necessary. Leaders redistribute the ammunition.

(14) The squad leader reports to the platoon leader when the room is clear.

d. **Reasons for Modifying the Entry Technique.** Although this technique is an effective procedure for clearing a room, leaders may be required to modify the existing action to meet their current situation. Some example reasons and methods of modifying the technique are shown in Table 3-1.

REASON	METHOD
Objective rooms are consistently small.	Clear with two or three men.
Shortage of personnel.	Clear in teams of two or three.
Enemy poses no immediate threat.	One or two men search each room to ensure no enemy or noncombatants are present.
No immediate threat, and speed is of the essence	One man visually searches each room.

Table 3-1. Reasons and methods for modifying entry techniques.

e. **Three- and Two-Man Teams.** When full four-man teams are not available for room clearing three- and two-man teams can be used. Figures 3-32 (below) and 3-33 (page 3-34) show the points of domination and sectors of fire for a three-man clearing team. Figures 3-34 and 3-35 (pages 3-34 and 3-35) show the same thing for a two-man team. Leaders should use the entry technique *blueprint* when modifying their techniques.

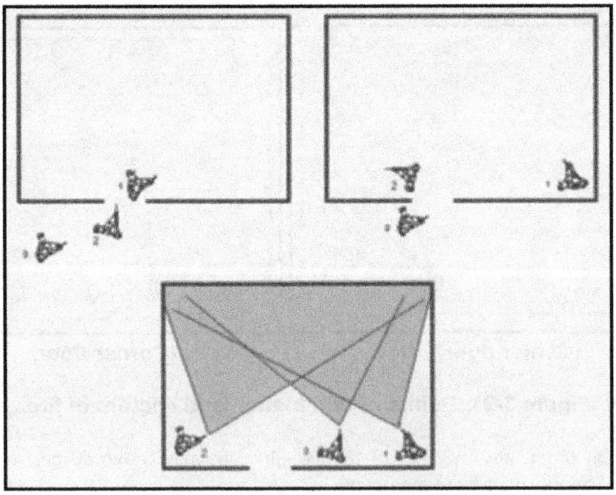

Figure 3-32. Points of domination and sectors of fire (three-man team, center door).

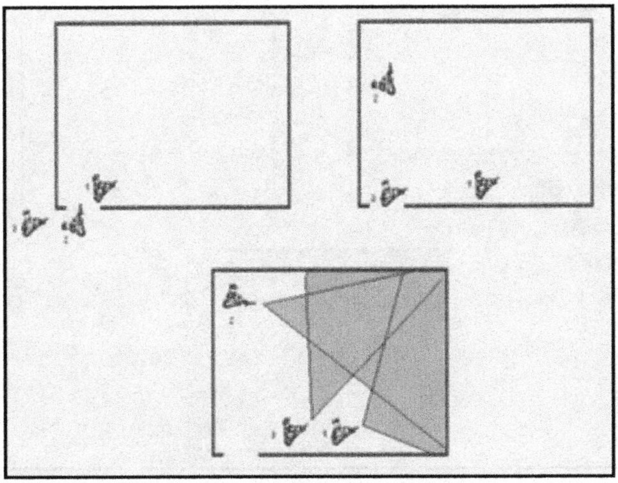

Figure 3-33. Points of domination and sectors of fire (three-man team, corner door).

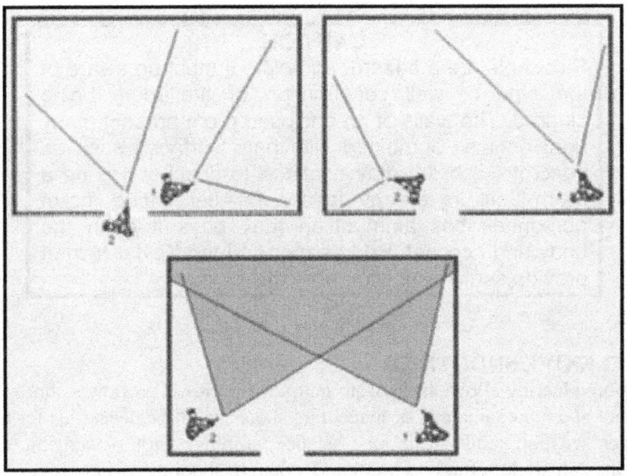

Figure 3-34. Points of domination and sectors of fire (two-man team, center door).

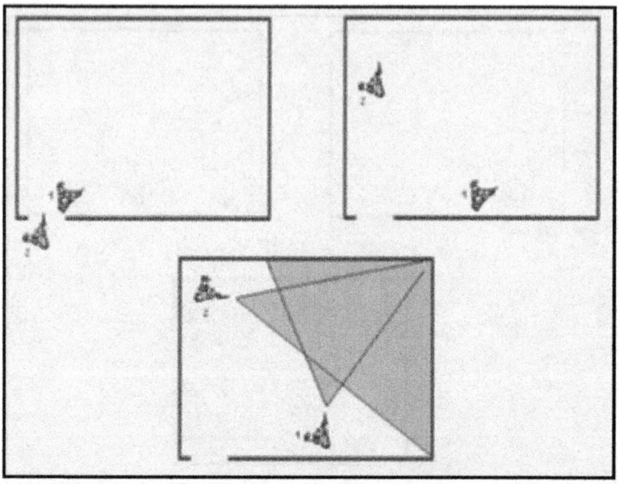

Figure 3-35. Points of domination and sectors of fire (two-man team, corner door).

> **CAUTION**
>
> Ricochets are a hazard. All soldiers must be aware of the type of wall construction of the room being cleared. The walls of an enclosed room present many right angles. Combined with hard surfaces such as concrete, a bullet may continue to ricochet around a room until its energy is spent. After hitting threat personnel, ball ammunition may pass through the body and ricochet. Body armor and the Kevlar helmet provide some protection from this hazard.

3-23. REFLEXIVE SHOOTING

Precision room clearing allows little or no margin for error. Too slow a shot at an enemy, too fast a shot at a noncombatant, or inaccurate shots can all be disastrous for the clearing team. Proper weapon ready technique, stance, aiming, shot placement, and trigger manipulations constitute reflexive shooting. Reflexive shooting techniques are used by all members of the fire team, to include M203 and M249 gunners.

a. **Weapon Ready Positions.** The two weapon ready positions are low ready and high ready (Figure 3-36, page 3-36).

(1) ***Low Ready Position.*** The butt of the weapon is placed firmly in the pocket of the shoulder with the barrel pointed down at a 45-degree angle. This position is the safest

carry position. It should be used by the clearing team while inside the room, except when actually entering and clearing.

(2) ***High, Ready Position.*** The butt of the weapon is held under the armpit, with the barrel pointed slightly up, keeping the front sight assembly under the line of sight but within the gunner's peripheral vision. To engage a target, the gunner pushes the weapon out as if to bayonet the target. When the weapon leaves the armpit, he slides it up into the firing shoulder. This technique is used when moving in a single file.

Figure 3-36. Ready positions for the M16A2.

b. **Stance.** Feet are about shoulder-width apart. Toes are pointed to the front (direction of movement). The firing side foot is slightly staggered to the rear of the non-firing side foot. Knees are slightly bent and the upper body is leaned slightly forward. Shoulders are square and pulled back, not rolled over or slouched. The head is up and both eyes are open. When engaging targets, the gunner holds the weapon with the butt in the pocket of his shoulder.

c. **Aiming with Iron Sights.** The four aiming techniques all have their place during combat in urban areas, but the aimed quick-kill technique is the one most often used in precision room clearing.

(1) ***Slow Aimed Fire.*** This technique is the most accurate. It consists of taking up a steady, properly aligned sight picture and squeezing off rounds. It is normally used for engagements beyond 25 meters or when the need for accuracy overrides speed.

(2) ***Rapid Aimed Fire.*** This technique features an imperfect sight picture in which windage is critical but elevation is of lesser importance. When the front sight post is in line with the target, the gunner squeezes the trigger. This technique is used against targets out to 15 meters and is fairly accurate and very fast.

(3) ***Aimed Quick Kill.*** This technique consists of using a good spot weld and placing the front sight post flush on top of the rear peep sight. It is used for very quick shots out to 12 meters. Windage is important, but elevation is not critical with relation to the target. This technique is the fastest and most accurate. With practice, soldiers can become deadly shots at close range.

(4) ***Instinctive Fire.*** This technique is the least desirable. The gunner focuses on the target and points the weapon in the target's general direction, using muscle memory to compensate for lack of aim. This technique should be used only in emergencies.

d. **M68 Close Combat Optic**. The M68 close combat optic (CCO) is an excellent close combat aiming system when used properly. Remember, the M68 is not a telescope sight.

(1) ***Aimed Fire.*** This technique requires looking through the CCO with both eyes open and focusing on the target. An optical illusion places a red aiming dot in front of the firer. The dot is placed on the target then the target is engaged with fire. The aiming dot does not have to be centered in the optic. The CCO is used in the same manner at all ranges. Therefore, there is no distinction between slow aimed fire, rapid aimed fire, and aimed quick kill techniques.

(2) ***Instinctive Fire.*** This technique remains the same with the CCO.

e. **Trigger Manipulation.** Rapid, aimed, semiautomatic fire is the most effective method of engaging targets during precision room clearing. As each round is fired from the aimed quick-kill position, the weapon's recoil makes the front sight post move in a small natural arc. The gunner should not fight this recoil. He should let the weapon make the arc and immediately bring the front sight post back onto the target and take another shot. This two-shot combination is known as firing a *controlled pair*. Soldiers must practice a *controlled pair* until it becomes instinctive. Clearing team members continue to fire *controlled pairs* until the target goes down. If there are multiple targets, team members engage with a controlled pair and then return to reengage any enemy left standing or still trying to resist.

f. **Shot Placement.** In precision room clearing, enemy soldiers must be incapacitated immediately. Shots that wound or are mortal but do not incapacitate the target instantaneously are better than misses but may allow the enemy to return fire. While a solid *head-shot* is expected to instantaneously incapacitate the enemy, a target area of 5 by 8 inches may be difficult to hit when moving rapidly in a low crouch position.

(1) Members of clearing teams should concentrate on achieving solid, well-placed shots (controlled pairs) to the upper chest, then to the head (Figure 3-37, page 3-38). This shot placement increases the first round hit probability and allows for a second round incapacitating shot.

(2) This engagement technique is more reliable than attempting *head-shots* only and is easy for soldiers to learn, having been taught previously to aim at center of mass.

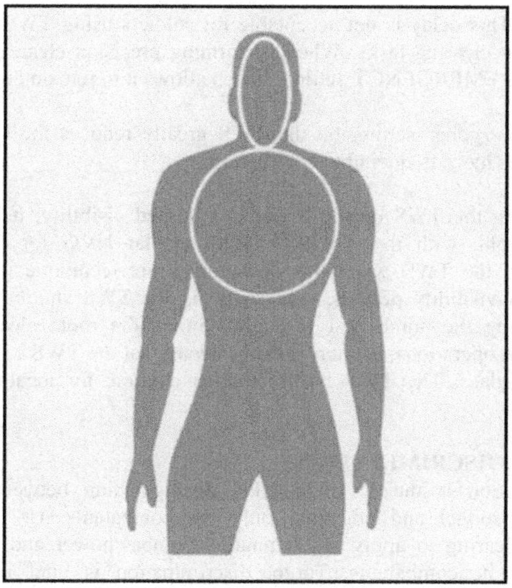

Figure 3-37. Lethal to incapacitating shot placement.

g. **Reflexive Shooting Techniques During Limited Visibility.** Reflexive shooting techniques are also used during periods of limited visibility.

(1) ***Visible Illumination.*** When using flashlights or other visible illumination, treat all engagements as day engagements and use the applicable technique as described above. Bright light shone into the enemy's eyes can limit his effectiveness; also, be aware that a flashlight marks your location as well.

(2) ***AN/PAQ-4 and AN/PEQ-2 Aiming Lights.*** When using IR aiming lights in conjunction with night vision goggles (NVGs), use the instinctive fire technique to point the weapon at the target while activating the aiming light. This technique should place the aiming dot within the field of view of the NVGs and on or near the target. Adjust placement of the aiming dot onto the target and fire. Note that target discrimination is more difficult when using NVGs. IR illumination provided by flashlights with IR filters, or the illuminator that is integral with the PEQ-2, can aid in target identification and discrimination. IR illumination is also required inside buildings when there is no ambient light.

(3) ***AN/PAS-13 Thermal Weapons Sight.*** The thermal weapons sight (TWS) offers some distinct advantages over IR viewers. It does not require any ambient light and does not *bloom* out when encountering a sudden light source. However, its weight and bulk are a disadvantage when performing reflexive firing techniques. With the sight in the ON position, the TWS has a power saving feature that turns off the viewer after a period of inactivity. The soldier reactivates the sight by placing his eye against the rubber eyecup. When reactivated, it takes a few seconds for the sight to cool itself down enough to

regain an image. This delay is not acceptable for soldiers using TWS while conducting room and building clearing tasks. When performing precision clearing tasks, the TWS must remain in the EMERGENCY setting, which allows it to remain continuously active.

NOTE: The *emergency* setting on the TWS greatly reduces the battery life, which requires more frequent battery changes.

(4) When using the TWS during periods of limited visibility, it is best to use the PAQ-4 aiming light, with the AN/PVS-14 Monocular NVG for reflexive shooting engagements. Use the TWS when the slow aimed fire technique is appropriate. For daytime and high visibility periods, soldiers using the TWS should not be placed on *point*, or be among the numbers 1 through 3 men of a room clearing team. When employed in urban operations, soldiers must be aware that the TWS cannot detect targets through window glass. The TWS is effective in daytime for locating targets hidden in shadows.

3-24. TARGET DISCRIMINATION

Target discrimination is the act of quickly distinguishing between combatant and noncombatant personnel and engaging only the combatants. US forces engage in precision room clearing to apply discriminating combat power and limit unnecessary casualties among noncombatants. Target discrimination is vital in precision room clearing. If there are no noncombatants then there is less of a need for selective engagements. However, even if an area is known to be free of noncombatants, other soldiers moving through the area may be mistaken as enemy and engaged unless clearing team members are disciplined and well-trained in fire control and target discrimination. Even with well-trained, disciplined soldiers, precision room clearing can result in unintentional casualties among noncombatants. Commanders must recognize this and take steps to relieve the stress it causes soldiers.

3-25. MOVEMENT WITHIN A BUILDING

When operating under precision conditions, movement techniques may be modified based on the room clearing technique being used. The terrain, the enemy situation, visibility, and the likelihood of contact dictate movement techniques.

a. **Individual Movement.** When moving within a building, the soldier avoids silhouetting himself in doors and windows (Figure 3-38, page 3-40). When moving in hallways, he never moves alone—he always moves with at least one other soldier for security. The soldier should try to stay 12 to 18 inches away from walls when moving; rubbing against walls may alert an enemy on the other side, or, if engaged by an enemy, ricochet rounds tend to travel parallel to a wall.

Figure 3-38. Movement within a building.

b. **Hallway Clearing Techniques.** The clearing team must always be alert. Team members provide security at the breach point and to the rear. Inside buildings they provide security laterally down corridors, and upward if near stairs or landings. The two basic techniques for moving down hallways are shown in Figure 3-39. Hallway intersections are dangerous areas and should be approached cautiously (Figures 3-40 and 3-41, pages 3-42 through 3-44).

(1) ***Serpentine.*** The serpentine technique is used in narrow hallways. The number 1 man provides security to the front. His sector of fire includes any enemy soldiers who appear at the far end of the hall or from any doorways near the end. The number 2 and number 3 men cover the left and right sides of the number 1 man. Their sectors of fire include any soldiers who appear suddenly from nearby doorways on either side of the hall. The number 4 man, normally carrying the M249, provides rear protection against any enemy soldiers suddenly appearing behind the clearing team.

(2) ***Rolling T.*** The rolling-T technique is used in wide hallways. The number 1 and number 2 men move abreast, covering the opposite side of the hallway from the one they are walking on. The number 3 man covers the far end of the hallway from a position behind the number 1 and number 2 men, firing between them. Once again, the number 4 man provides rear security.

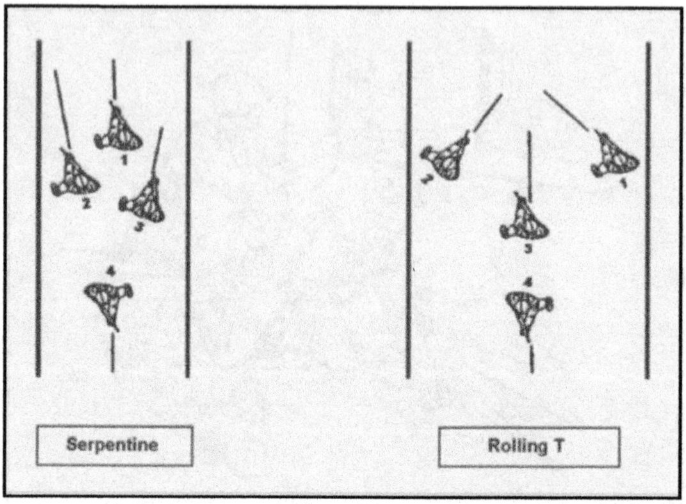

Figure 3-39. Hallway clearing techniques.

(3) ***Clearing "T" Intersections.*** Figure 3-40 (page 3-42) depicts the fire team's actions upon reaching a hallway "T" intersection when approaching from the *base* of the "T". The fire team is using the serpentine formation for movement.

- The team configures into a 2-by-2 formation with the numbers 1 and 2 men left, and the 3 and 4 men right. (When clearing a right-hand corner, use the left-handed firing method to minimize exposure.)
- The numbers 1 and 3 men move to the edge of the corner and assume a low crouch or kneeling position. On signal, the numbers 1 and 3 men simultaneously turn left and right, respectively.
- At the same time, the numbers 2 and 4 men step forward and turn left and right, respectively maintaining their (high) position. (Sectors of fire interlock and the *low/high* positions prevent soldiers from firing at another.)
- Once the left and right portions of the hallway are clear, the fire team resumes the movement formation.

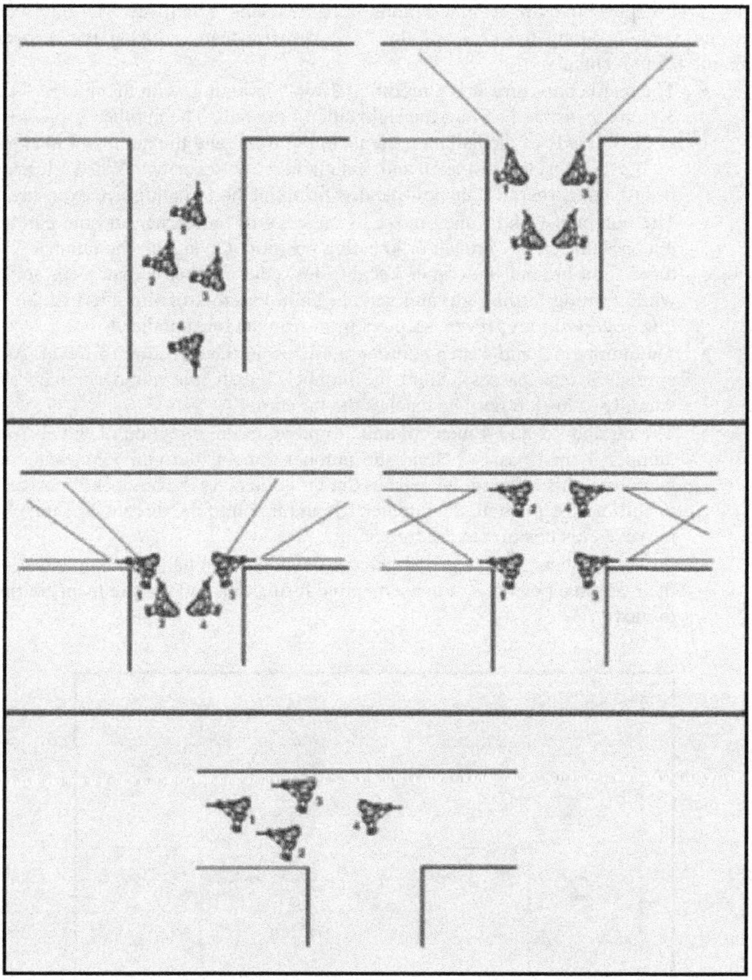

Figure 3-40. T-shaped hallway intersection clearing positions.

Figure 3-41 depicts the fire team's actions upon reaching a hallway "T" intersection when approaching along the *cross* of the "T". The fire team is using the serpentine formation for movement.

- The team configures into a modified 2-by-2 formation with the numbers 1 and 3 men abreast and toward the right side of the hall. The number 2 man moves to the left side of the hall and orients to the front, and the number 4 man shifts to the right side (his left) and maintains rear security. (When clearing a right-hand corner, use the left-handed firing method to minimize exposure.)
- The numbers 1 and 3 men move to the edge of the corner and the number 3 man assumes a low crouch or kneeling position. On signal, the number 3 man turns right around the corner keeping low, the number 1 man steps forward while turning to the right and staying high. (Sectors of fire interlock and the *low/high* positions prevent soldiers from firing at one another.)
- The numbers 2 and 4 men continue to move in the direction of travel. As the number 2 man passes behind the number 1 man, the number 1 man shifts laterally to his left until he reaches the far corner.
- The numbers 2 and 4 men continue to move in the direction of travel. As the number 4 man passes behind the number 3 man, the number 3 man shifts laterally to his left until he reaches the far corner. As the number 3 man begins to shift across the hall, the number 1 man turns into the direction of travel and moves to his position in the formation.
- As the numbers 3 and 4 men reach the far side of the hallway, they too assume their original positions in the serpentine formation, and the fire team continues to move.

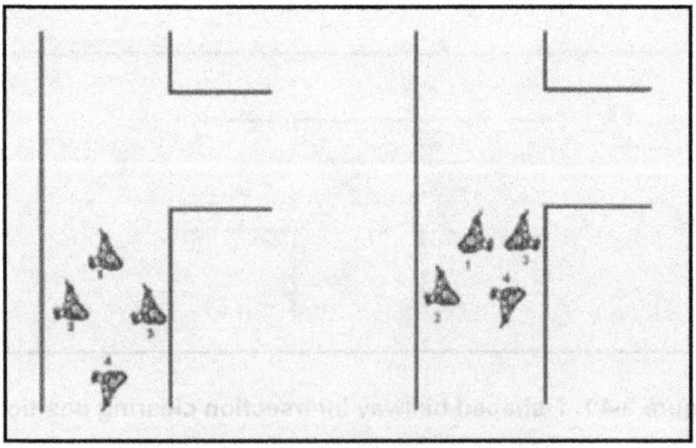

Figure 3-41. Hallway junction clearing.

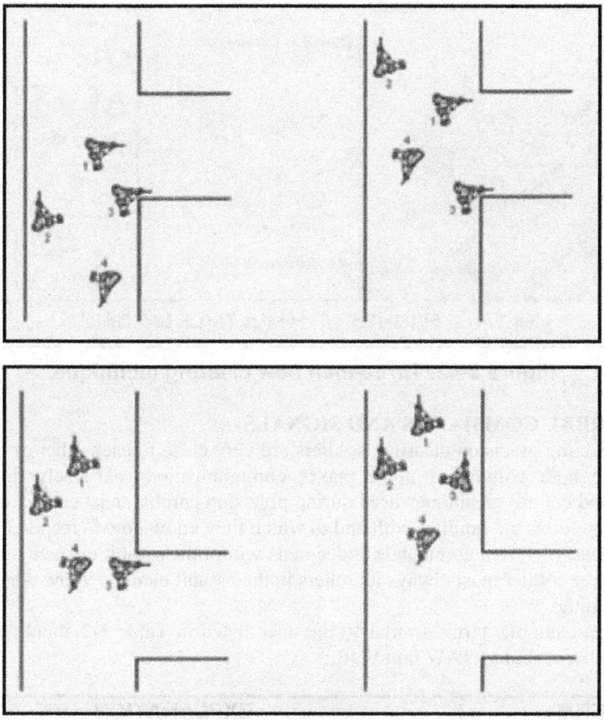

Figure 3-41. Hallway junction clearing (continued).

c. **Clearing Stairwells and Staircases.** Stairwells and staircases are comparable to doorways in that they create a *fatal funnel*; however, the danger is intensified by the three-dimensional aspect of additional landings. The ability of the squad or team to conduct the movement depends upon which direction they are traveling and the layout of the stairs. Regardless, the clearing technique follows a basic format:

- The squad leader designates an assault element to clear the stairs.
- The squad or team maintains 360-degree, three-dimensional security in the vicinity of the stairs.
- The squad leader then directs the assault team to locate, mark, bypass and or clear any obstacles or booby traps that may be blocking access to the stairs.
- The assault element moves up (or down) the stairways by using either the two-, three-, or four-man flow technique, providing overwatch up and down the stairs while moving. The three-man variation is preferred (Figure 3-42).

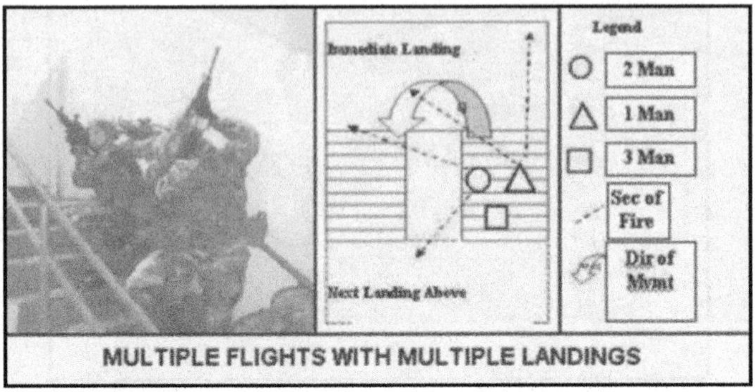

Figure 3-42. *Three-man flow* clearing technique.

3-26. VERBAL COMMANDS AND SIGNALS

When conducting precision clearing, soldiers are very close to each other as they engage targets. The high volume of noise makes communications extremely difficult. The command and control techniques used during precision combat must consist of terms and actions that soldiers are familiar with and to which they know how to respond.

a. The use of verbal commands and signals within the assault element are extremely important. The soldier must always let others in the assault element know where he is and what he is doing.

b. As an example, terms similar to the ones listed in Table 3-2 should be a part of each soldier's vocabulary IAW unit SOP.

TERM	EXPLANATION
"STATUS!"	Signal by an element leader that requires all members to report whether their sectors are clear and if they are prepared to continue the mission.
"CLEAR!"	Signal given by individuals to report their sector is clear.
"UP!"	Signal given by individuals to report they are ready to continue the mission (weapon loaded, equipment accounted for).
"ROOM CLEAR!"	Signal from team leader to team members, squad leader, and follow-on teams that the room is secure and cleared.
"COMING OUT!"	Signal given by an individual or team that they are about to exit a room.
"COME OUT!"	Reply given by security element or follow-on team that it is safe to exit the room.
"COMING IN!"	Signal given by an individual who is about to enter an occupied room.
"COME IN!"	Reply given by an occupant of a room stating it is safe to enter.
"COMING UP (DOWN)!	Signal given by an individual or team that is about to ascend or descend a stairway.

Table 3-2. Verbal commands and signals.

TERM	EXPLANATION
"COME UP (DOWN)!"	Reply given by security element that it is safe to ascend or descend a stairway.
"MAN DOWN!"	Signal given when an individual has been wounded or injured and cannot continue his mission.
"SHORT ROOM!"	Signal given by either the number 1 man or the number 2 man to indicate a small room, and that all team members should not enter.
"GRENADE!"	A command given by any soldier, when an enemy grenade has been thrown. All soldiers need to take immediate actions. Although difficult, the soldier should identify the location of the grenade, if possible.
"GO LONG!"	A command given by one member of the team to tell another team member to take up security farther into the room or farther down a hallway.
"GUN DOWN"	A signal given when an individual's weapon has malfunctioned and is being corrected.
"GUN UP"	A signal given when an individual has corrected a malfunction and is ready for action.
"RELOADING"	A signal given when an individual is reloading any weapon system. This signal is followed by "GUN UP" when ready.

Table 3-2. Verbal commands and signals (continued).

NOTE: The use of loud verbal commands may reveal to the enemy the location and immediate intent of friendly forces. Although code words may be substituted, they can be heard and used by enemy forces if friendly forces use them too loudly.

3-27. SAFETY AND FORCE PROTECTION

Precision clearing is high risk, and even training for it can be hazardous. Only well-trained, disciplined soldiers are able to execute these techniques successfully.

a. Leaders at all levels must enforce safe handling of weapons and demolitions. The concern that individual soldiers not be injured in accidents is essential to mission accomplishment. Unintentional and unsafe weapons fire or detonation of explosives or munitions can jeopardize the mission of the clearing team and subsequently the entire unit.

b. Soldiers engaged in precision clearing should wear all their protective equipment.

(1) Soft body armor, such as the standard Army-issue Kevlar vest, is effective in preventing death or serious injury from high-velocity fragments that strike the torso area. Although the Kevlar protective vest is effective, flexible, and relatively comfortable, it is not designed to stop bullets. As a rule, soft body armor stops some low-power handgun rounds but not rifle or carbine ammunition.

(2) Some versions of hard body armor stops almost any round fired at it. They tend to be heavy and stiff, but they have proven effective during precision clearing. If a commander knows his unit is going to conduct lengthy precision room clearing, he requests a special issue of threat level III or IV protective equipment. This equipment is excellent, but soldiers must train and rehearse wearing it before they enter combat.

All precision clearing is tiring, and soldiers wearing threat level III or IV protection tire or overheat more quickly.

(3) The standard Army Kevlar helmet and ballistic protective eyeglasses have also been proven to significantly reduce casualties during precision room clearing.

(4) Hard plastic knee and elbow protectors are available on special request. They are useful, especially during prolonged search and clear operations. They prevent injury from rubble and broken glass when kneeling or prone.

c. Detailed knowledge of weapons and munitions effects is important to the safety of members of the clearing team, as well as to mission accomplishment. Most interior building walls do not stop rifle fire. Fragments from grenades often penetrate interior walls. Standard home furnishings or office furniture offer little protection from high-velocity rounds. Excessive amounts of demolitions used to breach a wall may knock it down instead, perhaps even bring the roof of the building down also.

CAUTION

Goggles or ballistic eye protection should always be worn to protect soldiers from debris caused by explosives, tools, weapons, grenades, and so forth.

Section IV. FIGHTING POSITIONS

Whether a unit is attacking, defending, or conducting retrograde operations, its success or failure depends on the ability of the individual soldier to place accurate fire on the enemy with the least exposure to return fire. Consequently, the soldier must immediately seek and use firing positions properly.

3-28. HASTY FIGHTING POSITION

A hasty fighting position is normally occupied in the attack or the early stages of the defense. It is a position from which the soldier can place fire upon the enemy while using available cover for protection from return fire. The soldier may occupy it voluntarily or he may be forced to occupy it due to enemy fire. In either case, the position lacks preparation before occupation. Some of the more common hasty fighting positions in an urban area are: corners of buildings, behind walls, windows, unprepared loopholes, and the peak of a roof.

a. **Corners of Buildings.** The soldier must be capable of firing his weapon both right- and left-handed to be effective around corners.

(1) A common error made in firing around corners is firing from the wrong shoulder. This exposes more of the soldier's body to return fire than necessary. By firing from the proper shoulder, the soldier can reduce exposure to enemy fire.

(2) Another common mistake when firing around corners is firing from the standing position. The soldier exposes himself at the height the enemy would expect a target to appear, and risks exposing the entire length of his body as a target for the enemy.

b. **Walls.** When firing from behind walls, the soldier must fire around cover and not over it (Figure 3-43, page 3-48).

Figure 3-43. Soldier firing around cover.

c. **Windows.** In an urban area, windows provide convenient firing ports. The soldier must avoid firing from the standing position since it exposes most of his body to return fire from the enemy and could silhouette him against a light-colored interior beyond the window. This is an obvious sign of the soldier's position, especially at night when the muzzle flash can easily be observed. In using the proper method of firing from a window (Figure 3-44), the soldier is well back into the room to prevent the muzzle flash from being seen, and he is kneeling to limit exposure and avoid silhouetting himself.

Figure 3-44. Soldier firing from window.

d. **Loopholes.** The soldier may fire through a hole created in the wall and avoid windows (Figure 3-45). He stays well back from the loophole so the muzzle of the weapon does not protrude beyond the wall, and the muzzle flash is concealed.

Figure 3-45. Soldier firing from loophole.

e. **Roof.** The peak of a roof provides a vantage point for snipers that increases their field of vision and the ranges at which they can engage targets (Figure 3-46). A chimney, a smokestack, or any other object protruding from the roof of a building can reduce the size of the target exposed and should be used.

Figure 3-46. Soldier firing from peak of a roof.

f. **No Position Available.** When the soldier is subjected to enemy fire and none of the positions mentioned above are available, he must try to expose as little of himself as possible. The soldier can reduce his exposure to the enemy by lying prone as close to a building as possible, on the same side of the open area as the enemy. To engage the soldier, the enemy must then lean out the window and expose himself to return fire.

g. **No Cover Available.** When no cover is available, the soldier can reduce exposure by firing from the prone position, by firing from shadows, and by presenting no silhouette against buildings.

3-29. PREPARED FIGHTING POSITION

A prepared firing position is one built or improved to allow the soldier to engage a particular area, avenue of approach, or enemy position, while reducing his exposure to return fire. Examples of prepared positions include barricaded windows, fortified loopholes, sniper positions, antiarmor positions, and machine gun positions.

a. The natural firing port provided by windows can be improved by barricading the window, leaving a small hole for the soldier's use. Materials torn from the interior walls of the building or any other available material may be used for barricading.

(1) When barricading windows, avoid barricading only the windows that are going to be used as firing ports. The enemy can soon determine that the barricaded windows are fighting positions.

(2) Also avoid neat, square, or rectangular holes that are easily identified by the enemy. A barricaded window should not have a neat, regular firing port. The window should keep its original shape so that the position of the soldier is hard to detect. Firing from the bottom of the window gives the soldier the advantage of the wall because the firing port is less obvious to the enemy. Sandbags are used to reinforce the wall below the window and to increase protection for the soldier. All glass must be removed from the window to prevent injury to the soldier. Lace curtains permit the soldier to see out and prevent the enemy from seeing in. Wet blankets should be placed under weapons to reduce dust. Wire mesh over the window keeps the enemy from throwing in hand grenades.

b. Although windows usually are good fighting positions, they do not always allow the soldier to engage targets in his sector.

(1) To avoid establishing a pattern of always firing from windows, an alternate position is required such as in an interior room and firing through a rubbled outer wall (Figure 3-47), or a prepared loophole (Figure 3-48). The prepared loophole involves cutting or blowing a small hole into the wall to allow the soldier to observe and engage targets in his sector.

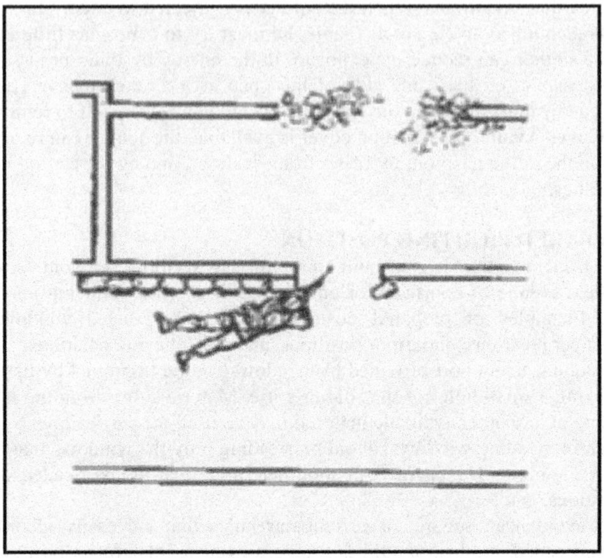

Figure 3-47. Interior room position.

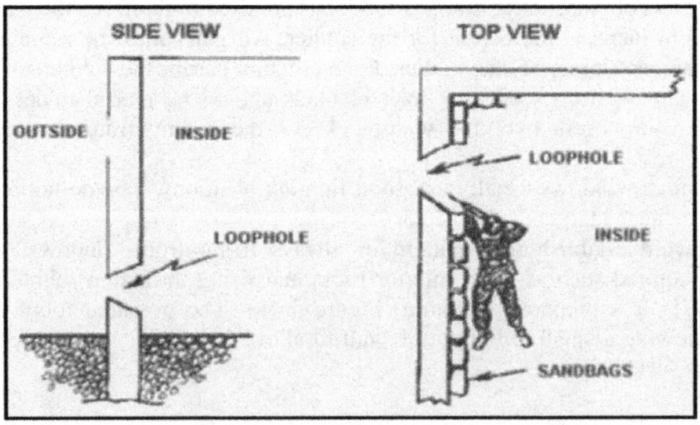

Figure 3-48. Prepared loophole.

(2) Sandbags are used to reinforce the walls below, around, and above the loophole (Figure 3-49, page 3-52). Two layers of sandbags are placed on the floor under the soldier to protect him from an explosion on a lower floor (if the position is on the second floor or higher). A wall of sandbags, rubble, furniture, and so on should be constructed to the rear of the position to protect the soldier from explosions in the room.

(3) A table, bedstead, or other available material can provide overhead cover for the position. This cover prevents injury to the soldier from falling debris or explosions above his position.

Figure 3-49. Cut-away view of a sandbag reinforced position.

(4) The position should be camouflaged by knocking other holes in the wall, making it difficult for the enemy to determine which hole the fire is coming from. Siding material should be removed from the building in several places to make loopholes less noticeable.

(5) Because of the angled firing position associated with loopholes, primary and supplementary positions can be prepared using the same loophole (Figure 3-50). This procedure allows the individual to shift his fire onto a sector that was not previously covered by small arms fire.

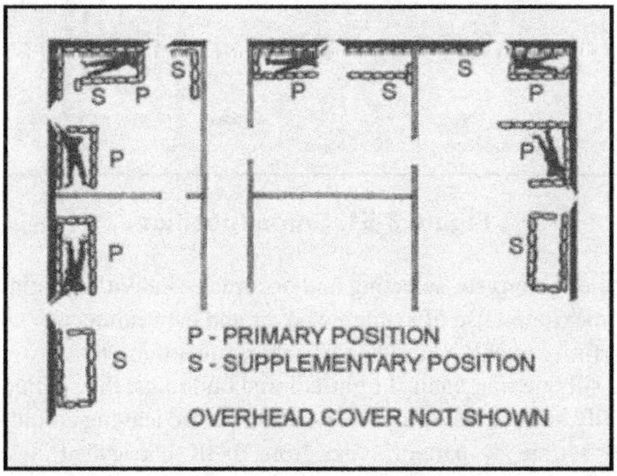

Figure 3-50. Loopholes with primary and supplementary positions.

c. A chimney or other protruding structure provides a base from which a sniper position can be prepared. Part of the roofing material is removed to allow the sniper to fire around the chimney. He should stand inside the building on the beams or on a platform with only his head and shoulders above the roof (behind the chimney). Sandbags placed on the sides of the position protect the sniper's flanks.

d. When the roof has no protruding structure to provide protection, the sniper position should be prepared from underneath on the enemy side of the roof (Figure 3-51). The position is reinforced with sandbags, and a small piece of roofing material should be removed to allow the sniper to engage targets in his sector. The missing piece of roofing material should be the only sign a position exists. Other pieces of roofing should be removed to deceive the enemy as to the true sniper position. The sniper should be invisible from outside the building and the muzzle flash must be hidden from view.

Figure 3-51. Sniper position.

e. Some considerations for selecting and occupying individual fighting positions are:

- Make maximum use of available cover and concealment.
- Avoid firing over cover; when possible, fire around it.
- Avoid silhouetting against light-colored buildings, the skyline, and so on.
- Carefully select a new fighting position before leaving an old one.
- Avoid setting a pattern; fire from both barricaded and non-barricaded windows.
- Keep exposure time to a minimum.
- Begin improving your hasty position immediately after occupation.

- Use construction material that is readily available in an urban area.
- Remember that positions that provide cover at ground level may not provide cover on higher floors.

f. In attacking an urban area, the recoilless AT weapon and ATGM crews may be hampered in choosing firing positions due to the backblast of their weapons. They may not have enough time to knock out walls in buildings and clear backblast areas. They should select positions that allow the backblast to escape such as corner windows where the round fired goes out one window and the backblast escapes from another. When conducting defensive operations the corner of a building can be improved with sandbags to create a firing position (Figure 3-52).

Figure 3-52. Corner firing position.

g. The rifle squad during an attack on and in defense of an urban area is often reinforced with attached antitank weapons. The rifle squad leader must be able to choose good firing positions for the antitank weapons under his control.

h. Various principles of employing antitank weapons have universal applications such as: making maximum use of available cover; trying to achieve mutual support; and allowing for the backblast when positioning recoilless weapons, TOWs, Dragons, Javelins, and AT4s.

i. Operating in an urban area presents new considerations. Soldiers must select numerous alternate positions, particularly when the structure does not provide cover from small-arms fire. They must position their weapons in the shadows and within the building.

j. AT4s and Javelins firing from the top of a building can use the chimney for cover (Figure 3-53). The rear of this position should be reinforced with sandbags but should not interfere with backblast area.

Figure 3-53. A recoilless weapon crew firing from a rooftop.

k. When selecting firing positions for recoilless weapons and ATGMs, make maximum use of rubble, corners of buildings, and destroyed vehicles to provide cover for the crew. Recoilless weapons and ATGMs can also be moved along rooftops to obtain a better angle to engage enemy armor. When buildings are elevated, positions can be prepared using a building for overhead cover (Figure 3-54, page 3-56). The backblast under the building must not damage or collapse the building or injure the crew. See Chapter 7.

NOTE: When firing from a slope, ensure that the angle of the launcher relative to the ground or firing platform is not greater than 20 degrees. When firing within a building, ensure the enclosure is at least 10 feet by 15 feet, is clear of debris and loose objects, and has windows, doors, or holes in the walls for the backblast to escape.

Figure 3-54. Prepared positions using a building for overhead cover.

l. The machine gun can be emplaced almost anywhere. In the attack, windows and doors offer ready-made firing ports (Figure 3-55). For this reason, the enemy normally has windows and doors under observation and fire, which should be avoided. Any opening in walls created during the fighting may be used. Small explosive charges can create loopholes for machine gun positions (Figure 3-56). Regardless of what openings are used, machine guns should be in the building and in the shadows.

Figure 3-55. Emplacement of machine gun in a doorway.

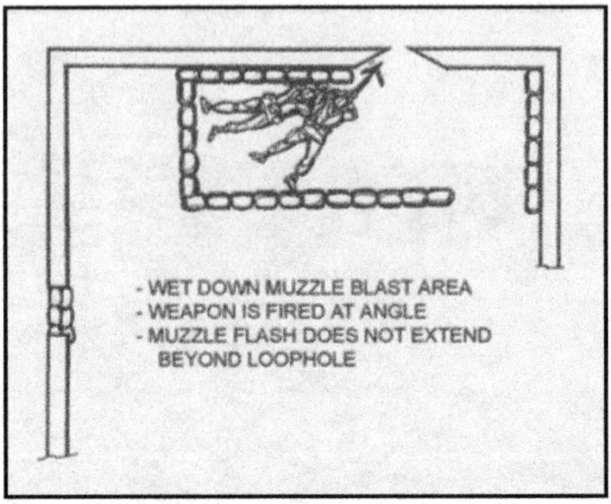

Figure 3-56. Use of a loophole with a machine gun.

m. Upon occupying a building, soldiers' board up all windows and doors. By leaving small gaps between the slots, soldiers can use windows and doors as good alternate positions.

n. Loopholes should be used extensively in the defense. They should not be constructed in any logical pattern, nor should they all be at floor or tabletop level. Varying their height and location makes them hard to pinpoint and identify. Dummy loopholes, knocked off shingles, or holes cut that are not intended to be used as firing positions aid in the deception. Loopholes located behind shrubbery, under doorjambs, and under the eaves of a building are hard to detect. In the defense, as in the offense, a firing position can be constructed using the building for overhead cover.

o. Increased fields of fire can be obtained by locating the machine gun in the corner of the building (Figure 3-57, page 3-58), in the cellar (Figure 3-58, page 3-58), or sandbagged under a building (Figure 3-59, page 3-59). Available materials, such as desks, overstuffed chairs, couches, and other items of furniture, should be integrated into the construction of bunkers to add cover and concealment.

Figure 3-57. Corner machine gun bunker.

Figure 3-58. Machine gun position in cellar.

Figure 3-59. Sandbagged machine gun emplacement under a building.

p. Although grazing fire is desirable when employing the machine gun, it may not always be practical or possible. Where destroyed vehicles, rubble, and other obstructions restrict the fields of grazing fire, the gun can be elevated to where it can fire over obstacles. Firing from loopholes on the second or third story may be necessary. A firing platform can be built under the roof and a loophole constructed (Figure 3-60). Again, the exact location of the position must be concealed. Camouflage the position by removing patches of shingles, over the entire roof.

Figure 3-60. Firing platform built under roof.

3-30. TARGET ACQUISITION

Urban areas provide unique target acquisition challenges to units. Buildings mask movement and the effects of direct and indirect fires. The rubble from destroyed buildings, along with the buildings themselves, provides cover and concealment for attackers and defenders, making target acquisition difficult. Urban areas often favor the defender's ability to acquire targets so this makes offensive target acquisition extremely important, since the side that fires first may win the engagement. Target acquisition must be continuous, whether a unit or soldier is halted or moving. The six steps of target acquisition, search, detection, location, identification, classification, and confirmation are no different in an urban environment than anywhere else but are usually performed at a much faster pace.

a. **Search.** Using all senses during the search step enhances the detection capabilities of all soldiers on the urban battlefield. The techniques of patrolling and using observation posts apply in urban as well as in wooded or more open terrain. These techniques enable units to search for and locate the enemy. Soldiers searching the urban battlefield for targets should employ target acquisition devices. These devices can include binoculars, image intensification devices, thermal sights, ground surveillance radar (GSR), remote sensors (REMs), platoon early warning systems (PEWS), and field expedient early warning devices. Several types of devices should be used since no single device can meet every need of a unit.

(1) ***Observation.*** Observation duties must be clearly given to squad members to ensure 360 degrees and three-dimensional security as they move. This security continues at the halt. Soldiers soon recognize the sights, smells, sounds and so forth, associated with their urban battlefield and can soon distinguish targets.

(2) ***Movement.*** Stealth should be used when moving in urban areas since there are often short distances between attackers and defenders. Hand and arm signals should be used until contact is made. The unit should stop periodically to look and listen. Routes should be carefully chosen so that buildings and piles of rubble can be used to mask the unit's movement.

(3) ***Movement Techniques.*** Techniques are basically the same as in open terrain (traveling, traveling overwatch, bounding overwatch). When a unit is moving and enemy contact is likely, the unit must use a movement technique with an overwatching element. This principle applies in urban areas as it does in other kinds of terrain except that in urban terrain, the overwatching element must observe both the upper floors of buildings and street level.

(4) ***Observation Posts.*** The military aspects of urban terrain must be considered in selecting observation posts (OPs). OPs can be positioned in the upper floors of buildings, giving soldiers a better vantage point than at street level. Leaders should avoid selecting obvious positions, such as water towers or church steeples that attract the enemy's attention (Figure 3-61).

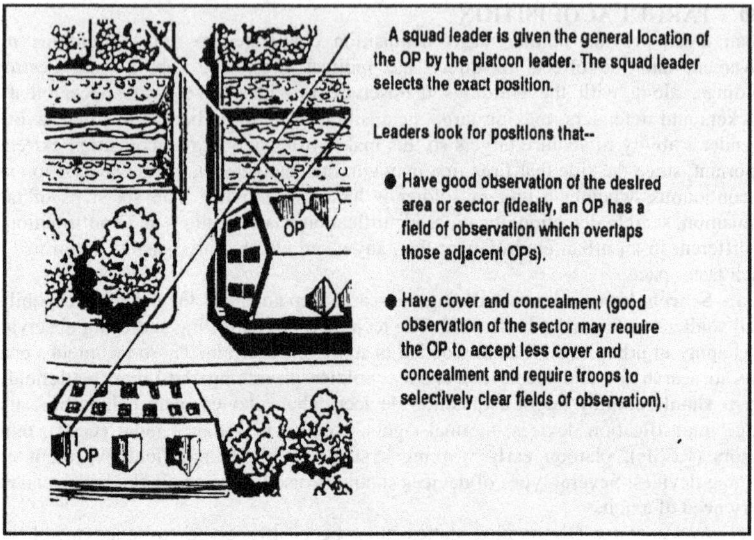

Figure 3-61. Selection of OP location.

b. **Detection.** Personnel, weapons, and vehicles have distinguishing signatures. Soldiers must recognize signatures so they can acquire and identify targets. This is extremely important in the urban battlefield, where one or more senses can be degraded. For example, soldiers operating in an urban area where smoke is used as an obscurant will have their sense of sight degraded, since they may not be able to see through the smoke with the naked eye. Their sense of smell and breathing is also affected. Some considerations are:

- Soldiers must look for targets in areas where they are most likely to be employed. Squad leaders must place OPs where they are most likely to see targets.
- Odors from diesel fuel, gasoline, cooking food, burning tobacco, after-shave lotion, and so forth reveal enemy and friendly locations.
- Running engines, vehicles, and soldiers moving through rubble-covered streets can be heard for great distances. Vehicles driven in urban areas produce more noise than those moving through open terrain. Soldiers moving through rubble on a street or in the halls of a damaged building create more noise than in a wooded area.
- Sounds and smells can aid in acquiring targets at night since they transmit better in the cooler, damper night air.
- Dust and noise created by the firing of some weapons such as a tank main gun can be seen and smelled.
- Irregularly shaped objects that do not conform to the surrounding area stand out.

- Abnormal reflections or flashes from movement of optics or metal can be seen.
- Voices can often be heard at long distances, with the sound reflecting off of structures.
- Shadows can be seen day or night.
- When scanning multistory buildings, soldiers may have to scan up as well as out (three-dimensional scanning).

c. **Location.** In an urban environment, determining the target location can be difficult. The cover and concealment provided by buildings and rubble can provide the enemy with an advantage that is not easily overcome. After the enemy is detected or contact is made, soldiers must visualize the situation from the enemy's viewpoint. This visualization helps the soldier determine where the likely enemy position is. At that point, the suspected enemy position should be suppressed, consistent with the ROE.

d. **Identification.** Being able to identify potential targets as quickly as possible after they are detected gives soldiers the advantage during urban combat. As a minimum, identification must determine if the potential target is friend, foe, or, a noncombatant. Correct identification is the key to preventing fratricide. Soldiers must know and understand the ROE. Soldiers must know what to engage and what not to engage.

e. **Classification.** To determine an appropriate method of dealing with a target, the soldier must determine the danger it represents. It requires quick decisions as targets are observed and occurs virtually simultaneously with identification. Situational awareness is vitally important. Multiple targets must be classified from most dangerous to least dangerous and engaged starting with the most dangerous.

f. **Confirmation.** This rapid verification of the initial identification and classification of the target is the final step of target acquisition. Identification, classification, and confirmation are done simultaneously.

3-31. DEFENSE AGAINST FLAME WEAPONS AND INCENDIARY MUNITIONS

Incendiary ammunition, special weapons, and the ease with which incendiary devices can be constructed from gasoline and other flammables make fire a threat during urban operations. During defensive operations, fighting fire should be a primary concern. Steps must be taken to reduce the risk of a fire that could make a chosen position indefensible.

a. Soldiers should construct positions that do not have large openings. These positions should provide as much built-in cover as possible to prevent penetration by incendiary ammunition. All unnecessary flammable materials should be removed including ammunition boxes, furniture, rugs, newspapers, curtains, and so on. Electricity and gas coming into the building must be shut off.

b. A concrete block building, with concrete floors and a tin roof, is an ideal place for a position. However, most buildings have wooden floors or subfloors, wooden rafters, and wooden inner walls, which require improvement. Inner walls should be removed and replaced with blankets to resemble walls from the outside. Sand should be spread 2 inches deep on floors and in attics to retard fire.

c. All available fire-fighting gear is pre-positioned so it can be used during actual combat. For the individual soldier such gear includes entrenching tools, helmets, sand or earth, and blankets. These items are supplemented with fire extinguishers.

d. Fire is so destructive that it can easily overwhelm personnel regardless of precautions. Soldiers should plan routes of withdrawal so a priority of evacuation from fighting positions can be established. This procedure allows soldiers to exit through areas that are free from combustible material and provide cover from enemy direct fire.

e. The confined space and large amounts of combustible material in urban areas can influence the enemy to use flame weapons or incendiary munitions. Two major first-aid considerations are burns and smoke inhalation. These can easily occur in buildings and render the victim combat ineffective. Although there is little defense against flame inhalation and lack of oxygen, smoke inhalation can be reduced by wearing the individual protective mask. Medics and combat lifesavers should be aware of the withdrawal plan and should be prepared to treat and evacuate burn and smoke inhalation casualties.

f. Offensive operations also require plans for fighting fire since the success of the mission can easily be threatened by fire. Poorly planned use of incendiary munitions can make fires so extensive that they become obstacles to offensive operations. The enemy may use fire to cover his withdrawal and to create obstacles and barriers to the attacker. Intentional flame operations, in an urban area, are difficult to control and may undermine mission success.

g. When planning offensive operations, the attacker must consider the effects of all weapons and munitions. Targets are chosen during the initial planning to avoid accidentally destroying critical facilities within the urban area. When planning flame operations in an urban area, priorities must be established to determine which critical installations (hospitals, power stations, radio stations, and historical landmarks) should have primary fire-fighting support.

h. Every soldier participating in the attack must be ready to deal with fire. The normal fire-fighting equipment available includes the entrenching tool, helmet (for carrying sand or water), and blankets (for snuffing out small fires).

3-32. DEFENSE AGAINST ENHANCED FLAME WEAPONS

Combat operations in Afghanistan, Chechnya, and Bosnia saw the increased use of enhanced flame weapons in an urban environment. While these weapons have been in existence for some time, US forces have not had much experience (after Vietman) in the use of and defense against them. Because future threats may use these weapons against US forces, this paragraph explains what enhanced flame weapons are and how to defend against them.

a. **Enhanced Flame Weapons.** These types of weapons primarily rely on blast, flame and concussion to inflict damage, rather than explosively driven projectiles, fragments, or shaped charges. The Russians found these weapons to be especially effective in Chechnya because they produced casualties without fragmentation and shrapnel. As Chechens would "hug" Russian units to negate the use of Russian firepower, Russians would use directed blast weapons against enemy personnel and positions to minimize fratricide due to ricochets, shrapnel, and fragmentation.

(1) ***Types of Enhanced Flame Weapons.*** There are two types of these weapons, though their effects are the same. Fuel air explosives (FAE) are the older generation of blast weapons. FAE rely on distributing fuel in the air and igniting it. Casualties are primarily produced by fuel exploding and burning in the air. The newer generation blast weapons are referred to as volumetric or thermobaric. They throw out explosives from a

warhead into a larger volume and use oxygen to ignite as a single event. This technique provides more reliable and controllable effects than FAE. Thermobaric weapons cause a tremendous blast in a confined space, such as a room or small building—the larger the volume of the weapon, the larger the blast effect. Many of these weapons are shoulder fired and are operated by a single gunner (Figure 3-62). Some shoulder fired blast weapons have tandem warheads that consist of a shaped charge followed by a Thermobaric munition (Figure 3-63). Currently, there are no thermobaric weapons in the US inventory, but are under research and development as a possible replacement for the M202A2 (Flash).

Figure 3-62. Russian RPO-A SHMEL, a shoulder fired thermobaric weapon.

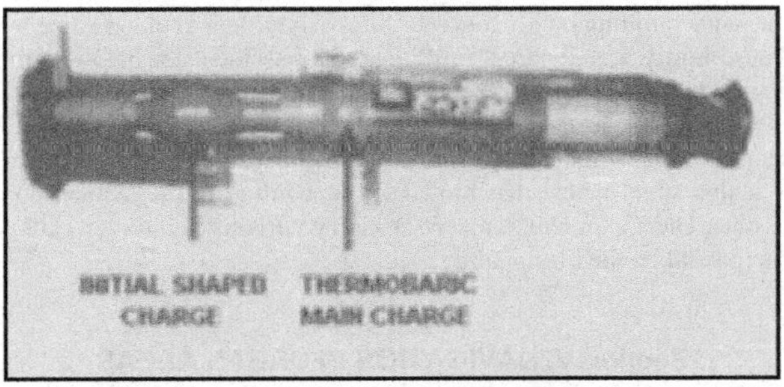

Figure 3-63. Russian RShG-1, tandem warhead.

(2) ***Effects of Enhanced Flame Weapons.*** These types of weapons are characterized by the production of a powerful fireball (flame temperatures of up to 1,200 degrees centigrade) together with a relatively long duration pressure wave. The fireball, and its associated dust storm, damages exposed skin and eyes over a wider radius than the blast effect. Most physical damage is caused by the heave and the push of the blast wave. This blast wave can collapse brick or block-built structures. Therefore, internal injuries to vital

organs and internal bleeding are common blast effects to personnel. Such weapons are particularly effective against fortified positions such as buildings. Confined spaces enhance the blast effect and, unlike fragments, blast and flame can travel around corners and down passages such as hallways or tunnels. Since blast pressure falls off rapidly in the open, much shorter minimum safety distances are possible and assault troops can be relatively close (to within 40 meters depending on the size of the munition) when many of these weapons are employed.

b. **Defensive Measures.** Using materials that absorb its energy or block its path can reduce the lethality of the blast/flame wave. The best protection is to isolate personnel from the wave; however, this procedure may not be possible in many tactical situations. Balance should be struck between protecting soldiers and not hampering their ability to fight or protect themselves from other threats. The first step is to prevent the munition from entering a structure by providing a physical barrier. If that is not possible, then the next step is to minimize damage from the weapons by weakening and isolating their effect. Another consideration is to make enhanced flame weapon gunners' priority targets for snipers or selected marksmen.

(1) ***Personnel.*** Personal injury can be minimized if soldiers wear a balaclava or similar garment to protect the face, goggles to protect the eyes from flash and flying dust and debris, and leather gloves to protect the hands.

(2) ***Armored Vehicles.*** If vehicles are *buttoned up*, the crew is protected against blast/flame damage; however, antennas, external components, and optics suffer varying degrees of damage. Tandem warheads pose a greater threat to armored vehicles.

c. **Fighting Positions.** Fixed fortifications, such as concrete bunkers or heavy-clad framed buildings, provide good protection against enhanced flame weapons detonating near the outside of the structure. Hastily prepared fighting positions or prepared fighting positions in lighter clad or framed buildings are more susceptible to blast effects. Unframed masonry buildings with concrete floors should be avoided since a falling floor is likely to cause injury to personnel. Fighting from basements or below ground positions or from prepared strong points in heavy-clad framed buildings provide additional protection. To reduce blast effects within a structure, unused openings inside buildings should be sealed to block the blast/flame wave path, while exterior openings should be left open or sealed with panels that blow off, depending on the tactical considerations, allowing the blast energy an exit route. Wet heavy curtains hung over exits, entries, and firing ports help weaken the blast energy.

Section V. NAVIGATION IN URBAN AREAS

Urban areas present a different set of challenges involving navigation. Deep in the city core, the normal terrain features depicted on maps may not apply—buildings become the major terrain features and units become tied to streets. Fighting in the city destroys buildings and the rubble blocks streets. Street and road signs are destroyed during the fighting if defenders do not remove them. Operations in subways and sewers present other unique challenges. Maps and photographs are available to help the unit overcome these problems. The global positioning system (GPS) can provide navigation assistance in urban areas.

3-33. MILITARY MAPS

The military city map is a topographical city map delineating streets and showing street names, important buildings, and other urban elements. The scale of a city map can vary from 1:25,000 to 1:50,000 depending on the importance and size of the city, density of detail, and intelligence information.

a. Special maps, prepared by supporting topographic engineers, can assist units in navigating in urban areas. These maps have been designed or modified to give information not covered on a standard map, which includes road and bridge networks, railroads, urban areas, and electric power fields. They can be used to supplement military city maps and topographical maps. Products that can be developed by the National Imagery Mapping Agency (NIMA) can be specifically tailored for the area of operations.

b. Once in the urban area, soldiers use street intersections as reference points much as they use hills and streams in rural terrain. City maps supplement or replace topographic maps as the basis of navigation. These maps enable units moving in the urban area to know where they are and to move to new locations even though streets have been blocked or a key building destroyed.

c. Techniques such as compass reading and pace counting can still be used, especially in a city where street signs and buildings are not visible. The presence of steel and iron in the urban environment may cause inaccurate compass readings. Sewers must be navigated much the same way. City sewer departments maintain maps providing the basic layout of the sewer system. This information includes directions the sewer lines run and distances between manhole covers. Along with basic compass and pace count techniques, such information enables a unit to move through the city sewers.

d. Helicopters can assist units in moving to objectives. An OH-58D assisting with a laser or an IR searchlight can be a useful technique.

e. Operations in an urban area adversely affect the performance of sophisticated electronic devices such as GPS and data distribution systems. These systems function the same as line-of-sight communications equipment. They cannot determine underground locations or positions within a building. These systems must be employed on the tops of buildings, in open areas, and down streets where obstacles do not affect line-of-sight readings.

f. City utility workers are assets to units fighting in urban areas. They can provide maps of sewers and electrical fields, and information about the city, which is especially important with regard to the use of the sewers. Sewers can contain pockets of highly toxic methane gas. City sewer workers know the locations of these danger areas and can advise a unit on how to avoid them.

3-34. GLOBAL POSITIONING SYSTEMS

Most GPS use a triangulation technique using satellites to calculate their position. Preliminary tests have shown that small urban areas, such as villages, do not affect GPS. Large urban areas with a mixture of tall and short buildings cause some degradation of most GPS. This effect may increase as the system is moved into the interior of a large building or taken into subterranean areas.

3-35. AERIAL PHOTOGRAPHS

Current aerial photographs are excellent supplements to military city maps and can be substituted for a map. A topographic map, military map, or city map could be obsolete. A recent aerial photograph shows changes that have taken place since the map was made, which could include destroyed buildings and streets that have been blocked by rubble as well as enemy defensive preparations. More information can be gained by using aerial photographs and maps together than using either one alone. Whenever possible, the aerial photos or satellite imagery should be acquired during the noon hour to minimize the amount of shadowing around structures.

Section VI. CAMOUFLAGE

To survive and win in combat in urban areas, a unit must supplement cover and concealment with camouflage. To properly camouflage men, vehicles, and equipment, soldiers must study the surrounding area and make positions look like the local terrain.

3-36. APPLICATION

Only the material needed for camouflaging a position should be used since excess material could reveal the position. Material must be obtained from a wide area. For example, if defending a cinderblock building, do not strip the front, sides, or rear of the building to camouflage a position.

a. Buildings provide numerous concealed positions. Armored vehicles can often find isolated positions under archways or inside small industrial or commercial structures. Thick masonry, stone, or brick walls offer excellent protection from direct fire and provide concealed routes.

b. After camouflage is completed, the soldier inspects positions from the enemy's viewpoint. He makes routine checks to see if the camouflage remains natural looking and actually conceals the position. If it does not look natural, the soldier must rearrange or replace it.

c. Positions must be progressively camouflaged, as they are prepared. Work should continue until all camouflage is complete. When the enemy has air superiority, work may be possible only at night. Shiny or light-colored objects attracting attention from the air must be hidden.

d. Shirts should be worn since exposed skin reflects light and attracts the enemy.

e. Camouflage face paint is issued in three standard, two-tone sticks. When face-paint sticks are not available, burnt cork, charcoal, or lampblack can be used to tone down exposed skin. Mud should be used as a last resort since it dries and peels off, leaving the skin exposed.

3-37. USE OF SHADOWS

Buildings in urban areas throw sharp shadows, which can be used to conceal vehicles and equipment (Figure 3-64, page 3-68). Soldiers should avoid areas not in shadows. Vehicles may have to be moved periodically as shadows shift during the day. Emplacements inside buildings provide better concealment.

Figure 3-64. Use of shadows for concealment.

a. Soldiers should avoid the lighted areas around windows and loopholes. They are better concealed if they fire from the shadowed interior of a room (Figure 3-65).

b. A lace curtain or piece of cheesecloth provides additional concealment to soldiers in the interior of rooms if curtains are common to the area. Interior lights are prohibited.

Figure 3-65. Concealment inside a building.

3-38. COLOR AND TEXTURE

Standard camouflage pattern painting of equipment is not as effective in urban areas as a solid, dull, dark color hidden in shadows. Since repainting vehicles before entering a urban area is not always practical, the lighter sand-colored patterns should be subdued with mud or dirt.

a. The need to break up the silhouette of helmets and individual equipment exists in urban areas as it does elsewhere. Burlap or canvas strips are a more effective camouflage than foliage (Figure 3-66). Predominant colors are normally browns, tans, and grays rather than greens, but each camouflage location should be evaluated.

Figure 3-66. Helmet camouflaged with burlap strips.

b. Weapons emplacements should use a wet blanket (Figure 3-67), canvas, or cloth to keep dust from rising when the weapon is fired.

Figure 3-67. Wet blankets used to keep dust down.

c. Command posts and logistical emplacements are easier to camouflage and better protected if located underground. Antennas can be remoted to upper stories or to higher buildings based on remote capabilities. Field telephone wire should be laid in conduits, in sewers, or through buildings.

d. Soldiers should consider the background to ensure they are not silhouetted or skylined, but blend into their surroundings. To defeat enemy urban camouflage, soldiers should be alert for common camouflage errors such as:

- Tracks or other evidence of activity.
- Shine or shadows.
- An unnatural color or texture.
- Muzzle flash, smoke, or dust.
- Unnatural sounds and smells.
- Movement.

e. Dummy positions can be used effectively to distract the enemy and make him reveal his position by firing.

f. Urban areas afford cover, resources for camouflage, and locations for concealment. The following basic rules of cover, camouflage, and concealment should be adhered to:

- Use the terrain and alter camouflage habits to suit your surroundings.
- Employ deceptive camouflage of buildings.
- Continue to improve positions. Reinforce fighting positions with sandbags or other fragment- and blast-absorbent material.
- Maintain the natural look of the area.
- Keep positions hidden by clearing away minimal debris for fields of fire.
- Choose firing ports in inconspicuous spots when available.

CHAPTER 4
OFFENSIVE OPERATIONS

"From 1942 to the present, shock units or special assault teams have been used by attackers (and often by defenders) with great success. These assault teams are characterized by integration of combined arms. Assault teams typically contain Infantry with variable combinations of armor, artillery, or engineers."

Technical Memorandum 5-87
Modern Experience in City Combat
US Army Human Engineering Laboratory
March, 1987

Section I. OFFENSIVE CONSIDERATIONS

Offensive operations in urban areas are based on offensive doctrine modified to conform to the urban terrain. Urban combat also imposes a number of demands that are different from other field conditions such as combined arms integration, fires, maneuver, and use of special equipment. As with all offensive operations, the commander must retain his ability to fix the enemy and maneuver against him. Offensive UO normally have a slower pace and tempo than operations in other environments. Unlike open terrain, units cannot maneuver quickly, even when mounted. Missions are more methodical. Brigades must be prepared to operate independently or within a division or joint task force (TF). The brigade and its subordinate battalion TFs must also be prepared to conduct different missions simultaneously. For example, a battalion may establish checkpoints in one section of a city and clear enemy in another section simultaneously.

4-1. REASONS FOR ATTACKING URBAN AREAS

Reasons for attacking urban areas include the following:

a. The results of the commander and staff's estimate may preclude bypassing as an option. The mission itself may dictate an attack of an urban area.

b. Cities control key routes of commerce and provide a tactical advantage to the commander who controls them. Control of features, such as bridges, railways, and road networks, can have a significant outcome on future operations. The requirement for a logistics base, especially a port or airfield, may play a pivotal role during a campaign.

c. The political importance of some urban areas may justify the use of time and resources to liberate it. Capturing the city could deal the threat a decisive psychological blow and or lift the moral of the people within the city.

(1) The tactical situation may require the enemy force to be contained.

(2) The urban area itself may sit on dominating terrain that would hinder bypassing for combat support (CS) and combat service support (CSS) elements.

(3) The enemy within that urban area may be able to interdict lines of communications even though the terrain around an urban area may facilitate its bypass.

4-2. REASONS FOR NOT ATTACKING URBAN AREAS

Conversely, reasons for not attacking urban areas include the following:

a. The commander may decide to bypass if he determines no substantial threat exists in the urban area that could interdict his unit's ability to accomplish its mission. The commander's intent may dictate speed as essential to the mission. Since combat in an urban area is time consuming, the commander may choose to bypass the urban area to save time.

b. During the estimate process, the commander and staff may realize a sufficient force is not available to seize and clear the urban area, or enough forces are available to accomplish the mission but cannot be logistically supported. If the tactical situation allows, the commander should avoid attacks on urban areas.

c. The urban area may be declared an *open city* to prevent civilian casualties or to preserve cultural or historical sites. An open city, by the law of land warfare, is a city that cannot be defended or attacked. The defender must immediately evacuate the open city and cannot distribute weapons to the city's inhabitants. The attacker assumes administrative control of the city and must treat its citizens as noncombatants in an occupied country.

4-3. TROOP REQUIREMENTS

Due to the nature of combat in urban areas, more troops are normally needed than in other combat situations. This situation is due to the number of requirements placed upon units, soldier fatigue, controlling civilians, and evacuation of casualties.

a. Because of the need to clear buildings and provide security, the number of troops required to accomplish an offensive mission is much greater. Some forces must be left behind in a building once it has been cleared to prevent enemy forces from repositioning or counterattacking friendly forces. Commanders and staffs need to be keenly aware that attacking units will effectively lose manpower from assault elements as they secure rooms and floors. They must ensure that the proper force ratios exist to conduct the missions assigned to subordinate units.

b. Commanders must also consider soldier fatigue. Fighting in urban areas is physically demanding and quickly tires a force. Commanders must plan for the relief or rotation of their forces before they reach the point of exhaustion. This situation is facilitated through proper task organization and maintenance of adequate reserves.

c. Additional forces may be needed to deal with noncombatants in the urban area. These forces must protect the noncombatants, provide first aid, and prevent them from interfering with the tactical plan.

d. Fighting in an urban area may result in a greater number of friendly casualties. The greater the restrictions on firepower, the less suppressive fire can be used, and the more the individual soldier is exposed to enemy fire. MEDEVAC/CASEVAC must be planned and subordinate units designated to conduct this task.

4-4. FIRES AND MANEUVER

As in other terrain, units conduct penetrations, envelopments, turning movements, and frontal attacks. Unlike open terrain, commanders cannot maneuver their units and attachments quickly due to the close, dense environment. Clearing buildings and looking for antiarmor ambushes, snipers, and booby traps degrade the ability of subordinate units to maneuver. Due to the dense environment and its effects on weapon systems, the synchronization of combat power is one of the commander's main challenges. Offensive operations need to be planned in detail, with subordinate elements given specific instructions and on order missions. Maintaining situational awareness assist in overcoming the inability to maneuver quickly.

a. **Indirect Fires.** The fire support plan may require extensive air and artillery bombardment to precede the ground attack on an urban area. Supporting fire suppresses the defender's fire, restricts his movement, and may destroy his position. However, indirect fire in urban areas with heavily clad construction creates rubble, which can be used for cover but may restrict the movements of attacking troops. For that reason, the artillery preparation should be short and violent. Assaulting troops must follow the artillery fire closely to exploit its effect on the defenders. While the supporting fire suppresses the enemy, maneuver units move near the coordinated fire line (CFL). As the attacking force assaults the objective, fires are lifted or shifted to block enemy withdrawal or to prevent the enemy from reinforcing their position.

(1) Prior coordination is critical to determine the techniques and procedures for communication, target identification, and shifting of fires. Consideration must be given to the noncombatants, houses of worship, medical centers, schools, public services, and historical monuments. The fire support plan can include integrating tanks, Infantry weapons, artillery, and dismounted direct and indirect fires.

(2) Indirect fire is planned to isolate objectives, to prevent reinforcement and resupply, to neutralize known and suspected command and observation posts, and to suppress enemy defenders. Most indirect fires are high-angle in urban terrain.

(3) Mortars are the most responsive indirect fires to hit targets of opportunity at the close ranges typical of combat in urban areas. Forward observers move with the forward units to adjust fire on targets as requested by the supported troops.

b. **Direct Fires.** Direct-fire is the most effective fire support in urban areas. Once a target can be located in a building, one or two direct-fire rounds can accomplish what entire salvos of indirect fire cannot. The best direct-fire support is provided by Bradley fighting vehicles (BFVs) but can also be provided by tanks and or howitzers. (See Chapter 7 for specific weapons effects.) Tanks and howitzers may create rubble and building and street damage that could restrict movement for the attacking force.

(1) Tanks may support by fire when lead units are seizing a foothold. During the attack of an urban area, tanks overwatch the Infantry's initial assault until an entry into the area has been secured. Tanks are supported by Infantry organic weapons to suppress enemy strongpoints while they move into overwatch positions. Commanders employ tanks to take advantage of the long range of their main gun. This procedure is usually achieved with tanks employed outside the urban area, for the duration of the attack to cover high-speed mounted avenues of approach, especially during the isolation phase. Tanks may also support Infantry in the urban area as an assault and support weapon. In both cases, Infantry must protect tanks.

(2) In house-to-house and street fighting, tanks and or BFVs move down streets protected by the Infantry, which clears the area of enemy ATGM weapons. Tanks and BFVs in turn support the Infantry by firing their main guns and machine guns to destroy enemy positions. Tanks are the most effective weapon for heavy fire against structures

and may be used to clear rubble with dozer blades (Figure 4-1). The BFV can provide sustained, accurate suppressive fires with its 25-mm gun.

(3) Large-caliber artillery rounds that are shot by direct fire are effective for destroying targets in buildings. If available, self-propelled 155-mm howitzers can use direct fire to destroy or neutralize bunkers, heavy fortifications, or enemy positions in reinforced concrete buildings (Figure 4-2). The self-propelled 155-mm can be used to clear or create avenues of approach. The 105-mm artillery can be used in this role but are not the preferred artillery pieces used in offensive UO. When artillery is used in the direct fire role, it must be close to the Infantry for security against enemy ground attack. Prior coordination must be accomplished so the bulk of the field artillery unit's shells are switched to High Explosive (HE). (See Chapter 7, paragraph 7-12; Chapter 10, paragraph 10-9; and Chapter 12, paragraph 12-2.)

Figure 4-1. Tank in direct fire supported by Infantry.

Figure 4-2. Artillery in direct-fire role.

(4) Tanks, self-propelled artillery, and BFVs are vulnerable in urban areas because streets and alleys provide ready-made fire lanes for defenders. Motorized traffic is restricted, canalized, and vulnerable to ambush and close-range fire. Tanks are at a further disadvantage because their main guns cannot be depressed sufficiently to fire into basements or elevated to fire into upper floors of buildings at close range (Figure 4-3).

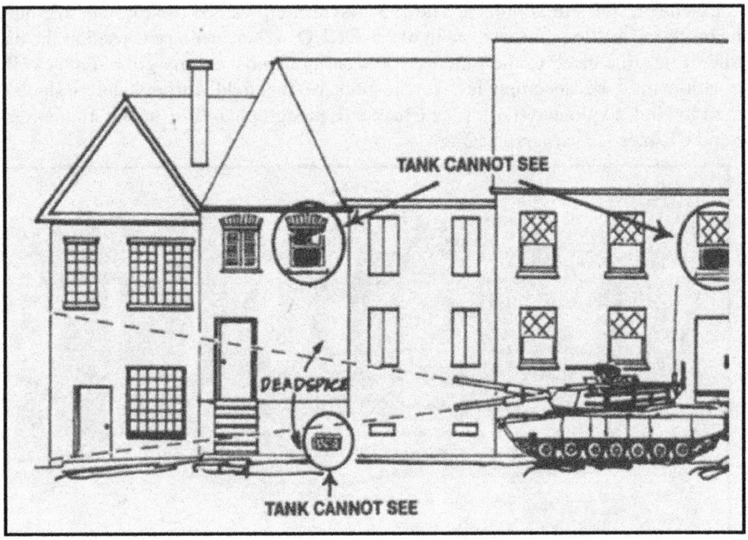

Figure 4-3. Tank dead space.

(5) Direct-fire systems organic to Infantry battalions—mainly ATGMs and recoilless weapons, such as the AT4, are initially employed to support the seizure of a foothold. Then, if necessary, they are brought forward to fight enemy armor within the town. Antitank weapons are not as effective as tank rounds for neutralizing targets behind walls. They neutralize a target only if that target is located directly behind the point of impact. ATGMs are at a greater disadvantage because of their 65-meter arming distance and the possibility of their guiding wires becoming caught on ground clutter. These factors limit employment in close engagements like those in urban areas.

(6) Snipers are a valuable asset during urban operations. They must be equipped with effective observation devices and placed in a key area to be effective. In situations where the ROE permit the use of destructive force, snipers can be used as part of the support element to provide accurate, long-range fires. Depending on the commander's concept, snipers can be employed in the counter-sniper role or assigned priority targets. If a restrictive ROE is in effect, snipers may be used to prevent collateral damage. Snipers can also overwatch breaching operations and call for indirect artillery fires. (For more

information on the offensive employment of snipers, see FM 71-2, FM 7-20, TC 23-14, and Chapter 6.)

c. **Maneuver.** The first phase of the attack should be conducted when visibility is poor. Troops can exploit poor visibility to cross open areas, gain access to rooftops, infiltrate enemy areas, and gain a foothold. If the attack must be made when visibility is good, units should consider using smoke to conceal movement.

(1) The formation used in an attack depends on the width and depth of the zone to be cleared, the character of the area, anticipated enemy resistance, and the formation adopted by the next higher command.

(2) Lead companies may have engineers attached for immediate support. Tasks given to the engineers may include:

- Preparing and using explosives to breach walls and obstacles.
- Finding and exploding mines in place or helping remove them.
- Clearing barricades and rubble.
- Cratering roads and other countermobility measures.

(3) When the unit is involved in clearing, bypassing buildings increases the risk of attack from the rear or flank. A single building may be an objective for a rifle squad, or if the building is large, for a rifle platoon or company. When the commander's concept is based on speed or when conducting a hasty attack, a battalion may be directed not to clear its entire zone.

(4) The reserve should be mobile and prepared for commitment. The reserve can stay close to forward units because of the available cover in urban areas. Battalion reserves normally follow one to two blocks to the rear of the lead company. A company reserve, if available, follows within the same block so it can immediately influence the attack. A unit with a reserve mission may be called upon to perform one or more of the following tasks:

- Attacking from another direction.
- Exploiting an enemy weakness or friendly success.
- Clearing bypassed enemy positions.
- Securing the rear or a flank.
- Maintaining contact with adjacent units.
- Supporting or counterattacking by fire.

(5) The battalion reconnaissance scout platoon is normally employed to reconnoiter the battalion's flanks and rear. Its capability for reconnaissance and security is somewhat reduced in urban areas. The reconnaissance/scout platoon can also help isolate a village or small town. They must be prepared to dismount and enter buildings for reconnaissance or for setting up OPs. Infantry platoons and squads conduct reconnaissance patrols and man OPs to supplement the reconnaissance/scout platoon effort.

(6) Security in an urban area presents special problems. All troops must be alert to an enemy that may appear from the flanks, from above, or from subterranean areas.

d. **Movement**. Moving from building to building or between buildings present a problem to units conducting offensive operations. Historical examples, recent operations in Somalia, and the Russian experience in Grozny have shown that most casualties can be expected during movement from building to building and down streets. Therefore, during mission analysis, commanders and staffs should plan operations in such a manner that allow subordinate elements to take maximum advantage of covered and concealed routes within the urban area. Additionally, commanders and staffs must carefully analyze which buildings must be isolated, suppressed, and obscured, consistent with the ROE, as well as using armored assets as shields for maneuver elements.

(1) In movement down narrow streets, or down wider streets with narrow paths through the debris, Infantry should move ahead of the tanks clearing the buildings on each side. Personnel movement across open areas must be planned with a specific destination in mind. Street intersections should be avoided, since they are normally used as engagement areas. Suppression of enemy positions and smoke to cover Infantry movement should also be included in the fire support plan. When needed, tanks move up to places secured by the Infantry to hit suitable targets. When an area is cleared, the Infantry again moves forward to clear the next area. Tanks and Infantry should use the traveling overwatch movement technique and communicate with tank crews by using arm-and-hand signals and radio.

(2) For movement down wider streets, Infantry platoons normally have a section of attached tanks with one tank on each side of the street. Single tanks should not be employed. Other tanks of the attached tank platoon should move behind the Infantry and fire at targets in the upper stories of buildings. In wide boulevards, commanders may employ a tank platoon secured by one or more Infantry platoons. The Infantry can secure the forward movement of the lead tanks, while the trailing tanks overwatch the movement of the lead units.

(3) If an Infantry unit must travel along streets that are too narrow for mutual tank support, the tanks travel in single file for support. The tanks move and fire to cover each other's approach while the Infantry provides ATGM fire from buildings as necessary.

(4) Tanks may drive inside buildings or behind walls for protection from enemy antitank missile fire where feasible. Buildings are cleared by the Infantry first. Ground floors are checked to ensure they support the tank and there is no basement into which the tank could fall. When moving, all bridges and overpasses are checked for mines, booby traps, and load capacity. Specific Infantry elements are assigned to protect specific tanks.

4-5. LIMITATIONS

Commanders attacking an urban area must recognize some important limitations in the use of available assets.

a. **Indirect Fires.** Normally, the use of indirect fires is much more restricted in urban areas than in open terrain. Consideration must be given to the effects of the indirect fire on the urban area and noncombatants. This procedure is especially true when extremely restrictive ROE are in effect. When indirect fires are authorized, they must be fired in greater mass to achieve the desired effect. When units are performing multiple missions, indirect fire supporting one element can easily cause casualties in adjacent elements. The rubbling caused by massive indirect fires adversely affect a unit's ability to maneuver during the attack.

b. **Noncombatants.** If there are noncombatants intermingled with combatants, the ability to use all available firepower may be restricted.

c. **Night Vision Devices.** Commanders and leaders must consider the effect that city lights, fires, and background illumination have on night vision devices. These elements may limit the effectiveness of night vision goggles (NVGs) and make thermal imagery identification difficult.

d. **Communications.** Communications equipment may not function to its maximum effectiveness because of the density in building construction. Intelligent use of graphic control measures, understanding the commander's intent, and maintaining situational awareness at all levels become more important to mission accomplishment.

Section II. MISSION, ENEMY, TERRAIN, TROOPS, TIME, CIVIL FACTORS

The planning, preparation, and conduct of offensive operations in an urban area are the same as all other offensive operations and must be based on the mission, enemy, terrain, troops, time, and civil (METT-TC) factors. Commanders must focus on the synchronization of maneuver forces and the fire support plan to accomplish the assigned mission. Combat support (CS) and combat service support (CSS) play a critical role in the offense. (See Chapters 13 and 14 for further details on CS and CSS.)

4-6. MISSION

The commander and staff must receive, analyze, and understand the mission before beginning planning. The conditions of the operation; either precision or high intensity; the ROE; and the desired end-state must be clearly understood and stated. As stated earlier, brigades and battalions may be required to conduct different missions simultaneously. A detailed discussion of the urban environment's effects on the planning process is in Chapter 2, Urban Analysis. Additional considerations that are specific to offensive operations are discussed below. When conducting this analysis, commanders and staff must consider the higher level commander's intent. For example, a brigade must determine if supporting efforts are needed to shape the battlefield prior to the main effort executing its mission. A battalion must determine if a mission given to them means clearing every building within an area, block by block, or if the seizure of key terrain only requires clearing along the axis of advance.

a. In certain circumstances, subordinate units may secure rather than clear buildings. Normally, clearing means entering and searching each building to kill, capture, or force the withdrawal of the threat in the zone of action or objective area as well as leaving security to prevent reoccupation of cleared buildings. This procedure may not be feasible due to the nature of the mission and should be made clear when orders are issued. Clearing requires a systematic search of every room. Securing means a search of selected areas and preventing the occupation or reoccupation of the area by the threat and questioning of noncombatants, if present.

b. Commanders and staffs must also consider how and where the unit is postured in order to conduct follow-on missions to facilitate higher echelon missions, and influences the missions that are given to subordinate units.

c. When the battalion is involved in clearing operations, bypassing buildings increases the risk of attack from the rear or flank unless planned support isolates and suppresses those buildings.

d. A battle may transition quickly from precision to high intensity conditions. The transition can be caused by enemy actions. Commanders must be prepared to request changes in ROE for certain areas or buildings. Indications of an enemy-forced change of ROE (and a change from precision conditions to high intensity) include:

- The requirement to breach multiple obstacles.
- The use of booby traps by the enemy.
- The requirement to use repetitive explosive breaching to enter a building; and rooms.

4-7. ENEMY

The unique factor the commander must determine to complete the IPB process is the type threat he is attacking—conventional, unconventional or other, such as gangs, factional elements, or organized criminals. The type of threat determines how the unit task-organizes and how combat power is synchronized to accomplish the mission. (See Chapter 2, Section IV.)

4-8. TERRAIN AND WEATHER

Offensive operations must be tailored to the urban environment based on a detailed analysis of each urban terrain setting, its types of urban areas, and existing structural form. (See Chapter 2, Sections II and III and FM 34-130 for details of urban terrain analysis.) Commanders and subordinate leaders must incorporate the following special planning considerations for an urban environment when conducting an offensive operation.

a. Alternates for military maps that do not provide enough detail for urban terrain analysis or reflect the underground sewer system, subways, underground water system, mass transit routes, and utility generation.

b. Natural terrain surrounding the urban area.

c. Key and decisive terrain (stadiums, parks, sports fields, school playgrounds, public buildings, and industrial facilities).

d. Confined spaces limiting observation, fields of fire and maneuver, which also prevents the concentration of fires at critical points.

e. Covered and concealed routes to and within the urban area.

f. Limited ability to employ maximum combat power due to the need to minimize damage and rubbling effects.

g. A greater demand for ammunition and rations, thus imposing unusual strains on logistics elements.

h. Problems with conducting effective reconnaissance during conventional operations. (Reconnaissance by force becomes the most effective reconnaissance means. This method involves probing a defense with successively larger units until the enemy positions are disclosed and successfully attacked. During unconventional operations, the opposite is true. Reconnaissance and security are easily accomplished by both sides and may be difficult to prevent.)

4-9. TROOPS AVAILABLE

Troop density for offensive missions in urban areas can be as much as three to five times greater than for similar missions in open terrain. Urban operations may require unique task organizations. Commanders must consider providing assets where they are needed to accomplish specific tasks. All phases of mission execution must be considered when developing task organization. Changes in task organization may be required to accomplish different tasks during mission execution. Task organizations could very well change as conditions and missions change. For example, high intensity offensive operations probably require different task organizations from precision offensive operations. Likewise, task organizations change as mission transitions from offense to stability and support and vice-versa. (See brigade, battalion, company, and platoon sections [Sections V, VI, VII, VIII] for specific task organizations and troop considerations.)

4-10. TIME AVAILABLE

Combat in urban areas has a slower tempo and an increased use of methodical, synchronized missions. Additionally, a brigade or battalion may find itself planning different operations simultaneously. For example, a task force may have the mission to conduct offensive missions in one part of the brigade's AO and another battalion may be conducting stability missions in yet another part of the brigade's AO. In planning UO, the commander and staff must take these factors into account. Plans must also take into account that more time is required for clearing buildings, blocks, or axes of advance due to the density of urban terrain and that troops tire more quickly because of stress and additional physical exertion caused by the environment. More time must be allowed for thorough reconnaissance and subordinate unit rehearsals. Allocating time for rehearsals is especially important when units are not habitually used to working with each other.

4-11. CIVIL CONSIDERATIONS

The commander and staff must understand the composition, activities, and attitudes of the civilian population, to include the political infrastructure, within the urban area. Various options are available to the commander to control the impact of civilians on the operation such as screening civilians, prohibiting unauthorized movement, diverting or controlling refugee movements, and evacuating. Understanding the urban society requires comprehension of—

- Living conditions.
- Cultural distinctions.
- Ethnicity.
- Factions.
- Religious beliefs.
- Political affiliation and grievances.
- Attitude toward US forces (friendly, hostile, neutral).

See Chapter 13, Section III for additional considerations.

Section III. COMMAND AND CONTROL

Urban operations require centralized planning and decentralized execution. Therefore the staff must develop a detailed plan that synchronizes the BOS in order to meet the commander's intent and also provide subordinate units with the means to accomplish the mission.

4-12. COMMAND

Subordinate units require mission-type orders that are restrictive in nature. Commanders should use detailed control measures to facilitate decentralized execution. Increased difficulties in command, control, and communications from higher headquarters demand increased responsibility and initiative from subordinate leaders. Understanding of the commander's intent two levels up by all leaders becomes even more important to mission accomplishment in an urban environment.

4-13. CONTROL

Control of the urban battlefield is difficult. In urban areas, radio communications are often less effective than field telephones and messengers. Units often fight without continuous communications, since dependable communications are uncertain. Pyrotechnic signals are hard to see because of buildings and smoke. The high noise level of battles within and around buildings degrades voice alerts. Voice communication can also signal the unit's intention and location to the enemy. Graphic control measures common to other tactical environments are also used in urban combat. These and other control measures ensure coordination throughout the chain of command, enhance the mission, and thus prevent fratricide. Thorough rehearsals and detailed backbriefs also enhance control. It is also important that subordinate leaders clearly understand the commander's intent (two levels up) and the desired mission end state in order to facilitate control. Commanders should consider using the executive officer (XO), the S3, and other staff members to control certain portions of the fight, when the commander's attention needs to be focused elsewhere.

a. **Radio Communications.** Radio communications in urban areas pose special problems to tactical units. Communications equipment may not function properly because of the massive construction of buildings and the environment. In addition to the physical blockage of line of sight transmissions, there is also the interference from commercial power lines, absorption into structures and the presence of large quantities of metal in structures. Leaders should consider these effects when they allocate time to establish communications. Unit SOPs become much more important in urban terrain. The time needed to establish an effective communications system might be greater in urban areas. Leaders should consider the following techniques when planning for radio communications:

- Emplace radios and retransmission sites on the upper floors of buildings. Radio antennas should blend in with the building structure so as not to be easily identifiable to the enemy.
- Construct field expedient antennas to enhance capabilities.
- RTOs should utilize an earpiece to keep their hands free in order to write messages and use their weapon to defend themselves.
- Use windows and holes in walls to extend antennas for better communications.
- Open doors and windows to enhance the flow of FM signals.

b. **Other Types of Communications.** Wire laid at street level is easily damaged by rubble and vehicle traffic. Also, the noise of urban combat is much louder than in other areas, making sound or verbal signals difficult to hear.

- Develop and utilize other nonverbal signals. Use color-coded signaling devices per unit SOP. Marking areas as the unit moves is a key to success. (See Appendix I.)
- If possible, lay wire through buildings for maximum protection.
- Use existing telephone systems. Telephones are not always secure even though many telephone cables are underground.
- Use messengers at all levels since they are the most secure means of communications.

c. **Graphic Control Measures.** The use of detailed graphic control measures is critical to mission accomplishment and fratricide avoidance in urban terrain. Phase lines can be used to report progress or to control the advance of attacking units. Limits of advance should be considered. Principal streets, rivers, and railroad lines are suitable phase lines or limits of advance. Examples are shown below.

(1) When attacking to seize a foothold, a battalion normally assigns each company a sector or a group of buildings as its first objective. When an objective extends to a street, only the near side of the street is included in the objective area. Key buildings or groups of buildings may be assigned as intermediate objectives. The battalion's final objective may be a group of buildings within the built-up area, key terrain, or nodes, depending on the brigade's mission. To simplify assigning objectives and reporting, all buildings along the route of attack should be identified by letters or numbers (Figure 4-4). Mixing numbers and letters may help differentiate between blocks as an attack progresses.

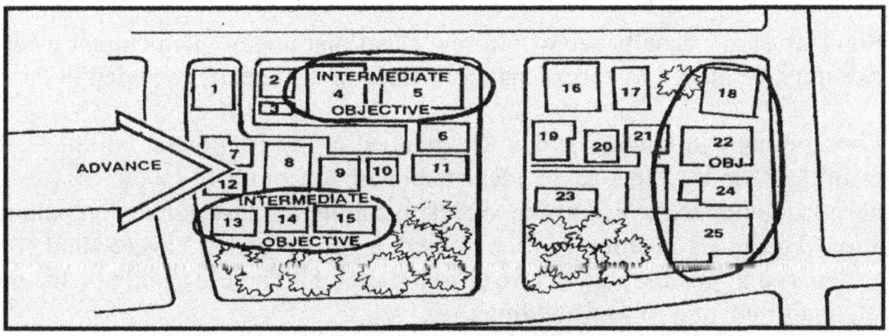

Figure 4-4. Example of a numbering system.

(2) Phase lines can be used to report progress or to control the advance of attacking units (Figure 4-5). Phase lines should be on the near side of the street or open area. In systematic clearing, a unit may have the mission to clear its zone of action up to a phase line or limit of advance. In that case, the commander chooses his own objectives when assigning missions to his subordinate units.

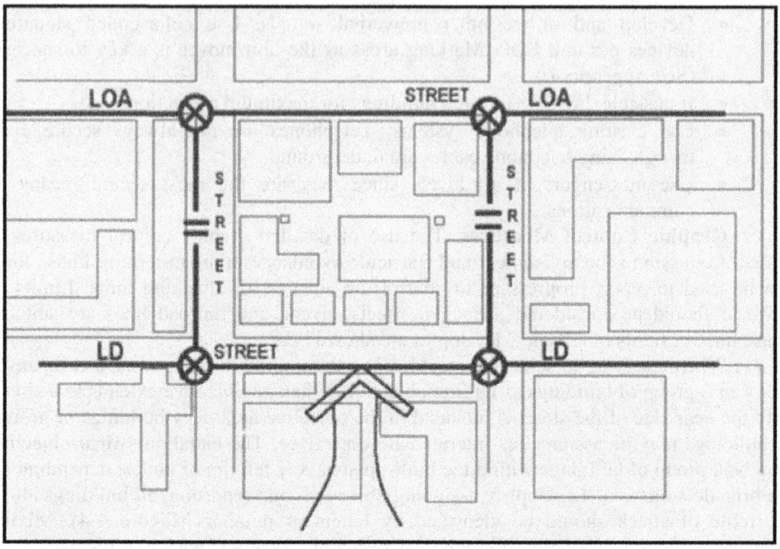

Figure 4-5. Boundaries and phase lines.

(3) Boundaries are usually set within blocks so that a street is included in the zone. Boundaries must be placed to ensure that both sides of a street are included in the zone of one unit.

(4) Checkpoints and contact points are planned at street corners, buildings, railway crossings, bridges, or any other easily identifiable urban feature.

(5) Forward units may occupy an attack position for last-minute preparation and coordination. The attack position is often behind or inside the last covered and concealed position, such as a large building, before crossing the LD. The LD should be the near side of a street, a rail line, or a row of buildings.

(6) A unit's assigned frontage for the attack of a built-up area depends on the size of buildings and the resistance anticipated. A company normally attacks on a one- to two-block front, and a battalion on a two- to four-block front, based on city blocks averaging 175 meters in width.

4-14. FOCUS ON THE THREAT

During the mission analysis, the plan should focus on the factors of METT-TC. Make the plan enemy-oriented instead of terrain-oriented. Use terrain factors to defeat the threat—do not attack buildings for the sake of seizing buildings, attack buildings to defeat the threat. Considerations include, but are not limited to, the following:

a. Thorough evaluation of the urban area's related terrain and threat may take much longer than other environments. This time factor also affects friendly planning efforts.

b. Determine the threat's location, strength, and capabilities. Develop a plan that defeats his direct and indirect fire systems.

c. Focus the axis of advance on the threat's weaknesses while maintaining adequate force protection measures. When possible employ multiple and supporting axes of advance.

d. Divide the objective area into manageable smaller areas that facilitate battalion TF maneuver.

e. Isolate the objective area and establish a foothold at the point of entry. The location chosen for the foothold must allow for expansion.

f. The brigade and battalion maneuver plans directly affect the company schemes of maneuver. Every company within the brigade must know what enemy targets will be engaged by brigade and battalion assets.

4-15. COMMANDER'S CRITICAL INFORMATION REQUIREMENTS

The commander's critical information requirements (CCIR) directly affect his decisions and dictate the successful execution of tactical operations. The staff must develop the components of CCIR that facilitate the commander's ability to make decisions that impact the plan during urban operations. Logical deductions are that essential elements of friendly information (EEFI) should address the enemy commander's priority intelligence requirements (PIR) and friendly forces information requirements (FFIR) should be items that cause the commander to make decisions that impact the plan. The following are examples of PIR, EEFI, and FFIR that would be more likely to help the commander in an urban environment.

a. **PIR**. These are intelligence requirements that a commander has anticipated and have stated priority in task planning and decision making. Examples include:

- Where are the threat command posts?
- What are the most likely threat infiltration routes into the area of operations?
- What streets and alleys restrict movement of friendly armored and wheeled vehicles?
- Where are the likely threat strong points and engagement areas?
- What is the threat air defense capability against Army aviation assets?

b. **EEFI.** These are critical aspects of a friendly operation that, if known by the threat, would subsequently compromise, lead to failure, or limit success of the operation and, therefore, must be protected from detection. Examples include:

- Is the unit command net vulnerable to intercept, direction finding, and electronic attack?
- Is the unit vulnerable to HUMINT collection and sabotage by local nationals?
- Where are the supply routes/LOC most vulnerable to ambush and snipers?
- Are friendly troop concentrations and movement under threat observation?

c. **FFIR.** This requirement is information the commander and staff need about the friendly forces available for the operation. Examples include:

- Scouts captured or compromised.
- Main bridge locations along the ground route that have been blown.
- OPORD compromised.
- Loss of cryptographic equipment.
- Expected personnel and equipment replacements that did not arrive.

4-16. REHEARSALS

After developing a thorough, well-synchronized plan, commanders should require subordinate units to conduct combined arms rehearsals and include all phases of the operation. When conducted properly, combined arms rehearsals identify potential problems in the synchronization of the plan between maneuver, combat support, and combat service support elements. Rehearsals provide a means for units that seldom operate together to train collective skills. Carefully consider where rehearsals are conducted within the brigade AO. It is preferable to conduct rehearsals on urban terrain similar to the objective area.

Section IV. OFFENSIVE FRAMEWORK AND TYPES OF ATTACKS

This section discusses the framework that is used and the types of attacks that are conducted during offensive UO.

4-17. OFFENSIVE FRAMEWORK

Figure 4-6 on page 4-16 depicts the operational framework of brigade urban offensive operations. The brigade commander's primary responsibility is to set the conditions for tactical success for his subordinate units. Whenever possible, close combat by maneuver units is minimized and brigades attempt to move from assess to transition. At the brigade level and below, offensive operations often take the form of either a hasty or deliberate attack. Both hasty and deliberate attacks are characterized by as much planning, reconnaissance, and coordination as time and the situation permit. Battalions and below conduct those attacks executing the tasks shown in Figure 4-6, page 4-16. The elements of offensive operations are not phases. There is no clear line of distinction that delineates when the brigade moves from one element to another. Properly planned and executed operations involve all four elements. They may be conducted simultaneously or sequentially, depending on the factors of METT-TC. During offensive operations, the brigade commander seeks to:

- Synchronize precision fires (lethal and non-lethal effects) and information operations.
- Isolate decisive points.
- Use superior combat power to destroy high pay-off targets.
- Use close combat, when necessary, against decisive points.

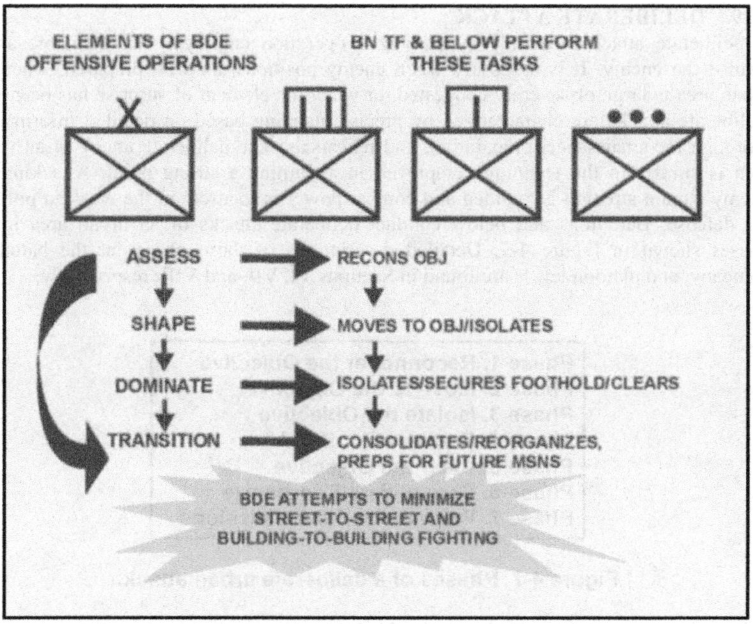

Figure 4-6. Offensive urban operational framework.

4-18. HASTY ATTACK

Battalions and companies conduct hasty attacks as a result of a movement to contact, a meeting engagement, or a chance contact during a movement; after a successful defense or part of a defense; or in a situation where the unit has the opportunity to attack vulnerable enemy forces. When contact is made with the enemy, the commander immediately deploys; suppresses the enemy; attacks through a gap, flank, or weak point; and reports to his higher commander. The preparation for a hasty attack is similar to that of a deliberate attack, but time and resources are limited to what is available. The hasty attack in an urban area differs from a hasty attack in open terrain because the terrain makes command, control, communications and massing fires to suppress the enemy difficult.

a. In urban areas, incomplete intelligence and concealment may require the maneuver unit to move through, rather than around, the friendly unit fixing the enemy in place. Control and coordination become critical to reduce congestion at the edges of the urban area.

b. On-order missions, be-prepared missions, or fragmentary orders may be given to a force conducting a hasty attack so it can react to a contingency once its objective is secured.

4-19. DELIBERATE ATTACK

A deliberate attack is a fully synchronized operation employing all available assets against the enemy. It is necessary when enemy positions are well prepared, when the urban area is large or severely congested, or when the element of surprise has been lost. Deliberate attacks are characterized by precise planning based on detailed information, thorough reconnaissance, preparation, and rehearsals. The deliberate attack of an urban area is similar to the technique employed in assaulting a strong point. Attacking the enemy's main strength is avoided and combat power is focused on the weakest point of his defense. Battalions and below conduct deliberate attacks of an urban area in the phases shown in Figure 4-7. Detailed descriptions of these phases at the battalion, company, and platoon levels are found in Sections VI, VII, and VIII, respectively.

Phase 1. Reconnoiter the Objective
Phase 2. Move to the Objective
Phase 3. Isolate the Objective
Phase 4. Secure a Foothold
Phase 5. Clear the Objective
Phase 6. Consolidate/Reorganize
Phase 7. Prepare for Future Missions

Figure 4-7. Phases of a deliberate urban attack.

Section V. BRIGADE OFFENSIVE OPERATIONS

A brigade may be assigned an objective that lies within an urban area, and may conduct the full range of offensive operations within a single large city or in an AO that contains several small villages and towns.

4-20. TASK ORGANIZATION

Proper task organization is essential for successful execution of offensive UO.

a. During UO, the brigade is often augmented with additional assets, which may include aviation, engineers, signal, smoke and or decontamination, ADA, MI, counterintelligence, MP, public affairs, PSYOP, civil affairs, translators, and LRS assets, when available. The brigade may also receive additional mechanized Infantry or armor. A sample Infantry brigade task organization is shown at Figure 4-8 on page 4-18. Actual task organizations are METT-TC dependent. How the brigade commander task-organizes so that the BOS can be synchronized is of critical importance to tactical success.

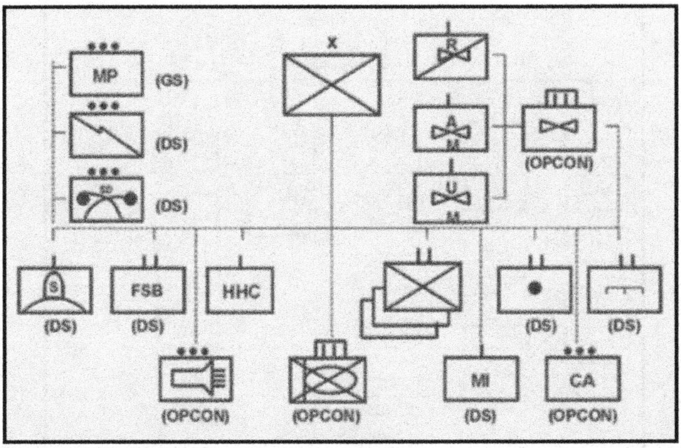

Figure 4-8. Sample UO task organization for an Infantry brigade.

NOTE: The task organization shown in Figure 4-8 would be essentially the same for light, airborne, and air assault Infantry brigades. Heavy brigades would differ based on the composition of their Table of Organization and Equipment (TOE).

b. Urban operations may require unique task organizations. Figure 4-9 depicts a sample brigade task organization for offensive operations, showing units under brigade control, and subordinate task forces necessary to accomplish decisive and shaping operations, specifically, the main and supporting efforts and the brigade reserve. Commanders must consider providing assets where they are needed to accomplish specific tasks. All phases of mission execution must be considered when developing task organization. Changes in task organization may be required to accomplish different tasks during mission execution. Task organizations could very well change from shape through transition.

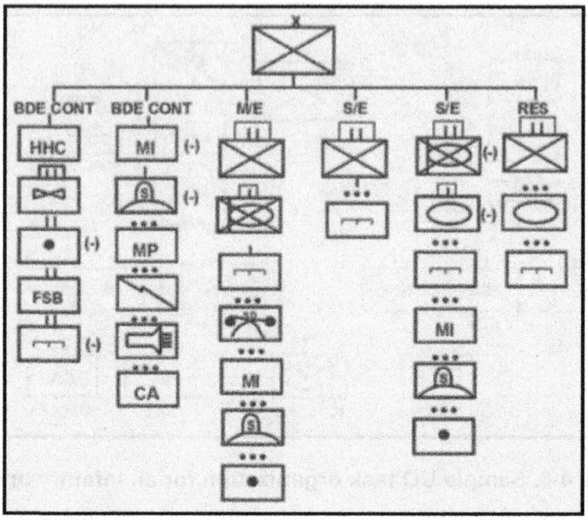

Figure 4-9. Sample brigade task organization for offensive UO.

NOTE: Figure 4-9 also depicts two field artillery platoons that have been given DS missions to provide direct fire support to the main and the supporting attacks.

4-21. ASSESS

Brigades primarily assess the urban environment using the military decision-making process (MDMP); intelligence preparation of the battlefield (IPB) acts as a key tool in that process. IPB is combined with the following:

- Division or joint task force (JTF) reconnaissance efforts and other shaping operations.
- Reconnaissance efforts of brigade units.
- Results of previous operations that impact current operations.

a. An important step in mission analysis is to determine the essential tasks. Combat power is applied precisely at decisive points, and other portions of the urban area are isolated to the extent necessary to ensure they do not adversely influence the UO. Operations are conducted both sequentially and simultaneously, as appropriate. Specific tasks to subordinates may address the full spectrum of Army operations, and the brigade will likely be conducting support, stability, and combat operations simultaneously. The complexity of UO may require simultaneous full spectrum operations down to company level.

b. The brigade commander and staff must determine, during assessment, whether the shaping efforts of higher headquarters are sufficient for the brigade to accomplish its missions or whether additional shaping efforts are required—for example, isolation of nodes or other key terrain. Additionally, the brigade commander and staff must assess whether the shaping efforts of higher headquarters permit them to move directly to domination and or transition.

4-22. SHAPE

Brigades normally shape the area of operations through isolation. Isolation is defined as a tactical task to seal off (both physically and psychologically) an enemy from his sources of support, to deny an enemy freedom of movement, and prevent an enemy unit from having contact with other enemy forces. During isolation, the brigade commander sets the conditions for tactical success. Implied in this step are the thorough reconnaissance of the objectives and movement of subordinate units to positions of tactical advantage. The brigade commander must carefully determine the extent and the manner in which his forces can isolate the objectives. The factors of METT-TC determine how the brigade will isolate the objective psychologically and physically. Only areas essential to mission success are isolated.

a. **Psychological Isolation of the Objective.** Isolation begins with the efforts of the division and corps psychological and civil affairs operations to influence enemy and civilian actions. The brigade commander should consider using PSYOP teams to broadcast appropriate messages to the threat and to deliver leaflets directing the civilian population to move to a designated safe area. These actions must be coordinated with the overall PSYOP plan for the theater and must not sacrifice surprise. By themselves, PSYOP are seldom decisive. They take time to become effective and often their effects are difficult to measure until after the actual attack, but they have usually proven to be successful. Under some METT-TC conditions, they have achieved results far outweighing the effort put into them.

b. **Sensors and Reconnaissance Units.** One of the more common methods of isolation involves the use of a combination of sensors and reconnaissance units along avenues of approach to detect enemy forces as they attempt to enter or leave the objective area. The brigade can engage these enemy forces with indirect fires, aerial fires, or a combination of the two, consistent with the ROE. This technique may be effective in detecting and stopping large enemy units from entering or leaving, but the cover and concealment the urban area provides make it difficult to totally seal off the urban objective. To be successful, this technique requires skillful reconnaissance units and responsive fires. It may not be possible for the brigade to observe all avenues of approach, and enemy units may escape detection by infiltrating or exfiltrating. It may be difficult to distinguish between enemy and friendly personnel and noncombatants moving in and out of the urban area. Indirect fires may cause unacceptable damage to key parts of the urban area.

c. **Snipers.** In certain situations that require precise fire, snipers can provide an excellent method of assisting in isolating key areas. Skillful application of snipers can provide lethal fire while simultaneously minimizing collateral damage and noncombatant casualties. Snipers can also be used to observe and report enemy activity and to call for and adjust indirect fire.

d. **Combination of Assets.** The most effective method of isolating an urban objective is probably the use of a combination of sensors, reconnaissance elements, and maneuver forces. The brigade can move platoons and companies into positions where they can dominate avenues of approach with observation and direct fires. Smaller urban areas with clearly defined boundaries make this method easier to accomplish. Larger urban areas may prevent a maneuver force from gaining access to a position from which to stop enemy movement into the objective area.

e. **Use of Fires and Smoke.** In some instances, where the ROE permit, indirect and aerial fires may be the only available or appropriate method of isolation. This technique is the most destructive; it demands large amounts of ammunition, and it may only last for short periods of time. Brigade fire planners can improve the effectiveness of this technique by careful selection of high pay-off targets and use of precision munitions. Mortar and light artillery fires falling onto large buildings are not as effective in preventing enemy movement as fires falling into open areas. Targeting them against larger avenues, parks, and other open areas force the enemy to move within buildings. Artillery and aerial fires can be directed against buildings that the enemy is using for movement and observation. This method slows and impedes enemy movement, but not stop it. It can also hinder enemy supply efforts and make it difficult to reinforce units under attack. Targeting obvious choke points, such as bridges or main road junctions, can also assist in the isolation effort. Smoke can be used to isolate the objective from enemy observation, but it is difficult to predict what smoke does in an urban area.

NOTE: Multiple flat polished surfaces in an urban area may degrade laser use, thereby rendering some weapon systems useless. Close coordination must occur between maneuver and fire support planners in order to obtain the desired effects of laser-guided precision munitions. Also, obscuration rounds may cause uncontrolled fires in the city and must be carefully planned.

4-23. DOMINATE

The brigade uses all combined arms available, consistent with the ROE, to defeat or destroy the enemy at decisive points and achieve the desired end-state of the mission. The brigade seeks to dominate the enemy through well-planned isolation and skillful use of combined arms. The brigade commander seeks to minimize the amount of street to street and house to house fighting that must be performed by battalions.

4-24. TYPES OF OFFENSIVE OPERATIONS

The brigade conducts the same types of offensive operations as it would on open terrain. (See FMs 7-30 and 71-3.) Techniques that may be more applicable during urban offensive operations are discussed in the following paragraphs. These techniques are applicable to all forms of offensive maneuver and would be determined by METT-TC factors.

a. **Movement to Contact, Search and Attack Technique.** Figure 4-10, page 4-22, depicts a brigade conducting a movement to contact in an urban area using the search and attack technique. This technique is used when knowledge of the enemy is unclear and contact is required. It is normally employed against a weak enemy force that is disorganized and incapable of massing strength against task forces (for example, urban insurgents or gangs). The brigade divides the AO into smaller areas and coordinates the movement of battalions through the brigade AO. In the example shown in Figure 4-10, the enemy is found and fixed during isolation and finished during domination. During a mission of this type, the urban environment makes it difficult for conventional Infantry forces to find, fix, and finish the enemy. For example, movement of units may become canalized due to streets and urban *canyons* created by tall buildings. The application of firepower may become highly restricted based on the ROE. The use of HUMINT in this type of action becomes increasingly more important and can be of great assistance during the *find* portion of the mission. (Table 4-1 shows the advantages and disadvantages of search and attack.)

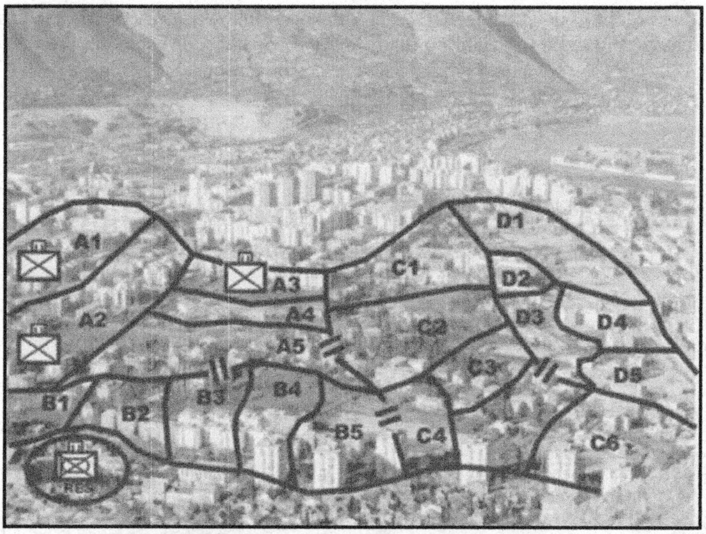

Figure 4-10. Search and attack technique.

TECHNIQUE	ADVANTAGES	DISADVANTAGES
Search and Attack	Requires enemy to fight in multiple directions. Increases maneuver space and flexibility.	Difficult to command and control. Difficult to provide CS and CSS. Difficult to provide for mutual support of maneuver forces. Find/fix/finish forces are challenged/limited.

Table 4-1. Advantages and disadvantages of search and attack.

b. **Attack on a Single Axis.** If the brigade must mass combat power in order to conduct a deliberate attack against an enemy strongpoint, an attack on a single axis may be considered. This technique would be used when the axis of advance is not well defended by the enemy. Figure 4-11 depicts a brigade conducting an attack on a single axis on OBJ GOLD. In the example shown, the lead task force (TF) has the mission of conducting a supporting attack to seize OBJ 22 and facilitate passage of the second the TF through OBJ 22. The second TF conducts the main attack to seize and clear OBJ 21 with an on order mission to seize OBJ 23. A third TF follows in reserve. In the example shown below, the brigade would normally receive assistance in isolating the objective. (Table 4-2 shows the advantages and disadvantages of an attack on a single axis.)

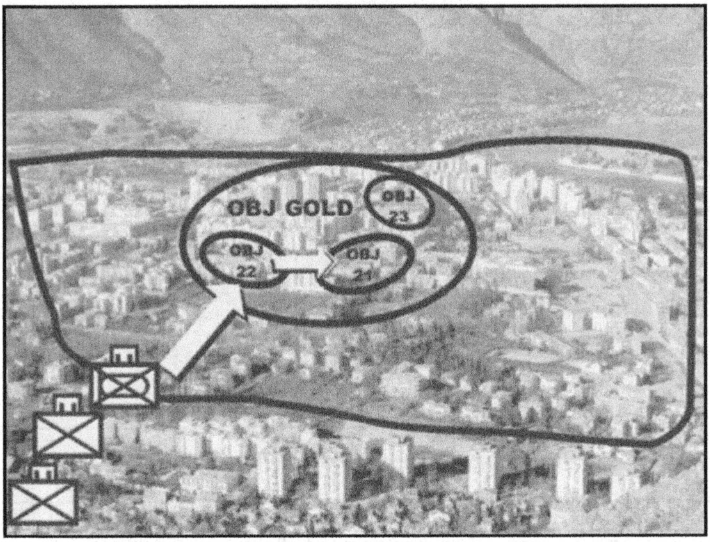

Figure 4-11. Attack on a single axis.

TECHNIQUE	ADVANTAGES	DISADVANTAGES
Attack on a Single Axis	Facilitates command and control. Limited combat power to the front. Concentrates combat power at a critical point.	Limits manuever. Presents denser target to the enemy. Presents a single threat to the enemy. Reduces flexibility.

Table 4-2. Advantages and disadvantages of an attack on a single axis.

c. **Attack on Multiple Axes.** If enemy defenses are more robust and the brigade commander wishes to force the enemy to fight in multiple directions, an attack on multiple axes can be considered.

(1) Figure 4-12 depicts a brigade conducting the same attack on OBJ GOLD using multiple axes. In this case, a battalion TF (air assault) conducts an air assault on OBJ C and then conducts a supporting attack to seize OBJ D. A second TF conducts a supporting attack to seize OBJ B, with a third conducting the main attack to seize and clear OBJ A. The supporting attacks isolate OBJ A. (Table 4-3 shows the advantages and disadvantages of an attack on multiple axes.) Synchronization of BOS is crucial to ensure the massing of effects at the critical points and to prevent the isolation and piecemeal destruction of smaller elements separated by the structures in the urban area.

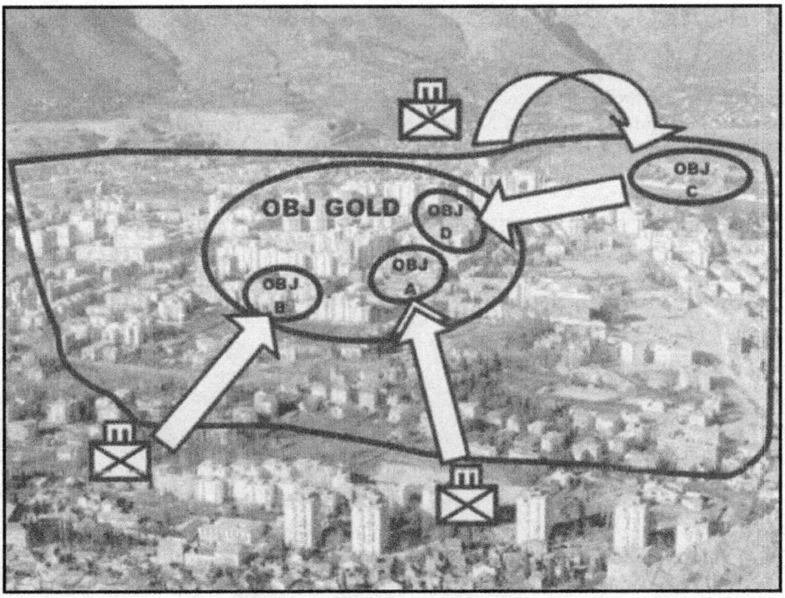

Figure 4-12. Attack on multiple axes.

TECHNIQUE	ADVANTAGES	DISADVANTAGES
Attack on Multiple Axes	Better distributes combat power. Requires the enemy to fight in multiple directions. Increases maneuver space and flexibility.	More difficult to command and control. More difficult to provide CS and CSS.

Table 4-3. Advantages and disadvantages of an attack on multiple axes.

(2) Figure 4-13 depicts an attack on multiple axes on different terrain. In this situation the brigade has the mission to seize OBJ ZULU (OBJs DOG, RAT, and CAT). The brigade commander has decided to attack on multiple axes with two battalion task forces conducting supporting attacks to seize OBJs DOG and RAT in order to isolate OBJ CAT. The brigade main attack seizes and clears OBJ CAT.

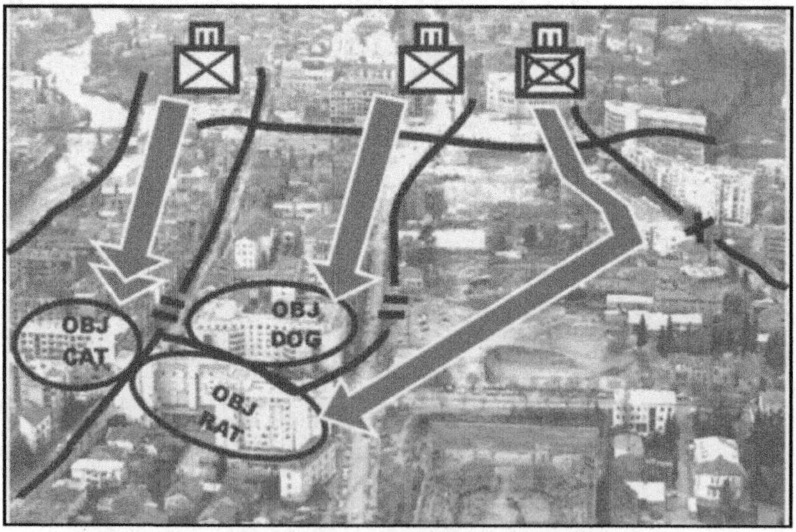

Figure 4-13. Attack on multiple axes, different terrain.

d. **Cordon and Attack.** The brigade may find itself in a position where it may physically isolate a large portion of an urban objective. The brigade commander may also determine that he can force the enemy out of his positions and out into more open areas where he can be engaged by direct and indirect fires. In this case, the cordon and attack technique may be considered. A cordon is a type of isolation. Cordon is a tactical task given to a unit to prevent withdrawal from or reinforcement of a position. Cordon implies seizing or controlling key terrain and or mounted and dismounted avenues of approach. Figure 4-14 depicts a brigade attacking to seize and clear OBJ EAGLE using the cordon and attack technique. One task force (four company teams) cordons OBJ EAGLE by occupying battle positions. (A cordon may also be accomplished using ambushes, roadblocks, checkpoints, OPs, and patrols.) The example in Figure 4-14 shows one TF seizing and clearing OBJ EAGLE and another as the brigade reserve. Skillful application of fires and other combat multipliers may also defeat the enemy when this technique is used and minimize or preclude close combat. (Table 4-4 lists the advantages and disadvantages of cordon and attack.

Figure 4-14. Cordon and attack.

NOTE: In the example shown in Figure 4-14, the battle positions are oriented to place fires on the enemy leaving OBJ EAGLE and to prevent his withdrawal from the objective area. The factors of METT-TC determine how the battle positions are oriented and what the mission end-state will be. Additional direct fire control measures, such as TRPs and engagement areas, as well as indirect fire control measures can focus fires and assist in canalizing the enemy into desired areas.

TECHNIQUE	ADVANTAGES	DISADVANTAGES
Cordon and Attack	Concentrates combat power. Provides mutual support of maneuver forces.	Sequencing the cordon can be difficult. Considerable combat power can be committed to the cordon.

Table 4-4. Advantages and disadvantages of cordon and attack.

e. **Fix and Bypass.** A brigade may find itself in a position where it is conducting operations near an urban area that needs to be bypassed. In certain situations the enemy may have to be fixed prior to the brigade's bypassing the urban area. Figure 4-15 depicts a brigade conducting a limited offensive action to fix the enemy with a small force and bypass the urban area with the bulk of the brigade's combat power. If entering the urban area is unavoidable or force protection requirements force the brigade to attack the urban area, the fix and bypass technique may be considered. (Table 4-5 on page 4-28 lists the advantages and disadvantages of fix and bypass.) It is preferable to completely avoid the urban area if it is eventually bypassed. During the planning process, routes are chosen so that close combat in the urban area can be avoided. Also, the brigade may be able to fix the enemy with fires and avoid having to enter the urban area.

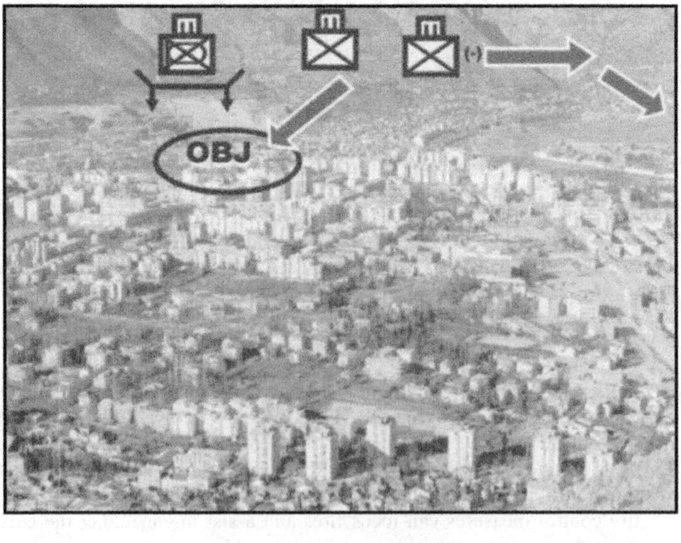

Figure 4-15. Fix and bypass.

TECHNIQUE	ADVANTAGES	DISADVANTAGES
Fix and Bypass	Avoids urban area. Facilitates freedom of action.	Requires the brigade to separate, commit, and support part of its force. Fixing force may become isolated and cut off.

Table 4-5. Advantages and disadvantages of fix and bypass.

f. **Multiple Nodal Attacks.** The brigade may be given the mission to attack multiple nodes either simultaneously or sequentially. This mission is characterized by rapid attacks followed by defensive operations. The enemy situation must permit the brigade to divide its forces and seize key nodes. Multiple attacks such as this require precise maneuver and supporting fires. This mission may be given to a brigade before an anticipated stability operation, or to isolate an urban area for other units that are going to conduct offensive operations inside the urban area. Figure 4-16 depicts a brigade conducting multiple nodal attacks. This technique is used to deny the enemy the use of key infrastructure. Use of this technique may also require designated rapid response elements in reserve in the event that enemy forces mass and quickly overwhelm an attacking battalion. The duration of this attack should not exceed the brigade's self-sustainment capability. (Table 4-6 lists the advantages and disadvantages of multiple nodal attacks.)

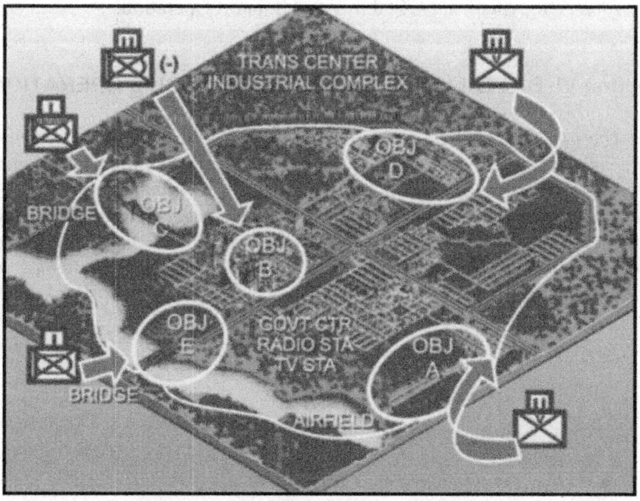

Figure 4-16. Multiple nodal attacks.

TECHNIQUE	ADVANTAGES	DISADVANTAGES
Multiple Nodal Attacks	Presents multiple threats to the enemy. Increases maneuver space and flexibility.	Difficult to command and control. Difficult to provide CS and CSS. Difficult to provide for mutual support of maneuver forces. Difficult to sequence.

Table 4-6. Advantages and disadvantages of multiple nodal attacks.

4-25. TRANSITION

During transition, the brigade continues to use all CS and CSS assets consistent with the mission end-state and ROE to move from offensive operations to stability and or support operations in order to return the urban area back to civilian control. During this step, the roles and use of SOF, CSS, and CS units, such as civil affairs (CA), PSYOP, medical, and MPs become more important with the requirements to maintain order and stabilize the urban area. Subordinate task forces and other brigade units consolidate, reorganize, conduct area protection and logistical missions, and prepare for follow-on missions. The brigade staff prepares to transition from being a *supported* force to being the *supporting* force. (See Chapter 14, Stability and Support Operations.)

Section VI. BATTALION TASK FORCE OFFENSIVE OPERATIONS

The battalion plan of action was as follows: one platoon of Company "F," with a light machine gun section, would stage the initial diversionary attack. It would be supported by two tanks and two tank destroyers, who were instructed to shoot at all or any suspected targets. Observation posts had been manned on a slag pile to support the advance with 81-mm mortar fire...The platoon action was to be the first step...to reduce the town of Aachen.

...the remainder of our zone of action...would be cleared by Companies "F" and "G," who would execute a flanking attack, jumping off abreast of each other through the area secured by the Company "F" platoon...Preparatory fire by medium artillery was to be planned...Mortar observers would accompany each company...Tanks and tank destroyers were assigned to each company...

LTC Darrel M. Daniel
Commander, 2nd Bn, 26th In Rgt
October, 1944, Battle of Aachen

This section discusses tactics, techniques, and procedures (TTP) and considerations that battalion task forces can employ to conduct independent UO or to conduct operations as part of larger brigade UO. The TTP described in this section can apply to all types of battalion task forces, with modifications made for the assets available.

4-26. TASK ORGANIZATION

As with brigade UO, battalion task forces (TF) may require unique task organizations. For example, UO provide one of the few situations where Infantry and armor elements may be effectively task-organized below platoon levels. Battalion commanders must consider providing assets where they are needed to accomplish specific tasks. All phases of mission execution must be considered when developing task organization. Changes in task organization may be required to accomplish different tasks during mission execution. Figure 4-17 depicts a sample task organization for a light Infantry TF conducting an offensive UO that consists of a main effort, two supporting efforts, and a reserve.

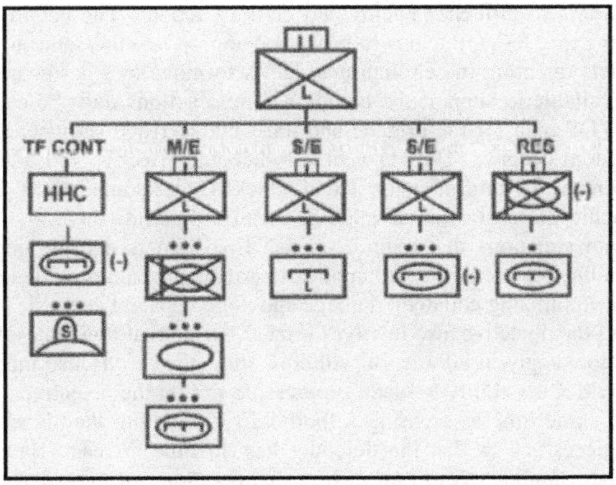

Figure 4-17. Sample offensive task organization.

NOTE: The task organization shown may change after the assault when the TF reorganizes for follow-on missions.

4-27. DELIBERATE ATTACK

Because companies or company teams may become isolated during the attack, the TF commander should attach some support elements to ensure the success of his plan. Armored vehicles (tanks, BFVs, self-propelled artillery) attached to light units must have their own logistics packages. Tanks and BFVs can be used to clear or isolate hardened targets protected by buildings or rubble. Engineers can neutralize obstacles hindering the attack. The TF commander plans to conduct a deliberate attack by performing the following actions.

a. **Reconnoiter the Objective.** This method involves making a physical reconnaissance of the objective with battalion assets and those of higher headquarters, as the tactical situation permits. It also involves making a map reconnaissance of the objective and all the terrain that affects the mission, as well as the analysis of aerial imagery, photographs, or any other detailed information about the buildings or other urban terrain the battalion is responsible for. Additionally, any human intelligence (HUMINT) collected by reconnaissance and surveillance units, such as the battalion reconnaissance platoon, snipers, and so forth, should be considered during the planning process.

b. **Move to the Objective.** This method may involve moving through open and or urban terrain. Movement should be made as rapidly as possible without sacrificing security. Movement should be made along covered and concealed routes and can involve moving through buildings, down streets, in subsurface areas, or a combination of all three. Urban movement must take into account the three-dimensional aspect of the urban area.

c. **Isolate the Objective.** Isolation begins with the efforts of SOF units controlled by higher headquarters to influence enemy and civilian actions. The battalion commander should consider using PSYOP teams to broadcast appropriate messages to the threat and to deliver leaflets directing the civilian population to move to a designated safe area, if the units are available to support the battalion. These actions must be coordinated with the overall PSYOP plan for the brigade and must not sacrifice surprise. By themselves, PSYOP are seldom decisive. They take time to become effective and often their effects are difficult to measure until after the actual attack. Under some METT-TC conditions, PSYOP have achieved results far outweighing the effort put into them.

(1) In certain situations that require precise fire, snipers can provide an excellent method of isolating key areas. Skillful application of snipers can provide lethal fire while simultaneously minimizing collateral damage and noncombatant casualties.

(2) Isolating the objective also involves seizing terrain that dominates the area so that the enemy cannot supply, reinforce, or withdraw its defenders. It also includes selecting terrain that provides the ability to place suppressive fire on the objective. (This step may be taken at the same time as securing a foothold.) If isolating the objective is the first step, speed is necessary so that the defender has no time to react. Battalions may be required to isolate an objective as part of brigade operations, or may be required to do so independently (Figure 4-18, page 4-32). Depending on the tactical situation, companies within the battalion may isolate an objective by infiltration and stealth.

(3) Cordon is a tactical task given to a unit to prevent withdrawal from or reinforcement of a position. A cordon is a type of isolation. It implies seizing or

controlling key terrain and or mounted and dismounted avenues of approach. Figure 4-18 depicts a brigade attacking to seize and clear OBJ EAGLE using the cordon and attack technique. One battalion TF (four company teams) cordons (isolates) OBJ EAGLE by occupying battle positions. (A cordon may also be accomplished through use of ambushes, roadblocks, checkpoints, OPs, and patrols.) Skillful application of fires and other combat multipliers may also defeat the enemy when this technique is used and minimize or preclude close combat. In the example shown in Figure 4-18, the battle positions are oriented to place fires on the enemy leaving OBJ EAGLE and to prevent his withdrawal from the objective area. The factors of METT-TC determine how the battle positions are oriented and what the mission end-state will be. Additional direct fire control measures, such as TRPs and engagement areas, as well as indirect fire control measures, can focus fires and assist in canalizing the enemy into desired areas.

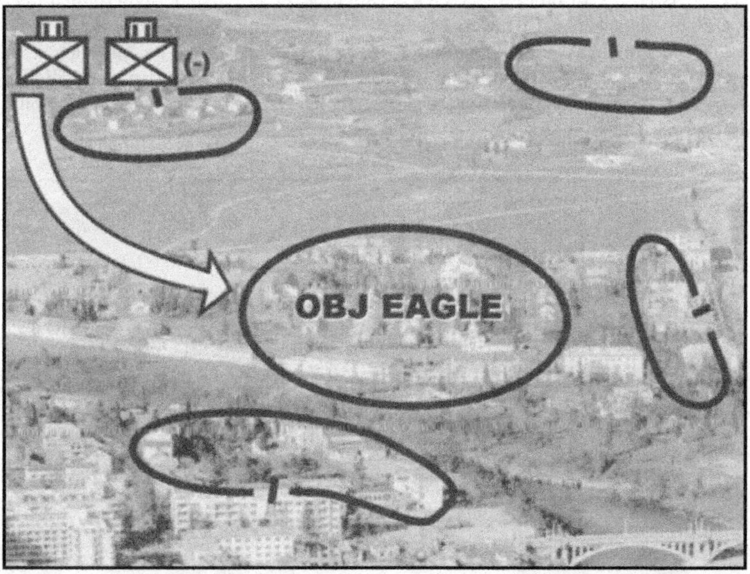

Figure 4-18. Isolation of an urban area by an Infantry battalion using the cordon technique.

NOTE: Combat experience and recent rotations at the CTCs have shown that many casualties can be sustained when moving between buildings, down streets, and through open areas to enter a building either to gain a foothold or to clear it. One purpose of isolation at the company and battalion levels must be to dominate the area leading to the points of entry to protect assaulting troops entering the building from effective enemy fire. This technique is accomplished by using direct and indirect fires and obscurants, maintaining situational awareness, and exercising tactical patience prior to movement.

d. **Secure a Foothold.** Securing a foothold involves seizing an intermediate objective that provides cover from enemy fire and a location for attacking troops to enter the urban area. The size of the foothold is METT-TC dependent and is usually a company intermediate objective. In some cases a large building may be assigned as a company intermediate objective (foothold).

(1) As a company attacks to gain a foothold, it should be supported by suppressive fire and smoke. In the example shown in Figure 4-19, the center TF conducts a supporting attack to seize OBJ DOG. (In the brigade scheme of maneuver, the TF on the left conducts the main attack to seize and clear OBJ CAT, and the TF on the right conducts a supporting attack to seize OBJ RAT. The seizure of OBJs RAT and DOG isolates OBJ CAT). In order to seize OBJ DOG the TF commander determined that two intermediate objectives were necessary.

(2) One company secures a foothold in OBJ Y. As a follow-on mission, the same company seizes OBJ Z and supports the battalion main effort by fire, or facilitates the passage of another company through OBJ Y to seize OBJ Z to support the battalion main effort by fire.

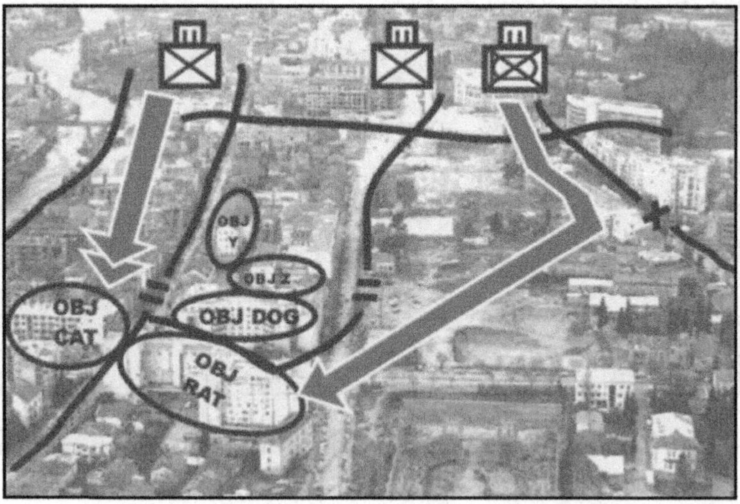

Figure 4-19. Securing a foothold, battalion attack.

e. **Clear an Urban Area.** Before determining to what extent the urban area must be cleared, the factors of METT-TC must be considered. The ROE affect the TTP subordinate units select to move through the urban area and clear individual buildings and rooms. The commander may decide to clear only those parts necessary for the success of his mission if—

- An objective must be seized quickly.
- Enemy resistance is light or fragmented.
- The buildings in the area have large open areas between them. In this case, the commander would clear only those buildings along the approach to his objective, or only those buildings necessary for security.

An Infantry battalion may have a mission to systematically clear an area of all enemy. Through detailed analysis, the commander may anticipate that he will be opposed by strong, organized resistance or will be in areas having strongly constructed buildings close together. Companies may be assigned their own AO within the battalion sector in order to conduct systematic clearing (Figure 4-20).

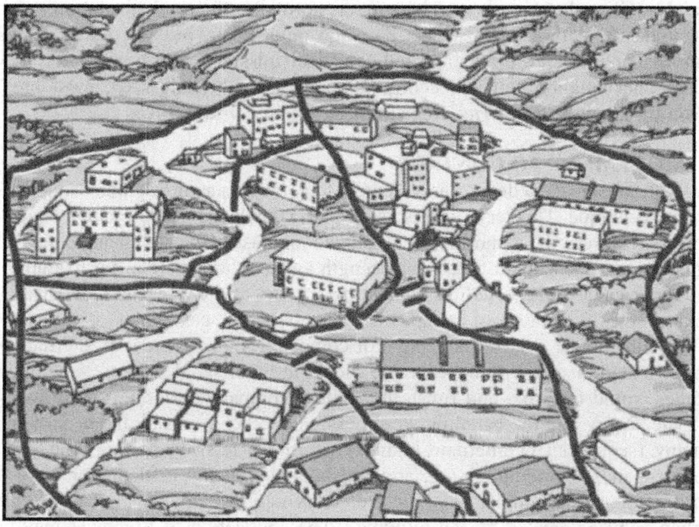

Figure 4-20. Systematic clearance within assigned areas.

f. **Consolidate/Reorganize and Prepare for Future Missions.** Consolidation occurs immediately after each action. Reorganization and preparation for future missions occurs after consolidation. Many of these actions occur simultaneously.

(1) Consolidation provides security and facilitates reorganization, and allows the battalion to prepare for counterattack. Rapid consolidation after an engagement is extremely important in an urban environment. The assault force in a cleared building must be quick to consolidate in order to repel enemy counterattacks and to prevent the enemy from infiltrating back into the cleared building. After securing a floor, selected members of the assault force are assigned to cover potential enemy counterattack routes to the building. Priority must be given to securing the direction of attack first.

(2) Reorganization actions (many occurring simultaneously) prepare the unit to continue the mission. The battalion prepares to continue the attack, prepares for future missions, and prepares for the possible transition to stability and support operations.

NOTE: Friendly force situational awareness is significantly improved in digitally equipped units through the use of Force XXI Battalion Command Brigade and Below (FBCB2) assets.

g. **Transition.** During transition, the battalion continues to use all CS and CSS assets consistent with the mission end-state and ROE to move from offensive operations to stability and or support operations in order to return the urban area to civilian control. During this step, the roles and use of SOF, CS, and CSS units, such as civil affairs (CA), PSYOP, medical, and MPs, become more important with the requirements to maintain order and stabilize the urban area. These assets normally support the battalion's transition efforts under brigade control. The battalion and other brigade units consolidate, reorganize, conduct area protection and logistical missions, and prepare for follow-on missions. The battalion staff, in coordination with the brigade staff, must prepare to transition from being a *supported* force to being the *supporting* force.

4-28. MOVEMENT TO CONTACT

Figure 4-21 on page 4-36 depicts a movement to contact in an urban area using the search and attack technique. This technique is used when knowledge of the enemy is unclear and contact is required. It is normally employed against a weak enemy force that is disorganized and incapable of massing strength against the battalion; for example, urban insurgents or gangs. The battalion divides its portion of the AO into smaller areas and coordinates the movement of companies. The battalion can either assign sectors to specific companies or control movement of companies by sequential or alternate bounds within the battalion sector. In the example shown in Figure 4-21 on page 4-36, individual companies would find, fix, and finish the enemy (company sectors), or they would find and fix the enemy and the battalion would assign another company the task of finishing the enemy (sequential or alternate bounds). During a mission of this type, the urban environment makes finding, fixing, and finishing the enemy difficult for conventional Infantry forces. For example, movement of units may become canalized due to streets and urban *canyons* created by tall buildings. The application of firepower may become highly restricted based on the ROE. The use of HUMINT in this type of action becomes increasingly more important and can be of great assistance during the *find* portion of the mission.

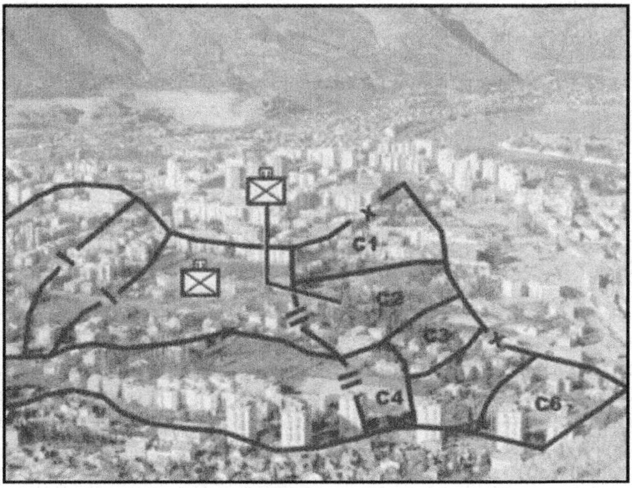

Figure 4-21. Search and attack technique.

4-29. INFILTRATION

The following example describes the actions of an Infantry battalion conducting an infiltration. With some modification, it could also apply to a dismounted mechanized Infantry battalion.

a. The outskirts of an urban area may not be strongly defended. Its defenders may have only a series of antiarmor positions, security elements on the principal approach, or positions blocking the approaches to key features in the town. The strongpoints and reserves are deeper in the urban area.

b. A battalion may be able to seize a part of the urban area by infiltrating platoons and companies between those enemy positions on the outskirts. Moving by stealth on secondary streets by using the cover and concealment of back alleys and buildings, the battalion may be able to seize key street junctions or terrain features, to isolate enemy positions, and to help following units pass into the urban area. Such an infiltration should be performed when visibility is poor and no civilians are in the area. Bypassing enemy strongpoints may result in flank and rear security problems for the infiltrating battalion. Bypassed units may become a counterattack force or cut lines of communications, if not isolated. Planning should include securing all mounted and dismounted avenues of approach from the bypassed enemy strongpoints to ensure their isolation.

c. The Infantry battalion is organized into infiltration companies with appropriate attachments and a reserve consistent with METT-TC. Each company should have an infiltration lane that allows stealthy infiltration by company or smaller size units. Depending on the construction of the urban area and streets, the infiltration lane may be 500 to 1,500 meters wide.

d. The infiltrating companies advance stealthily on foot using available cover and concealment. Mortar and artillery fire can be used to divert the enemy's attention and cover the sound of infiltrating troops.

e. Armored vehicles and antiarmor weapons are positioned to cover likely avenues of approach for enemy armored vehicles. The battalion commander may position antiarmor weapons to cover the likely avenues of approach, if no BFVs or tanks are available. The reconnaissance platoon and antiarmor company screen the battalion's more vulnerable flanks. Also, the antiarmor company can support by fire if the situation provides adequate support by fire positions.

f. As the companies move into the urban area, they secure their own flanks. Security elements may be dropped off along the route to warn of a flank attack. Engineers assist in breaching or bypassing minefields or obstacles encountered. Enemy positions are avoided but reported.

g. The infiltrating companies proceed until they reach their objective. At that time, they consolidate and reorganize and arrange for mutual support. They patrol to their front and flanks, and establish contact with each other. The company commander may establish a limit of advance to reduce chances of enemy contact or to ensure safety from friendly forces.

h. If the infiltration places the enemy in an untenable position and he must withdraw, the rest of the battalion is brought forward for the next phase of the operation. If the enemy does not withdraw, the battalion must clear the urban area before the next phase of the operation (Figure 4-22).

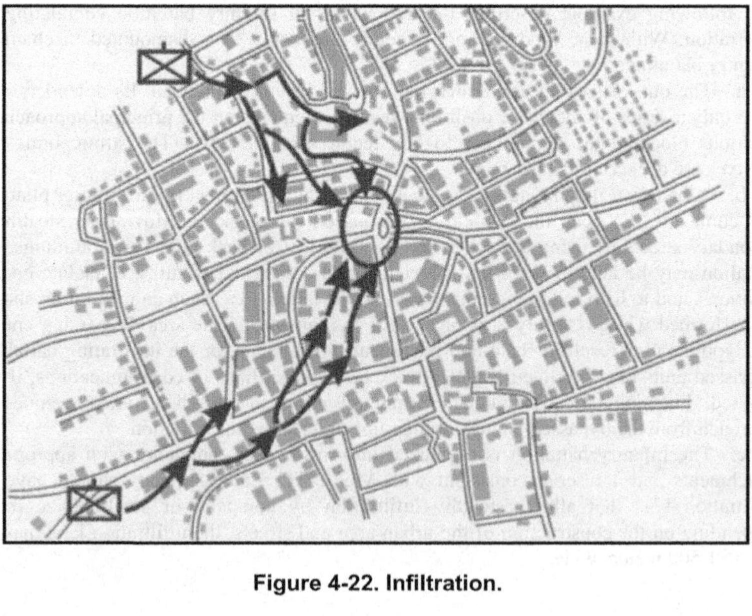

Figure 4-22. Infiltration.

4-30. ATTACK OF A VILLAGE

The battalion may have to conduct either a hasty or deliberate attack of a village that is partially or completely surrounded by open terrain. (Figure 4-23 depicts a TF conducting such an attack.) After the factors of METT-TC have been considered, the tactical tasks discussed in paragraph 4-27 are performed (specifically, reconnoiter the objective, move to the objective, isolate the objective, secure a foothold, clear the objective, and consolidate and reorganize and or prepare for future missions). In the example shown in Figure 4-23, two companies and or company teams isolate the village, and a company team secures a foothold and enters and clears the village.

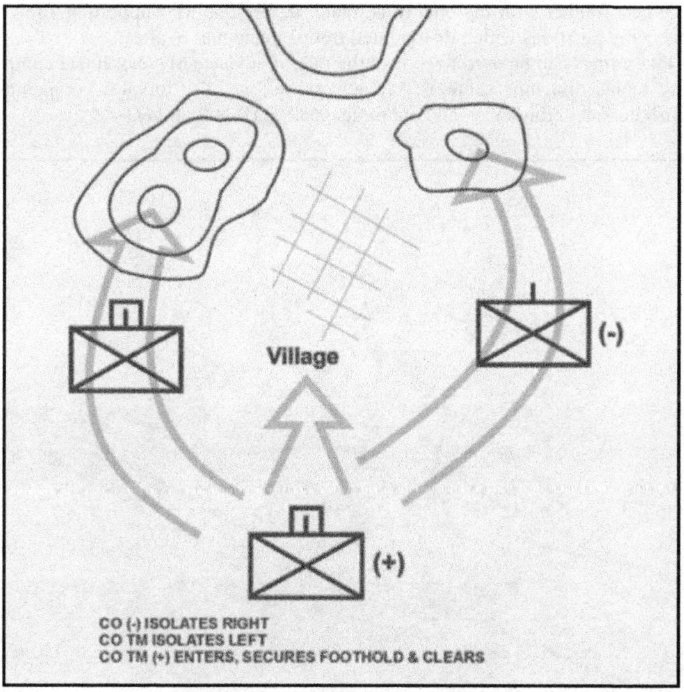

Figure 4-23. Attack of a village.

4-31. ROUTE SECURITY AND CLEARANCE

An Infantry battalion may have to clear buildings to secure a route through a city. How quickly the battalion can clear the buildings depends on enemy resistance and the size and number of the buildings. The battalion deploys companies/company teams IAW with METT-TC factors. Figure 4-24 shows three companies abreast clearing routes in sector. The enemy situation must permit the battalion to deploy its subordinate units.

This mission would not normally be executed against well prepared enemy defenses in depth. In outlying areas, the forward units proceed by bounds from road junction to road junction. Other platoons provide flank security by moving down parallel streets and by probing to the flanks.

a. Depending on the required speed and enemy situation, the Infantry may either move mounted or dismounted. The platoons move down the widest streets, avoiding narrow streets. Each BFV section overwatches the squad to its front, keeping watch on the opposite side of the street. Sections provide their wingman with mutual support. Combat vehicles providing overwatch should be secured by dismounted troops. The rest of the Infantry should stay mounted to maximize speed and shock effect until required to dismount due to the enemy situation or upon reaching the objective.

b. When contact with the enemy is made, tanks support. Supporting fire fixes and isolates enemy positions which dismounted troops maneuver to attack.

c. Phase lines can be used to control the rate of advance of subordinate companies or company teams and other action. At each phase line, the forward companies might reestablish contact, reorganize, and continue clearing (Figure 4-24).

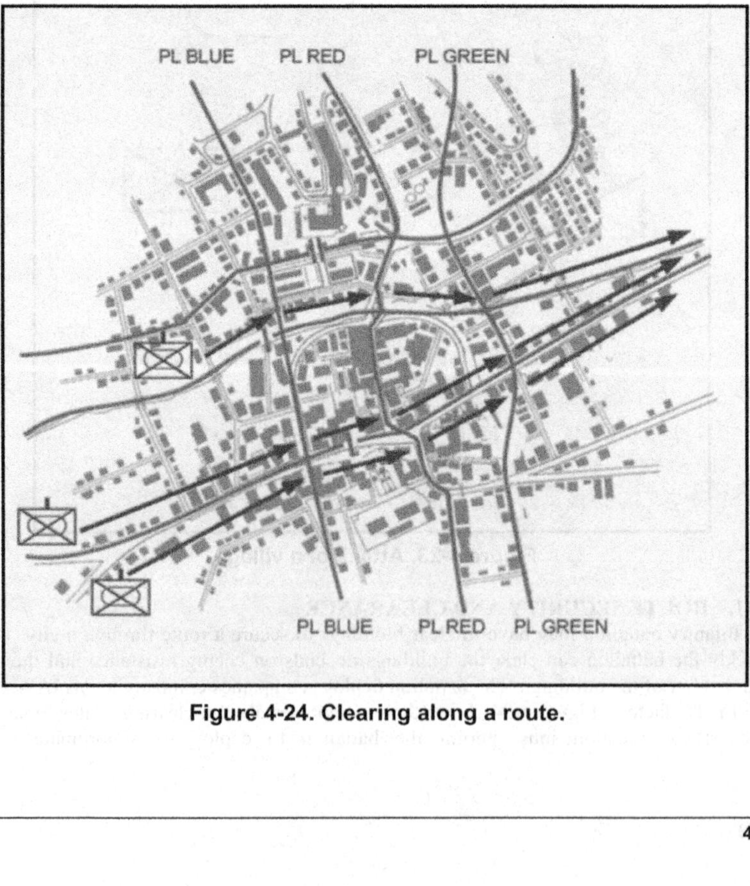

Figure 4-24. Clearing along a route.

4-32. NODAL ATTACK

The battalion may be given the mission to seize key nodes as part of a brigade operation. (See Figures 4-25 and 4-26.) In certain situations, the battalion may be required to seize nodes independently. This mission is characterized by rapid attacks followed by defensive operations. The enemy situation must permit the attacking force to divide its forces and seize key nodes. Multiple attacks, as depicted in Figures 4-25 and 4-26, require precise maneuver and supporting fires. This mission may be given to a battalion before an anticipated stability and or support operation, or to isolate an urban area for other units that will be conducting offensive operations inside the urban area. Figure 4-25 depicts a brigade conducting multiple nodal attacks. Figure 4-26 depicts a battalion TF executing its assigned mission. This technique is used to deny the enemy key infrastructure. Use of this technique may also require designated rapid response elements in reserve in the event that enemy forces mass and quickly overwhelm an attacking battalion. Normally the reserve is planned at brigade level. Battalions executing a nodal attack independently needs to plan for a designated rapid response reserve element. The duration of this attack should not exceed the battalion's self-sustainment capability.

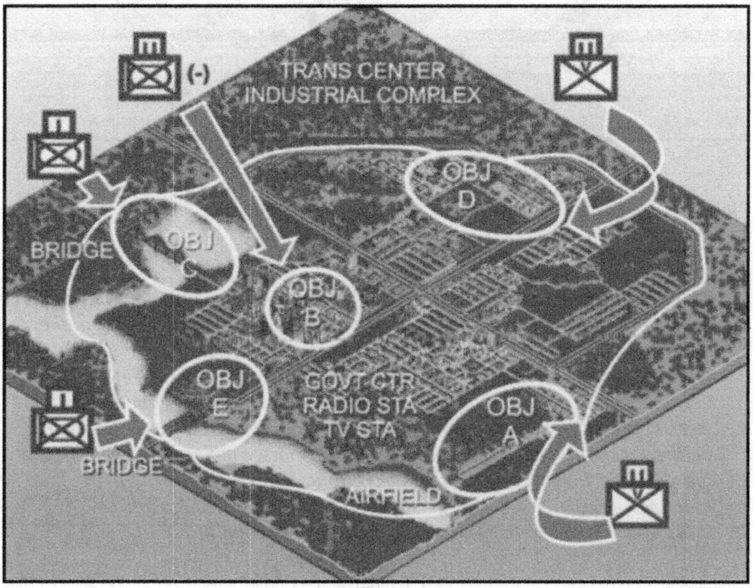

Figure 4-25. Brigade scheme of maneuver, nodal attack.

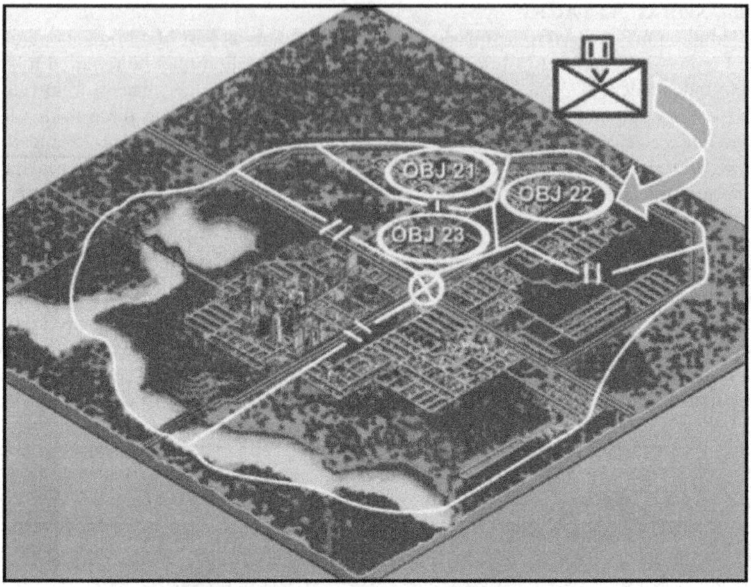

Figure 4-26. Battalion nodal attack.

Section VII. COMPANY TEAM ATTACK OF AN URBAN AREA

"We finally reached the front of the company where the lead APC was stopped and learned that the fire was coming from the large hotel on the left side of the street, about 50 meters to the front of the lead platoon. I guided the MK 19 HMMWV up onto a steep sidewalk so the gunner could get an effective shot and told him to watch my M16 tracer rounds and to work the building from top to bottom. I fired several tracers into the hotel; he fired a spotting round into one of the top story windows and then fired the grenade launcher on automatic, hitting every single window in the building. The effects were devastating. Concrete fragments flew everywhere, and one or two Somalis fell out of the building."

CPT Charles P. Ferry
Mogadishu, October 1993:
Personal Account of a Rifle Company XO
Infantry Magazine, Sep-Oct 94

This section discusses tactics, techniques, and procedures (TTP) and considerations that company teams can employ to conduct independent UO or to conduct operations as part of larger battalion TF UO. The TTP described in this section can apply to all types of company teams, with modifications made for the assets available.

4-33. TASK ORGANIZATION

The company commander normally task-organizes his company into two elements: an assault element and a support element. The support element may be given a number of tasks that are conducted on order or simultaneously; specifically, support by fire, isolate the objective, and conduct other support functions. The tactical situation dictates whether or not separate elements need to be task-organized in order to conduct these support missions. The mission to breach is METT-TC dependent and may be given to the assault or support element; or a separate element may be formed to conduct this task. If available, engineers are usually task-organized into the element that performs the breach. The size and composition of the elements are determined by METT-TC. If the company is part of a battalion operation, the company could be given the mission to conduct one or more of the tasks mentioned above. If conducting an urban attack independently, the company team performs both assault and support tasks.

a. **Assault Element.** The purpose of the assault element is to kill, capture, or force the withdrawal of the enemy from an urban objective. The assault element of a company team may consist of one or more platoons usually reinforced with engineers, BFVs, and possibly tanks. Building and room clearing are conducted at the platoon and squad level. The assault element must be prepared to breach to gain entry into buildings.

b. **Support Element.** The purpose of the support element is to provide any support that may be required by the assault element. The support element at company level normally consists of the company's organic assets (platoons, mortars, and antitank weapons), attachments, and units that are under the OPCON of the company commander. This assistance includes, but is not limited to, the following:

- Suppressing and obscuring enemy within the objective buildings and adjacent structures.
- Isolating the objective buildings with observation and direct or indirect fires to prevent enemy withdrawal, reinforcement, or counterattack.
- Breaching walls en route to and in the objective structure.
- Destroying or suppressing enemy positions with direct fire weapons.
- Securing cleared portions of the objective.
- Providing squads to assume assault element missions.
- Providing resupply of ammunition, explosives, and personnel.
- Evacuating casualties, EPWs, and noncombatants.

c. **Reserves.** Companies fighting in urban terrain may not be able to designate a reserve, based on the number of troops required to conduct offensive operations. A platoon may be detached from the company to form a battalion reserve. The company reserve, if one is designated, should be mobile and prepared for commitment. Because of the available cover in urban areas, the reserve can stay close to forward units. The reserve normally follows within the same block so that it can immediately influence the attack. The size of the reserve is METT-TC dependent, but at company level, the reserve normally consists of a squad, detached from an organic platoon, or attached elements. In addition to the tasks discussed in FM 7-10, Chapter 4, the reserve may be called upon to perform one or more of the following tasks based on the commander's priority of commitment:

- Assuming the mission of the assault element.
- Clearing bypassed enemy positions.
- Moving behind the assault element to provide security in cleared buildings, allowing the assault element to continue to move.

d. **Breaching Element.** At the company level, breaching is normally conducted by the assault element. However, a separate breaching element may be created and a platoon may be given this mission and task organized accordingly. The purpose of breaching is to provide the assault element with access to an urban objective. Breaching can be accomplished using explosive, ballistic, thermal, or mechanical methods. Ballistic breaching includes using direct fire weapons; mechanical breaching includes the use of crowbars, axes, saws, sledgehammers, or other mechanical entry devices. Thermal breaching is accomplished through the use of a torch to cut metal items such as door hinges. Attached engineers, or a member of the assault element who has had additional training in mechanical, thermal, ballistic, and explosive breaching techniques, may conduct the breach.

e. **Sample Task Organizations.** Task organization of the company varies based on the factors of METT-TC and the ROE.

(1) ***Light Infantry Task Organization.*** An Infantry company conducting this mission might task-organize as follows:

Assault Two rifle platoons and one rifle platoon(-) reinforced with engineers (attached to the platoons).
Reserve A squad from one of the platoons.
Support The company AT weapons, 60-mm mortar section, and M240 machine guns. (Other support provided by the battalion task force.)

(2) ***Light/Heavy Task Organizations.*** Different METT-TC factors might produce the following light/heavy task organizations:

Example 1:

Assault Two rifle platoons, each reinforced with engineers.
Reserve One rifle platoon.
Support BFV platoon and the company AT weapons and 60-mm mortar section. (Other support provided by the battalion task force.)

Example 2:

Assault Two rifle platoons reinforced with engineers.
Reserve One rifle platoon.
Support One tank platoon. The company AT weapons and 60-mm mortar section.

Example 3:

Assault Two rifle platoons, each with engineers. One tank section OPCON to an Infantry platoon.

Reserve One rifle platoon.

Support A tank section and the company AT weapons under the tank platoon leader's control. The company 60-mm mortar section. (All available direct and indirect fire weapons should be used to isolate objective buildings. Direct fire down streets and indirect fire in open areas between buildings to help in the objective isolation.)

NOTE: The company commander may use the company executive officer, tank platoon leader, BFV platoon leader, or first sergeant to control the support element, as the task organization and situation dictate. Based on METT-TC factors, a BFV platoon can perform any of the missions described above (assault, support, reserve). Unit integrity should be maintained at the platoon level. If the tactical situation requires the employment of sections, it should be for a limited duration and distance.

4-34. DELIBERATE ATTACK

At the company level, a deliberate attack of an urban area usually involves the sequential execution of the tactical tasks below.

a. **Reconnoiter the Objective.** This method involves making a physical reconnaissance of the objective with company assets and those of higher headquarters, as the tactical situation permits. It also involves a map reconnaissance of the objective and all the terrain that affects the mission, to include the analysis of aerial imagery, photographs, or any other detailed information about the building or other urban terrain, which the company is responsible for. Additionally, any human intelligence (HUMINT) collected by reconnaissance and surveillance units, such as the battalion reconnaissance platoon, snipers, and so forth, should be considered during the planning process.

b. **Move to the Objective.** This method may involve moving the company tactically through open and or urban terrain. Movement should be made as rapidly as possible without sacrificing security. Movement should be made along covered and concealed routes and can involve moving through buildings, down streets, subsurface areas, or a combination of all three. Urban movement must take into account the three-dimensional aspect of the urban area.

c. **Isolate the Objective.** Isolating the objective involves seizing terrain that dominates the area so that the enemy cannot supply, reinforce, or withdraw its defenders. It also includes selecting terrain that provides the ability to place suppressive fire on the objective. (This step may be taken at the same time as securing a foothold.) If isolating the objective is the first step, speed is necessary so that the defender has no time to react. Companies may be required to isolate an objective as part of a battalion operation or may be required to do so independently. Depending on the tactical situation, an Infantry company may isolate an objective by infiltration and stealth.

d. **Secure a Foothold.** Securing a foothold involves seizing an intermediate objective that provides cover from enemy fire and a location for attacking troops to enter the urban area. The size of the foothold is METT-TC dependent and is usually a company intermediate objective. In some cases a large building may be assigned as a company intermediate objective (foothold). As the company attacks to gain a foothold, it should be supported by suppressive fire and smoke.

e. **Clear an Urban Area.** Before determining to what extent the urban area must be cleared, the factors of METT-TC must be considered. The ROE influence the TTP platoons and squads select as they move through the urban area and clear individual buildings and rooms.

(1) The commander may decide to clear only those parts necessary for the success of his mission if—

- An objective must be seized quickly.
- Enemy resistance is light or fragmented.
- The buildings in the area have large open areas between them. In this case, the commander would clear only those buildings along the approach to his objective, or only those buildings necessary for security. (See Figure 4-26.)

(2) A company may have a mission to systematically clear an area of all enemy. Through detailed analysis, the commander may anticipate that he will be opposed by a strong, organized resistance or will be in areas having strongly constructed buildings close together. Therefore, one or two platoons may attack on a narrow front against the enemy's weakest sector. They move slowly through the area, clearing systematically from room to room and building to building. The other platoon supports the clearing units and is prepared to assume their mission.

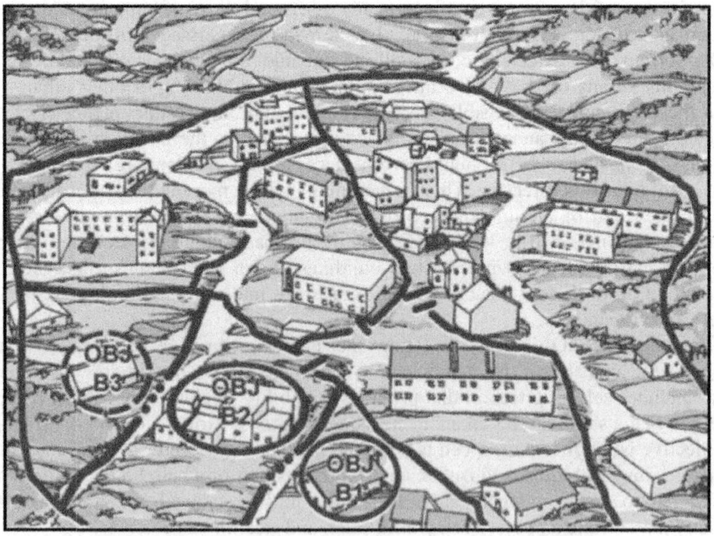

Figure 4-26. Clearing selected buildings within sector.

f. **Consolidate/Reorganize and Prepare for Future Missions.** Consolidation occurs immediately after each action. Consolidation is security and allows the company to prepare for counterattack and to facilitate reorganization. It is extremely important in an urban environment that units consolidate and reorganize rapidly after each engagement. The assault force in a cleared building must be quick to consolidate in order to repel enemy counterattacks and to prevent the enemy from infiltrating back into the cleared building. After securing a floor, selected members of the assault force are assigned to cover potential enemy counterattack routes to the building. Priority must be given to securing the direction of attack first. Those soldiers alert the assault force and place a heavy volume of fire on enemy forces approaching the building. Reorganization occurs after consolidation. Reorganization actions prepare the unit to continue the mission; many actions occur at the same time.

(1) ***Consolidation Actions.*** Platoons assume hasty defensive positions after the objective has been seized or cleared. Based upon their specified and implied tasks, assaulting platoons should be prepared to assume an overwatch mission and support an assault on another building, or another assault within the building. Commanders must ensure that platoons guard enemy mouseholes between adjacent buildings, covered routes to the building, underground routes into the basement, and approaches over adjoining roofs.

(2) ***Reorganization Actions.*** After consolidation, the following actions are taken:

- Resupply and redistribute ammunition, equipment, and other necessary items.
- Mark the building to indicate to friendly forces that the building has been cleared.
- Move support or reserve elements into the objective if tactically sound.
- Redistribute personnel and equipment on adjacent structures.
- Treat and evacuate wounded personnel.
- Treat and evacuate wounded EPW and process remainder of EPW.
- Segregate and safeguard civilians.
- Re-establish the chain of command.
- Redistribute personnel on the objective to support the next phase or mission.

(3) ***Prepare for Future Missions.*** The company commander anticipates and prepares for future missions and prepares the company chain of command for transition to defensive and or stability and support missions.

NOTE: Friendly force situational awareness is significantly improved in digitally equipped units through the use of Force XXI Battle Command Brigade and below (FBCB2) assets.

4-35. ISOLATE AN URBAN OBJECTIVE

Infantry companies isolate an urban objective to prevent reinforcement of, or a counterattack against, the objective and to kill or capture any withdrawing enemy forces. When planning the isolation, commanders must consider three-dimensional and in-depth isolation of the objective (front, flanks, rear, upper stories, rooftops, and subsurface). All available direct and indirect fire weapons, to include attack helicopters and CAS, should be employed, consistent with the ROE. Isolating the objective is a key factor in facilitating the assault and preventing casualties. The company may perform this mission as the support element for a battalion operation, or it may assign the task to its own internal support element for a company attack. In certain situations, companies may be required to isolate an objective or an area for special operations forces or for stability/support operations. When possible, the objective should be isolated using stealth and or rapid movement in order to surprise the enemy. Depending on the tactical situation, companies may use infiltration in order to isolate the objective. Likely tasks include, but are not limited to, the ones described below.

NOTE: Combat experience and recent rotations at the CTCs have shown that many casualties can be sustained when moving between buildings, down streets, and through open areas in order to gain entry into a building either to gain a foothold or to clear it. One of the purposes of isolation at the company level must be to dominate the outside area that leads to the point of entry in order to allow assaulting troops to enter the building without receiving effective fire from the enemy. This method is accomplished by the effective use of direct and indirect fires, obscurants, maintaining situational awareness, and exercising tactical patience prior to movement.

a. **Isolating the Objective (Battalion Attack).** A company may isolate the objective as the support element for a battalion operation. When a company is given this mission, the objective is normally a larger structure, a block, or a group of buildings. The company commander task-organizes his platoons and assigns them support by fire positions based on the factors of METT-TC. In addition to isolating the objective, the company (support element) may be given additional tasks that will be conducted on order or at the same time. Examples of these additional tasks include assuming assault element missions, securing cleared buildings, handling noncombatants and EPWs, and CASEVAC.

b. **Isolating the Objective (Company Attack).** When a company conducts an attack, the task organization and tasks given to the company support element is determined by the factors of METT-TC. If the company conducts an attack, the objective can be a building, a block or group of buildings, a traffic circle, or a small village (Figure 4-27, page 4-48). Emphasis must be placed on suppressing or neutralizing the fires on and around the objective. Figure 4-27 depicts an infantry company with tanks assaulting Buildings (BLDG) 41 and 42. In order to secure a foothold and clear BLDGs 41 and 42, the commander has assigned a platoon to support by fire and suppress the enemy squad in BLDG 11 and the medium machine gun in BLDG 21. A tank section suppresses the light machine gun in BLDG 51 and assists in the suppression of BLDG 11. Another platoon supports by fire and suppresses any enemy fire from BLDGs 31, 41, and 42. The company's third platoon, positioned in buildings behind the support element, acts as the assault element to clear BLDGs 41 and 42. In this manner, three-dimensional isolation of the objective (BLDGs 41 and 42) is accomplished.

NOTE: All buildings within the support element's sector of fire were numbered to facilitate command and control.

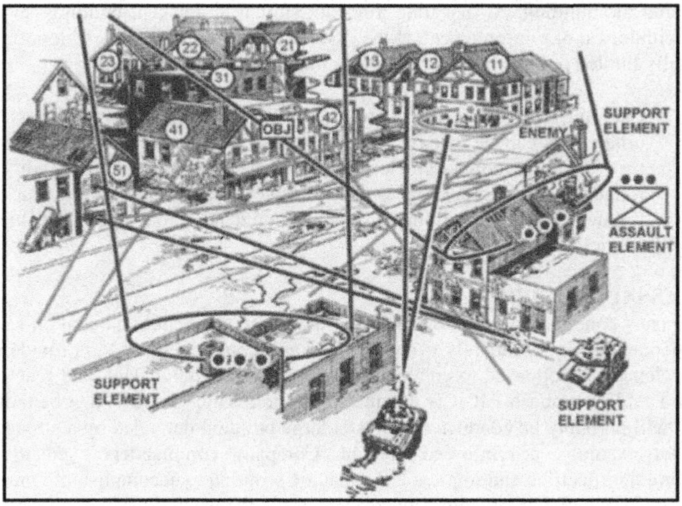

Figure 4-27. Isolating an urban objective.

c. **Tasks.** The company commander isolates the objective with direct and indirect fires before and during the assault element's execution of its mission. The company will—

- Suppress known, likely, and suspected enemy targets, consistent with the ROE, with direct and indirect fire weapons. Under restrictive ROE, suppression may be limited only to actual enemy locations.
- Cover mounted avenues of approach with antiarmor weapons.
- Cover dismounted avenues of approach with automatic weapons.
- Control key terrain near or adjacent to the objective in order to prevent the enemy from reinforcing his positions, withdrawing, or counterattacking.
- Be prepared to move to other locations in order to suppress enemy fires and neutralize enemy positions as the assault element performs its tasks.

(1) Company commanders must give specific instructions to subordinate leaders concerning where to place fires in support of the assault element. For example, from TRP 1 to TRP 2, along the third and second floor windows on the east side of Building 21, shift fires to the west side of the objective from TRP 1 to TRP 4 when the green star cluster is seen, and so on. Once suppressive fires on the objective begin, they normally increase and continue until masked by the advancing assault element. Suppressive fires may or may not be used from the beginning of the assault depending on the ROE. Targets can be marked and identified with tracer rounds; M203 smoke, HE, or illumination rounds; voice and arm-and-hand signals; laser pointers; or similar devices.

(2) The precise well-placed volume of fire, as opposed to a volume of fire, suppresses the enemy. The volume of fire and types of weapons employed is ROE dependent. Once masked, fires are shifted to upper or lower windows and continued until the assault force

has entered the building. At that time, fires are shifted to adjacent buildings to prevent enemy withdrawal or reinforcement. If the ROE are restrictive, the use of supporting fires is normally limited to known enemy locations that have engaged the unit.

NOTE: Care must be taken in urban areas when WP, ILLUM, or tracers are used since urban fires can be caused. Care must also be exercised, if sabot rounds are used by the armored vehicles, based on the its penetration capability. Sabot rounds can penetrate many walls and travel great distances to include passing through multiple buildings, creating unintended damage, casualties, and fratricide.

4-36. ASSAULT A BUILDING

The company conducts this mission as part of the assault element of a battalion task force or independently. (Independently is defined here as a company having to provide its own support element, as opposed to conducting an operation without flank and rear support, such as a raid or ambush.) If it is conducted as the assault element of a battalion task force, it will probably be conducted against a large building defended by a strong enemy force; for example, a reinforced platoon. Company commanders need to clearly understand the specified and implied tasks that are required to accomplish the mission, as well as the brigade/battalion commanders' intent and the desired mission end-state. This procedure allows the company commander to task-organize and issue specific missions to his subordinate elements as to which floors and rooms to clear, seize, or bypass. As an example, Figure 4-28, on page 4-50, depicts an Infantry TF assigned the mission of clearing the objectives in its sector (DOG and TAIL). Company B has been given the TF supporting effort of seizing and clearing OBJ TAIL. The company commander has decided to assign an intermediate objective (WING) to 1st platoon. 3d platoon is the support element with the mission of isolating WING (1st and 2d squads) and providing one squad to act as the company reserve (3d squad). 2d platoon has the mission of passing through 1st platoon, which will mark a passage lane and seize TAIL.

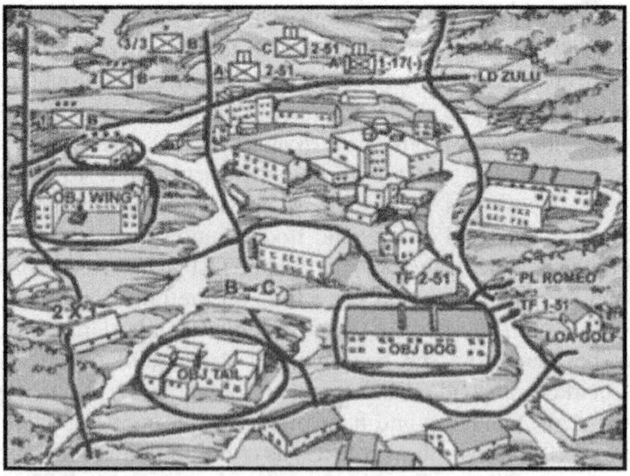

Figure 4-28. Assault of a building.

a. **Execution.** Platoons should move by bounds by floor when clearing a multistory building. This procedure permits troops to rest after a floor has been cleared. It is likely that platoons are required to leave security on floors and in cleared rooms and also facilitate the passage of another platoon in order to continue the assault. The assault element must quickly and violently execute its assault and subsequent clearing operations. Once momentum has been gained, it is maintained to prevent the enemy from organizing a more determined resistance on other floors or in other rooms. If platoons come across rooms/hallways/stairwells that are barricaded with furniture or where obstacles have been placed, they should first attempt to bypass the barricade or obstacle and maintain the momentum of the attack. If they cannot bypass the barricade or obstacle, security should be placed on it, it should be checked for booby traps, and should then be reduced. Also, sealing doors and floors may be an option in order to maintain momentum. Subordinate leaders should continue the momentum of the assault, yet not allow the operation to become disorganized.

b. **Ammunition and Equipment.** METT-TC factors and the ROE determine how the assault element is equipped and armed. The assault element carries only a fighting load of equipment and as much ammunition as possible, especially grenades (fragmentation, smoke, concussion, and stun consistent with the building construction and the ROE). The support element maintains control of additional ammunition and equipment not immediately needed by the assault element. An often-overlooked munition in an urban battle is the light antitank weapon such as the M72 LAW and the AT4. Soldiers can use these for a variety of purposes such as suppressing a manned position or supporting the breaching or assault elements. Resupply should be pushed to the assault element by the support element. Commanders must carefully manage the soldier's load during the assault. Normally, ammunition, water, special assault weapons/equipment, and medical supplies/litters are the only items carried in the assault. Attached or OPCON tank or BFV platoons should also configure their ammunition load to support their mission, consistent with the ROE.

c. **Assault Locations.** The assault may begin from the top or bottom of the building.

(1) ***Top Entry.*** Entry at the top and fighting downward is the preferred method of clearing a building. This method is only feasible, however, when access to an upper floor or rooftop can be gained by ladder; from the windows or roofs of adjoining, secured buildings; or when enemy air defense weapons can be suppressed and troops can be transported to the rooftops by helicopter.

(2) ***Bottom Entry.*** Entry at the bottom is common and may be the only option available. When entering from the bottom, breaching a wall is the preferred method because doors and windows may be booby trapped and covered by fire from inside the structure. If the assault element must enter through a door or window, entry from a rear or flank position is preferred. Under certain situations, the ROE may not permit the use of certain explosives, therefore entry through doors and windows may be the only option available. Armored vehicles can be especially useful in supporting bottom entry.

d. **Breaching.** Squads and platoons will have to conduct breaching. Engineers may be attached to the unit responsible for breaching. Depending on the factors of METT-TC, company commanders may need to designate specific breaching locations or delegate the task to platoon leaders. The ROE also influences whether mechanical, thermal, ballistic, or explosive breaching is used. For example, if BFVs are attached to the company and the

ROE permit their use, they can breach the wall by main-gun fire for the initial-entry point.

e. **Assault Tasks.** Once inside the building, the priority tasks are to cover the staircases and to seize rooms that overlook approaches to the building. These actions are required to isolate enemy forces within the building and to prevent reinforcement from the outside. The assault element clears each room on the entry floor and then proceeds to clear the other floors to include the basement. If entry is not made from the top, consideration may be given to rushing/clearing and securing a stairwell and clearing from the top down, if the tactical situation permits. If stairwell use is required, minimize their use and clear them last. If there is a basement, it should be cleared as soon as possible, preferably at the same time as the ground floor. The procedures for clearing a basement are the same as for any room or floor, but important differences do exist. Basements may contain entrances to tunnels such as sewers and communications cable tunnels. These should be cleared and secured to prevent the enemy from infiltrating back into cleared areas.

DANGER

A SAFETY CONSIDERATION FOR CLEARING BUILDINGS IS THE HIGH PROBABILITY OF RICOCHET.

f. **Suppressive Fires During the Assault.** The support element provides suppressive fire while the assault element is systematically clearing the building. It also provides suppressive fire on adjacent buildings to prevent enemy reinforcements or withdrawal. Suppressive fire may consist of firing at known and suspected enemy locations; or, depending on the ROE, may only include firing at identified targets or returning fire when fired upon. The support element destroys or captures any enemy trying to exit the building. The support element must also deal with civilians displaced by the assault. Armored vehicles are useful in providing heavy, sustained, accurate fire.

g. **Clearing Rooms.** Company commanders must ensure that clearing platoons carry enough room marking equipment and plainly mark cleared rooms from the friendly side IAW unit SOP. (See Appendix I.) Also, if the operation occurs during limited visibility, marking must be visible to friendly units. The support element must understand which markings will be employed and ensure that suppressive fires do not engage cleared rooms and floors. Maintaining situational awareness concerning the location of the assault teams and which rooms/floors have been cleared is imperative and a key command and control function for the company commander. Radios can be consolidated, if necessary, with priority going to the squads and platoons clearing rooms. When exiting cleared buildings friendly troops should notify supporting elements using the radio or other preplanned signals.

4-37. ATTACK OF A BLOCK OR GROUP OF BUILDINGS

A company team normally attacks a block or group of buildings as part of a battalion task force. To attack a block or a group of buildings, a company team may need to be reinforced with BFVs or tanks and engineers, consistent with the ROE and the enemy situation.

a. **Execution.** The execution of this mission is characterized by platoon attacks supported by both direct and indirect fires. Success depends on isolating the enemy positions which often become platoon objectives, suppressing enemy weapons, seizing a foothold in the block, and clearing the block's buildings room by room.

b. **Direct Fire Weapons.** BFVs, tanks, machine guns, and other direct fire support weapons fire on the objective from covered positions, consistent with the ROE. These weapons should not be fired for prolonged periods from one position. The gunners should use a series of positions and displace from one to another to gain better fields of fire and to avoid being targeted by the enemy. Direct fire support tasks can be assigned as follows:

(1) Machine guns fire along streets and into windows, doors, mouseholes, and other probable enemy positions. ROE may restrict firing only to known enemy locations.

(2) BFVs, tanks, and antitank weapons fire at enemy tanks and other armored vehicles can also provide a countersniper capability due to their range and target acquisition capability.

(3) Tanks fire at targets protected by walls and provide protection against enemy tanks, as required.

(4) BFVs may be used to create breaches with the 25-mm gun and TOW.

(5) Riflemen engage targets of opportunity.

c. **Obscuration and Assault.** Before an assault, the company commander should employ smoke to conceal the assaulting platoons. He secures their flanks with direct fire weapons and by employment of the reserve, if necessary. Concealed by smoke and supported by direct fire weapons, an assaulting platoon attacks the first isolated building. The assault element utilizes the cover of suppressive fires to gain a foothold. The company commander must closely coordinate the assault with its supporting fire so that the fire is shifted at the last possible moment. The squads and platoons then clear each designated building. After seizing the block, the company consolidates and reorganizes to repel a counterattack or to continue the attack. Periods of limited visibility may provide the best conditions to attack, especially if NVGs provide the company a technological advantage over the threat.

NOTE: Obscuration rounds may cause uncontrolled fires in the city and must be carefully planned.

4-38. HASTY ATTACK

A company team may find itself moving to an urban area or conducting a movement to contact with a mission of clearing a village of enemy. The following discussion provides a technique for conducting a hasty attack on a village. The company commander makes a quick assessment of the factors of METT-TC and reacts appropriately to support the higher level commander's intent.

a. **Establish Support.** If attached or OPCON, tanks, BFVs, MK19s or M2HBs mounted on HMMWVs, and TOWs assume support-by-fire positions from which they can fire on the village, prevent the enemy from withdrawing, and destroy any reinforcements (support element functions). If these assets are not available, then the company commander moves Infantry elements into position to accomplish the same tasks. The company's 60-mm mortar and AT sections also provide fire support. Armored vehicles can reposition during the assault, if necessary, to gain better fields of fire and provide better support.

b. **Assault the Village.** The rifle platoons assault from a covered route so as to hit the village at a vulnerable point (Figure 4-29). As the platoons approach the village, smoke is employed to screen their movement and supporting fires are shifted. Once the platoons close on the village, they clear the buildings quickly, consistent with the ROE, and consolidate. The company is then ready to continue operations.

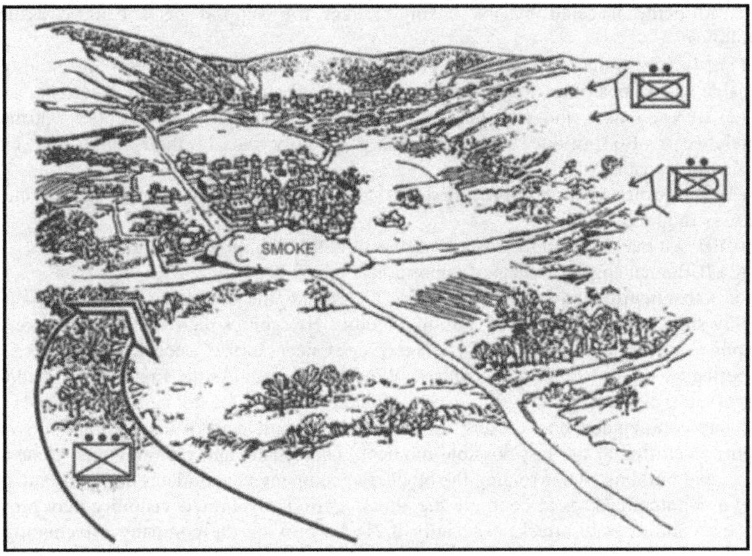

Figure 4-29. Hasty attack of a village.

4-39. MOVEMENT TO CONTACT AND RECONNAISSANCE

In a fast-moving situation, a company team may have to conduct a movement to contact through an urban area to fix enemy forces (Figure 4-30, page 4-54). Similarly, a company team may have to reconnoiter such a route to prepare for a battalion task force attack or other mission. This type of reconnaissance is accomplished with a company team. It is preferable to conduct this mission with tanks and or BFVs. The actual task organization will be determined by the factors of METT-TC.

Figure 4-30. Movement to contact through an urban area.

a. **Tempo.** These operations are characterized by alternating periods of rapid movement to quickly cover distances and much slower movement for security. The speed of movement selected depends on the terrain and enemy situation.

b. **Execution.** An infantry company without support from tanks or BFVs would conduct travelling overwatch or bounding overwatch along urban routes. In open areas where rapid movement is possible due to terrain, a tank section should lead, if available. In closer terrain, the infantry should lead while overwatched by the tanks. Another infantry platoon and the other tank section should move on a parallel street. Artillery fire should be planned along the route. Engineers accompany the lead platoon on the main route to help clear obstacles and mines.

c. **Danger Areas.** The company should cross danger areas (crossroads, bridges, and overpasses, and so forth) by a combination of actions:

- Between danger areas, the company moves with the infantry mounted, or rapidly on foot, when contact is not likely.
- When enemy contact is likely, the company moves to clear enemy positions or to secure the danger area. Tanks and other combat vehicles support infantry.

d. **Axis of Advance.** In peripheral areas, this advance should be on one axis with the lead unit well forward and security elements checking side streets as they are reached. In the city core, this operation is conducted as a coordinated movement on two or three axes for more flank security.

e. **Enemy Positions.** Enemy positions can be either destroyed by the company itself or, if the need for speed is great, bypassed, reported, and left to following units if the situation allows.

f. **Coordination.** The company commander must ensure that the actions of platoons and attached or OPCON elements are coordinated. Situational awareness must be maintained in a rapidly moving or changing environment. The company commander reports all information collected to the battalion task force.

4-40. SEIZURE OF KEY URBAN TERRAIN

A traffic circle, bridge or overpass that spans a canal, a building complex, or, in some cases, the population itself are examples of key urban terrain. Therefore, seizing such terrain intact and securing it for friendly use is a likely mission for a company team. The discussion below describes the TTP for seizing and controlling a bridge and seizing a traffic circle.

a. **Seizure of a Bridge.** For this mission (Figure 4-31), a company team should perform the following actions.

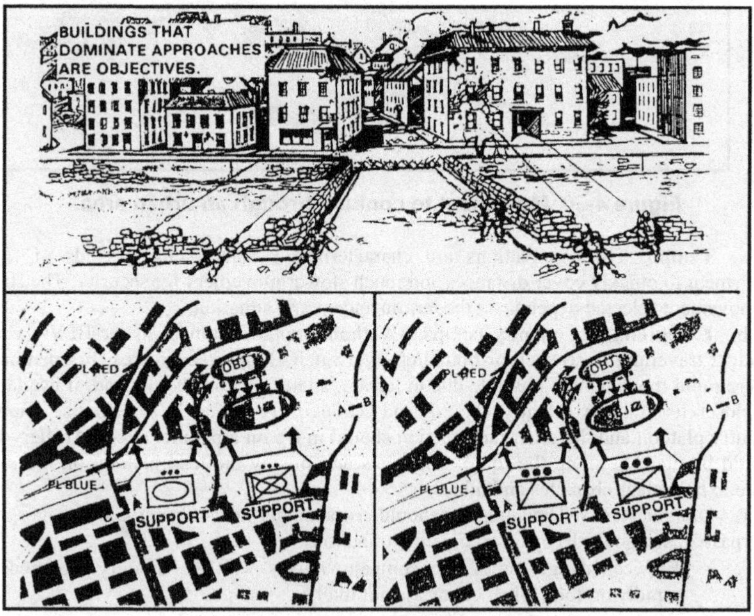

Figure 4-31. Seizure of a bridge.

(1) ***Clear the Near Bank.*** The first step in seizing a bridge is to clear the buildings on the near bank that overwatch the bridge and the terrain on the far side. The commander must find out which buildings dominate the approaches to the bridge. Buildings that permit him to employ anti-tank weapons, machine guns, and riflemen are cleared while supporting fire prevents the enemy from reinforcing his troops on the far bank and keeps enemy demolition parties away from the bridge.

(2) ***Suppress.*** Suppress enemy weapons on the far bank with direct and indirect fire. In suppressing the enemy's positions on the far bank, priority is given to those positions from which the enemy can fire directly down the bridge. Tanks, BFVs, TOWs, and machine guns mounted on HMMWVs are effective in this role. TOWs, Dragons, Javelins, and AT4s can be used against enemy tanks covering the bridge. Use screening smoke to limit enemy observation. All suppression must be consistent with the ROE.

(3) ***Assault.*** Seize a bridgehead (buildings that overwatch and dominate the bridge) on the far bank by an assault across the bridge. The objectives of the assaulting platoons are buildings that dominate the approaches to the bridge on the far side. One or two platoons assault across the bridge using all available cover while concealed by smoke. In addition to a frontal assault across the bridge, other routes should be considered. They are supported by the rest of the company and any attached and OPCON forces. Once on the other side, they call for the shifting of supporting fire and start clearing buildings. When the first buildings are cleared, supporting fire is lifted and or shifted again and the assault continues until all the buildings in the objective area are cleared.

(4) ***Clear the Bridge.*** Secure a perimeter around the bridge so that the engineers can clear any obstacles and remove demolitions from the bridge. The company commander may expand his perimeter to prepare for counterattack. Once the bridge is cleared, tanks, BFVs, and other support vehicles are brought across to the far bank.

b. **Seizure of a Traffic Circle.** A company may have to seize a traffic circle either to secure it for friendly use or to deny it to the enemy (Figure 4-32). This operation consists of seizing and clearing the buildings that control the traffic circle, and bringing direct-fire weapons into position to cover it. After gathering all available intelligence on the terrain, enemy, and population, the commander takes the following steps:

- Isolates the objective.
- Seizes and or clears the buildings along the traffic circle.
- Consolidates and prepares for counterattack.

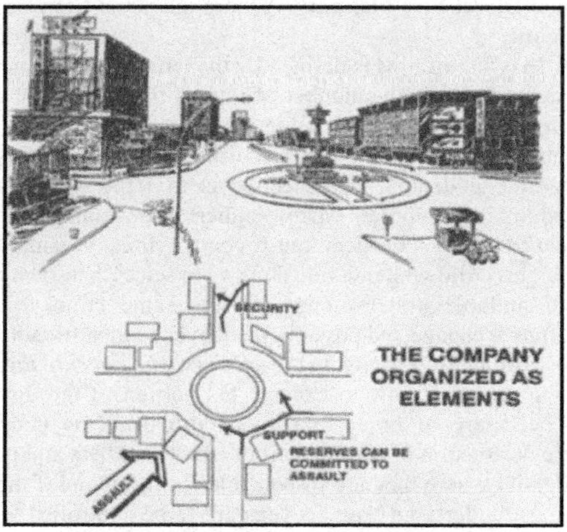

Figure 4-32. Seizure of a traffic circle.

(1) ***Troop Safety.*** Friendly troops should not venture into the traffic circle until it is secure. A traffic circle is a natural kill zone.

(2) ***Task Organization.*** The company should be organized with assault, support, and reserve elements based on the factors of METT-TC and the ROE.

(a) *Assault.* Seizes and or clears the terrain (buildings) that influence the objective. (For example, two rifle platoons, reinforced with engineers.)

(b) *Support.* Isolates the traffic circle and provides security. This element can be mounted (preferred) or dismounted; for example, an infantry platoon with a section of BFVs. Provides direct fire support for the assault element. The element could consist of tanks, BFVs, TOWs, MK 19s, or M2HBs mounted on HMMWVs, occupying a support-by-fire position.

(c) *Reserve.* Reinforces the assault element on order. (Normally a detached squad from one of the rifle platoons.)

(3) ***Flexibility.*** At various stages in this operation, roles may change. For example, the assault element may clear buildings until the support element can no longer support it. Then the reserve can be committed to the assault. It may also occur that one of the assault elements finds itself in a better position to isolate the traffic circle and becomes the support element. At that time, the isolating (support) element would become part of the assault element. The support element may also have to reposition to continue support.

4-41. DIRECT FIRE PLANNING AND CONTROL

One of the company commander's primary responsibilities will be to plan and control direct fires during the attack of an urban objective. The information below applies to a company isolating an objective for either a battalion or company attack. Direct fire support can be very resource intensive. Commanders must ensure that enough ammunition is available to support their fire plans.

a. **Principles of Direct Fire Planning.** A thorough direct fire plan will adhere to the principles stated below.

(1) ***Mass fires.*** Massing of fires is defined by the terminal effect on the enemy, not by the number of systems firing or the number of rounds fired. Mass must not be confused with volume of fires. Massing fires is achieved by placing accurate fires on multiple targets at the same time. This method means firing at enemy targets in or outside of buildings laterally and in depth. The objective is to force the enemy to respond to multiple threats and to kill or suppress enemy soldiers or positions.

(2) ***Leaders control fires.*** Leaders must control fires to simultaneously engage different priority targets. Allowing individual crews to select their own priority target will probably result in multiple systems engaging the same target while leaving other dangerous targets free to engage and possibly maneuver against friendly units.

(3) ***Fire plans must be understood by the soldiers who execute them.*** It is imperative that every soldier understands how to execute his portion of the direct fire plan. This understanding is necessary in order to avoid fratricide. This understanding is also necessary to ensure destruction or suppression of enemy soldiers and positions. A soldier must be able to identify where they are responsible for firing and if there is an enemy to engage. Then he must understand how his fires are to be controlled and directed during

the course of the fight. Ensuring terms are commonly understood assists all involved. Exchanging SOPs, chalk talks, terrain models, and rehearsals assist in understanding.

(4) ***Focus fires.*** Focusing fires means accurately directing fires to hit specific targets, points, or areas, and is the most difficult task of controlling fires. The commander focuses fires by clearly conveying instructions (either preplanned or hasty) to direct the fires of the individual platoons on specific targets or areas that support his plan for distribution. Platoons must be able to recognize the point at which to focus their fires. Failure to do so will result in different units/assets in the support element engaging the same targets, while others are not engaged. Recognizable control measures allow the support element to focus fires (see paragraph d). OPORDs and rehearsals must paint the visual picture of how the commander wants the fires focused and what the platoons will see to focus their fires.

(5) ***Distribute fires.*** Distributing fires is the process of engaging different enemy threats simultaneously to avoid overkill by multiple systems engaging the same targets and to degrade the enemy's ability to deal with single threats one at a time. Proper distribution ensures critical targets are engaged first and the enemy is engaged three dimensionally. The following points should be emphasized:

a. Avoid target overkill. Minimize engaging targets that are already destroyed or suppressed.

b. Use each weapon system in its best role. Different weapons systems and ammunition types have specific characteristics that maximize their capability to kill or suppress specific enemy weapons systems at different ranges. For example, an AT 4 can be used to suppress an enemy sniper position, but it will probably not destroy the position itself.

c. Destroy the most dangerous targets first. Proper focus, distribution, and firing first are the keys to maximizing this principle.

d. Concentrate on enemy crew-served weapons and combat vehicles. This method deprives the enemy of his ability to use his fire support weapons against friendly troops.

e. Take the best shots and expose only those weapons systems actually needed to fire in order to maximize the probability of hitting and killing enemy targets, and to protect friendly forces as long as possible.

(6) ***Shift fires.*** Shifting fires is the process of re-focusing weapons systems to change the distribution of fires as targets are destroyed or as the situation changes, for example, the introduction of new forces on the battlefield. At the company level, this method is accomplished by shifting the fires of the support element and focusing them on new targets. This fire may be used to isolate, suppress, prevent counterattack/reinforcement, and so forth.

(7) ***Rehearse the fire plan.*** The most important part of any operation that requires soldiers to shoot their weapons is the fire plan. Every fire plan must be rehearsed; for example, what is the fire plan and how is it executed in each phase: isolation/gaining a foothold, breaching, assault of the building. A rehearsed fire plan enhances execution, prevents fratricide, identifies shortcomings, and works to synchronize the operation.

b. **Questions to Answer.** When the direct fire plan is complete, the commander should be able to answer the following questions:

- How does the fire plan help achieve success at the decisive point?
- What is the company mission and the desired effect of our fires?

- Is the fire plan consistent with the ROE?
- Where are combat vehicles or other dangerous weapons systems?
- Which course of action has the enemy selected?
- What are the PIR to determine the enemy's actions?
- Where are we going to kill or suppress the enemy?
- From where will we engage him?
- Which enemy weapons do we want to engage first?
- How will we initiate fires with each weapon system?
- Which weapons will fire first? What will each engage? What are the engagement criteria?
- What is the desired effect of fires from each unit in the support element?
- How will we distribute the fires of platoons to engage the enemy three dimensionally?
- What will the support element focus their fires on? (How will the support element units know where to engage? Will they be able to see and understand the control measures?)
- How will we mass fires to deal with multiple enemy threats and achieve the desired volume of fire?
- Where will leaders be positioned to control fires; how will we focus fires on new targets?
- How will we deal with likely enemy reactions to our fires?
- Does the plan avoid overkill; use each weapon system in its best role; concentrate on combat vehicles, take the best shots, expose only those friendly weapons needed, destroy the most dangerous targets first?
- Have my fires been massed to achieve suppression, obscuration, and security needs of the breach?
- Will the fires be masked by buildings or assault element movement?

c. **Fire Commands.** Fire commands are verbal orders used to control direct fires. They are standard formats that rapidly and concisely articulate the firing instructions for single or multiple engagements. They can be given over the radio or landline to control fires. At company level fire commands must control the fires of multiple elements with different weapons systems. Fire commands should concentrate on ensuring that the support element is accurately focused and understands its portion of fire distribution. Platoon leaders generally give these commands after the company commander gives the order to initiate fires. A general format includes:

- Alert (call sign).
- Weapon ammunition (optional, METT-TC dependent).
- Target description.
- Location or method to focus fires.
- Control pattern technique (optional, METT-TC dependent).
- Execution (my command, your command, event).

Sample Fire Command:

"Tango 27 (PSG) This is Tango 16 (PL), over"- Alert
"7.62mm, 40mm, and AT 4s" – Weapons/ammunition
"Windows and Door" – Target description
"OBJ 4; White; A1, B1, C1" – Location
"Fire" – Execution.

d. **Direction of Assault Technique of Direct Fire Control.** In this technique, the company commander assigns building numbers in a consistent pattern in relation to the direction of assault. In the example shown in Figure 4-33 on page 4-62, the commander numbered the buildings consecutively, in a counterclockwise manner. Further, the sides of the buildings were color coded consistently throughout the objective area (WHITE=direction of assault side; GREEN=right side; BLACK=rear side; RED=left side; BLUE=roof). An odd-shaped building is also shown. Note that a *four-sided* concept was retained to minimize confusion. Further designations of WHITE 1, WHITE 2, WHITE 3, and so on from left to right can be added to specify which wall will be engaged. Apertures on the buildings are also labeled consecutively using rows and columns, as shown. In the example, "OBJ 4, WHITE, window A1" is the lower left-hand window on the direction of assault side of OBJ 4. All designations are labeled in relation to the direction of assault. (See FM 34-130 for additional information on building shapes and structural labeling.)

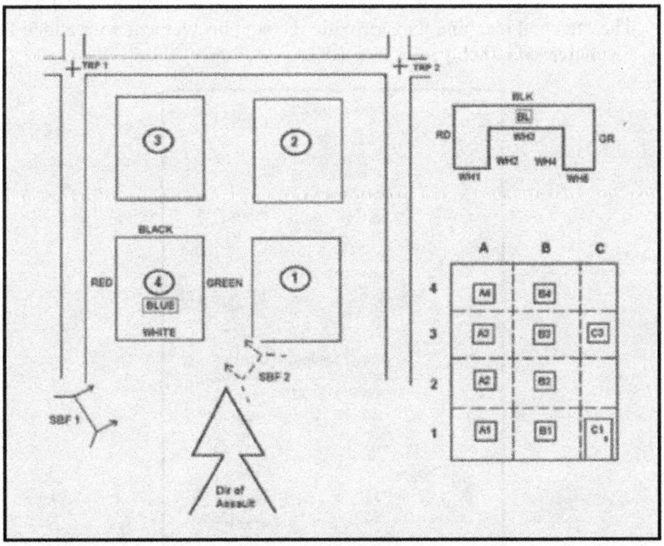

Figure 4-33. Direction of assault technique of direct fire control.

Section VIII. PLATOON ATTACK OF AN URBAN AREA

The Infantry platoon will normally conduct offensive tasks as part of a company mission. However, there may be times that the platoon will be required to perform an independent offensive operation in support of the main effort. This section discusses how the platoon conducts the various tasks as part of a company operation or as an independent mission.

4-42. TASK ORGANIZATION (PLATOON ATTACK OF A BUILDING)

The platoon leader will normally organize his platoon into at least two elements: an assault element consisting of two rifle squads, and a support element consisting of the platoon's crew-served weapons and one rifle squad as the support or reserve (Figure 4-34). If engineers are not available, he can designate a breaching team from within either the assault or the support element or, depending on the situation, he may task organize a separate breach element. The size and composition of these elements are determined by the mission given, the number of troops available, the type and size of the objective building, whether the adjacent terrain provides open or covered approaches, and the organization and strength of the enemy defenses. As part of a company operation, the platoon will be part of either the assault element or the support element.

- As part of the company's assault element, the platoon would organize into three assault squads with two assault teams each, and will attach the machine guns to the company support element.
- As the part of the company's support element, the platoon may be organized into three support squads with machine guns and antiarmor weapons attached. The attached machine guns provide the support element with added firepower for increased lethality.

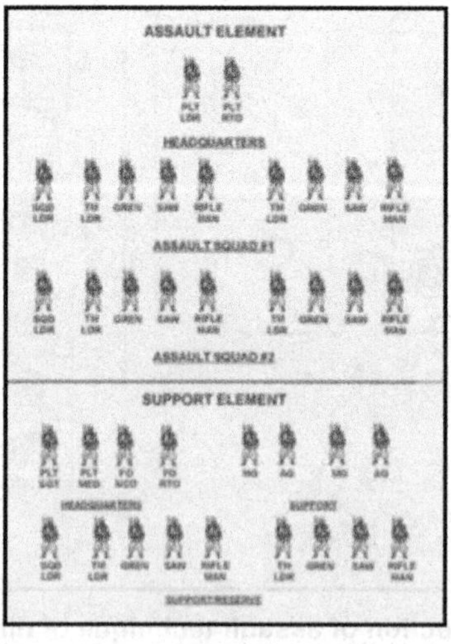

Figure 4-34. Platoon organization.

a. **Assault Element.** The purpose of the assault element is to kill, capture, or force the withdrawal of the enemy from an urban objective and to seize key terrain. The assault element of a platoon may consist of one, two, or three squads. Squad leaders will normally organize their two fire teams into two assault teams or, in special circumstances, the squad may be kept as a single assault element.

NOTE: Clearing techniques are designed to be executed by the standard four-man fire team. This method does not mean that all four members must enter a room to clear it. Because of the confined spaces typical of building/room clearing operations, units larger than squads quickly become awkward and unmanageable. When shortages of personnel demand it, two- and three-man teams can conduct room-clearing operations, but four-man teams are best suited. Using fewer personnel adds to the combat strain and greatly increases the risks to the team. For clearing large open buildings, such as hangars or warehouses, it may be necessary to commit two squads at the same time using a bounding overwatch movement technique to effectively cover the entire structure and provide force protection.

b. **Support Element.** The purpose of the support element (except for the medic) is to provide immediate suppressive fire support to enable the assault element to close with the enemy. Suppressive fires must be closely controlled to avoid excessive expenditure of ammunition and prevent fratricide. The support element is normally controlled by the platoon sergeant or a senior squad leader and normally consists of the platoon's crew-served weapons, light and medium antitank weapons systems, forward observer team, platoon medic, and any personnel not designated as part of the assault element (Figure 4-35). The support element provides both direct and indirect fire support and other assistance to advance the assault element. This support includes, but is not limited to, the following:

- Suppressing enemy weapons systems and obscuring the enemy's observation within the objective building and adjacent structures.
- Isolating the objective building with direct and indirect fires to prevent enemy withdrawal, reinforcement, or counterattack.
- Obscuring enemy observation of obstacles en route to the objective and at the entry point of the objective during breaching operations.
- Destroying or suppressing enemy positions with direct fire weapons.
- Engaging armored vehicles.
- Securing cleared portions of the objective.
- Providing replacements for the assault element.
- Providing the resupply of ammunition and pyrotechnics.
- Bringing up specific equipment that the assault element could not carry in the initial assault.
- Treating and evacuating casualties, prisoners, and civilians.

NOTE: The platoon sergeant must be prepared to rapidly evacuate wounded from the objective area to the company casualty collection point (CCP). The use of ground ambulances may be impeded by rubble in the streets, barricades, and demolition of roads; therefore, litter teams could be used extensively. Also, snipers can affect medical evacuation from forward positions.

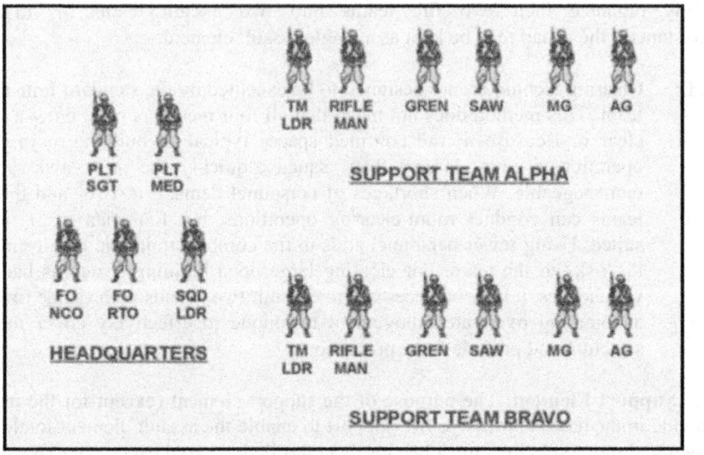

Figure 4-35. Platoon support element with squad integrated.

c. **Breaching Element.** The purpose of the breaching element is to clear and mark lanes through obstacles during movement, providing the assault element with access to an urban objective. The platoon leader organizes the force to ensure breaching elements are designated. One technique is to assign one fire team from the assault element as the breaching element. Alternatively, the breach can be conducted using an attached engineer or any member of the assault or support element who has had additional breach training.

4-43. MOVEMENT IN URBAN TERRAIN

As the lead element for the company when conducting movement, the platoon must be prepared to react to contact.

a. Platoon members must be ready to take cover and return fire immediately. They must also be alert for any signs or indications of the enemy and report promptly.

b. The rate of movement is controlled by the lead element based on the density of the urban terrain and enemy threat. In outlying or lightly defended areas, a mechanized infantry platoon may proceed along streets mounted, but send dismounted squads forward to reconnoiter key terrain (bridges, intersections or structural choke points).

c. Enemy action against the unit may come in the way of an ambush along a street, enfilade fire down the street, sniper fire from upper stories of buildings, or artillery and mortar fire when canalized. For protection from those types of threats, the platoon should move through buildings, along walls and other forms of cover, use tanks, BFV's, as well as indirect and direct fire weapons to overwatch and support movement.

d. The platoon moves using a lead maneuver element (one squad on narrow streets and two squads on wide streets). These squads will move forward along the streets using buildings for cover when possible. They will scout danger areas and close with the enemy. An overwatching element (the rest of the platoon and the supporting weapons) will follow securing the flanks and rear while providing support to the point element. At any time the platoon leader may choose to rotate the point squad with an overwatching squad (Figure 4-36).

Figure 4-36. Movement down a street.

4-44. ATTACKING IN URBAN TERRAIN

As the culminating effort of a planned (deliberate) attack, or a result of a movement to contact, a meeting engagement or a chance contact during movement, the platoon may be required to be part of a company attack or conduct a platoon attack on an urban area or building.

a. The attack involves isolating the building to prevent the reinforcing or withdrawal of its defenders (normally planned at company level); suppressing the enemy with BFVs, tanks, machine gun and mortar fire; entering the building at the least defended point; and clearing the building. There must be close coordination between the isolation/support elements and the assault elements.

b. As the lead element in the company movement formation when a chance contact is made with the enemy (hasty attack), the platoon takes the following actions:

(1) Forward squad (or squads) will immediately return fire, get down, seek cover and suppress the enemy.

(2) Those squads not in direct fire contact will provide supporting fire with individual and crew-served weapons (to include tanks and BFVs, if attached). Engage known, then suspected, enemy positions.

(3) Provide a situation report to the commander.

(4) The commander will either direct the platoon to establish a support by fire position in order to allow another platoon to assault or, if the threat is small and disorganized, he will direct the platoon in contact to conduct a platoon attack of the enemy position (subparagraph c, below).

c. When conducting a deliberate attack of an urban objective there are three steps that must be considered, planned and coordinated in order to achieve success.

- Isolate the objective.
- Enter the building (secure a foothold).
- Clear the building (room by room, floor by floor).

(1) Isolation of the objective requires the seizing of dominant terrain in order to cut off enemy routes for reinforcing, supplying, or facilitating the withdrawal of its defenders. The intent is to completely dominate what comes and goes within the objective area and provide early warning for the assault element (Figure 4-37).

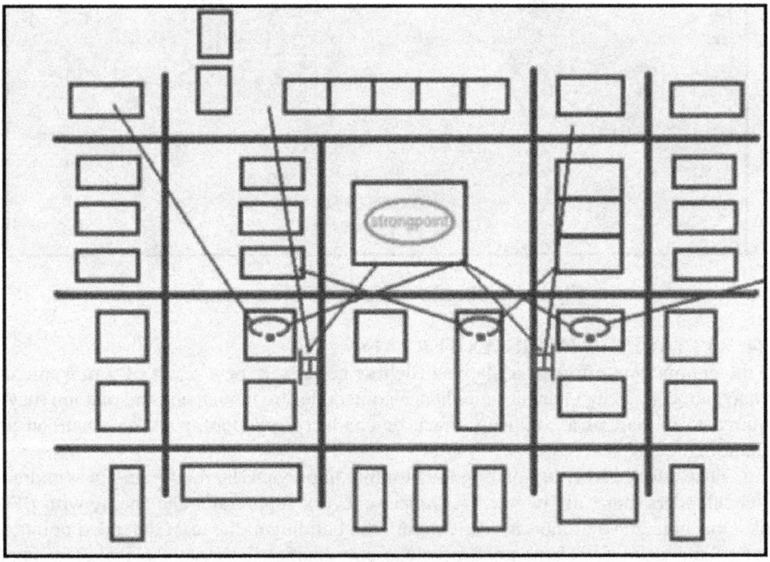

Figure 4-37. Infantry platoon with two tanks as support element, isolating the objective.

(2) As the assault element for the company, the platoon (or platoons) is responsible for entering and clearing the objective building. This method may involve creating a breach into the building and securing a foothold as well as killing, capturing or forcing the withdrawal of all enemy personnel within the structure. Squads and teams perform room clearing. The squad leader controls the maneuver of the two fire teams as they clear along hallways, stairways, and in rooms. (See Chapter 3.) The platoon leader alternates

the squads as required, and maintains momentum, and ensures resupply of ammunition and water.

d. If a platoon is conducting an assault of a building independently, it should be organized with an assault element and a support element (Figure 4-38). The assault element, usually led by the platoon leader, normally consists of two squads with two fire teams each. The support element, usually controlled by the platoon sergeant, normally consists of one rifle squad equipped with antitank weapons, two medium machine gun crews, and attached forward observers. The support element must designate individuals to provide flank and rear security. In addition to its own support element, BFVs, tanks, and other company assets can support the platoon.

NOTE: Isolation of the surrounding area in conducted by the rest of the company.

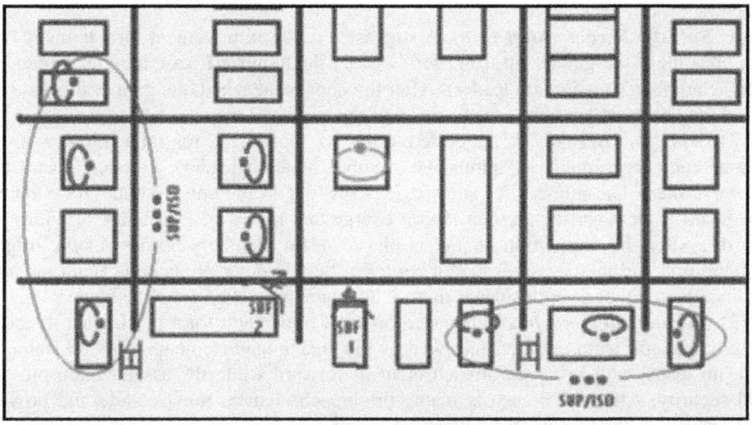

Figure 4-38. Platoon attack of a building with two platoons isolating.

4-45. PLATOON ASSAULT OF A BUILDING

The assault force, regardless of size, must quickly and violently execute the assault and subsequent clearing operations. Once momentum has been gained, it is maintained to deny the enemy time to organize a more determined resistance on other floors or in other rooms. The small unit leaders are responsible for maintaining the momentum of the assault, controlling movement, yet not allowing the operation to become disorganized. Enemy obstacles may slow or stop forward movement. Leaders must maintain the momentum by rapidly creating a breach in the obstacle, or by redirecting the flow of the assault over or around the obstacles.

a. **Approaches.** All routes to the breach and or entry point are planned in advance. The best route is confirmed and selected during the leaders' reconnaissance. The route should allow the assault element to approach the breach (entry) point from the enemy's blind side, if possible.

b. **Order of March.** The assault team's order of march to the breach point is determined by the method of breach and their intended actions at the breach (entry) point. This preparation must be completed prior to or in the last covered and concealed location before reaching the breach (entry) point. Establishing an order of march is done to aid the team leader with C2 and to minimize exposure time in open areas and at the entry point. An order of march technique is to number the assault team 1, 2, 3, and 4. The number 1 man should always be responsible for frontal/door security. If the breach has been conducted prior to their arrival the assault team quickly moves through the breach (entry) point. If a breach has not been made prior to their arrival at the breach (entry) point, and depending on the type of breach to be made, the team leader conducts the breach himself or signals forward the breach man/element. One option is to designate the squad leader as the breach man. If the breach man is part of the assault team, he is normally the last of the four men to enter the building or room. This method allows him to transition from his breaching task to his combat role.

(1) ***Ballistic Breach (Shot Gun)***. A suggested order of movement for a ballistic (shot gun) breach has the gunner up front, followed by the number 1 man, number 2 man, and then the number 3 man (team leader). After the door is breached, the gunner moves to the rear of the lineup and assumes the position of the number 4 man.

(2) ***Explosive Breach***. A suggested order of movement for an explosive breach without engineer support is; number 1, number 3 (team leader), number 2, and then number 4 man. The number 1 man provides security at the entry point. The number 3 man (team leader) carries the demolition charge and places it. Number 4 provides rear security. After the demolition charge is placed, team members re-form in the original configuration and take cover around a corner or behind other protection. Team members can line up on either or both sides if there is adequate protection from the blast.

(3) ***Mechanical Breach***. A suggested order of movement for a mechanical breach is the initial assault team in order, followed by the breach man/element. At the breach point the team leader will bring the breach element forward while the assault team provides local security. After the breach is made, the breach element moves aside and provides local security as the assault team enters the breach.

c. **Security.** Because of the three-dimensional threat associated with urban terrain, the assault element must maintain 360-degree security during movement to the breach (entry) point. If the assault element is to stop in the vicinity of the breach (entry) point to wait for the breach element to complete its task, the support element must maintain suppressive fire to protect the assault element.

d. **Assault Locations.** Entry at the top and fighting downward is the preferred method of clearing a building. This method forces the defenders down and out of the building where the support element can engage them. This method is only feasible, however, when access to an upper floor or rooftop can be gained from the windows or roofs of adjoining, secured buildings. Rooftops are treated as danger areas when surrounded by higher buildings from which enemy forces could engage the assault element. Helicopters should land only on those buildings that have a roof structure that can support their weight. If the structure cannot support the helicopter, soldiers can dismount as the helicopter hovers a few feet above the roof. Troops then breach the roof or common walls to gain entrance into the building. (If using explosives on the rooftop, ensure cover is available to the soldiers.) They may use ropes or other means to enter the lower floors through the holes created.

NOTE: Soldiers should consider the use of devices and techniques that allow them upper level access without using interior stairways. These devices and techniques include, but are not limited to, adjacent rooftops, fire escapes, portable ladders, and various soldier-assisted lifts.

e. **Support Element.** The support element isolates the building with direct and indirect fires to support the assault element's move to the breach point. The support element covers mounted avenues of approach with antiarmor weapons, covers dismounted avenues of approach with automatic weapons, and suppresses enemy fires and neutralizes enemy positions to enable the breach team and assault element to move into position. The location of adjacent units must be considered in the emplacement of supporting fires.

(1) The support element uses smoke to obscure the movement of the breach team and assault element to the building. If possible, the smoke obscuration is maintained until the assault element has entered the building.

(2) Depending upon the ROE, just before the rush of the assault element, the support element increases suppressive fires on the objective and continues until masked by the advancing assault element. Once masked, fires are shifted to upper or lower windows and continued until the assault element has entered the building. At that time, fires are shifted to adjacent buildings to prevent enemy withdrawal or reinforcement.

(3) If the ROE are very restrictive, the use of supporting fires may be restricted to known enemy locations that have engaged the unit.

(4) The support element must also deal with civilians displaced by the assault, EPWs, and casualties.

4-46. CONSOLIDATION AND REORGANIZATION

The squad and platoon will conduct consolidation and reorganization immediately after each action where soldiers are engaged and ammunition is expended. Consolidation is the action taken by the squad or platoon to ensure its security, to prepare for a counterattack by the enemy, and to prepare to continue the mission. Consolidation in an urban environment must be quick in order to repel enemy counterattacks and to prevent the enemy from infiltrating back into cleared buildings or floors. After securing a floor (bottom, middle, or top), selected members of the unit are assigned to cover potential enemy counterattack routes to the building. Priority must be given initially to securing the direction of attack. Security elements alert the unit and place a heavy volume of fire on enemy forces approaching the unit. Reorganization occurs after consolidation. These actions prepare the unit to continue the mission by ensuring key leadership positions are filled and important weapon systems are manned. Many reorganization actions occur simultaneously during the consolidation of the objective.

a. **Consolidation Actions.** Squads assume hasty defensive positions to gain security immediately after the objective has been seized or cleared. Squads that performed missions as assault elements should be prepared to assume an overwatch mission and to support another assault element. Units must guard all avenues of approach leading into their area. These may include:

- Enemy mouse-holes between adjacent buildings.
- Covered routes to the building.
- Underground routes into the basement.
- Approaches over adjoining roofs.

b. **Reorganization Actions.** After consolidation, leaders ensure the following actions are taken:

- Resupply and redistribute ammunition.
- Mark buildings to indicate to friendly forces that they have been cleared.
- Treat and evacuate wounded personnel. Once the objective area is secure, begin evacuating noncombatants then enemy wounded.
- Process EPWs.
- Segregate and safeguard noncombatants.
- Reestablish the chain of command.

c. **Continuation of the Assault.** If the unit is going to continue with its original mission, its "be prepared/on order" mission, or receives a new mission, it must accomplish the following:

(1) The momentum must be maintained. Keeping momentum is a critical factor in clearing operations. The enemy is not allowed to move to its next set of prepared positions or to prepare new positions.

(2) The support element pushes replacements, ammunition, and supplies forward to the assault element.

(3) Security for cleared areas must be established IAW the OPORD or TACSOP.

(4) The support element must displace forward to ensure that it is in place to provide support to the assault element, such as isolation of the new objective.

CHAPTER 5
DEFENSIVE OPERATIONS

"[Captain] Liebschev prepared his defenses with extraordinary thoroughness, choosing only to defend the northern half of the town. The southern half was turned into a nightmare of trapped and mined houses some of which were blown into the streets to form road blocks and others were blown up to clear arcs of fire. All his strong points were linked by what is best described as 'mouse holing' from house to house. All approaches to the defended sector were either heavily mined or under concealed enfilade fire. The main approach into the town square was left attractively unobstructed....The 2nd Canadian Brigade was given the task of clearing a way through the town and was forced to fight its way from house to house on not more than a 250-yard front. Every building, when taken, had to be occupied to stop the Germans infiltrating back into it again after the leading troops had passed on. The fighting was at such close quarters that artillery support was impossible..."

Extracted from The Battle for Italy
By General W. G. F. Jackson

Section I. DEFENSIVE CONSIDERATIONS

Full spectrum operations require that units be prepared to defend in urban areas. Before making a decision to defend urban areas, commanders at all levels should consider the issues discussed in this chapter.

5-1. REASONS FOR DEFENDING URBAN AREAS

The worldwide increase in urban sprawl has made it virtually impossible for forces conducting operations to avoid cities and towns. For various reasons, these areas must be defended.

a. Certain urban areas contain strategic industrial, transportation, or economic complexes that must be defended. Capitals and cultural centers may be defended for strictly psychological or national morale purposes even when they do not offer a tactical advantage to the defender. Because of the sprawl of such areas, significant combat power is required for their defense. The decision to defend these complexes is made by political authorities or the theater commander.

b. The defenders' need to shift and concentrate combat power, and to move large amounts of supplies over a wide battle area may require retention of vital transportation centers. Since most transportation centers serve large areas, the commander must defend the urban area to control such centers.

c. Most avenues of approach are straddled by small towns every few kilometers and must be controlled by defending forces. These areas can be used as battle positions or strongpoints. Blocked streets covered by mortar and or artillery fire can canalize attacking armor into mined areas or zones covered by antiarmor fire. If an attacker tries to bypass an urban area, he may encounter an array of tank-killing weapons. To clear such an area, the attacker must sacrifice speed and momentum, and expend many resources. A city or town can easily become a major obstacle.

d. A well-trained force defending an urban area can inflict major losses on a numerically superior attacker. The defender can conserve the bulk of his combat power so it is available for use in open terrain. The defenders remaining in urban areas perform an economy-of-force role.

e. Aerial photography, imagery, and sensory devices cannot detect forces deployed in cities. Well-emplaced CPs, reserves, CSS complexes, and combat forces are hard to detect.

5-2. REASONS FOR NOT DEFENDING URBAN AREAS

Reasons for not defending urban areas include the following.

a. The location of the urban area does not support the overall defensive plan. If the urban area is too far forward or back in a unit's defensive sector, is isolated, or is not astride an enemy's expected avenue of approach, the commander may choose not to defend it.

b. Nearby terrain allows the enemy to bypass on covered or concealed routes. Some urban areas, mainly smaller ones, are bypassed by main road and highway systems.

c. Structures within the urban area do not adequately protect the defenders. Extensive areas of lightly built or flammable structures offer little protection. Urban areas near flammable or hazardous industrial areas, such as refineries or chemical plants, should not be defended because of increased danger of fire to the defenders.

d. Dominating terrain is close to the urban area. If the urban area can be dominated by an enemy force occupying this terrain, the commander may choose to defend from there rather than the urban area. This applies mainly to small urban areas such as a village.

e. Better fields of fire exist outside the urban area. The commander may choose to base all or part of his defense on long-range fields of fire outside an urban area. This applies mainly to armor-heavy forces defending sectors with multiple, small, urban areas surrounded by open terrain, such as agricultural areas with villages.

f. The urban area has cultural, religious, or historical significance. The area may have been declared an "open city" in which case, by international law, it is demilitarized and must be neither defended nor attacked. The attacking force must assume civil administrative control and treat the civilians as noncombatants in an occupied country. The defender must immediately evacuate and cannot arm the civilian population. A city can be declared open only before it is attacked. The presence of large numbers of noncombatants, hospitals, or wounded personnel may also affect the commander's decision not to defend an urban area.

5-3. GENERAL CONSIDERATIONS

The basic fundamentals of defense do not change in an urban environment. In urban combat, the defender does possess key advantages over the attacker. The defender can shape the battlefield by maximizing the natural restrictions and obstacles found in the restrictive terrain of the urban environment. US forces may not wish to inflict collateral damage on the urban terrain they are defending but the very nature of conducting an urban defense may lead to high-intensity conditions on the urban battlefield and to extensive collateral damage. Typically, US forces should not expect enemy forces attacking in urban terrain to be bound by restrictive ROE and should therefore not expect to accrue any of the advantages that a defender might have if the attacker is restricted in the application of force.

Section II. MISSION, ENEMY, TERRAIN, TROOPS AND TIME AVAILABLE, CIVIL CONSIDERATIONS

The defense of an urban area should be organized around key terrain features, buildings, and areas that preserve the integrity of the defense and provide the defender ease of movement. The defender must organize and plan his defense considering factors of mission, enemy, terrain, troops and time available, and civil considerations (METT-TC). Procedures and principles for planning and organizing the defense of an urban area are the same as for other defensive operations. In developing a defensive plan, the defender considers METT-TC factors with emphasis on fire support, preparation time, work priorities, and control measures. Planning for the defense of an urban area must be detailed and centralized. As in the offense, execution is decentralized as the battle develops, and the enemy forces assault the buildings and rooms. Therefore, it is imperative that all leaders understand the mission end-state and the commanders' intent, two levels up.

5-4. MISSION

Commanders and leaders must receive, analyze, and understand the mission before they begin planning. They may receive the mission as a FRAGO or as a formal OPORD, and must analyze all specified and implied tasks. Depending on mission requirements, an infantry unit at brigade and battalion level must be prepared to defend as part of a larger force or independently; companies and below normally defend as part of a larger force. Mission analysis for defense in urban terrain will essentially be the same as for other defensive operations. Detailed IPB is essential and must include building construction; routes, including underground systems; civilian communications; and utilities. (See Appendix G.) A hasty defense may be conducted in any of the defensive situations described in this chapter, immediately after offensive operations, or when a higher state of security is warranted during stability operations or support operations. The major difference between a hasty defense and a deliberate defense is in the amount of time for preparation. Similar to offensive operations, units must be prepared to transition to offensive or stability and support missions, and back.

5-5. ENEMY

Units must also analyze the type of enemy force they may encounter. If the attacker is mostly dismounted infantry, the greatest danger is allowing him to gain a foothold. If the attacker is mostly armor or mounted motorized infantry, the greatest danger is that he will mass direct fire and destroy the defender's positions. If the threat is primarily asymmetrical, force protection measures must be enhanced. (See Chapter 2 for detailed discussion of urban threat evaluation.)

5-6. TERRAIN AND WEATHER

Specific defensive considerations are discussed in this paragraph. Terrain in urban areas is three-dimensional; the defender must make use of the entire battle space:

- Surface (ground level, for example streets and parks).
- Supersurface (buildings, both interior and exterior).
- Subterranean (subways and sewers).

Analysis of all man-made and natural terrain features is critical when planning to defend in urban terrain. The type of urban area in which it will be operating affects the unit's defensive plan.

a. **Observation and Fields of Fire.** Although concealment and cover will be plentiful, observation will be limited. Attacking forces generally advance by crossing streets and open areas between buildings where they are exposed to fires from concealed positions.

(1) ***Weapons and Range.*** Units must position weapons to obtain maximum effect and mutual supporting fire. This allows for long-range engagements out to the maximum effective ranges. FOs should be well above street level to adjust fires on the enemy at maximum range. Observed fire will be very difficult in densely constructed areas. Fires and FPFs should be preplanned and, if possible and ROE permit, preregistered on the most likely approaches to allow for their rapid shifting to threatened areas.

(2) ***Limited Visibility.*** Units can expect the attacker to use limited visibility conditions to conduct necessary operations to sustain or gain daylight momentum. The following should be considered:

- Unoccupied areas that can be observed and covered by fire during daylight may have to be occupied or patrolled at night.
- Remote sensors and early warning devices should be employed in dead space and on avenues of approach.
- The artificial illumination available in urban terrain should be considered for use during the defense.

Responding to night probes with direct fire weapons should be avoided, as this gives away the location of the positions.

b. **Cover and Concealment.** Battle positions should be prepared using the protective cover of walls, floors, and ceilings. Units will continue to improve positions using materials on hand. Units prepare the terrain for movement between positions and can reduce exposure by—

- Using prepared breaches through buildings.
- Moving through reconnoitered and marked subterranean systems.
- Using trenches.
- Using the concealment offered by smoke and darkness to cross open areas.

c. **Obstacles.** An urban area is by its very nature an obstacle and or an obstruction. The series of man-made structures inherent in urban terrain canalizes and impedes an attack.

d. **Key Terrain.** Key terrain is any place where seizure, retention, or control affords a marked advantage to either enemy or friendly forces. Primary examples of key terrain are ports, airfields, bridges over canals or rivers, building complexes, or parks. Urban areas are unusual in that the population of the area itself may be considered key terrain. The identification of key terrain allows the defender to select his defensive positions and assists in determining the enemy's objectives. A special kind of key terrain is the nodes that are found in urban areas. These include governmental centers, power distribution facilities, and communication hubs. These nodes may have to be protected by the defender from asymmetrical as well as conventional threats.

(1) ***Villages.*** Villages are often on choke points in valleys, dominating the only high-speed avenue of approach through the terrain (Figure 5-1). If the buildings in such a village are well constructed and provide good protection against both direct and indirect fires, a formidable defense can be mounted by placing a company in the town, while controlling close and dominating terrain with other battalion task force elements.

Figure 5-1. Village.

(2) ***Strip Areas.*** Strip areas consist of houses, stores, and factories and are built along roads or down valleys between towns and villages (Figure 5-2). They afford the defender the same advantages as villages. If visibility is good and enough effective fields of fire are available, a unit acting as a security force need occupy only a few strong positions spread out within the strip. This will deceive the enemy, when engaged at long ranges, into thinking the strip is an extensive defensive line. Strip areas often afford covered avenues of withdrawal to the flanks once the attacking force is deployed and before the security force becomes decisively engaged.

Figure 5-2. Strip area.

(3) ***Towns and Small Cities.*** Small forces can gain an advantage in combat power when defending a small city or town (Figure 5-3) that is a choke point if it places tanks, BFVs, TOWs, Javelins, and Dragons on positions dominating critical approaches, when facing a predominantly armored enemy. To deny the enemy the ability to bypass the town or city, the defending force must control key terrain and coordinate with adjacent forces. Reserve forces should be placed where they can quickly reinforce critical areas. Obstacles and minefields assist in slowing and canalizing the attacker.

Figure 5-3. Towns and small cities.

(4) ***Large Cities.*** In large cities, units must consider that the terrain is restrictive due to large buildings that are normally close together (Figure 5-4). This situation requires a higher density of troops and smaller defensive sectors than in natural open terrain. Units occupy defensive frontages about one-third the size of those in open areas.

Figure 5-4. Large cities.

e. **Avenues of Approach.** The defender must not only consider the surface (streets, boulevards, parks) avenues of approach into and out of the urban area, but also supersurface (interior and exterior of buildings) and subterranean avenues of approach. The defender normally has the advantage. He knows the urban area and can move rapidly from position to position through buildings and underground passages. Control of these above- and below-ground avenues of approach becomes more critical when the defense of nodes must be oriented against terrorism and sabotage.

5-7. TIME AVAILABLE

Units must organize and establish priorities of work, depending upon the time available. Many tasks can be accomplished simultaneously, but priorities for preparation should be in accordance with the commander's order. A sample priority of work sequence follows:

- Establish security and communications.
- Assign sectors of responsibility and final protective fires.
- Clear fields of fire.
- Select and prepare initial fighting positions.
- Establish and mark routes between positions.
- Emplace obstacles and mines.
- Improve fighting positions.

5-8. TROOPS AVAILABLE

The defensive employment of troops in urban areas is governed by all METT-TC factors and on the ROE. The defender has a terrain advantage and can resist the attacker with much smaller forces.

5-9. CIVIL CONSIDERATIONS

See Chapter 13, Section III for more information on civil considerations.

Section III. DEFENSIVE FRAMEWORK AND ORGANIZATION

This section discusses the defensive framework and organization used during the planning and execution of defensive UO.

5-10. DEFENSIVE FRAMEWORK

Similar to offensive operations, the brigade will be the primary headquarters that will be task-organized to conduct defensive urban operations. The brigade can conduct the full range of defensive operations within a single urban area or in an AO that contains several small towns and cities using the elements shown in the defensive urban operational framework in Figure 5-5, page 5-8. The elements are similar to those in offensive operations in that the brigade commander attempts to set the conditions for tactical success. Isolation of the brigade by the enemy is avoided through security operations; defensive missions are assigned subordinate task forces in order to achieve the commander's intent and desired end-state; and then the brigade transitions to stability and or support operations. During urban defensive operations, the transition to stability and support operations may not be clear to the soldiers conducting the operations. Commanders must offset this tendency with clear mission type orders and updated ROE.

Again, as in offensive operations, the elements are not phases. They may occur simultaneously or sequentially. Well planned and executed defensive operations will have all four elements present. During defensive operations the brigade commander seeks to:

- Avoid being isolated by the enemy.
- Defend only the decisive terrain, institutions, or infrastructure.
- Conduct counter or spoiling attacks to retain the initiative.

Battalion TFs and below conducts defensive operations by conducting counterreconnaissance missions and patrols (shaping/avoiding isolation); assigning battle positions or sectors to subordinate units (dominating); and consolidating/reorganizing and preparing for follow-on missions (transitioning).

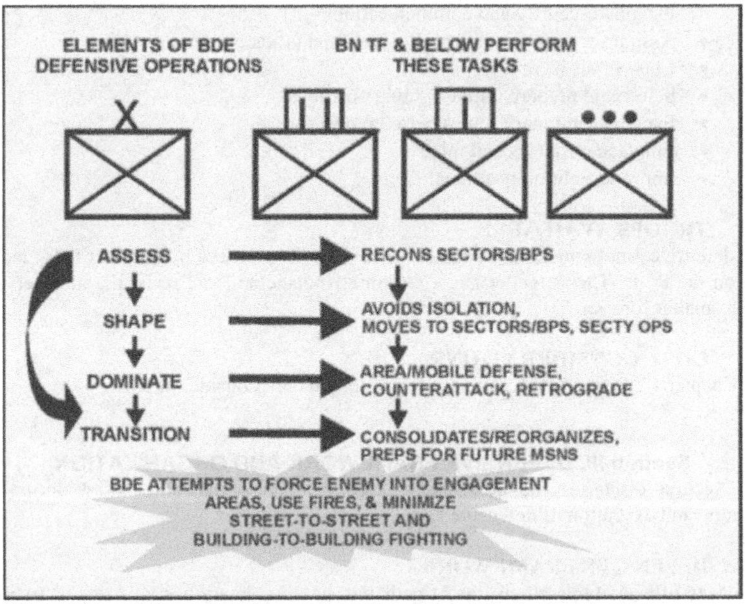

Figure 5-5. Defensive urban operational framework.

5-11. COMMAND AND CONTROL

In all defensive situations, commanders should position themselves well forward so that they can control the action. In urban terrain, this is even more critical due to obstacles, poor visibility, difficulty in communication, and intense fighting. Other key leaders may be placed in positions to report to the commander and to make critical, time-sensitive decisions.

a. **Graphic Control Measures.** The use of graphic control measures to synchronize actions become even more important to mission accomplishment in an urban environment (Figure 5-6). Phase lines can be used to report the enemy's location or to control the advance of counterattacking units. Principal streets, rivers, and railroad lines are suitable phase lines, which should be clearly and uniformly marked on the near or far side of the street or open area. Checkpoints aid in reporting locations and controlling movement. Contact points are used to designate specific points where units make physical contact. Target reference points (TRPs) can facilitate fire control. Many of these points can be designated street intersections. These and other control measures ensure coordination throughout the chain of command.

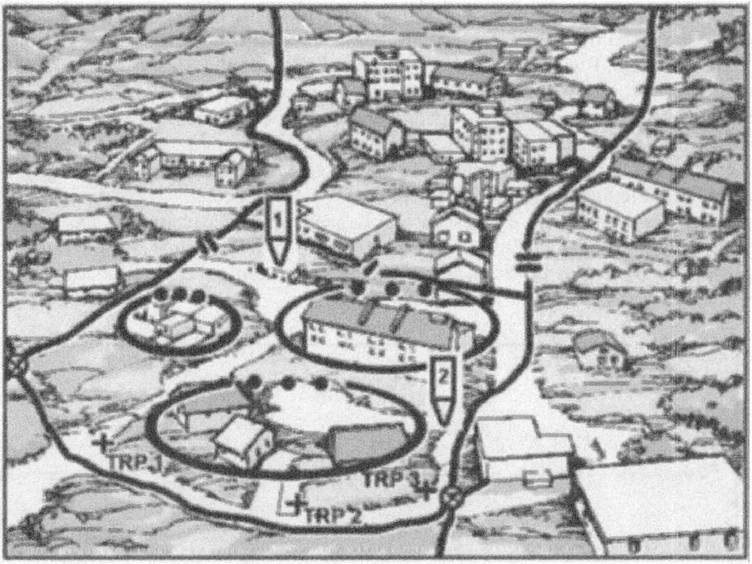

Figure 5-6. Graphic control measures.

b. **Command Post Facilities.** Command post (CP) facilities should be located underground, if possible, or in solidly constructed buildings. Their vulnerability requires all-round security. Since each facility may have to secure itself, it should be near the reserve for added security. When collocated with another unit, command post facilities may not need to provide their own security. Also, a simplified organization for command posts is required for ease of movement. Since rubble often hinders movement of tracked and wheeled vehicles, the CP must be prepared to backpack communications and other needed equipment for operations. Identification of alternate CP locations and routes to them must also be accomplished.

c. **Communications Restrictions.** Radio communications is initially the primary means of communication for controlling the defense of an urban area and for enforcing security. Structures and a high concentration of electrical power lines may degrade radio communication in urban areas. Wire is emplaced and used as the primary means of communications as time permits. However, wire can be compromised if interdicted by the enemy. Messengers can be used as another means of communication. Visual signals may also be used but are often not effective because of the screening effects of buildings and walls. Signals must be planned, widely disseminated, and understood by all assigned and attached units. Increased battle noise makes the effective use of sound signals difficult.

5-12. ORGANIZATION AND PREPARATION OF THE DEFENSE

The defensive organization described in this paragraph will likely be used against a conventional enemy force that may threaten US forces with mechanized and dismounted Infantry supported by other combined arms. This defensive organization may also occur in a brigade area of operation (AO) where there are multiple threats. For example, one part of the AO may require linear features; other parts may require the use of other defensive techniques, such as a perimeter defense, against different types of threats in the same brigade AO. METT-TC factors and the ROE determine how units plan, prepare, and execute the defense. The defense is organized into three areas—the security force area, main battle area, and rear area (Figure 5-7). Units defending in urban areas may have missions in any one of these areas, depending on the nature of the operation. Infantry units are well suited to conduct defensive operations in close urban terrain where engagement ranges will be short, where there is abundant cover and concealment, and where the enemy's assault must be repelled.

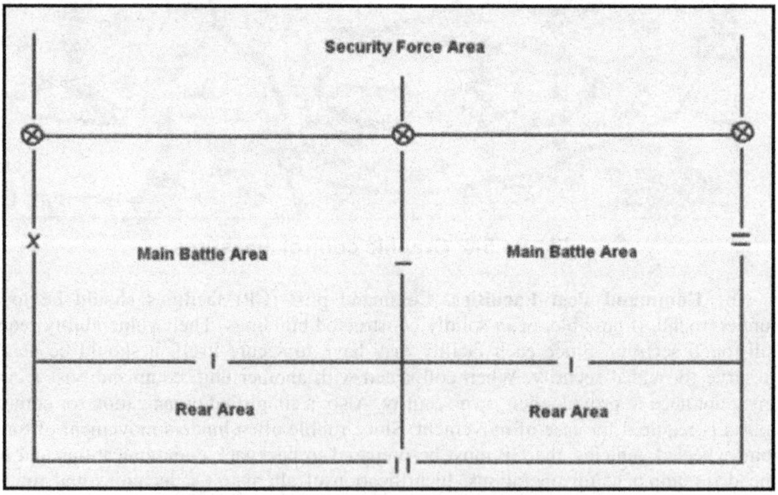

Figure 5-7. Organization of the defense.

a. **Patterns of Defense.** Of the two patterns of defense, area and mobile, the area defense will probably be the pattern most used since many of the reasons for defending on urban terrain are focused on retaining terrain. The mobile defense pattern is more focused on the enemy and the commander may decide to use it based on his estimate of the situation. Many defenses may include a combination of both. In large urban areas, the concept of defensive operations may be mobile and exploit depth, with the defender concentrating on moving forces from key terrain features or buildings to other similar features. The defender must seek to disrupt the enemy throughout all phases of battle.

b. **General Concept of the Defense.** Planning of the defense must be detailed and centralized while execution is decentralized. In an urban area, the defender must take advantage of inherent cover and concealment afforded by the urban terrain. He must also consider restrictions to the attacker's ability to maneuver and observe. By using the terrain and fighting from well-prepared and mutually supporting positions, a defending force can inflict heavy losses on, delay, block, or fix a much larger attacking force. The defense of an urban area should be organized around key terrain features, buildings, and areas that preserve the integrity of the defense and that provide the defender ease of movement. The defender must organize and plan his defense by considering obstacles, avenues of approach, key terrain, observation and fields of fire, cover and concealment, (OCOKA) and the considerations in this paragraph. Detailed knowledge of the terrain by the defender allows him to force an attacker to expend exorbitant amounts of time, supplies, equipment, and manpower.

(1) ***Reconnaissance.*** To obtain the detailed knowledge that they need, the commanders and staffs need to conduct a reconnaissance of the defensive area. The amount of time spent and the level of detail obtained will vary greatly between a deliberate defense and a hasty defense. The defender must identify the following:

- Positions that enable him to place suppressive fires on the enemy.
- Covered and concealed routes for friendly elements to move between positions (subways and sewers).
- Structures that dominate large areas.
- Areas such as parks, boulevards, rivers, highways, and railroads where antiarmor weapons have fields of fire.
- Firing positions for mortars.
- Command locations that offer cover, concealment, and ease of command and control.
- Protected storage areas for supplies.

(2) ***Security Operations.*** The defensive battle normally begins with a combined arms force conducting security operations well forward of the main body. Operations consist of security, reconnaissance, and counterreconnaissance tasks. Counterreconnaissance missions to support these operations employ ambushes, mines, obstacles, deception, security patrols, OPs, indirect fires, camouflage, demonstrations, and other measures to destroy or deceive the enemy's reconnaissance elements. Again, urban areas are well suited for infantry counterreconnaissance operations because of the abundance of cover and concealment that permits infantry to move by stealth.

c. **Main Battle Area**. The decisive battle is usually fought in the main battle area (MBA). Depending on the threat, units can deploy on the forward edges of the urban area or in battle positions in depth. In either case, the defense is made stronger by including forces that are defending on close terrain or on the flanks into the defensive scheme.

(1) ***Size of Battle Positions.*** The size and location of battle positions within the area of operations depends mainly on the type of enemy encountered and the ability to move between positions to block threatened areas. It may be desirable to place small antiarmor elements, secured by infantry, on the forward edges while the main defense is deployed in depth.

(2) ***Considerations.*** Defensive positions on the forward edge of a city or town should:

- Provide early warning of the enemy's advance.
- Engage the enemy at long range.
- Deceive the enemy as to the true location of the defense.

(3) ***Sectors.*** Depending on the factors of METT-TC, units may also assign sectors to defend instead of battle positions. In certain instances, the units may employ both. Sectors would normally be assigned when blocks and streets provide a grid type pattern and boundaries can be clearly delineated. (See FMs 7-20 and 7-10 for detailed information on when to assign either or both.)

(4) ***Frontages.*** Infantry units will normally occupy less terrain in urban areas. For example, an infantry company, which might occupy 1,500 to 2,000 meters in open terrain, is usually restricted to a frontage of 300 to 800 meters in urban areas. The density of buildings and rubble and street patterns will dictate the frontage of the unit (Table 5-1).

UNIT	FRONTAGES	DEPTHS
Battalion or Battalion TF	4 to 8 blocks	3 to 6 blocks
Company or Company Team	2 to 4 blocks	2 to 3 blocks
Platoon	1 to 2 blocks	1 block
NOTE: An average city block has a frontage of about 175 meters. These minimum figures apply in areas of dense, block-type construction; multistory buildings; and underground passages.		

Table 5-1. Approximate frontages and depths in large urban areas.

(5) ***Selection of Buildings.*** Buildings that add most to the general plan of defense are chosen for occupation. Mutual support between these positions is vital to prevent the attacker from maneuvering and outflanking positions, making them untenable. Buildings chosen for occupation as defensive positions should:

- Offer good protection.
- Have strong floors to keep the structure from collapsing under the weight of debris.
- Have thick walls.
- Be constructed of nonflammable materials (avoid wood).
- Be strategically located (corner buildings and prominent structures).
- Be adjacent to streets, alleys, vacant lots, and park sites. These buildings usually provide better fields of fire and are more easily tied in with other buildings.
- Be covered by friendly fire and offer good escape routes.

(6) ***Occupation of Positions.*** See paragraph 5-12 and Chapter 3, Section IV.

(7) ***Obstacles.*** Obstacles are easily constructed in an urban area. An urban area itself is an obstacle since it canalizes and impedes an attack. Likely avenues of approach should be blocked by obstacles and covered by fire (Figure 5-8, page 5-14). Units must hinder or prevent enemy maneuver without interfering with its own maneuver elements. Therefore, the battalion usually detonates cratering charges at key street locations on order. Mines are laid on the outskirts of the urban area or the sector and along routes the unit will not use. Barriers and obstacles are normally emplaced in three belts, consistent with the ROE. All avenues of approach (three-dimensional) must be denied. Units must not overlook the use of field-expedient materials, such as cars, light poles, and so on, or the emplacement of command-detonated antipersonnel mines and antitank mines. Commanders must clearly understand the ROE and what they will be permitted to emplace. When necessary, obstacles can be emplaced without mines and covered by fire.

(a) *First Belt.* The first obstacle belt is at the nearest buildings across from and parallel to the main defensive position (MDP). This belt consists of wire and improvised barriers to include: building interiors, subterranean avenues of approach, and exterior areas, such as open areas, danger areas, and dead space. The barriers and obstacles are covered by long-range fires. This belt impedes enemy movement, breaks up and disorganizes attack formations, and inflicts casualties and is protective in nature.

(b) *Second Belt.* The second obstacle belt is placed between the first belt and the MDP buildings, but out of hand grenade range from defensive positions. It impedes movement, canalizes the enemy into the best fields of fire, breaks up attack formations, and inflicts casualties. This belt is not meant to stop enemy soldiers permanently. It should be constructed efficiently to give the most benefit—not to be an impenetrable wall. It consists mainly of wire obstacles, improvised barriers, road craters, and mine fields. It should include command-detonated Claymores. Triple-strand concertina is placed along the machine gun final protective line (FPL), as marked earlier IAW unit SOP, to slow the enemy on the FPL and to allow the machine gun to be used effectively.

(c) *Third Belt.* The third obstacle belt is the defensive position's denial belt. It consists of wire obstacles placed around, through, and in the defensive buildings and close-in mine fields as well as in subterranean accesses. It impedes and complicates the enemy's ability to gain a foothold in the defensive area. Command-detonated Claymores should be used extensively. Claymores should be placed so as not to cause friendly casualties when detonated.

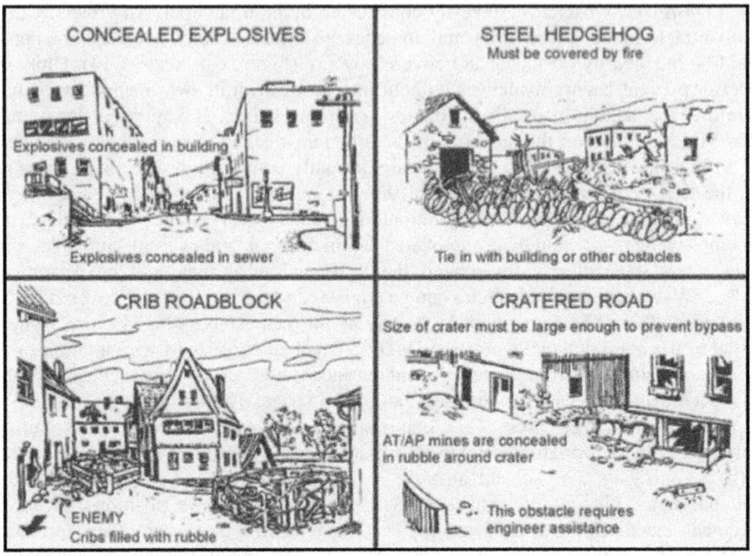

Figure 5-8. Example of urban obstacles.

(8) ***Rubbling.*** If they have the authority and the ROE permit, commanders also decide if buildings should be rubbled to increase fields of fire. However, rubbling the buildings too soon or rubbling too many may disclose exact locations and destroy cover from direct fire. Because rubbling may take more resources than are available to units, careful consideration of available resources must be made prior to rubbling. Additionally, care must be taken not to rubble areas that are necessary to support operations, such as MSRs. Buildings are normally rubbled with engineer assistance; engineers will usually employ explosives and engineer equipment to accomplish this task. If available, armored vehicles can be used to rubble buildings.

(9) ***Fire Hazards.*** The defender's detailed knowledge of the terrain permits him to avoid areas that are likely to be fire hazards. All urban areas are vulnerable to fire, especially those with many wooden buildings. The defender can deliberately set fires—

- To disrupt and disorganize the attackers.
- To canalize the attackers into more favorable engagement areas.
- To obscure the attacker's observation.

Likewise, the enemy may cause fires to confuse, disrupt, or constrain friendly forces and efforts. Units should anticipate this possibility and ensure that fire-fighting equipment is on hand when conducting these types of operations. Battalion S4s can move sand and water to buildings. The S5 can coordinate for usage of local fire-fighting equipment. Defensive positions should not be located atop known gas lines, oil storage tanks, or other highly flammable areas.

(10) ***Actions on Contact.*** When enemy forces enter and maneuver to seize initial objectives, the defender should employ all available fires to destroy and suppress the direct-fire weapons that support the ground attack. Tanks and enemy APCs should be engaged as soon as they come within the effective range of antiarmor weapons. As the enemy attack develops, the actions of small-unit leaders assume increased importance. Squad and platoon leaders are often responsible for fighting independent battles. Thus, it is important that all leaders understand their commander's concept of the defense. Situational awareness must be maintained and where the enemy's efforts are likely to result in a gaining a foothold, violent counterattacks must be employed to deny him access into the MBA.

(11) ***Employment of a Reserve.*** The unit defensive plan must always consider the employment of a reserve. The reserve force should be prepared to counterattack to regain key positions, to block enemy penetrations, to protect the flanks, or to assist by fire in the disengagement and withdrawal of positions. During urban combat, a reserve force—

- Normally consists of infantry.
- Must be as mobile as possible.
- May be a company or platoon at battalion level; a squad or platoon at company level.
- May be supported by tanks or other armored vehicles.
- Should be positioned as close as possible to the place where it is anticipated being employed.

(12) ***Counterattacks.*** All elements should be prepared to counterattack. The best counterattack force is a small, infantry-heavy element supported by BFVs and or tanks, if available. They should be prepared to counterattack to regain key positions, to block enemy penetrations, to provide flank protection, and to assist by fire the disengagement and withdrawal of endangered positions. It is especially important for enemy footholds to be repelled violently. When an element is committed to counterattack to reinforce a unit, it may be attached to the unit in whose sector the counterattack is taking place. Otherwise, the counterattack becomes the main effort. This makes coordination easier, especially if the counterattack goes through the unit's positions.

(13) ***Defense During Conditions of Limited Visibility.*** Commanders can expect the attacker to use conditions of limited visibility to conduct operations to sustain or gain daylight momentum.

(a) Commanders should employ the following measures to defend against attacks during limited visibility:

- Defensive positions and crew-served weapons should be shifted from an alternate position or a hasty security position just before dark to deceive the enemy as to the exact location of the primary position.
- Unoccupied areas between units, which can be covered by observed fire during daylight, may have to be occupied, blocked, or patrolled during limited visibility. Early warning devices and obstacles need to be installed.
- Radar, remote sensors, and night observation devices should be emplaced to cover streets and open areas. Thermal imagery devices, such as the one found on the TOW weapon system, are excellent for observation during limited visibility.

- Noise-making devices, tanglefoot tactical wire, and LP/OPs should be positioned on all avenues of approach for early warning and to detect infiltration.
- Artificial illumination should be planned, to include the use of street lamps, stadium lights, pyrotechnics, visible and IR ILLUM, and so forth.
- Indirect fire weapons, grenade launchers, and hand grenades should be used when defenses are probed to avoid disclosure of defensive positions.
- Tank and BFV platoons must know the locations of friendly positions. The use of thermal recognition signals and markers can help decrease the possibility of fratricide.

(b) Commanders should initiate FPFs through the use of a planned signal. Crew-served weapons, armored vehicle-mounted weapons if available, and individual riflemen fire within their assigned sectors. Grenades and command-detonated mines should be used to supplement other fires as the enemy approaches the positions.

(c) Defenders should move to daylight positions before BMNT. Buildings should be marked from the friendly side IAW unit SOP in order to facilitate movement. Armored vehicles can be used to cover the movement of friendly troops.

d. **Rear Area.** Units may be deployed in the rear area to protect CSS elements and to defend high payoff assets, lines of communications, C2 nodes, and other key locations. Units will employ the tactics, techniques, and procedures (TTP) discussed in Sections V, VI, and VII.

5-13. PRIORITIES OF WORK

Priorities of work in during defensive operations in urban areas are the same as other defensive operations. Specific considerations for a defense on urbanized terrain are discussed in this paragraph.

a. **Establish Security.** Units should quickly establish all-round security by placing forces on likely avenues of approaches. The level of security (50 percent, 30 percent, and so forth) is determined by METT-TC factors. The reconnaissance and counterreconnaissance plan should be emphasized. While security is being established, civilians located within the defensive area need to be identified and evacuated.

b. **Assign Areas of Responsibility.** Boundaries define sectors of responsibility. They include areas where units may fire and maneuver without interference or coordination with other units. Responsibility for primary avenues of approach should never be split. In areas of semidetached construction, where observation and movement are less restricted, boundaries should be established along alleys or streets to include both sides of a street in a single sector. Where buildings present a solid front along streets, boundaries may have to extend to one side of the street. Battle positions should also be specifically assigned, as required by METT-TC. Commanders and leaders should specify which buildings comprise the battle positions or strongpoints. Positions should be clearly designated so that no doubt remains as to which elements will have responsibility for occupation or control.

c. **Clear Fields of Fire.** In urban areas, units may need to rubble certain buildings and structures to provide greater protection and fields of fire to the defender (see paragraph 5-12c(8), Rubbling). If the ceiling of a lower-story room can support the weight of the rubble, collapsing the top floor of a building before the battle starts may afford better protection against indirect fires. Rubbling an entire building can increase the fields of fire and create an obstacle to enemy movement. Planning must be extensive so that rubbled buildings will not interfere with planned routes of withdrawal or counterattack. Vehicles may also have to be moved to clear fields of fire.

d. **Select and Prepare Initial Fighting Positions.** Units should select positions in depth. Units should prepare positions as soon as troops arrive and continue preparing as long as positions are occupied. Enemy infiltration or movement sometimes occurs between and behind friendly positions. Therefore, each position must be organized for all-round defense. The defender should also:

(1) Make minimum changes to the outside appearance of buildings where positions are located.

(2) Screen or block windows and other openings to keep the enemy from seeing in and tossing in hand grenades. This must be done so that the enemy cannot tell which openings the defenders are behind.

(3) Remove combustible material to limit the danger of fire. Fires are dangerous to defenders and create smoke that could conceal attacking troops. For these reasons, defenders should remove all flammable materials and stockpile fire-fighting equipment (water, sand, and so forth). The danger of fire also influences the type of ammunition used in the defense. Tracers or incendiary rounds should not be used extensively if threat of fire exists.

(4) Turn off electricity and gas. Both propane and natural gas are explosive. Natural gas is also poisonous, displaces oxygen, and is not filtered by a protective mask. Propane gas, although not poisonous, is heavier than air. If it leaks into an enclosed area, it displaces the oxygen and causes suffocation. Gas mains and electricity should be shut off at the facility that serves the urban area.

(5) Locate positions so as not to establish a pattern. Units should avoid obvious firing locations like towers and buildings prohibited for use by the Law of Land Warfare, such as churches.

(6) Camouflage positions.

(7) Reinforce positions with all materials available such as mattresses, furniture, and so forth. The S4 will have to arrange for as much protective material as possible. Caution should be taken as mattresses and fabric furniture are flammable. Drawers and cabinets should be filled with earth or sand to provide cover. Vehicles, such as trucks or buses can be placed over positions outside buildings. Flammable fluids should be drained. Other flammables, such as seats should be removed, and the gas tank filled with water.

(8) Block stairwells and doorways with wire or other material to prevent enemy movement. Create holes between floors and rooms to allow covered and concealed movement within a building.

(9) Prepare range cards, fire plans, and sector sketches.

(10) Look at how basements may be used. If grazing fire can be achieved from basement widows, emplace machine guns in basements. When basements are not used, they should be sealed to prevent enemy entry.

(11) Cache resupply of ammunition, water, and medical supplies.

e. **Establish Communications.** Commanders should consider the effects of urban areas on communications when they allocate time to establish communications. Line-of-sight limitations affect both visual and radio communications. Wire laid at street level is easily damaged by rubble and vehicle traffic. The noise of urban area combat is much louder than in other areas, making sound signals difficult to hear. Therefore, the time needed to establish an effective communications system in urban terrain may be greater than in other terrain. Units should consider the following techniques when planning for communications:

- Emplace line of sight radios and retransmission sites on the upper floors of buildings.
- Use existing telephone systems. However, telephones are not secure even though many telephone cables are underground.
- Use messengers at all levels since they are the most secure means of communications.
- Lay wire through buildings for maximum protection, if the assets are available.

f. **Emplace Obstacles and Mines.** To save time and resources in preparing the defense, commanders must emphasize using all available materials (automobiles, railcars, rubble) to create obstacles. Civilian construction equipment and materials must be located and inventoried. This equipment can be used with engineer assets or in place of damaged equipment. Coordination must be made with proper civilian officials before use.

(1) Engineers must be able to provide advice and resources as to the employment of obstacles and mines. The principles for employing mines and obstacles do not change in the defense of an urban area; however, techniques do change. For example, burying and concealing mines in streets are hard due to concrete and asphalt. Consider placing mines in sandbags so they cannot be seen and also using fake mines placed in sandbags in order to deceive the enemy.

(2) FASCAM may be effective on the outskirts of an urban area or in parks; however, in a city core, areas may be too restrictive. Mines and obstacles must be emplaced consistent with the ROE. Any antipersonnel mines must be command-detonated. Riot control agents may be employed to control noncombatant access into defensive areas, if permission is granted by the National Command Authority (NCA).

g. **Improve Fighting Positions.** When time permits, all positions, to include supplementary and alternate positions, should be reinforced with sandbags and provided overhead cover. Attached engineers can help in this effort by providing advice and assisting with construction.

h. **Establish and Mark Routes Between Positions.** Reconnaissance by all defending elements will assist in route selection for use by defenders moving between positions. Movement is crucial in fighting in urban areas. Early selection and marking of routes adds to the defender's advantages.

Section IV. BRIGADE DEFENSIVE OPERATIONS

This section discusses planning considerations and provides tactics and techniques for the planning of brigade defensive UO.

5-14. DEFENSIVE PLANNING

In planning a defense in an urban area, the brigade staff must identify the following:

- Positions and areas that must be controlled to prevent enemy infiltration.

- Sufficient covered and concealed routes for movement and repositioning of forces.
- Structures and areas that dominate the urban area.
- Areas such as parks and broad streets that provide fields of fire for tanks and antiarmor weapons.
- Position areas for artillery assets.
- C2 locations.
- Protected areas for CSS activities.
- Suitable structures that are defensible and provide protection for defenders.
- Contingency plans in the event that the brigade must conduct breakout operations.
- Plans for rapid reinforcement.

a. Units defending in urban areas must prepare their positions for all around defense. The brigade must employ aggressive security operations that include surveillance of surface and subsurface approaches. The brigade must constantly patrol and use OPs and sensors to maintain effective security. Special measures must be taken to control possible civilian personnel who support the enemy or enemy combatants who have intermixed with the local population. Consideration must also be given to the protection of non-combatants that remain in the AO, and contingency actions in the event that the situation deteriorates and requires their evacuation.

b. Defensive fire support in urban operations must take advantage of the impact of indirect fires on the enemy before he enters the protection of the urban area. Fire support officers at all levels must coordinate and rehearse contingencies that are inherent to nonlinear fire support coordination measures and clearance of fires. Mutually supporting observation plans for daylight and periods of limited visibility must account for the degradation of lasers in well-lit urban areas. The brigade fire support officer also plans and coordinates nonlethal capabilities for the brigade (see Chapter 10). Civil affairs and PSYOP assets should be coordinated with the appropriate command and control warfare/information operations headquarters.

5-15. INTEGRATING THE URBAN AREA INTO THE DEFENSE

The brigade may also integrate villages, strip areas, and small towns into the overall defense, based on higher headquarters' constraints and applicable ROE (Figure 5-9, page 5-20). A defense in an urban area or one that incorporates urban areas normally follows the same sequence of actions and is governed by the principles contained in FM 7-30, Chapters 5 and 6. When defending large urban areas, the commander must consider that the terrain is more restrictive due to buildings that are normally close together. This requires a higher density of troops and smaller AOs than in open terrain. The brigade normally assigns task force AOs and may use phase lines, control measures, or other positions to position forces in depth.

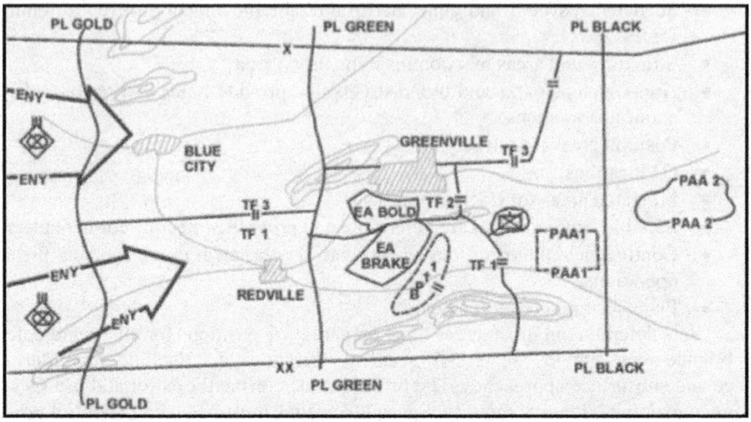

Figure 5-9. Integrating urban areas into a defense.

5-16. NODAL DEFENSE

Figure 5-10 depicts a transitional situation where the brigade moves from an offensive to a defensive or stability operation. The brigade mission may contain METT-TC factors that require varying defensive techniques by the subordinate battalion TFs under the brigade's control. Considerations in a situation such as this include:

a. **Task Organization.** TFs may very well have to be task-organized differently to conduct the specific missions assigned by the brigade commander. The task organization required for the defensive or stability operation will probably be different from the task organization used in an offensive operation.

b. **Symmetrical/Asymmetrical Threats.** The brigade will likely respond to both symmetrical and asymmetrical threats within the area of operations. The defensive techniques chosen by subordinate battalion TFs should be capable of responding to the specific threats in their respective AOs.

c. **Boundary Changes.** Again, based on the commander's intent and the brigade's defensive scheme of maneuver, boundary changes may be required in order to give battalion's more or less maneuver space.

d. **ROE Modification.** The ROE may require modification based on the type of mission to be conducted. The ROE may become more or less restrictive based on METT-TC factors. Commanders and leaders must insure that the ROE are clearly stated and widely disseminated at the beginning and conclusion of each day.

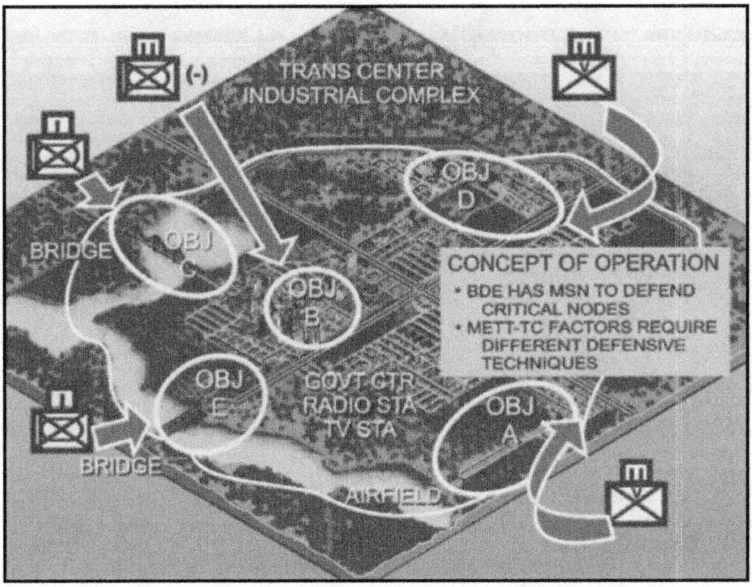

Figure 5-10. Nodal defense, transitional situation.

Figure 5-11 on page 5-22 depicts a nodal defense where TFs employ varying defensive techniques in order to achieve the brigade commander's desired end-state. The brigade commander's intent is to safeguard the key nodes that were seized during the offensive action in order to eventually return the infrastructure of this particular urban area back to civilian control. A combination of sectors, battle positions, strong points, roadblocks, checkpoints, security patrols, and OPs could be employed throughout the brigade AO. Figure 5-11 depicts the changed TF task organizations, the extended boundaries, and directed brigade OPs.

NOTE: TF operational graphics were drawn in order to provide an example of a possible technique that may be employed within the brigade AO in order to meet the brigade commander's intent. For example, the TF defending the transportation center has elected to use a perimeter defense for inner security and has assigned the attached mechanized infantry company team the mission to conduct outer security by means of a screen and manning the designated brigade OP.

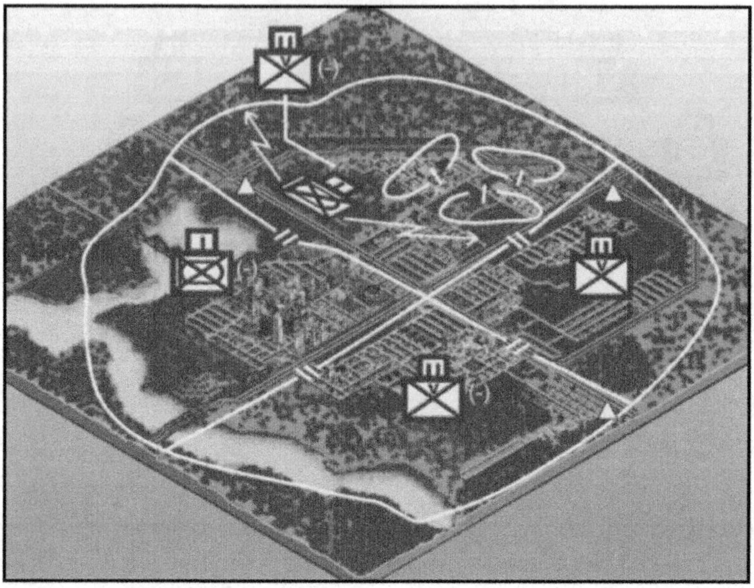

Figure 5-11. Nodal defense, varying defensive techniques.

Section V. BATTALION DEFENSIVE OPERATIONS

This section discusses planning considerations and provides tactics and techniques for the planning of battalion TF defensive UO.

5-17. EMPLOYMENT OF COMBAT AND COMBAT SUPPORT ASSETS

This paragraph will concentrate on the employment of combat and combat support assets at the battalion level. Once the battalion has decided where to defend, it should select company battle positions or sectors that block or restrict the enemy's ability to maneuver and control key areas. The battalion needs to plan two levels down at the platoon level where the battle will be fought. The frontage for a platoon is about one to two city blocks long. Platoons can occupy about three small structures or one larger two- or three-story building (Table 5-1 on page 5-12 and Figure 5-6 on page 5-9), depending on METT-TC factors. Companies may be tasked to detach a platoon to act as the battalion reserve.

a. **Mortar Platoon.** The battalion mortar platoon may be initially positioned forward in support of the security area. After withdrawal of security forces, it is positioned to support the entire battalion. (See Chapter 12 and Appendix K for detailed information concerning mortar employment in urban terrain.) Mortars at the battalion level are employed to maximize the effect of their high-angle fires. They should be used to engage:

- Enemy overwatch positions.
- Enemy infantry before they seize a foothold.
- Targets on rooftops.
- Enemy reinforcements within range.

b. **AT Weapons.** The commander will give the AT platoon missions that support the defensive scheme of maneuver based on the capabilities and limitations of the system and the type of threat that the battalion will face. For example, battalion defending against conventional threats that have armored vehicles will most likely give the AT platoon missions that primarily defend against armored threats. Battalions defending against asymmetrical threats will most likely give the AT platoon missions that will enhance force protection.

c. **Scout Platoon.** Depending on the situation and terrain, the battalion scout platoon may provide a security force forward of the battalion to give early warning of enemy activity. Alternately, the scout platoon may be used to screen a flank or the rear.

d. **Employment of Tanks and BFVs.** The battalion should employ tanks and BFVs to take advantage of their long-range fires and mobility. Urban areas restrict the mobility of tanks and BFVs and make them vulnerable to enemy infantry antiarmor weapons.

(1) When tanks and BFVs are employed in the defense of an urban area, infantry should be positioned to provide security against close antitank fires and to detect targets for the armored vehicles. Tanks and BFVs should be assigned engagement areas in support of the defensive scheme of maneuver. BFVs may be placed along the forward edge of the area in order to engage enemy armored vehicles. Friendly armored vehicles can also be placed in positions to the rear of the buildings and interior courtyards where their weapon systems can provide added rear and flank security. Combat vehicles are assigned primary, alternate, and supplementary positions as well as primary and secondary sectors of fire. They should be positioned in defilade behind rubble and walls or inside buildings for movement into and out of the area. Armored vehicles can also be used for resupply, CASEVAC, and rapid repositioning during the battle. BFVs can also provide a mobile reserve. Tank or BFV elements should be placed OPCON to a light infantry battalion rather than attached. A tank or BFV element attached or OPCON will have to be divided up within the defensive area to take advantage of the fires available to this asset. BFVs and antitank weapons should supplement tank fires. Tanks and BFVs may be—

- Positioned on the edge of the urban area in mutually supporting positions.
- Positioned on key terrain on the flanks of towns and villages.
- Used to cover barricades and obstacles by fire.
- Part of the reserve.

(2) Tanks and BFVs are normally employed as platoons. However, sections may be employed with light infantry platoons or squads based on METT-TC factors and identified engagement areas. This provides tanks and BFVs with the close security of the infantry. Tanks and BFVs provide the commander with a mobile force to respond quickly to enemy threats on different avenues of approach. They can also be effectively employed in counterattacks.

e. **Indirect Fire Support.** Fire planning must be comprehensive due to the proximity of buildings to targets, minimum range restrictions, repositioning requirements, and the ROE. Mortar and artillery fires are planned on top of and immediately around defensive positions for close support.

(1) ***Artillery.*** Artillery may be used as direct or indirect support. In the defense, artillery fire should be used to—

- Suppress and blind enemy overwatch elements.
- Disrupt or destroy an assault.
- Provide counterbattery fire.
- Support counterattacks.
- Provide direct fire when necessary.

(2) ***Fire Planning.*** Fire planning is conducted for urban areas in much the same manner as it is for other areas, taking into concern the limitations of the restrictive terrain. Consideration should be given to TRPs, covering obstacles, FPFs.

(3) ***Priorities of Fire.*** The commander should establish priorities of fire based on enemy avenues of approach and threat systems that present the greatest danger to the defense. For example, during the attacker's initial advance, tanks, BMPs, and overwatching elements are the greatest threat to the defense. In certain situations, enemy APCs may provide a larger threat than enemy tanks in an urban area; the APCs carry infantry, which can gain footholds in buildings. Artillery and mortar fires should suppress and destroy enemy ATGMs and overwatch positions and or elements. If enemy formations secure a foothold, priority is shifted to the destruction of enemy forces within the penetration.

(4) ***Control of Supporting Fires.*** As the enemy attack progresses in the city, fires are increased to separate infantry from supporting tanks and fighting vehicles. During this phase, friendly artillery concentrates on attacking infantry, counterfire missions, and the destruction of reinforcements that are approaching the city.

(5) ***Support of Counterattacks.*** When initiated, counterattacks are given priority of supporting fires. When artillery is firing the missions as mentioned above, it must remain mobile and be prepared to displace to preplanned positions to avoid enemy counterbattery fire.

f. **Employment of Engineers.** Normally, one engineer platoon or company supports a battalion or battalion task force. Engineers are employed under battalion control or attached to companies. Company commanders may be given an engineer squad to assist them in preparing the defense. The battalion commander and staff must consider engineer tasks that enhance survivability, mobility, and countermobility. The supporting engineers use C4 and other explosives to make firing ports, mouseholes, and demolition obstacles. Based upon priority of work, the battalion tells the attached or OPCON engineer element to assist each of the infantry companies preparing the village for defense and to execute their obstacle plan. The engineers' mission is to tell the infantrymen exactly where to place the demolitions and how much is needed for the desired effect. They assist in preparation of charges. Tasks that engineers can accomplish in the defense of an urban area include:

- Constructing obstacles and rubbling.
- Clearing fields of fire.
- Laying mines.
- Preparing mobility routes between positions.
- Preparing fighting positions.

g. **Air Defense Assets.** Air defense assets available to the commander, such as Stinger and Avenger, are normally employed to ensure all-round air defense. These assets are normally controlled at battalion level, however they may be placed under a company commander's control when METT-TC factors warrant that type of use. The lack of good firing positions for long-range air defense missile systems in some urban areas may limit the number of deployed weapons. In the defense, weapons systems may have to be winched or airlifted into positions. Rooftops and parking garages are good firing positions because they normally offer a better line-of-sight. Stingers and Avengers can be assigned the missions of protecting specific positions or of functioning in general support of the battalion.

h. **Battalion Trains/Service Support.** The battalion locates an area where the trains can be positioned near enough to provide support but far enough away to not get in the line of fire. (See Chapter 13.) A location is chosen near the main avenue of approach to ease resupply, recovery, and maintenance operations. Company trains are often collocated with the battalion trains. Ammunition expenditure is usually high when fighting in an urban area. To avoid moving around the village with ammunition resupply during the battle, ammunition should be stockpiled in each occupied platoon and squad position. Platoons should also stockpile firefighting equipment, drinking water, food, and first-aid supplies at each squad position. Other factors the battalion must consider are:

- Resupply.
- Medical evacuation.
- Firefighting.
- Security.

5-18. INTEGRATING URBAN AREAS INTO THE DEFENSE

The battalion may often integrate villages, strip areas, and small towns into the overall defense, based on higher headquarters' constraints and applicable ROE. (See Figure 5-12 on page 5-26.) A defense in an urban area, or one that incorporates urban areas, normally follows the same sequence of actions and is governed by the principles contained in Chapters 5 and 6 of FM 7-20. Specific TTP are discussed in paragraphs 5-19 through 5-22.

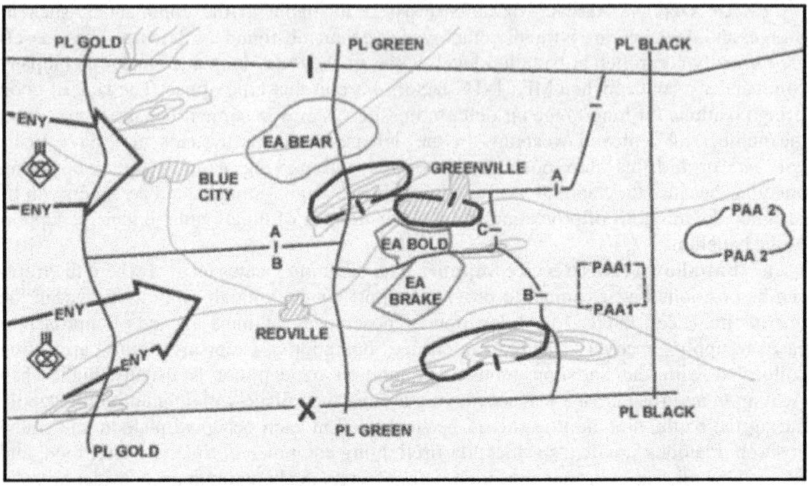

Figure 5-12. Integrating urban areas into the defense.

5-19. DEFENSE OF A VILLAGE

A battalion TF assigned a defensive sector that includes a village may incorporate the village as a strongpoint in its defense. This use of an urban area is most common when the village stands astride a high-speed avenue of approach or when it lies between two difficult obstacles. To incorporate such an area into its defense, the battalion TF must control the high ground on either side of the village to prevent the enemy from firing from those areas into the village.

a. The majority of the TF tanks and BFVs should be employed where maneuver room is the greatest on the key terrain to the flanks of the village. This is also where the TF BFVs should be employed. As the security force withdraws and companies and or teams assume the fight, BFVs can assume support by fire positions.

b. Although the battalion TFs disposition should prevent large enemy forces from threatening the rear and flanks of the village, the danger of small-unit enemy infiltration means the village must be prepared for all-round defense.

c. Engineers required for team mobility operations should stay with the company or company team in the village to provide continuous engineer support if the company team becomes isolated. The TF commander should centrally control engineer support for the rest of the TF. Engineer assets may be in DS of the other companies or company teams. The priority of barrier materials, demolitions, and mines should go to the company or company team in the village.

d. The TF commander should use any key terrain on the village flanks for maneuver to prevent the village's defense from becoming isolated. The strongpoints in the town should provide a firm location where the enemy can be stopped and around which counterattacks can be launched (Figure 5-13).

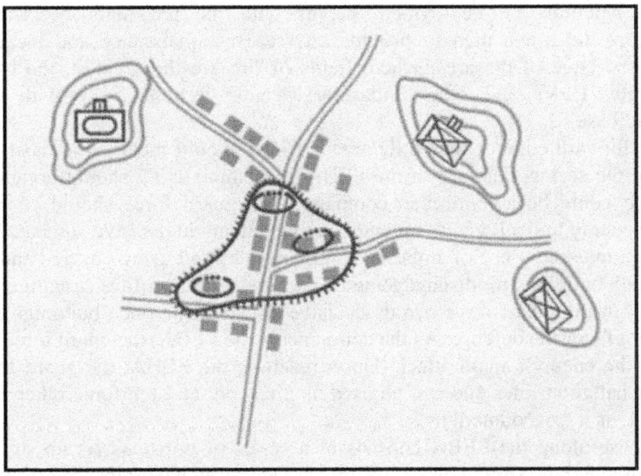

Figure 5-13. Battalion TF defense of a village.

5-20. DEFENSE IN SECTOR

A battalion TF may be given the mission of defending a sector in a city (Figure 5-14). The battalion should take advantage of the outlying structures to provide early warning and delay the enemy and take advantage of the tougher interior buildings to provide fixed defense. This defense should cover an area about 4 to 12 blocks square.

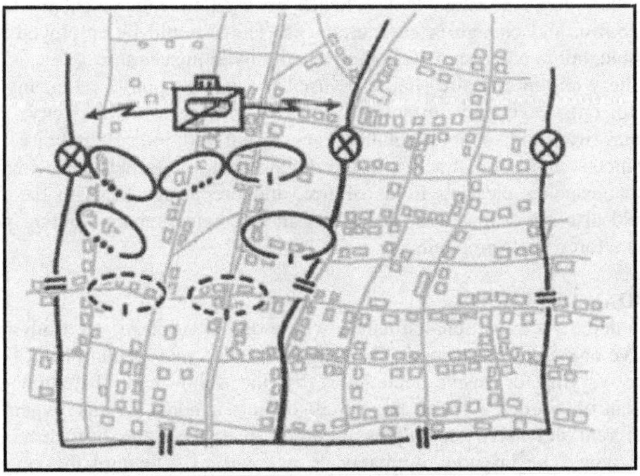

Figure 5-14. Defense in sector.

a. The battalion TF deployment begins with the reconnaissance/scout platoon reconnoitering the urban area to provide an area reconnaissance and location of the enemy. At the edge of the area, where fields of fire are the greatest, the battalion TF should deploy BFVs and other antiarmor weapon systems to provide long-range antiarmor defense.

b. The forward edge of the battle area (FEBA) should include the most formidable buildings in the sector. Forward of the FEBA, the battalion TF should organize a guard force, which could be a reinforced company. The guard force should concentrate on causing the enemy to deploy without engaging the enemy in decisive combat. This can be done through maximum use of ambushes and obstacles and using covered and concealed routes through buildings for disengagement. The guard force inflicts casualties and delays the enemy, but the guard force avoids decisive engagement since buildings beyond the FEBA do not favor the defense. As the action nears the FEBA, the guard force detects the location of the enemy's main attack. Upon reaching the FEBA, the guard force passes through the battalion lines and can be used as a reserve and reinforce other elements of the battalion, or it can counterattack.

c. Defense along the FEBA consists of a series of positions set up similar to that described in the company defense of the village (see paragraph 5-24). Key terrain features such as strong buildings, road junctions, and good firing positions should be the center of the strongpoint defense. Based on METT-TC considerations, the defense in sector may consist of either strongpoints or battle positions. Strongpoints located on or covering decisive terrain are extremely effective in the defense. Buildings should be prepared for defense as outlined in Chapter 3.

d. BFVs should be used to engage threat armored vehicles; to cover obstacles with fire; and to engage in counterattacks with tanks. They can also be used to transport casualties and supplies to and from the fight.

e. The battalion's attached tanks should be used to engage enemy tanks, cover obstacles by fire, and engage in counterattacks. They should be employed in platoons where possible, but in congested areas may be employed in sections.

f. Artillery and mortar fire should be used to suppress and blind enemy overwatch elements, to engage enemy infantry on the approaches to the city, to provide counterbattery fire, and to support counterattacks using both indirect and direct fire.

g. Engineers should be attached to the defending force to help in laying mines and constructing obstacles, clearing fields of fire, and preparing routes to the rear. These routes should also have obstacles. Engineers should help prepare fighting positions in support of the force in strongpoints.

5-21. NODAL DEFENSE

Figure 5-15 depicts a transitional situation where the battalion moves from an offensive to a defensive or stability operation. The brigade mission may contain METT-TC factors that require varying defensive techniques by the subordinate battalions under the brigade's control. Figure 5-16 on page 5-30 depicts a nodal defense where battalions employ different defensive techniques in order to achieve the brigade commander's desired end-state. The brigade commander's intent is to safeguard the key nodes that were seized during the offensive action in order to eventually return the infrastructure of this particular urban area back to civilian control. A combination of sectors, battle positions, strong points, roadblocks, checkpoints, security patrols, and OPs could be employed within the TF sector or AO. Figure 5-16 on page 5-30 depicts the changed TF task organizations, the extended boundaries, and directed brigade OPs. Considerations in a situation such as this include:

a. **Task Organization.** Companies may have to be task organized differently to conduct the specific missions assigned by the battalion or TF commander. The task organization required for the defensive or stability operation will probably be different from the task organization used in an offensive operation.

b. **Symmetrical/Asymmetrical Threats.** The battalion or TF will likely respond to both symmetrical and asymmetrical threats within the area of operations. The defensive techniques chosen by subordinate companies should allow them to respond to the specific threats in their respective AOs, battle positions, or sectors.

c. **Boundary Changes.** Again, based on the commander's intent and the battalion's or TF's defensive scheme of maneuver, boundary changes may be required in order to give companies more or less maneuver space.

d. **ROE Modification.** The ROE may require modification based on the type of mission to be conducted. The ROE may become more or less restrictive based on METT-TC factors. Commanders and leaders must ensure that the ROE are clearly stated and widely disseminated at the beginning and conclusion of each day.

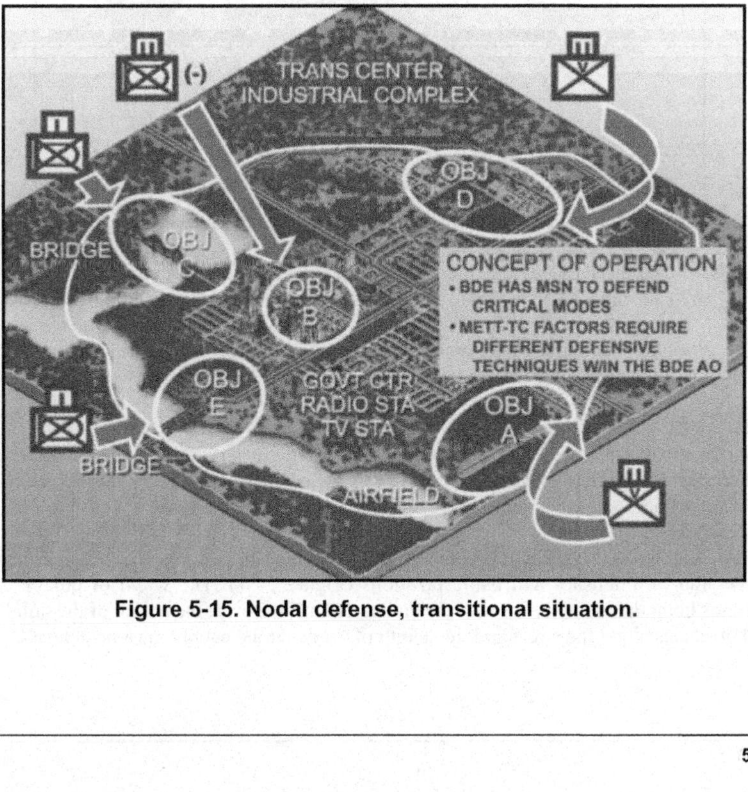

Figure 5-15. Nodal defense, transitional situation.

NOTE: In Figure 5-16, the northern TF defending the transportation center/industrial complex has decided to use a perimeter defense for inner security and has assigned the attached mechanized Infantry company the mission to conduct outer security by means of a screen and manning the designated brigade OP. Other TFs within the brigade AO may be required to use different defensive techniques.

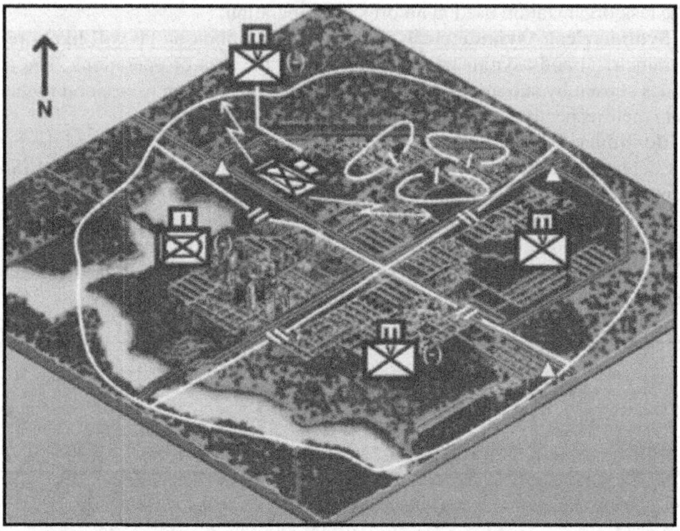

Figure 5-16. Nodal defense, different defensive techniques.

NOTE: The digital force has the potential to provide accurate threat information that can enhance situational awareness, which facilitates targeting and obstacle placement. JSTARS; GUARDRAIL; unmanned aerial vehicles, if present; and other reconnaissance assets will significantly improve the threat situational awareness and targeting capability of the unit.

5-22. DELAY

The purpose of a delay is to slow the enemy, cause enemy casualties, and stop the enemy (where possible) without becoming decisively engaged or bypassed. The delay can be oriented either on the enemy or on specified terrain such as a key building or manufacturing complex.

a. **Ambushes and Battle Positions.** A delay in an urban area is conducted from a succession of ambushes and battle positions (Figure 5-17). The width of the TF zone depends upon the amount of force available to control the area, the nature of the buildings and obstacles along the street and the length of time that the enemy must be delayed.

(1) ***Ambushes.*** Ambushes are planned on overwatching obstacles and are closely coordinated but they are executed at the lowest levels. The deployment of the TF is realigned at important cross streets. The ambushes can be combined with limited objective attacks on the enemy's flanks. These are usually effective in the edge of open spaces, parks, wide streets, and so on. Tanks and BFVs should execute these along with dismounted Infantry.

(2) ***Battle Positions.*** Battle positions should be placed where heavy weapons, such as tanks, BFVs, antiarmor weapons, and machine guns, will have the best fields of fire. Such locations are normally found at major street intersections, parks, and at the edge of open residential areas. Battle positions should be carefully and deliberately prepared, reinforced by obstacles and demolished buildings, and supported by artillery and mortars. They should be positioned to inflict maximum losses on the enemy and cause him to deploy for a deliberate attack.

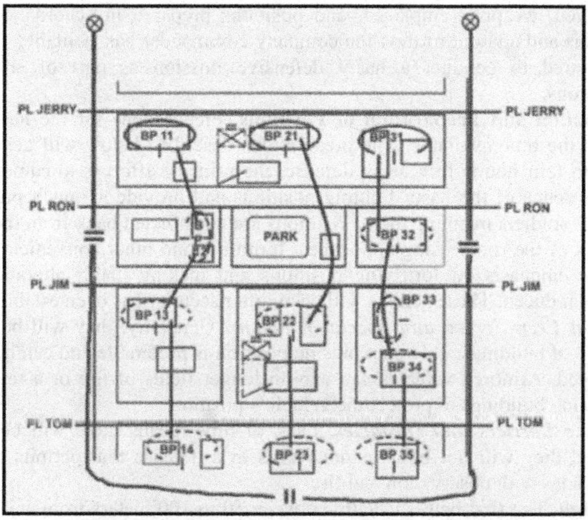

Figure 5-17. Battalion delay in an urban area.

b. **Two Delaying Echelons.** The TF is most effective when deployed in two delaying echelons, alternating between conducting ambushes and fighting from battle positions. As the enemy threatens to overrun a battle position, the company disengages and delays back toward the next battle position. As the company passes through the company to the rear, it establishes another battle position. Smoke and demolitions are used to aid in the disengagement. Security elements on the flank can be employed to prevent the enemy from out-flanking the delaying force. A small reserve can be used to react to unexpected enemy action and to conduct continued attacks on the enemy's flank.

c. **Engineers.** The engineer effort should first be centralized to support the preparation of battle positions and then decentralized to support the force committed to ambush.

Section VI. COMPANY DEFENSIVE OPERATIONS

This section discusses planning considerations and provides tactics and techniques for the planning of company team defensive UO.

5-23. HASTY DEFENSE

A very likely defensive mission for the Infantry company in urban terrain will be to conduct a hasty defense. This mission is characterized by reduced time for the preparation of the defense. All of the troop-leading procedures are the same. The priorities of work will basically be the same, but many will take place concurrently. Units will be deployed, weapons emplaced, and positions prepared in accordance with the mission analysis and amount of time the company commander has available. Companies must be prepared to conduct a hasty defensive mission as part of stability and support operations.

a. **Occupation and Preparation of Positions.** Preparations for the hasty defense will vary with the time available. The preparations described below will generally take between two to four hours. In a hasty defense, the primary effort is to camouflage and conceal the presence of the hasty fighting positions and provide as much protection as possible for the soldiers manning them. Positions are constructed back from the windows in the shadows of the room using appliances, furniture, and other convenient items and materials. The emphasis on fortifying positions and making major alterations to the environment is reduced. These actions will occur after security has been established.

(1) ***Position Crew-Served and Special Weapons.*** Generally, they will be employed from the inside of buildings, unless an outside position is preferable and can be protected and camouflaged. Armored vehicles can exploit longer fields of fire or a reverse slope engagement using buildings to protect the vehicle's position.

(2) ***Emplace Barriers and Obstacles.*** Lack of time means there will be two belts established and they will not be as extensive as in a defense that permits more time. Cover all obstacles with observation and fire.

(a) *First Belt.* The first belt is usually between 50 to 100 meters from and parallel to the defensive trace. It will normally consist of wire obstacles, improvised barriers, road craters, and minefields. For example, burning tires and trash have proven to be effective obstacles on urban terrain. Antitank and command detonated mines are used consistent with the ROE. This belt blocks, fixes, turns, or canalizes the enemy; disrupts attack formations; and inflicts casualties.

(b) *Second Belt.* The second belt is the denial belt. It consists of wire obstacles placed around, through, and in the defensive buildings and close-in mine fields as well as in subsurface accesses. It impedes and complicates the enemy's ability to gain a foothold in the defensive area. Command detonated Claymores are used extensively consistent with the ROE. Claymores are placed where they will not cause friendly casualties.

(c) *Field-Expedient Obstacles.* Field-expedient obstacles made from available materials, such as rubble, cars and light poles, should be employed.

(3) ***Prepare Positions.*** Squads and platoons prepare positions using whatever materials are available; for example, filling dressers or other furnishings with earth or other materials.

(4) ***Rehearsals.*** Conduct rehearsals with leaders and soldiers concerning the orientation of the defense, unit positions, location of crew served weapons, CASEVAC, resupply, execution of counterattack plans, withdrawal plan, and so on. One of the more important rehearsals to conduct is the synchronization of direct and indirect fires to accomplish the commander's intent.

(5) ***Movement Enhancement.*** There will not be much time to improve movement within the defense. Units should plan to use subsurface and supersurface (through buildings) routes. Priority should be given to removing obstructions to alternate positions and to the counterattack route.

(6) ***Communications.*** Check communications. Communications is initially radio. Plans are made for messengers, and routes improved for them. Wire is emplaced as an improvement to the defense as time and the terrain allows.

NOTE: The digital force has the potential to provide accurate threat information that can enhance situational awareness, which helps facilitate targeting and obstacle placement. JSTARS; GUARDRAIL; unmanned aerial vehicles, if present; and other reconnaissance assets will significantly improve the threat situational awareness and targeting capability of the unit.

b. **Improving the Defense.** As time permits, the following areas can be given consideration and prioritized in accordance with METT-TC.

- Sleep plan.
- Barrier and obstacle improvement.
- Improvement of primary and alternate positions.
- Preparation of supplementary positions.
- Additional movement enhancement efforts.
- Initiation of patrols.
- Improvement of camouflage.
- Maintenance/refueling.
- Continued rehearsals for counterattack and withdrawal.

5-24. DEFENSE OF A VILLAGE

An Infantry company may be given the mission to defend a village (Figure 5-18, page 5-34). Once the company commander has completed his reconnaissance of the village, he scouts the surrounding terrain and, with the information assembled, he develops his plan for the defense. One of his first decisions is whether to defend with his Infantry on the leading edge of the village or farther back within the confines of the village. Normally, defending on the leading edge will be more effective against an armor heavy force, where the defending company can take advantage of longer range observation and fields of fire. Defending in depth within the village will be more effective against a primarily Infantry heavy force, in order to deny the enemy a foothold. This decision will be based on the factors of METT-TC. This mission is usually characterized with the company defending an urban area that is surrounded by open terrain. The company may need to coordinate with adjacent units to plan for the defense or control of this terrain.

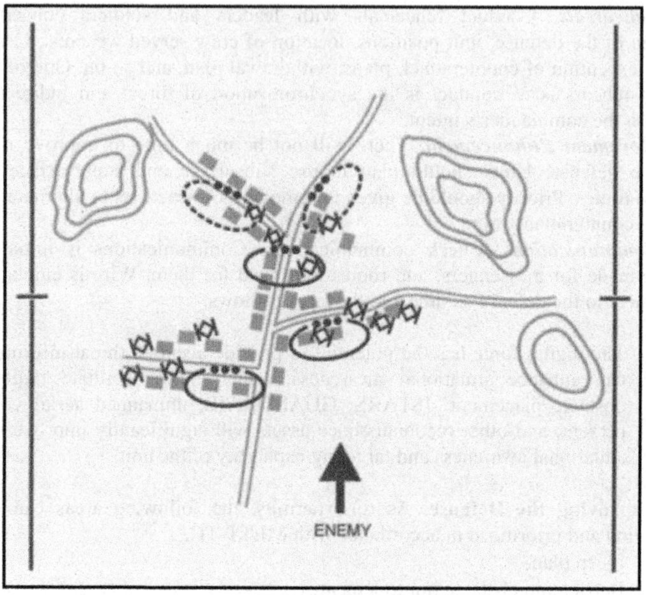

Figure 5-18. Company defense of a village.

a. **Influencing Factors.** Several factors influence the commander's decision. First, he must know the type of enemy that his company defends against. If the threat is mainly Infantry, the greater danger is allowing them to gain a foothold in the village. If the threat is armor or motorized Infantry, the greatest danger is that massive direct fire destroys the company's defensive positions. The company commander must also consider the terrain forward and to the flanks of the village from which the enemy can direct fires against his positions.

b. **Platoon Battle Positions.** Based on the mission analysis, platoons are normally given a small group of buildings in which to prepare their defense, permitting the platoon leader to establish mutually supporting squad-sized positions. This increases the area that the platoon can control and hampers the enemy's ability to isolate or bypass a platoon. A platoon may be responsible for the road through the village. The rest of the company is then positioned to provide all-round security and defense in depth.

c. **Company Mortars and Antitank** Weapons. A position for the company mortars must be chosen that protects mortars from direct fire and allows for overhead clearance. Antitank weapons are placed where they can engage targets at maximum ranges with alternate firing points. Infantry should protect antitank weapons.

d. **BFVs.** Based on METT-TC considerations, BFVs may be placed along the forward edge of the urban area to engage enemy armored vehicles. Friendly armored vehicles can also be placed in positions to the rear of the buildings and interior courtyards where their weapon systems can provide added rear and flank security. Combat vehicles are assigned primary, alternate, and supplementary positions as well as primary and secondary sectors of fire. They should be positioned in defilade behind rubble and walls or inside buildings for movement into and out of the area. Armored vehicles can also be used for resupply, CASEVAC, and rapid repositioning during the battle. BFVs can also provide a mobile reserve for the company. If a mechanized Infantry platoon is attached, it is controlled through its chain of command. If a mechanized Infantry section is attached, it can be controlled through the senior squad leader.

e. **Tanks.** If a tank platoon is available from the battalion task force, the company commander could place the tanks along the leading edge where rapid fire would complement the antitank weapons. The tank platoon leader should select exact firing positions and recommend engagement areas. If faced by enemy Infantry, the tanks move to alternate positions with the protection of friendly Infantry. These alternate positions allow the tanks to engage to the front as well as the flanks with as little movement as possible. Positions can be selected within buildings and mouseholes can be constructed. After they are withdrawn from the leading edge of the village, the tanks could provide a mobile reserve for the company.

f. **FPFs.** FPFs are planned to address the biggest threat to the company—the enemy's Infantry. When firing an FPF inside an urban area is required, mortars are more effective than artillery. This situation is true due to their higher angle of fall that gives them a greater chance of impacting on the street.

g. **Barriers and Obstacles.** Obstacles are easily constructed in an urban area. The company commander must stop enemy vehicles without interfering with his own movement in the village. Therefore, the company detonates cratering charges at key street locations on order. Mines are laid on the outskirts of the town and along routes the company will not use. Barriers and obstacles are normally emplaced in three belts. If attached or OPCON, the tank or BFV platoon leader can assist the commander by giving advice on where to place antivehicular obstacles.

h. **Engineers.** The supporting engineers use C4 and other explosives to make firing ports, mouseholes, and demolition obstacles. Based upon his priority of work, the commander tells the engineer squad leader to assist each of the Infantry platoons preparing the village for defense and to execute the company team's obstacle plan. The engineer squad leader's mission is to tell the Infantrymen exactly where to place the demolitions and how much is needed for the desired effect. He assists in preparation of charges. He also assists in the emplacement and recording of the minefields as well as the preparation of fighting positions.

i. **Communications.** To ensure adequate communications, redundant verbal and nonverbal communications are planned and checked. The company installs a wire net and develops a plan for pyrotechnic signals. Lay backup wire in case the primary lines are cut by vehicles, fires, or the enemy. The commander also plans for the use of messengers throughout the village.

5-25. DEFENSE OF A BLOCK OR GROUP OF BUILDINGS

An Infantry company operating in urban terrain may have to defend a city block or group of buildings in a core periphery or residential area. The company conducts this operation in accordance with the battalion task force's defensive scheme of maneuver. The operation should be coordinated with the action of security forces charged with delaying to the front of the company's position. The defense should take advantage of the protection of buildings that dominate the avenues of approaches into the MBA. This mission differs from defense of a village in that it is more likely to be conducted completely on urban terrain, without surrounding open terrain that characterizes the defense of a village. An Infantry company is particularly well suited for this type of mission, since the fighting will require the enemy to move Infantry into the urban area in order to seize and control key terrain. (See Table 5-1 on page 5-12.)

a. **Task and Purpose.** A well-organized company defense in an urban area—

- Defeats the enemy's attack on the streets and city blocks by using obstacles and fire.
- Destroys the enemy by ambush and direct fire from prepared positions within defensible buildings.
- Clears the enemy from footholds or remains in place for a counterattack.

b. **Reconnaissance and Security.** The execution of the mission will be more effective if the terrain is reconnoitered and obstacles and fire lanes are prepared. The LP/OPs should be supplemented by patrols, mainly during periods of limited visibility, and wire communications should be used. Platoons should be given the mission to provide one LP/OP in order to provide spot reports concerning the size, location, direction and rate of movement, and type of enemy assaulting the company sector or battle position.

c. **Task Organization.** METT-TC factors will determine how the company will be task organized to accomplish the mission. A possible task organization might be:

(1) ***Rifle Platoons.*** Three platoons (one platoon minus a squad) occupy the defensive sector.

(2) ***Reserve.*** Detached squad from one of the rifle platoons. The reserve should be given priority of commitment missions such as reinforcing the fires of the defense, reacting to a danger on the flank, or counterattacking to throw the enemy from a foothold. The biggest threat to the company is for the enemy to gain a foothold and use it to begin clearing buildings. Any foothold should be counterattacked and the enemy must be quickly and violently expelled.

(3) ***Fire Support.*** Company 60-mm mortar and antitank weapons.

(4) ***Company Control.*** An engineer squad, with priority to the company obstacle plan, then reverts to company reserve. Engineers should be controlled at company level. They construct obstacles, prepare access routes, and assist in preparing defensive positions. Additional attachments or OPCON units, such as BFVs and tanks may be placed under company control. For example, a BFV Infantry element can be used to defend a sector or battle position. The BFVs can stay under the control of the platoon sergeant and support by fire and or conduct other missions as determined by the company commander. A platoon or section of tanks attached or OPCON to the company should provide heavy direct-fire support, engage enemy tanks, and support counterattacks. An attached or OPCON tank platoon can initially attack by fire and then revert to a mobile

reserve role. The company executive officer can be used to control a reserve with multiple elements.

d. **Execution.** The defensive forces should ambush on the avenues of approach, cover the obstacles by fire, and prepare a strong defense inside the buildings. Counterattack forces should be near the front of the company sector in covered and concealed positions with an on order mission to counterattack. Rehearsals should be conducted both day and night. Counterattack forces should also be given specific instructions of what their actions will be after the enemy assault has been repelled; for example, stay in sector or revert back to reserve status.

5-26. DEFENSE OF KEY URBAN TERRAIN

An Infantry company may have to defend key urban terrain. This defense may be part of defensive operations or may be an adjunct mission to stability and support operations. In many cases, the mission is characterized by an unclear enemy situation and extremely restrictive ROE. The key terrain may be a public utility, such as gas, electrical, or water plants; a communications center, such as radio and or television; transportation center; a traffic circle; and so forth. When assigned a mission of this type, a company commander may often find his company having to defend a piece of terrain that he would rather not have to occupy. Often the facilities previously described are sited for their centrality of location and convenience and not for the defensibility of the terrain.

a. **Task Organization.** The factors of METT-TC will determine the task organization of the company. Figure 5-19, page 5-38, depicts an Infantry rifle company reinforced with an additional rifle platoon to defend the objective (water purification plant). Additional assets will be given to the company commander as they are requested or assigned, based on mission requirements and availability. In the situation depicted in Figure 5-19, the organic weapons of the Infantry company are sufficient to accomplish the mission. The only additional requirement was for another rifle platoon to defend the objective.

b. **Tasks.** In the situation shown in Figure 5-19, the company commander has determined that in order to properly defend the objective, he needs to deploy platoons on the defensible terrain available. Therefore, he is defending urban terrain (left), high ground (top), and low vegetated terrain (right, bottom). Additionally, it may be necessary to perform some of the tasks listed below:

- Provide inner and outer security patrols.
- Conduct counterreconnaissance.
- Establish LP/OPs.
- Establish checkpoints and roadblocks.
- Conduct civilian control and evacuation.
- Conduct coordination with local authorities.
- Prevent collateral damage.
- Supervise specific functions associated with operation of the facility, such as water purification tests, site inspections, and so forth.

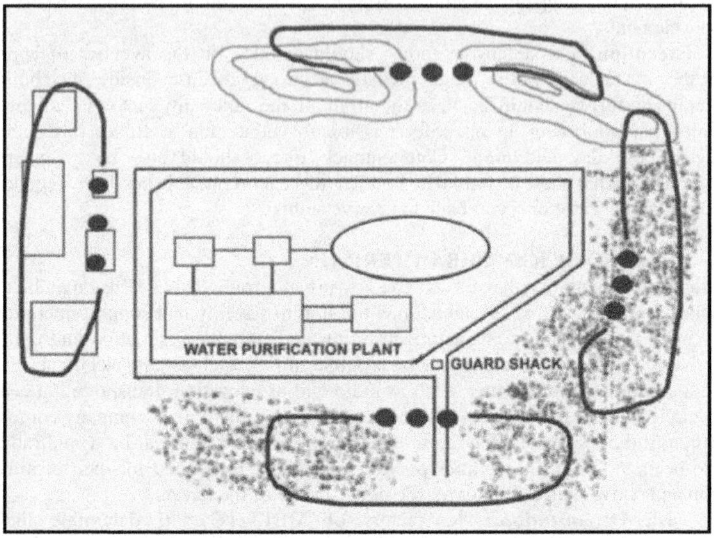

Figure 5-19. Perimeter defense of key terrain

c. **Execution.** The company commander will normally deploy platoons in a perimeter around the objective in order to dominate key terrain and cover the mounted and dismounted avenues of approach into the objective. Machine guns and antitank weapons will be emplaced to cover the dismounted and mounted avenues of approach into the objective, respectively. Wire obstacles will normally be used to restrict and deny entry into the objective area. Obstacles should be covered by fire and rigged with detection devices and trip flares. Antitank and command-detonated mines will be used consistent with the ROE. The company prepares to defend against a direct attack, such as a raid, or sabotage against key facilities within the objective, for example, water filtration system, pump station, and so forth. The commander makes an assessment as to the overall importance of the key facilities within the objective and prioritizes security requirements. The 60-mm mortar section is positioned to provide 360-degree fire support. The AT section is positioned to engage vehicular targets. If the threat does not require the employment of mortars or AT weapons, these sections are given other tasks.

NOTE: IBCT company assets will be positioned using the same considerations.

d. **Other Considerations.** Depending on the mission requirements and threat, the company commander may have to consider the need for the following.

- Artillery and attack helicopter support.
- ADA assets to defend against air attack.
- Engineer assets to construct obstacles.

- Interpreters to assist in the functioning of the facility and operation of the equipment.
- MP, civil affairs, and or PSYOP assets for civilian control and liaison/ coordination with local police and or authorities.
- BFVs or tanks to act as a mobile reserve or reaction force, or integrated into the company plan.

e. **Force Protection.** The company may be required to conduct a perimeter defense as part of force protection, such as defending a friendly base camp on urban terrain. The same techniques of establishing a perimeter defense would be used. The company maintains the appropriate level of security (100, 50, 30 percent, and so forth), consistent with the commander's plan and the enemy situation. Additional tasks may include:

- Setting up roadblocks and checkpoints.
- Searching individuals and vehicles prior to entry into the camp.
- Maintaining a presence as a show of force to the population outside the base camp.
- Conducting inner and outer security patrols.
- Clearing potential threats from any urban terrain that overwatches the base camp.
- Conducting ambushes to interdict any enemy forces moving towards the base camp.
- Restricting access to locations within the base camp. Conducting surveillance of these locations from within or from adjacent structures or positions.
- Conducting reaction force duties inside and outside the perimeter of the camp.

NOTE: See Chapter 14 for detailed information on roadblocks, checkpoints, and searches.

f. **Defense of a Traffic Circle.** An Infantry company may be assigned the mission of defending a key traffic circle in an urban area, or similar terrain, to prevent the enemy from seizing it or to facilitate movement of the battalion task force or other units (Figure 5-20, page 5-40).

(1) The company commander with this mission should analyze enemy avenues of approach into the objective and buildings that dominate those avenues. He should plan direct and indirect fires, consistent with the ROE, on to the traffic circle itself and on the approaches to it. He should also plan for all-round defense of the buildings that dominate the traffic circle to prevent encirclement. The company should prepare as many covered and concealed routes between these buildings as time permits. This makes it easier to mass or shift fires and to execute counterattacks.

(2) Obstacles can also deny the enemy the use of the traffic circle. Obstacle planning, in this case, must take into account whether friendly forces will need to use the traffic circle.

(3) Antitank weapons can fire across the traffic circle if fields of fire are long enough. Tanks should engage enemy armored vehicles and provide heavy direct-fire support for counterattacks. BFVs should engage enemy armored vehicles and provide direct fire to protect obstacles.

Figure 5-20. Defense of a traffic circle.

5-27. DEFENSE OF AN URBAN STRONGPOINT

A company may be directed to construct a strongpoint as part of a battalion defense (Figure 5-21). In order to do so, it must be augmented with engineer support, more weapons, and CSS resources. A strong point is defended until the unit is formally ordered out of it by the commander directing the defense. Urban areas are easily converted to strongpoints. Stone, brick, or steel buildings provide cover and concealment. Buildings, sewers, and some streets provide covered and concealed routes and can be rubbled to provide obstacles. Also, telephone systems can provide communications.

a. The specific positioning of unit in the strongpoint depends on the commander's mission analysis and estimate of the situation. The same considerations for a perimeter defense apply in addition to the following:

(1) Reinforce each individual fighting position (to include alternate and supplementary positions) to withstand small-arms fire, mortar fire, and artillery fragmentation. Stockpile food, water ammunition, pioneer tools, and medical supplies in each fighting position.

(2) Support each individual fighting position with several others. Plan or construct covered and concealed routes between positions and along routes of supply and communication. Use these to support counterattack and maneuver within the strongpoint.

(3) Divide the strongpoint into several independent, but mutually supporting, positions or sectors. If one of the positions or sectors must be evacuated or is overrun, limit the enemy penetration with obstacles and fires and support a counterattack.

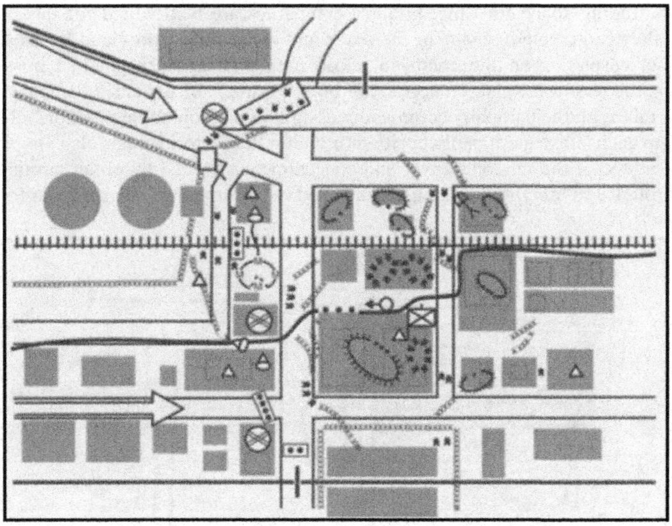

Figure 5-21. Urban strongpoint.

(4) Construct obstacles and minefields to disrupt and canalize enemy formations, to reinforce fires, and to protect the strongpoint from the assault. Place the obstacles and mines out as far as friendly units can observe them, within the strongpoint, and at points in between where they will be useful.

(5) Prepare range cards for each position and confirm them by fires. Plan indirect fires in detail and register them. Indirect fires should also be planned for firing directly on the strongpoint using proximity fuses.

(6) Plan and test several means of communication within the strongpoint and to higher headquarters. These are radio, wire, messenger, pyrotechnics, and other signals.

(7) Improve or repair the strongpoint until the unit is relieved or withdrawn. More positions can be built, routes to other positions marked, existing positions improved or repaired, and barriers built or fixed.

b. A strong point may be part of any defensive plan. It may be built to protect vital units or installations, as an anchor around which more mobile units maneuver, or as part of a trap designed to destroy enemy forces that attack it.

5-28. DELAY

The intent of a delay is to slow the enemy, cause casualties, and stop him, where possible, without becoming decisively engaged. This procedure is done by defending, disengaging, moving, and defending again. A company delay is normally conducted as part of the battalion task force's plan. The delay destroys enemy reconnaissance elements forward of the outskirts of the urban area, prevents the penetration of the urban area, and gains and maintains contact with the enemy to determine the strength and location of the main attack by trading space for time. Infantry companies are well suited for this operation, because they can take advantage of the cover and concealment provided by urban terrain and inflict casualties on the enemy at close range. Delays are planned by assigning platoon battle positions, platoon sectors, or both. Figure 5-22 depicts a company delay in urban terrain with the company commander assigning platoon battle positions. Routes are planned to each subsequent battle position or within the sector. Routes also are planned to take advantage of the inherent cover and concealment afforded by urban terrain, such as going through and hugging buildings, using shadows, subsurface areas, and so forth.

Figure 5-22. Company delay in an urban area.

a. The company's sector should be prepared with obstacles to increase the effect of the delay. Engineers prepare obstacles on main routes but avoid some covered and concealed routes that are known by the friendly troops for reinforcement, displacement, and resupply. These routes are destroyed and obstacles are executed when no longer needed.

b. Antiarmor weapon systems, tanks, and BFVs should be positioned on the outskirts of the urban area to destroy the enemy at maximum range. They should be located in defilade positions or in prepared shelters. They fire at visible targets and then fall back or proceed to alternate positions. Platoons should be assigned sectors from 100 to 300 meters (one to two blocks) wide. If available, they should be reinforced with sensors or GSRs, which can be emplaced on the outskirts or on higher ground to attain the maximum range in the assigned AO. Platoons delay by detecting the enemy early and inflicting casualties on him using patrols, OPs, and ambushes and by taking advantage of all obstacles. Each action is followed by a disengagement and withdrawal. Withdrawals occur on covered and concealed routes through buildings or underground. By day, the defense is dispersed; at night, it is more concentrated. Close coordination and maintaining situational awareness are critical aspects of this operation.

Section VII. PLATOON DEFENSIVE OPERATIONS

In urban areas, buildings provide cover and concealment, limit fields of observation and fire, and restrict the movement of troops and armored vehicles. This section covers the key planning considerations, weapons selection, preparations, and the construction of a platoon defensive position on urbanized terrain.

5-29. PLANNING THE DEFENSE

Planning the defense begins when the leader receives a mission or determines a requirement to defend such as during consolidation and reorganization after an assault. The leader must use terrain wisely and designate a point of main effort. He chooses defensive positions that force the enemy to make costly attacks or conduct time-consuming maneuvers to avoid them. A position that the enemy can readily avoid has no defensive value unless the enemy can be induced to attack it. The defense, no less than the offense, should achieve surprise. As platoon leaders conduct their troop-leading procedures, they also have to consider civilians, ROE, limited collateral damage, and coordination with adjacent units to eliminate the probability of fratricide. Maneuver, methods, and courses of action in establishing defensive positions in and around urbanized terrain are METT-TC intensive.

a. **Focus.** The squad's and platoon's focus for defending in an urban area is the retention of terrain. As with most defensive scenarios, the squad and platoon will defend as part of the company. The platoon will either be given a sector to defend or a battle position to occupy and the platoon leader must construct his defense within the constraints given to him. See Sections II and III for other planning considerations.

b. **Strongpoint.** One of the most common defensive tasks a platoon will be given during urban operations is to conduct a strongpoint defense of a building, part of a building, or a group of small buildings (see paragraph 5-27 and Figure 5-21). The platoon's defense is normally integrated into the company's mission. The platoon leader organizes the strongpoint defense by positioning personnel and their weapons systems to maximize their capabilities. Supporting fires are incorporated into the overall defensive plan to provide depth to the engagement area.

(1) The platoon leader organizes the defense into a series of individual, team, and squad fighting positions located to cover avenues of approach and obstacles, and to provide mutual support in order to repel the enemy advance. Snipers should be positioned to support the commander's intent and to allow for the opportunity to engage C2 and key targets.

(2) Depending on the length of the mission, the platoon should stockpile munitions (especially grenades), food and water, medical supplies, and fire-fighting equipment.

5-30. PRIORITIES OF WORK AND DEFENSIVE CONSIDERATIONS

A critical platoon- and squad-level defensive task during defensive urban operations is the preparation of fighting positions. General defensive considerations in urban terrain are similar to any other defensive operations. Fighting positions in urban areas are usually constructed inside buildings and are selected based on an analysis of the area in which the building is located, the individual characteristics of the building, and the characteristics of the weapons system.

a. **Priorities of Work.** The priorities of work are the same as those listed in paragraph 5-13. Specific considerations at platoon level are discussed below.

(1) Select key weapons and crew-served weapon positions to cover likely mounted and dismounted avenues of approach. To cover armored avenues of approach, position antiarmor weapons inside buildings with adequate space and ventilation for backblast (on upper floors, if possible, for long-range shots). Position machine guns/M249s to cover dismounted avenues of approach. Place them near ground level to increase grazing fires. If ground rubble obstructs grazing fires, place machine guns/M249s in the upper stories of the building. Ensure weapons are mutually supporting and are tied in with adjacent units.

(2) Ensure the position is free of noncombatants. Remove them from the area of operations before occupying the position.

(3) Clear fields of fire. Prepare loopholes, aiming stakes, sector stakes, and TRP markings. Construct positions with overhead cover and camouflage (inside and outside).

(4) Identify and secure subsurface avenues of approach (sewers, basements, stairwells, and rooftops).

(5) Stockpile ammunition, food, fire-fighting equipment, and drinking water.

(6) Construct barriers and emplace obstacles to deny the enemy any access to streets, underground passages, and buildings, and to slow his movement. Integrate barriers and or obstacles with key weapons. Cover all barriers and obstacles by fire (both direct and indirect) and or observation. (See Chapter 8 for more information concerning obstacles.)

(7) Improve and mark movement routes between positions as well as to alternate and supplementary positions. Improve routes by digging trenches, if possible; using sewers and tunnels; creating entry holes; and positioning ropes and ladders for ascending and descending.

b. **Considerations.** The following must be considered when establishing a defensive position.

(1) ***Security.*** The first priority is establishing all-around security. Each position should have at least one soldier providing security during all preparations.

(2) ***Protection.*** Select buildings that provide protection from direct and indirect fires. Reinforced concrete buildings with three or more floors provide suitable protection while buildings constructed of wood, paneling, or other light material must be reinforced to provide sufficient protection. One- and two-story buildings without a strongly constructed cellar are vulnerable to indirect fires and require construction of overhead protection for each fighting position. If possible, use materials gathered from the immediate area to build the overhead cover.

(3) ***Dispersion.*** A platoon position should not be established in a single building when it is possible to occupy two or more buildings that permit mutually supporting fires. A position without mutual support in one building is vulnerable to bypass, isolation, and subsequent destruction from any direction.

(4) ***Concealment.*** Do not select buildings that are obvious defensive positions (easily targeted by the enemy). If the requirements for security and fields of fire dictate the occupation of exposed buildings, the platoon will be required to add reinforcement materials to the building to provide suitable protection to the troops inside.

(5) ***Fields of Fire.*** To prevent isolation, individual and crew-served weapons positions should be mutually supporting and have fields of fire in all directions. When

clearing fields of fire, try to maintain the natural appearance of the surrounding area if possible. Removing objects that interfere with the gunner's field of vision may be necessary.

(6) ***Covered Routes.*** Defensive positions should have at least one covered and concealed route that allows resupply, medical evacuation, reinforcement, or withdrawal from the building without being detected, or at least provides protection from direct fire weapons. The route can be established using underground systems, communications trenches, or walls and buildings that allow covered movement.

(7) ***Observation.*** Positions in buildings should permit observation of enemy avenues of approach and adjacent defensive sectors. Upper stories offer the best observation but also attract enemy fire.

(8) ***Fire Hazard.*** If possible, avoid selecting positions in buildings that are obvious fire hazards. If these flammable structures must be occupied, reduce the danger of fire by wetting down the immediate area, laying an inch of sand on the floors, and providing fire extinguishers and fire fighting equipment. Ensure that each defender is familiar with the withdrawal routes and that they have the opportunity to rehearse their withdrawal using these planned routes in the event of fire.

(9) ***Tag Lines.*** Tag lines are a flexible handhold used to guide individuals along a route. Tag lines aid in navigation and movement when operating in confined spaces such as buildings, tunnel systems and caverns where visibility is limited and sense of direction can be lost. When preparing defensive positions inside buildings, tag lines can be run from each fighting position back to the command post, or along an egress route. These lines can be made of rope, string, cable, wire and so forth. The most effective item to be used as a tag line is WD-1A communications wire. Along with serving as a tag line it can be used as a primary means of communication between individual fighting positions and leader's positions.

(10) ***Time.*** Time is the one element in METT-TC that the platoon and its leaders have no control over. The most important factor to consider when planning the use of time is to provide subordinate leaders with two-thirds of all available time. The unit TACSOP provides the leaders with their priorities when time does not allow for detailed planning. The platoon will complete defensive preparation IAW the TACSOP and the commander's operational priorities.

c. **Preparation.** Preparation of the platoon's individual fighting positions will normally be conducted inside the buildings the platoon has been assigned to defend. As with all defensive positions, the leader's first task is to establish security. This will normally be in the form of an observation post located within the protection of the platoon's direct fire weapons. The OP should be manned with at least two personnel. Leaders then assign individual or two-man positions to adequately cover his sector. The squad leader will position himself to best control his squad. The platoon leader will designate the level of security to be maintained. The remaining personnel will continue to work preparing the defense. The leaders will continue to make improvements to the defense as time permits. (The preparation of fighting positions is discussed in detail in Chapter 3.)

d. **Other Typical Tasks.** Additional defensive preparation tasks may be required in basements, on ground floors, and on upper floors.

(1) ***Basements and Ground Floors.*** Basements require preparation similar to that of the ground floor. Any underground system not used by the defender that could provide enemy access to the position must be blocked.

(a) *Doors*. Unused doors should be locked or nailed shut, as well as blocked and reinforced with furniture, sandbags, or other field expedients.

(b) *Hallways*. If not required for the defender's movement, hallways should be blocked with furniture and tactical wire (Figure 5-23).

(c) *Stairs*. Unused stairs should be blocked with furniture and tactical wire, or removed. If possible, all stairs should be blocked (Figure 5-23), and ladders should be used to move from floor to floor and then removed.

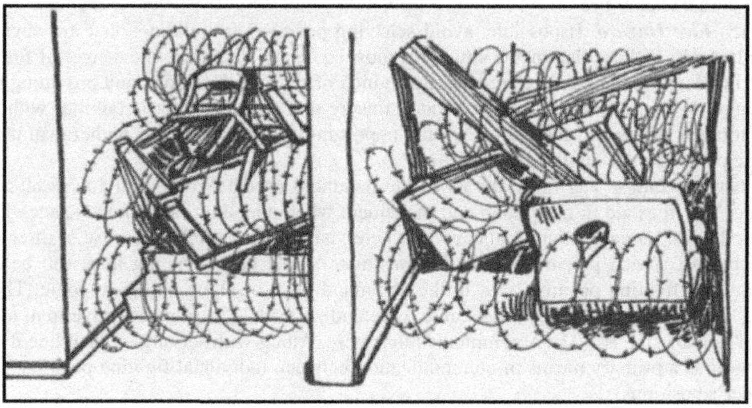

Figure 5-23. Blocking stairs and doorways.

(d) *Windows*. Remove all glass. Block unused windows with boards or sandbags to prevent observation and access.

(e) *Floors*. Make fighting positions in the floors. If there is no basement, fighting positions can give additional protection from heavy direct-fire weapons.

(f) *Ceilings*. Erect support for ceilings that otherwise would not withstand the weight of fortified positions or rubble from upper floors (Figure 5-24).

(g) *Unoccupied Rooms*. Block rooms not required for defense with tactical wire.

(2) ***Upper Floors.*** Upper floors require the same preparation as ground floors. Windows need not be blocked, but should be covered with wire mesh, canvas, ponchos, or other heavy material, to prevent grenades from being thrown in from the outside. The covering should be loose at the bottom to permit the defender to drop grenades.

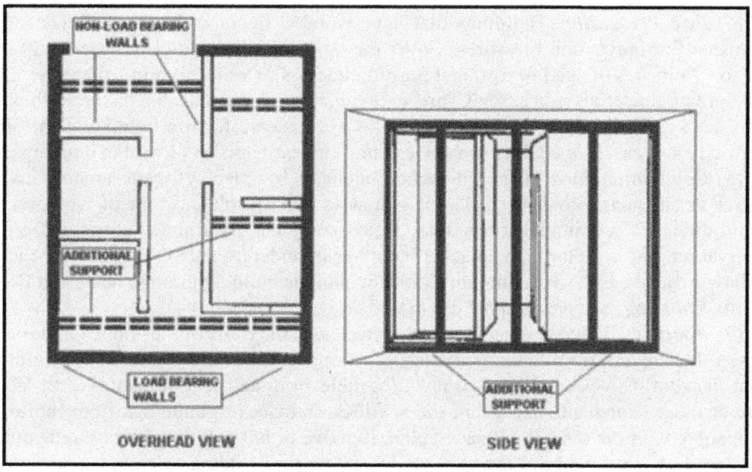

Figure 5-24. Reinforcing ceilings.

(3) ***Interior Routes.*** Routes are required that permit defending fire teams and squads to move within the building (Figure 5-25) to engage enemy forces from any direction. Plan and construct escape routes to permit rapid evacuation of a room or a building. Mouseholes should be made through interior walls to permit movement between rooms. Such holes should be marked to enable defenders to easily locate them during day and night conditions. Brief all personnel as to where the various routes are located. Conduct rehearsals so that everyone becomes familiar with the routes.

Figure 5-25. Movement routes within building.

(4) ***Fire Prevention.*** Buildings that have wooden floors and rafter ceilings require extensive fire prevention measures. Cover the attic and other wooden floors with about one to two inches of sand or dirt, and position buckets of water for immediate use. Place fire-fighting materials (dirt, sand, fire extinguishers, and blankets) on each floor for immediate use. Fill water basins and bathtubs as a reserve for fire fighting. Turn off all electricity and gas. If available, use any existing fire extinguishers found in buildings.

(5) ***Communications.*** Conceal radio antennas by placing them among civilian television antennas, along the sides of chimneys and steeples, or out of windows that would direct FM communications away from enemy early-warning sources and ground observation. Lay wire through adjacent buildings or underground systems or bury them in shallow trenches. Lay wire communications within the building through walls and floors.

(6) ***Rubbling.*** See paragraph 5-12c(8).

(7) ***Rooftops.*** Platoons must position obstacles on the roofs of flat-topped buildings to prevent helicopters from landing and to deny troops from gaining access to the building from the roof. Cover rooftops that are accessible from adjacent structures with tactical wire or other expedients and guard them. Block entrances to buildings from rooftops if compatible with the overall defensive plan. Remove or block the structure on the outside of a building that could aid the attacker in scaling the building to gain access to upper floors or to the rooftop.

(8) ***Obstacles.*** Position obstacles adjacent to buildings to stop or delay vehicles and infantry. To save time and resources in preparing the defense, platoon leaders must allow the use of all available materials, such as automobiles, railcars, and rubble, to create obstacles. Vehicles can be tied together by running poles through their windows. Leaders must supervise the construction of obstacles to ensure they are tied to buildings and rubble areas to increase effectiveness, and to canalize the enemy into engagement areas selected by the leader. Direct support engineers can provide advice and resources as to the employment of obstacles and mines.

(9) ***Fields of Fire.*** The field of fire is the area a weapon or group of weapons may cover effectively with fire from a given position. After the defensive positions are selected and the individuals have occupied their assigned positions, they will determine what clearance is necessary to maximize their field of fire. Leaders and individuals must view fields of fire from the fighting position and from the view of the enemy. Only selective clearing will be done to improve the field of fire. If necessary, the position will be relocated to attain the desired field of fire. Within the field of fire leaders will designate for each weapons system a primary and an alternate sector of fire. Each weapons system has unique requirements for its field of fire, and the platoon and squad leaders must ensure these requirements are met. Each position is checked to ensure that the fields of fire provide the maximum opportunity for target engagement and to determine any dead space within the sector of fire.

e. **Antitank Weapons Positions.** Employ antitank weapons in areas that maximize their capabilities in the urban area. The lack of a protective transport could require the weapon to be fired from inside a building, from behind the cover of a building, or from behind the cover of protective terrain. Leaders should make every effort to employ antitank weapons in pairs so that the same target can be engaged from different positions. Another consideration is security for the crew and system. This is necessary to allow the gunner to concentrate on locating and engaging enemy armor.

f. **Sniper Positions.** Snipers give the platoon a force multiplier by providing an overwatch capability and by engaging enemy C2 targets. Snipers normally operate in two-man teams, which provides the shooter with security and another set of eyes for observation and to locate and identify targets. Leaders should allow the snipers to select their own positions for supporting the defense. An effective sniper organization can trouble the enemy far more than its cost in the number of friendly soldiers employed. Snipers deploy in positions where they are not easily detected. and where they can provide the most benefit. (See Chapter 6.)

5-31. CONDUCT OF THE DEFENSE

The conduct of the defense in an urban area is similar to the conduct of the defense in any other environments.

a. **Occupy Positions.** After planning and preparing for the defense, the platoon moves to the defensive positions using prescribed movement techniques. To establish the defense the platoon will stop short of the actual site and conduct a reconnaissance to ensure the area is free of enemy or noncombatants, and to identify individual and crew served weapons positions. The platoon then establishes security and begins to occupy positions. Once the platoon has occupied, the priorities of work will be performed as established by the platoon leader.

b. **Locate the Enemy.** The platoon establishes and maintains OPs and conducts security patrols as directed by the commander. OPs, patrols, and individual soldiers look and listen using night vision devises, binoculars, and early warning systems to detect the enemy's approach.

c. **Action on Contact.** Once the enemy is detected, the platoon leader—

- Alerts the platoon sergeant, squad leaders and forward observer.
- Reports the situation to the company commander.
- If possible, calls in OP's.
- Initiates indirect fire mission when enemy is at maximum range.
- Initiates long-range direct fires on command.

d. **Fight the Defense.** Determining that the platoon can destroy the enemy from their current positions, the platoon leader—

- Continues with indirect and direct fire engagements.
- Controls fires using standard commands, pyrotechnics, and other prearranged signals.
- Initiates FPF as the enemy closes on the protective wire.

The platoon continues to defend until the enemy is repelled or ordered to disengage.

5-32. CONSOLIDATION AND REORGANIZATION

Once the enemy has been repelled, the order to consolidate and reorganize will be given by the platoon leader.

a. The platoon will—

- Reestablish security.
- Reman key weapons.
- Provide first aid and prepare to evacuate casualties.
- Repair damaged obstacles and replace mines and early warning devices.
- Redistribute ammunition and supplies.

- Relocate key weapons, and adjust positions for mutual support.
- Reestablish communications.
- Prepare for a renewed enemy attack.

b. Squad leaders provide ammunition, casualties and equipment (ACE) report to the platoon leader.

c. The platoon leader—

- Reestablishes the platoon chain of command.
- Provides a platoon ACE report to the commander.

d. The platoon sergeant coordinates for resupply and supervises casualty evacuation.

e. The platoon quickly reestablishes OP's, resumes patrolling and continues to improve the defense.

5-33. COUNTERATTACK

A platoon may be given the mission to counterattack in order to retake a defensive position or key point, to destroy or eject an enemy foothold, or to stop an enemy attack by hitting his flank and forcing him to stop his movement and establish a hasty defense.

a. A platoon counterattack is planned at company level to meet each probable enemy penetration. They must be well coordinated and aggressively executed. Counterattacks should be directed at the enemy's flank and supported with direct and indirect fires.

b. If tank support is available, it should be used to spearhead the counterattack. Tanks have the mobility, firepower, and survivability to quickly execute the counterattack mission. Tanks are ideally suited for destroying enemy armor, heavy weapons, and fortifications with their main gun and engaging enemy infantry with their coaxial machine gun. This capability will assist the infantry in executing their part of the mission.

c. The counterattack mission is planned and coordinated as part of the defensive operation.

(1) Considerations for counterattack planning may include, but are not limited to, the following:

- Location of friendly units.
- Location of noncombatants.
- Critical location in the defense that, if threatened, could collapse.
- Size and type of force required to defeat and eject the enemy.
- Where in the defense do we want the enemy to think he is successful?
- Who determines and initiates the execution of the counterattack?

(2) Control measures needed for the conduct of the counterattack include:

- Assembly area or blocking position.
- Start point, route, and release point, if necessary.
- Attack position.
- Line of departure or line of contact.
- Zone of action, direction of attack, and or axis of advance.
- Objective.
- Limit of advance.

5-34. DEFENSE AGAINST ARMOR

Urban terrain is well suited to an infantry's defense against mechanized infantry and armored forces. Mechanized infantry and armored forces will attempt to avoid the dense, canalizing urban areas but may be forced to pass through them. Well-trained infantry can inflict heavy casualties on such forces.

a. Urban areas have certain traits that favor antiarmor operations.

- Rubble in the streets can be used to block enemy vehicles, conceal mines, and cover and conceal defending infantry.
- The buildings restrict and canalize armor maneuver, fields of fire, and communications, reducing the enemy's ability to reinforce.
- Buildings provide cover and concealment for defending infantry.
- Rooftops, alleys, and upper floors provide good firing positions.
- Sewers, drains, and subways provide underground routes for infantry forces.

b. When preparing for antiarmor operations in urban areas leaders should:

(1) ***Choose a good engagement area.*** Enemy tanks should be engaged where most restricted in their ability for mutual support. The best way for infantrymen to engage tanks is one at a time, so they can destroy one tank without being open to the fires of another. Typical locations include narrow streets, turns in the road, "T" intersections, bridges, tunnels, split-level roads, and rubbled areas. Less obvious locations can include using demolitions or mines to create obstacles.

(2) ***Select good weapons positions.*** The best weapons positions are places where the tank is weakest and the infantry is most protected. A tank's ability to see and fire is limited, to the rear and flanks, if the tanks are buttoned up. Figure 5-26, on page 5-52, shows the weapons and visual dead space of a buttoned-up tank against targets located at ground level and overhead. The TRPs should be clearly visible through the gunner's sights and resistant to battle damage (for example, large buildings or bridge abatements, but not trees or cars). The leader of the antiarmor operation should specify what type of engagement should be used such as frontal, crossfire, or depth. Frontal fire is the least preferred since it exposes the gunner to the greatest probability of detection and is where armor is the thickest. (For more information on target engagement techniques, see FM 7-91 and or FM 23-1.)

Figure 5-26. Tanks cannot fire at close-range, street-level, and overhead targets.

(a) The best places to fire on tanks from the dismounted infantry perspectives are at the flanks and rear at ground level or at the top of tanks if the force is in an elevated position in a building (See Chapter 11 for minimum arming distance). A suitable antiarmor defense might be set up as shown in Figure 5-27.

(b) The best place to engage a tank from a flank is over the second road wheel at close range. This can be done using a corner so the tank cannot traverse the turret to counterattack.

(c) For a safe engagement from an elevated position, infantrymen should engage the tank from a range three times the elevation of the weapons.

(d) To engage at a longer range is to risk counterfire since the weapon's position will not be in the tank's overhead dead space. Overhead fire at the rear or flank of the tank is even more effective. Alternate and supplementary positions should be selected to enforce all-round security and to increase flexibility.

(3) ***Coordinate target engagement.*** The first task of the tank-killing force is to force the tanks to button up using all available direct and indirect fire because tanks are most vulnerable when buttoned up. The next task is to coordinate the fires of the antitank weapons so if there is more than one target in the engagement area, all targets are engaged at the same time.

c. Often armored vehicles are accompanied by infantry in built-up areas so antiarmor weapons must be supported by an effective all-round antipersonnel defense (Figure 5-28).

d. At a planned signal (for example, the detonation of a mine) all targets are engaged at the same time. If targets cannot be engaged simultaneously, they are engaged in the order of the most dangerous first. Although tanks present the greatest threat, threat armored personnel carriers (APCs) are also dangerous because their infantry can dismount and destroy friendly antiarmor positions. If the friendly force is not secured by several infantrymen, priority of engagement might be given to threat APCs.

Figure 5-27. A platoon's antiarmor defense.

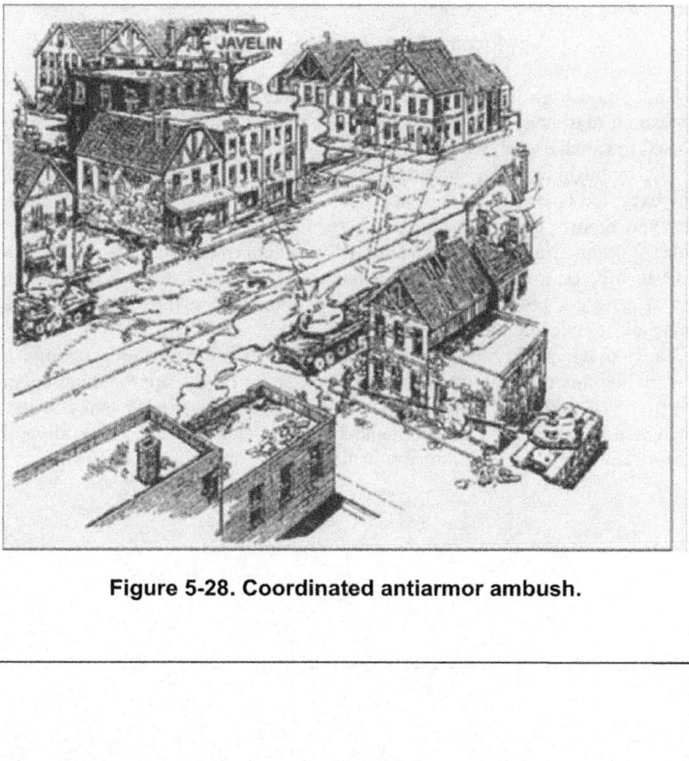

Figure 5-28. Coordinated antiarmor ambush.

5-35. CONDUCT OF ARMORED AMBUSH

A rifle company can use an attached tank platoon to conduct an armored ambush in a built-up area (Figure 5-29). To do so, the tank platoon should be reinforced with a BFV and one or two squads from the rifle company. The ambush can be effective against enemy armor if it is conducted in an area cleared and reconnoitered by friendly forces.

Figure 5-29. Armored ambush.

a. The ambushing tank platoon must know the area. The operation involves maneuver on a road network that is free of obstacles. Obstacles outside the ambush area can be used to canalize and delay the enemy.

b. The ambushing tanks should be located in a hide position about 1,000 meters from the expected enemy avenue of approach. A security post, located at a choke point, observes and reports the approach, speed, security posture, and activity of the enemy. This role is assigned to scouts, if available, or Infantrymen who use the BFV to move from OP to OP; or a series of dismounted OPs are established. When the enemy is reported at a trigger point or TRP, the tank platoon leader knows how much he must move his tanks to execute the ambush.

c. Tanks move quickly from their hide positions to firing positions, taking advantage of all available concealment. They try for flank shots on the approaching enemy at an average range of 300 to 400 meters. These ranges do not expose tanks to the enemy infantry. Once the enemy is engaged, tanks break contact and move to a rally point with close security provided by an infantry squad and moves to a new ambush site.

CHAPTER 6
SNIPER AND COUNTERSNIPER TACTICS, TECHNIQUES, AND PROCEDURES

Snipers have always played a large role in urban combat. They have been used to disrupt operations, inflict casualties, and tie down large numbers of troops searching for them. The lethality and accuracy of modern weapons, the three-dimensional aspect of urban battlefields, and the many alleyways, corridors, and rear exits available to a sniper make him a serious threat. Commanders and leaders at all levels must be aware of the value of employing snipers and the threat posed by enemy snipers. They must understand the effects a sniper can have on unit operations, and the steps by which he can be countered and his threat minimized. In this chapter, the term sniper is used to describe a trained sniper team or a single rifleman firing carefully aimed shots from short to long range.

Section I. EMPLOYMENT OF SNIPERS

The value of the sniper to a unit operating in an urban area depends on several factors. These factors include the type of operation, the level of conflict, and the rules of engagement (ROE). Where ROE allow destruction, the snipers may not be needed since other weapons systems available to a combined arms force have greater destructive effect. However, they can contribute to the fight. Where the ROE prohibit collateral damage, snipers may be the commander's most valuable tool. (See FM 7-20, FM 71-2, and TC 23-14 for more information.)

6-1. SNIPER CAPABILITIES

Sniper effectiveness depends in part on the terrain. Control is degraded by the characteristics of an urban area. To provide timely and effective support, the sniper must have a clear picture of the commander's concept of operation and intent.

a. Snipers should be positioned in buildings of masonry construction. These buildings should also offer long-range fields of fire and all-round observation. The sniper has an advantage because he does not have to move with, or be positioned with, lead elements. He may occupy a higher position that is to the rear or flanks and some distance away from the element he is supporting. By operating far from the other elements, a sniper avoids decisive engagement but remains close enough to kill distant targets threatening the unit. Snipers should not be placed in obvious positions such as church steeples and rooftops since the enemy often observes these and targets them for destruction. Snipers should not be positioned where there is heavy traffic because these areas invite enemy observation.

b. Snipers should operate throughout the area of operations, moving with and supporting the units as necessary. Some teams may operate independent of other forces and search for targets of opportunity, especially enemy snipers. The team may occupy multiple positions. A single position may not afford adequate observation for the entire team without increasing the risk of enemy detection. Separate positions must maintain mutual support.

c. Snipers may be assigned tasks such as:

- Killing enemy snipers (countersniper fire).
- Killing/destroying targets of opportunity. These targets may be prioritized by the commander and might include enemy snipers, key leaders, vehicle commanders, radio operators, sappers, and direct fire support/crew-served weapons crews, forward observers, radio telephone operators, protected equipment.
- Destroying key pieces of equipment or materials.
- Denying enemy access to certain areas or avenues of approach (controlling key terrain).
- Providing fire support for barricades and other obstacles.
- Maintaining surveillance of flank and rear avenues of approach (screening).
- Supporting local counterattacks with precision fire.
- Perform observation of key facility to gather intelligence.
- Perform observation to enhance general security measures.
- Call for and adjust indirect fires.

6-2. EMPLOYMENT CONSIDERATIONS

The characteristics of urban areas and the nature of urban warfare impact on both the effectiveness of the sniper weapons system and how the system may be employed. The sniper must consider the location of the target in relation to his position, whether he or the target is inside or outside the building. The sniper must also consider the structural configuration of the buildings in his area of operation. The basic classes of structures encountered in an urban area are concrete, masonry, and wood. However, any one building may include a combination of these materials. All buildings offer the sniper concealment, though the degree of protection varies with the materials used.

a. **Selecting Positions and Targets**. Commanders will provide the sniper with the general area of operation (building or group of buildings) in which to position himself, but the sniper selects the best position for his specific engagements. Sniper positions should cover:

- Obstacles.
- Roofs.
- Friendly routes.
- Likely avenues of approach.
- Gaps in the final protective fires.
- Dead space.
- Other areas that may provide the enemy an advantage.

The sniper also selects numerous alternate and supplementary positions to cover his areas of responsibility. The sniper should think three-dimensionally. Because the urban environment poses a three-dimensional battle space, the sniper should anticipate the threat from any direction at any time.

b. **Offensive Missions**. Offensive operations carry the fight to the enemy to destroy his capability and will to fight. The sniper can prove to be a key combat multiplier by neutralizing enemy targets that threaten the success of the attack. During offensive operations snipers—

- Conduct countersniper operations.
- Overwatch movement of friendly forces and suppress enemy targets that threaten the moving forces.
- Place precision fire on enemy crew-served weapons teams and into exposed apertures of bunkers.
- Place precision fire on enemy leaders, armored vehicle drivers or commanders, FOs, RATELOs, or other designated personnel.
- Place precision fire on small, isolated, bypassed elements.
- Place precision fire on enemy forces that are withdrawing or threatening a counterattack.
- Assist in screening a flank using supplemental fires.
- Dominate key terrain by controlling access with fires.

(1) To increase security and surprise, snipers must move covertly into position in an objective area long before the main attack forces arrive. Once the assault begins, snipers may need to reposition due to masking of fires by friendly forces. A detailed evaluation must be made when determining where and how the snipers would be most beneficial to the mission.

(2) Upon consolidation of forces on the objective area, the snipers may be displaced forward to new positions for security. These positions may not necessarily be on the objective. From these positions the snipers conduct observation and provide early warning to the friendly unit. They also provide precision fire against bypassed enemy positions, enemy counterattacks, or other enemy positions that could impede the unit's ability to exploit the success of the mission.

c. **Defensive Operations**. When employed properly, snipers can effectively enhance a commander's defensive fire plan. After analyzing the terrain, snipers should provide the commander with recommendations for their employment. Snipers can perform the following tasks during defensive operations:

- Cover obstacles, minefields, roadblocks and pre-positioned demolitions.
- Perform counterreconnaissance (terminate enemy reconnaissance efforts).
- Engage enemy OPs, exposed armored vehicle commanders and AT weapons teams.
- Engage enemy vehicles' optics to degrade vision and disrupt movement.
- Engage enemy crew-served weapons.
- Disrupt follow-on units with long-range precision fire.

(1) Snipers should be positioned to cover one or more avenues of approach into the defensive position. They can be used to enhance security, allowing the commander to concentrate his combat power against the most likely enemy avenue of approach. Snipers, with their optics for target acquisition and their long-range engagement capability, compliment the unit's machine guns. Snipers may also be used in an economy-of-force role to cover a dismounted enemy avenue of approach into positions the unit cannot cover.

(2) Multiple sniper teams can be positioned for surveillance and mutual support. If possible, they should establish positions in depth for continuous support during the fight. The sniper's rate of fire neither increases nor decreases as the enemy approaches. Specific targets are systematically and deliberately engaged—accuracy is never sacrificed for speed.

(3) When supporting a strongpoint defense, the sniper teams should be positioned outside the defensive position to allow for freedom of movement. Their primary mission is to conduct observation tasks or independent harassing engagements against enemy reconnaissance elements or induce the enemy to move into a predetermined engagement area.

d. **Stability and Support Operations**. Snipers are employed in stability and support operations during peacekeeping missions, which include, but are not limited to, urban guerrilla warfare and hostage situations.

(1) ***Urban Guerrilla Warfare***. The primary role of the sniper in an urban guerrilla environment is to dominate the area of operations by delivery of selective, aimed fire against specific targets as authorized by local commanders. Usually this authorization comes only when such targets are about to employ firearms or other lethal weapons against the peacekeeping force or innocent civilians. The sniper's other role, almost as important as his primary role, is the gathering and reporting of intelligence. Some of the specific tasks a sniper may be required to accomplish are:

- Engage dissidents/urban guerrillas involved in hijackings, kidnappings, hostage takings, and so on.
- Engage urban guerrilla snipers as opportunity targets or as part of a deliberate clearance operation.
- Covertly occupy concealed positions to observe selected areas.
- Record and report suspicious activity.
- Provide protection for other elements of the peacekeeping force, including firemen, repair crews, and so on.

(2) ***Hostage Situations***. Hostage situations will usually be delegated to special purpose units unless the tactical situation does not allow it. Snipers and their commanders must understand that even a well-placed shot may not always result in the instantaneous incapacitation of a terrorist. Even the best sniper armed with the best weapon and bullet combination cannot guarantee the desired results. An instantly fatal shot may not prevent the death of a hostage when muscle spasms in the terrorist's body trigger his weapon. As a rule, the sniper should be employed only when all other means of resolving the situation have been exhausted.

(a) *Command and Control*. Once the decision has been made by the local commander to employ the sniper, all command and control of his actions should pass to the sniper team leader. At no time should the sniper receive the command to fire from someone not in command. He should be given clearance to fire, and then he and his team leader decide when to engage the target. If more than one sniper team is used to engage one or more targets, it is imperative that the same rules of engagement apply to all the teams. However, communication between snipers will still be necessary. A *landline* or TA-312 phone loop, which is similar to a gun loop, can be used to communicate to all snipers. Landlines, however, do not allow for fluid mobility so radio communications is preferred.

(b) *Position Selection*. Generally, the selection of a firing position for a hostage situation is not much different from selecting a firing position for any other form of combat. The same principles and techniques apply. Remember, the terrain and the situation will dictate the sniper's selection of firing positions. Although the sniper is used as a last resort in a hostage situation, he should be positioned as early as possible. This will enable him to precisely determine the ranges to selected targets, positively identify hostages and terrorists within the target area, and select alternate firing positions for use should the situation change.

e. **Infiltration and Exfiltration**. One method the sniper may use when infiltrating an urban area is through the outskirts of the city or town. The outskirts are primarily residential and may not be heavily defended by enemy forces. Its defenders may be a series of antitank and antiaircraft positions, security elements covering main approaches, or positions blocking main avenues into the city center. Strongpoints and reserves are normally deeper in the city.

(1) As part of a larger force, the sniper moves with stealth along secondary streets, using the cover and concealment of back alleys and buildings. The sniper can assist the larger units in seizing key terrain and isolating enemy positions, allowing follow-on forces to enter the urban area. After the initial force has seized a foothold into the city the sniper teams may infiltrate and move into their areas of operation.

(2) Mortar and artillery fire may be used to suppress enemy observation and mask any sounds that may be made by an infiltrating element. Infiltration should be conducted during times of limited visibility and through areas that are free of civilians and domestic animals.

(3) Sniper teams may infiltrate into the city as part of a larger force during an airborne or air assault operation.

(4) When a sniper is conducting exfiltration from his area of operation, care must be taken to avoid detection. As with infiltration, stealth and use of all available cover and concealment must be maintained. Exfiltration should be conducted during hours of limited visibility to aid in avoiding detection. Special care must be taken to cover tracks as well.

(5) As operations in a specific area continue, sniper teams should vary the methods of infiltration and exfiltration in order to avoid the possibility of compromise.

6-3. COMMANDER'S RESPONSIBILITIES TO THE SNIPER

Operations in urban terrain require detailed intelligence. The commander (and or S2) should provide the sniper with the following materials and information for planning the operation:

a. **Tactical Maps and Aerial Photos**. Although tactical maps may not show all the man-made objects with enough detail for planning tactical operations in the urban area, they do provide a good representation of the surrounding terrain. Tactical maps should be supplemented with vertical and oblique aerial photographs.

b. **Civil Government and Local Military Information**. Considerable current information on practically all details of a city can be obtained from civil governments and local military forces:

- Large-scale city maps.
- Diagrams of underground sewer, utility, and transportation systems.
- Information on key public buildings and key personnel rosters.
- Information on the size and density of the population.
- Information on police and security capabilities.
- Information on civil defense, air-raid shelters, and fire-fighting capabilities.
- Information on utility systems, medical facilities, and multimedia and communications facilities.

c. **Clear Task and Purpose**. The mission assigned to a sniper team for a certain operation consist of the tasks that the commander wants the sniper team to accomplish, and the purpose (reason) for accomplishing the mission. The commander must decide how he wants his snipers to influence the urban battlefield. Then he must assign missions and tasks to achieve this effect. The commander must be sure to provide a prioritized target list so the sniper teams can plan effectively, and avoid involvement in sustained engagements.

(1) The commander may describe the effects or results he expects and allow the snipers to select key targets.

(2) The commander may prescribe specific types of targets. For example, if he wants to disrupt an enemy's defensive preparation he may task snipers to engage equipment operators and vehicle drivers. Or he may task them to engage soldiers preparing individual positions.

(3) The commander can also assign specific or key targets. These may include leaders, radio operators, ATGM gunners, armored vehicle commanders, or crew-served weapons crews.

Section II. COUNTERING THE URBAN SNIPER

In many host nations the "specially trained sniper" and the "trained marksman" are former rifle competitors, some of which may be Olympic-grade shooters. In Serbia one of the most successful snipers was a former Olympic contender. The shooters-snipers must not be underestimated, especially in regards to the range at which they can engage a target and obtain a hit.

6-4. TYPES OF ENEMY SNIPERS AND THEIR CAPABILITIES

The three general types of snipers are the specially trained and equipped individual, the trained marksman, and the civilian irregular. Each has different characteristics of operation and may be used to accomplish different purposes. Countermeasures effective against one type may be less effective against another.

a. **Specially Trained Sniper**. The most dangerous sniper is the individual who has been specially selected, trained, and equipped with a modern scope-mounted sniper rifle. These individuals are expert shots and are trained to select key individuals as their targets. They can hit at great range (sometimes out to 1,000 meters) and are skilled in avoiding detection. They are normally members of an organized, armed force and wear a standard uniform that may be modified to provide better camouflage. Their actions are carefully integrated into the overall plan of operation. This sniper is the most difficult to counter effectively. Until recently, there were not many potential adversaries of the US that could produce significant numbers of such individuals. Many armies in the world now have a renewed interest in snipers. More and more sniper training is taking place, with an increase of high-power rifles that are available at a reasonable cost on the world arms market. US forces can expect to see more and more trained snipers with improved weapons systems during future urban operations. Some of these may be equipped with rifles and night observation equipment that are among the best in the world. The US Army and its Western allies already have a relatively large number of this type sniper, as do several states of the former Soviet Union, and the Peoples Republic of China.

b. **Trained Marksman**. A trained marksman is a common sniper often found in urban combat. This sniper is a trained soldier, equipped with a standard issue weapon, who is an above-average shot. He normally has fair to good field craft skills and is difficult to detect in the urban environment. He may be employed singly or in teams to create confusion among friendly forces, cause casualties, or harass and disrupt the tempo of operations. He is often used by the enemy in an economy-of-force role as a rear guard or covering force, while the main enemy force withdraws. He may also be placed on the perimeter of a defended urban area to provide early warning of the approach of friendly forces and to disrupt and cause them to deploy early. The trained marksman is a dangerous foe. He can be found in fairly large numbers in the armies of many potential adversaries. He is normally a member of an organized, armed force and wears a standard uniform. He may, however, be a guerrilla fighter, in which case he may not wear a recognizable uniform but will normally carry his arms openly.

c. **Armed Irregular**. The third general type of sniper is the armed irregular. He may have little or no formal military training but may have experience in urban combat. He may or may not wear any distinguishing uniform and may even appear to be merely another of the thousands of noncombatants found in a large urban area. He may or may not carry his weapon openly and may go to great lengths to avoid identification as a sniper. His fires are normally not accurate, and he seldom deliberately targets specific individuals. His actions are not normally integrated into an overall enemy plan, although his attacks may be loosely coordinated with others in his general area. Although this type of sniper has the least ability to cause heavy losses among US forces, he has high value as an element of harassment, and in some stability and support situations he may achieve results far out of proportion to his actual ability to cause casualties.

d. **Range of Sniper Attacks**. The typical range for a sniper attack is 300 to 600 meters with medium-caliber rifles. Shots from 800 to 1,000 meters are the exception. Heavy sniper rifles (caliber .50, 12.7-mm, 14.5-mm, and 15-mm) with ranges of 1,200 to 1,500 meters are now available around the world. These heavy sniper rifles were originally intended as antimateriel weapons for stand-off attack against high-value targets, such as radar control vans, missiles, parked aircraft, and bulk fuel and ammunition storage sites. They are only marginally accurate enough for long-range shots against individual personnel. It is their ability to shoot through all but the heaviest shielding material, and their devastating effects, that make them valuable psychological weapons. The ability to shoot through common urban building materials makes these large weapons valuable as countersniper tools.

e. **Equipment Trends**. Several other equipment trends will result in a greater threat to US forces from urban snipers in the future.

(1) The quality and quantity of night observation devices sold on the world market is increasing daily. In the near future, even trained marksmen may be equipped with devices to allow accurate fires at night.

(2) The use of simple, direct-view optical sights on military rifles is increasing. Although not in the accuracy class of true sniper weapons, these sights make the trained marksman a much more dangerous foe. This is especially true within the shorter ranges (less than 200 meters) normally associated with combat in urban areas.

(3) Many armies are now buying simple but effective devices to either silence or suppress the muzzle blast of sniper weapons. These devices inhibit the task of

determining the location of a sniper. Although many of these devices significantly reduce the maximum effective range of the weapon, snipers can be very effective at less than 200 meters with these devices attached.

(4) The employment of heavy sniper rifles, such as the .50 caliber, has increased.

(5) The use of laser detection devices to detect, damage, degrade, or prevent the use of snipers has increased.

6-5. THE LAW OF LAND WARFARE APPLIED TO SNIPERS

Historically, units that suffered heavy and continual casualties from urban sniper fire and were frustrated by their inability to strike back effectively often have become enraged. Such units may overreact and violate the laws of land warfare concerning the treatment of captured snipers. This tendency is magnified if the unit has been under the intense stress of urban combat for an extended time. It is vital that commanders and leaders at all levels understand the law of land warfare and also understand the psychological pressures of urban warfare. It requires strong leadership and great moral strength to prevent soldiers from releasing their anger and frustration on captured snipers or civilians suspected of sniping at them.

a. The law of land warfare is not restricted to declared wars. It applies in all cases of armed conflict, such as many situations in stability and support operations. These laws and the legal orders of their superiors bind all US soldiers. Under the law, it is forbidden to kill, wound, or harm an enemy who, having laid down his arms or having no means of defense, has surrendered. A sniper who has been captured, or who has surrendered, must not be harmed. It does not matter how many friendly casualties he has caused or how long he waits before he surrenders.

b. Any sniper who wears the uniform of a belligerent, carries his arms openly, and conducts himself in accordance with the law and customs of warfare should be treated as a prisoner of war, not a criminal. An armed irregular who is part of an organized resistance movement, obeys the orders of a designated commander, carries his weapon openly, and obeys the laws and customs of war should also be accorded such treatment. A civilian who snipes at US forces without meeting these criteria can be detained by the military and tried by the appropriate court. Under no circumstances should a captured person be mistreated or killed in retaliation for sniping, regardless of how many casualties he may have caused.

c. In some stability and support situations, the ROE and the mandate under which the US forces are operating may severely restrict how much lethal counterforce can be used against snipers. Three principles govern the legal use of lethal force. The commander must—

- Make every effort to avoid causing unnecessary suffering.
- Use the minimum force necessary to accomplish the task at hand.
- Apply the type and degree of force using the rule of general proportionality.

d. ROE result from the law of war stretched over the situational template of a particular mission. It would violate the law of war, as well as most ROE, to respond to sniper fire with *massive, indiscriminate* return fire into an urban area if another less destructive tactic or weapon could be employed without greatly increasing risk to US forces. Whatever the situation, commanders and leaders must understand the ROE and ensure their soldiers follow them.

6-6. SNIPER AWARENESS

The first step in countering snipers is for commanders, leaders, and staff officers at all levels to be aware of the sniper threat. Although snipers may be more prevalent in some situations than in others, a sniper threat always exists in urban areas to some degree. Plans to counter the sniper threat and protect the friendly force from snipers must be integrated into the operation during the early stages of planning. Tactics and techniques must be taught to soldiers before they encounter sniper fire.

a. A careful METT-TC analysis and consultation with personnel familiar with the area may reveal the extent of the sniper threat. This is especially important during stability and support operations. Information on the local sniper threat may be obtained from:

- Host nation military, government, or police officials.
- US Embassy personnel.
- Allied special operations forces (SOF) or other allied forces.
- US SOF operating in the area.
- UN officials or other UN forces in the area.
- Nongovernment agency officials.
- Local militia members.
- Local civilians, including children.

b. Coordination with US snipers can identify specific areas and situations where enemy snipers may be operating and help the commander decide on countermeasures to be employed. In addition to the school-trained snipers assigned to the infantry battalion, there are several other sources of sniper expertise. Some examples of expert US snipers are:

- US SOF snipers such as special forces (SF) and rangers, sea-air-land teams (SEALs), and the US Army Sniper School.
- US law enforcement officials such as police special weapons and tactics (SWAT) teams, the Drug Enforcement Agency (DEA), the Federal Bureau of Investigation (FBI), or the Secret Service. (These are usually available only during domestic stability and support operations.)

6-7. PLANNING SNIPER COUNTERMEASURES

When planning sniper countermeasures, the commander and staff answers three basic questions. Their answers suggest the set of countersniper TTP that best fits the METT-TC conditions under which the unit is operating. Consideration of these questions assists the commander and staff in eliminating the TTP that are inappropriate to the situation.

a. **"What does the US commander want to accomplish, and which ROE govern his operations?"** If the commander's intent is to conduct combat operations, and if the ROE allow him to do this with the full range of his available firepower, he can either suppress and bypass the sniper or use the principle of fix and maneuver. He can use firepower to suppress and fix the sniper in position while maneuvering forces either to avoid the sniper and continue the mission or to close with and eliminate him.

(1) If the ROE limit the application of force, or the use of such force would create a large number of civilian casualties, the commander may be limited in his initial response to a sniper attack. In many stability and support situations, the key to success is

perseverance, restraint, and the use of minimum or appropriate force. The unlimited use of firepower in an urban area may undermine the legitimacy of the US force and work against the commander's ultimate intent.

(2) The right of self-defense is never denied US forces, but it may be limited. This is a difficult concept for soldiers to grasp, especially if they are taking fire from snipers. The time to explain it is *before* enemy contact, not *during*. Leaders must keep the commander's ultimate intent in mind when they plan and execute sniper countermeasures.

b. **"What does the enemy want to accomplish with his snipers, and what capabilities does he have to accomplish it?"** Enemy snipers may be striving for several goals:

(1) Defeat US forces. At the small-unit tactical level, this is possible for the specially trained sniper and possibly for the marksman.

(2) Force US forces to deploy, delaying them, breaking up their tactical tempo, and allowing the enemy to seize the initiative. Unless US forces react aggressively and counter the sniper's effects quickly, this goal is possible for all snipers.

(3) Harass US forces, exhausting them, and lowering morale. All snipers can do this, but most often the marksman and the irregular are used for this purpose.

(4) Kill a specific individual. This task is almost always assigned to a specially trained sniper. His target may be specific individuals by their function such as military officers, RATELOs, or armored vehicle commanders. During stability and support operations, the target may be political or community leaders, or classes of individuals such as members of the media, international aid workers, policemen, or civilians living in a contested area.

(5) Cause US casualties for political effect. As the information age progresses, potential adversaries will become more and more adept at manipulating the attitudes of the American public to turn them against US efforts in stability and support situations. One way to do this has been to cause US military casualties, regardless of their tactical effect, knowing the world media will transmit images to discourage Americans and lessen their support. The commander must determine the level of sniper to be countered and the type of weapons, ammunition, tactics, and night vision equipment available to the enemy. This information can be used to assess the expected range and lethality of enemy snipers, and will aid in identifying patterns to counter. It will also be useful to assess passive protective measures such as the likely effectiveness of body armor, light vehicle armor kits, screens, shields, and so forth.

c. **"What are the rules of engagement?"** There are three basic levels of ROE against snipers. The commander can modify each depending on the local situation.

(1) US troops use minimum force. This is common in many stability and support situations, especially during aid to domestic authorities, peacekeeping, noncombatant evacuations, and humanitarian relief.

(2) US troops use an equal or reasonable response to force used against them. This may be the situation in more violent peace enforcement operations.

(3) US troops use overwhelming force. This is the normal situation during combat in urban areas when the enemy poses a significant threat to US forces.

6-8. COUNTERSNIPER TACTICS, TECHNIQUES, AND PROCEDURES

Countersniper TTP by US forces involve two types of actions: *active* countermeasures and *passive* countermeasures. Each has its place, depending on the METT-TC conditions under which the unit is operating. Most sniper countermeasures are not new TTP for well-trained combat troops. They are simply common sense actions taken routinely while in a combat area to limit exposure to fire, conceal positions, move tactically, and respond to enemy contact. Some countermeasures are not routine and require additional training emphasis. No matter which TTP are employed, successful countersniper measures present leaders with a challenge to maintain unit discipline. The sniper has the initiative. Units must not implement countermeasures halfheartedly. To do so invites casualties from snipers who can wait hours for the moment a unit's guard is down.

a. **Active Countermeasures**. Active countermeasures either detect and destroy the sniper before he can fire, or engage and neutralize him after he fires. Active countermeasures include the use of the following:

(1) ***Observation Posts and Aerial Observers***. Observers can maintain a constant surveillance over potential sniper positions and detect snipers as they attempt to move into position for a shot. Once detected, snipers can be easily neutralized or forced to withdraw.

(a) Observation posts should have access to powerful spotting telescopes, medium-power binoculars, and night observation devices (thermal, if possible). Constantly scanning an area for the minute movements made by a well-trained sniper is exhausting. Therefore, personnel on OP duty should rotate frequently. However, a person who is intimately familiar with the area being scanned is most likely to notice a subtle change.

(b) As military and commercial lasers become more and more common, these devices may be used against US forces manning observation posts. Observers should be equipped with laser protective glasses, especially when using direct-view optical devices. Laser protective glasses, binoculars with laser filters, and indirect-view optics protect observers from most available laser systems.

(c) Aerial observers can operate from any of several platforms. The modernized OH-58, with its sophisticated night vision capability, and the AC 130 have excellent capabilities to detect individual snipers around US positions. Any of several unmanned aerial vehicles (UAVs), with their extended loiter time and video/night vision capability, can also be used effectively.

(2) ***Patrols***. Constant reconnaissance and security patrols around a unit's position hinder a sniper's getting into a firing position undetected. Small patrols are usually more effective.

(a) Like US sniper teams, enemy sniper teams are small and depend on stealth to approach a target along covered and concealed routes. Normally, they move to a hide or "shoot" position and remain there for long periods. These sniper teams are most effective when they have good fields of fire from 300 to 600 meters. At ranges of less than 300 meters, the sniper's movements and firing signature are more easily detected. A moving sniper who has been discovered by a small security patrol is at a great disadvantage. He lacks the firepower to fight a long engagement and is normally far from support or assistance.

(b) Small night security patrols using night vision devices can be very effective. Reconnaissance patrols should move by covered and concealed routes to good observation points, stop, observe, then move to another position. The patrol routes must vary, and a reaction force or supporting weapons must be ready if the patrol makes contact. Military working dogs and trained handlers can be useful in detecting enemy snipers. Dogs can quickly search large buildings for hidden enemy and can detect personnel at long range if downwind.

(c) In addition to reconnaissance patrols, small combat patrols are also effective. A variation of the ambush patrol is the stay-behind ambush. A small ambush element moves as part of a larger patrol and occupies its position without being observed. It then observes its kill zone, which may be very large if the element has a sniper team with it, and engages enemy snipers as they attempt to move into position.

(3) ***United States Snipers***. US snipers can be most effective as a counter to enemy snipers. Not only do they have expert knowledge of sniping and likely enemy hiding places, they can normally engage enemy marksmen and irregulars at a greater range than the enemy sniper can engage US forces. Their precision fires are also much less likely to cause civilian casualties than fires from other weapons. The commander must carefully consider whether the use of these scarce resources in such a purely defensive, reactive role is the best way to employ them. They may be more valuable inflicting casualties on enemy forces. In some stability and support operation situations, SOF sniper teams may be available. These highly trained teams are often equipped with special, long-range sniper weapons that can be used to dominate large areas around US forces.

(4) ***Unit Weapons***. If an enemy sniper engages a unit, it may be authorized to respond with fire from all its light weapons. In an urban area, the direction of enemy fire, especially from a single rifle shot, is often difficult to determine. If a unit can determine the general location of a sniper, it should return suppressive fire while maneuvering to engage the sniper from close range. This is not always successful because a well-trained sniper often has a route of withdrawal already chosen. Massive return of fire and immediate maneuver can be effective against short-range sniper fires, if the ROE permit this response. In high-intensity urban combat, they are often the best immediate responses. Exploding fragmentation rounds, such as 40-mm grenades from the M203 grenade launcher, are the most effective suppressors.

(5) O***vermatching Fire From Selected Weapons***. The use of overmatching return fires against snipers can be very effective in high-intensity or precision urban combat. The 25-mm cannon on the BFV is a powerful and accurate weapon that can penetrate deep into buildings with its APDS rounds. Fires from caliber .50 machine guns were effective against snipers during combat in Panama in 1989. Units reported the snipers seemed to be intimidated into inaction by the immediate return of heavy machine gun fire. In Somalia, immediate heavy fires from MK 19 automatic grenade launchers were often effective at stopping sniper fires from armed irregulars. Light or medium antitank weapons are also effective. Because of their accuracy, guided munitions such as the TOW, Hellfire, Dragon, or Javelin have the added advantage of limiting collateral damage. Tank cannon can also be used to respond to sniper fire, although the danger of collateral damage is greater because of the extreme penetration of the round.

(6) ***Preemptive Fires***. In high-intensity urban combat, preemptive fires can be used against likely sniper positions. This technique is more often used during offensive operations. It uses large amounts of ammunition but can be very effective for short

attacks. Fragmentation fires from artillery, mortars, and grenade launchers are best for suppressing snipers whose position has not yet been detected.

(7) ***Projected Smoke or Riot Control Agents***. Projected smoke that builds quickly is a good response to protect a unit from further casualties if engaged by an enemy sniper. It greatly limits his ability to acquire targets. The closer the smoke is placed to the sniper's location, the more effective it is. If the location of the sniper is unknown or cannot be reached by projected smoke, a smoke cloud established near the unit is still effective in reducing the sniper's chances of hitting a target. If the ROE permit, and permission has been granted for the use of riot control agents, they can be used effectively to reduce the sniper threat. Few snipers can deliver long-range, accurate fires while wearing protective masks.

b. **Passive Countermeasures**. Passive countermeasures prevent the sniper from acquiring a clear target or prevent his fires from causing casualties. Many passive countermeasures are not unique to countering enemy snipers. They are common sense actions taken by well-trained infantry units in a combat area to limit exposure and minimize casualties. Passive countersniper measures are rarely successful by themselves. They may be politically and psychologically effective in terms of reducing US casualties and the level of violence, but they are often ultimately counterproductive to the commander's main mission. They tend to isolate US forces, especially during stability and support operations, when a visible presence is often required. They tend to create a siege mentality, and they pass the initiative over to the sniper. Among the most common passive countermeasures are:

(1) ***Limit Exposure***. Consider the following when limiting exposure:

- Use covered and concealed routes.
- Avoid open plazas and intersections.
- Stay away from doorways and windows.
- Move along the side of the street, not down the center.
- Move in the shadows.
- Move dispersed, using traveling or bounding overwatch.
- Avoid lighted areas at night.
- Avoid being silhouetted against lights or the skyline.
- Move quickly across open areas that cannot be avoided.
- Remain crouched or prone behind cover or concealment whenever possible.
- If troops are riding in the cargo area of trucks, keep the canvas cargo cover mounted to screen them. (This countermeasure may not be appropriate if there is threat of ambush by enemy forces in addition to snipers.)
- Avoid gathering together in large groups in the open.
- Remain dispersed.
- Avoid wearing obvious badges of rank.
- Avoid exaggerated saluting or standing at attention for officers while in the open.

(2) ***Wear Protective Equipment***. The Kevlar helmet and protective vest will not always stop a sniper bullet, but they will significantly reduce the severity of wounds. They should be worn any time soldiers are exposed to potential sniper fire. In situations where dismounted movement across country is not required, request and issue soldiers

special, heavy protective vests that are actually bulletproof. All unit members should wear this protection.

(3) ***Use Armored Vehicles***. Whenever possible, move around the urban area in a protected vehicle with as little exposure as possible. Avoid open-sided cargo vehicles. Requisition or improvise vehicular armor against small-arms fire for all administrative and logistical vehicles.

(4) ***Erect Screens and Shields***. Use simple canvas or plastic screens to make a dangerous alleyway or street crossing much safer for foot traffic. Adapt screens on windows to allow vision out while hiding personnel inside. Use moveable concrete barriers to provide protection for personnel at static positions. Use common items, such as rubble-filled 55-gallon drums and sandbags, to provide cover.

(5) ***Deny the Enemy Use of Overwatching Terrain***. Either occupy such terrain with friendly forces or modify it to make it less useful to an enemy sniper. Pull down likely hiding places. Ensure all actions are in accordance with the laws and customs of war. Clear bushes and rubble. Board or brick up windows. Pile up earth and rubble in front of buildings to block lines of sight for snipers.

(6) ***Use Smoke Hazes or Smoke Screens to Obscure the Sniper's Field of View and Limit the Effectiveness of His Fires***. A clear atmosphere is required for accurate long-range sniping. Smoke hazes can be maintained over broad areas for long periods without significantly hindering friendly operations. Smoke screens can be created quickly and sustained for short periods so US forces can accomplish their objective.

CHAPTER 7
EMPLOYMENT AND EFFECTS OF WEAPONS

This chapter supplements the technical manuals and field manuals that describe weapons capabilities and effects against generic targets. It focuses on specific employment considerations pertaining to combat in urban areas, and it addresses both organic infantry weapons and combat support weapons.

7-1. EFFECTIVENESS OF WEAPONS AND DEMOLITIONS

The characteristics and nature of combat in urban areas affect the employment of weapons and the results they can achieve. Leaders at all levels must consider the following factors in various combinations.

a. **Surfaces.** Hard, smooth, flat surfaces are characteristic of urban targets. Rarely do rounds impact perpendicular to these flat surfaces; rather, they impact at some angle of obliquity, which reduces the effect of a round and increases the threat of ricochets. The tendency of rounds to strike glancing blows against hard surfaces means up to 25 percent of impact-fuzed explosive rounds may not detonate when fired onto rubbled areas.

b. **Engagement Ranges.** Engagement ranges are close. Studies and historical analyses have shown that only 5 percent of all targets are more than 100 meters away. About 90 percent of all targets are located 50 meters or less from the identifying soldier. Few personnel targets will be visible beyond 50 meters and engagements usually occur at 35 meters or less. Minimum arming ranges and troop safety from backblast or fragmentation effects must be considered.

c. **Engagement Times.** Engagement times are short. Enemy personnel present only fleeting targets. Enemy-held buildings or structures are normally covered by fire and often cannot be engaged with deliberate, well-aimed shots.

d. **Depression and Elevation.** Depression and elevation limits for some weapons create dead space. Tall buildings form deep canyons that are often safe from indirect fires. Some weapons can fire rounds to ricochet behind cover and inflict casualties. Target engagement from oblique angles, both horizontal and vertical, demands superior marksmanship skills.

e. **Reduced Visibility.** Smoke from burning buildings, dust from explosions, shadows from tall buildings, and the lack of light penetrating inner rooms all combine to reduce visibility and to increase a sense of isolation. Added to this is the masking of fires caused by rubble and man-made structures. Targets, even those at close range, tend to be indistinct.

f. **Risks from Friendly Fire.** Urban fighting often becomes confused melees with several small units attacking on converging axes. The risks from friendly fires, ricochets, and fratricide must be considered during planning. Control measures must be continually adjusted to lower the risks. Soldiers and leaders must maintain a sense of situational awareness and clearly mark their progress IAW unit SOP to avoid fratricide.

g. **Close Combat.** Both the shooter and target may be inside or outside buildings and they may both be inside the same or separate buildings. The enclosed nature of combat in urban areas means the weapon's effects, such as muzzle blast and backblast, must be considered as well as the round's impact on the target.

h. **Attacking Man-made Structures.** Usually man-made structures must be attacked before enemy personnel inside are attacked. Weapons and demolitions can be chosen for employment based on their effects against masonry and concrete rather than against enemy personnel.

i. **Modern Buildings.** Modern engineering and design improvements mean that most large buildings constructed since World War II are resilient to the blast effects of bomb and artillery attack. They may burn easily, but usually retain their structural integrity and remain standing. Once high-rise buildings burn out, they are still useful to the military and are almost impossible to damage further. A large structure can take 24 to 48 hours to burn out and become cool enough for soldiers to enter.

j. **Building Types.** The most common worldwide building type is the 12- to 24-inch brick building. Table 7-1 lists the frequency of occurrence of building types worldwide.

TYPE OF BUILDING	FREQUENCY OF OCCURRENCE (PERCENTAGE)
30-inch Stone	1
8- to 10-inch Reinforced Concrete	6.9
12- to 24-inch Brick	63
6-inch Wood	16
14-inch Steel and Concrete (Heavy Clad)	2
7-inch Steel and Concrete (Light Clad)	12

Table 7-1. Types of buildings and frequency of occurrence.

k. **Definitions.** The following definitions were determined based on the analyses of various studies relating to the size of "man-sized" holes and experimentation analyses from the MOUT advanced concepts technology demonstration (ACTD).

(1) ***Loophole.*** A loophole is a firing aperture (a minimum of 8 inches in diameter) made in a structure.

(2) ***Mousehole.*** A mousehole is an opening that is made to the interior or exterior of a structure (walls, floors, ceilings, roofs) to facilitate inter- and intra-building communications and movement. A mousehole is usually a minimum of 24 inches high by 30 inches wide in size.

(3) ***Breach Hole.*** A breach hole is an opening that is made in a structure using mechanical, ballistic, explosive, or thermal means to facilitate the entry of assault elements. A breach hole is normally 50 inches high by 30 inches wide in size. Breaches made through existing apertures, for example doors and windows, normally do not require additional size enhancement.

7-2. RIFLE, CARBINE, AND SQUAD AUTOMATIC WEAPON

The M16 rifle and the M4 carbine are the most common weapons fired in urban areas. These weapons, along with the M249 light machine gun, are used to kill enemy personnel, to suppress enemy fire and observation, and to penetrate light cover. Leaders can use tracer fire to designate targets for other weapons.

a. **Employment.** Close combat is the predominant characteristic of urban engagements. Riflemen must be able to hit small, fleeting targets from bunker apertures, windows, and loopholes. This requires pinpoint accuracy with weapons fired in the semiautomatic mode. Killing an enemy through an 8-inch loophole at a range of 50 meters is a challenge, but one that may be common in urban combat.

(1) When fighting inside buildings, rapid semiautomatic fire is used. To suppress defenders while entering a room, a series of rapid three-round bursts is fired at all identified targets and likely enemy positions. This technique is more effective than firing long bursts into a room with fully automatic fire. Soldiers should fire aimed shots from an underarm or shoulder position, not unaimed fire from the hip.

(2) When targets reveal themselves at short range inside buildings, the most effective engagement is the quick-fire technique with the weapon up and both eyes open. (See FM 23-9 for more detailed information.) Accurate, quick fire not only kills enemy soldiers but also gives the attacker fire superiority.

(3) Within urban areas, burning debris, reduced ambient light, strong shadow patterns of varying density, and smoke all limit the effect of night vision and sighting devices. The use of aiming stakes in the defense and the pointing technique in the offense, both using three-round bursts, are night firing skills required of all infantrymen. The individual laser aiming light can sometimes be used effectively with night vision goggles (NVGs). Any soldier using NVGs should be teamed with at least one soldier not wearing them.

b. **Weapon Penetration.** The penetration that can be achieved with a 5.56-mm round depends on the range to the target and the type of material being fired against. The M16A2, M4, and M249 achieve greater penetration than the older M16A1, but only at longer ranges. At close range, the weapons perform the same. Single 5.56-mm rounds are not effective against structural materials (as opposed to partitions) when fired at close range—the closer the range, the less the penetration.

(1) ***5.56 mm Maximum Penetration.*** For the 5.56-mm round, maximum penetration occurs at 200 meters. At ranges less then 25 meters, penetration is greatly reduced. At 10 meters, penetration by the M16 round is poor due to the tremendous stress placed on this high-speed round, which causes it to yaw upon striking a target. Stress causes the projectile to break up, and the resulting fragments are often too small to penetrate.

(2) ***Reduced Penetration.*** Even with reduced penetration at short ranges, interior walls made of thin wood paneling, Sheetrock, or plaster are no protection against 5.56-mm ball ammunition rounds. Common office furniture, such as desks and chairs, cannot stop these rounds, but a layer of books 18 to 24 inches thick can.

(3) ***Wood and Cinder Blocks.*** Wooden frame buildings and single cinder block walls offer little protection from 5.56-mm rounds. When clearing such structures, soldiers must ensure friendly casualties do not result from rounds passing through walls, floors, or ceilings.

(4) ***Armor-Piercing Rounds.*** Armor-piercing rounds are slightly more effective than ball ammunition in penetrating urban targets at all ranges. They are more likely to ricochet than ball ammunition when the target presents a high degree of obliquity.

c. **Protection.** The following common barriers in urban areas stop a 5.56-mm round fired at less than 50 meters:

- One thickness of well-packed sandbags.
- A 2-inch concrete wall (nonreinforced).
- A 55-gallon drum filled with water or sand.
- A small ammunition can filled with sand.

- A cinder block filled with sand (block will probably shatter).
- A plate glass windowpane at a 45-degree angle (glass fragments may be thrown behind the glass).
- A brick veneer.
- A car body (5.56-mm rounds penetrate but may not always exit).

d. **Wall Penetration.** Although most structural materials repel single 5.56-mm rounds, continued and concentrated firing can breach some typical urban structures (see Table 7-2).

(1) ***Breaching Masonry Walls.*** The best method for breaching a masonry wall is by firing short bursts (three to five rounds) in a U-shaped pattern. The distance from the gunner to the wall should be minimized for best results—ranges as close as 25 meters are relatively safe from ricochet. Ballistic eye protection, protective vest, and helmet should be worn.

(2) ***Ball and Armor-Piercing Ammunition.*** Ball ammunition and armor-piercing rounds produce almost the same results, but armor-piercing rounds are more likely to fly back at the shooter. The 5.56-mm round can be used to create either a loophole (about 7 inches in diameter) or a breach hole (large enough for a man to enter). When used against reinforced concrete, 5.56-mm rounds cannot cut the reinforcing bars.

TYPE	PENETRATION	ROUNDS (REQUIRED)
8-inch reinforced concrete	Initial Loophole	35 250
14-inch triple brick	Initial Loophole	90 160
12-inch cinder block with single-brick veneer	Loophole Breach hole	60 250
9-inch double brick	Initial Loophole	70 120
16-inch tree trunk or log wall	Initial*	1 to 3
12-inch cinder block (filled with sand)	Loophole	35
24-inch double sandbag wall	Initial*	220
3/8-inch mild steel door	Initial*	1
*Penetration only, no loophole.		

Table 7-2. Structure penetration capabilities of the 5.56-mm round against typical urban targets (range 25 to 100 meters).

7-3. MEDIUM AND HEAVY MACHINE GUNS (7.62-MM AND CALIBER .50)

In the urban environment, the Browning caliber .50 machine gun and the 7.62-mm M60 and M240B machine guns provide high-volume, long-range, automatic fires for the suppression or destruction of targets. They provide final protective fire along fixed lines and can be used to penetrate light structures—the caliber .50 machine gun is most effective in this role. Tracers from both machine guns are likely to start fires.

a. **Employment.** The primary consideration that impacts the employment of machine guns within urban areas is the limited availability of long-range fields of fire.

Although machine guns should be emplaced at the lowest terrain level possible, grazing fire at ground level is often obstructed by rubble.

(1) ***M2, Caliber .50 Machine Gun.*** The caliber .50 machine gun is often employed on its vehicular mount during both offensive and defensive operations. If necessary, it can be mounted on the M3 tripod for use in the ground role or in the upper levels of buildings. When mounted on a tripod, the caliber .50 machine gun can be used as an accurate, long-range weapon and can supplement sniper fires.

(2) ***M60/M240 Machine Guns.*** Medium machine guns are cumbersome, making them difficult to use inside while clearing a building. They are useful outside to suppress and isolate enemy defenders. If the gunner is unable to engage targets from the prone position, he can fire the M240B and the M60 from either the shoulder or the hip to provide a high volume of assault and suppressive fires. The use of the long sling to support the weapon and ammunition is preferred.

(3) ***Comparison.*** Medium machine guns are less effective against masonry targets than caliber .50 machine guns because of their reduced penetration power. The gun's availability and its lighter weight make it well suited to augment heavy machine gun fire. They can be used in areas where the caliber .50 machine guns cannot be positioned, or they can be used as a substitute when heavy machine guns are not available. The M60/M240B machine gun can be employed on its tripod to deliver accurate fire along fixed lines and then can quickly be converted to bipod fire to cover alternate fields of fire.

b. **Weapon Penetration.** The ability of the 7.62-mm and caliber .50 rounds to penetrate is also affected by the range to the target and type of material fired against. The 7.62-mm round is affected less by close ranges than the 5.56-mm; the caliber .50 rounds penetration is reduced least of all.

(1) At 50 meters, the 7.62-mm ball round cannot reliably penetrate a single layer of well-packed sandbags. It can penetrate a single sandbag layer at 200 meters, but not a double layer. The armor-piercing round does only slightly better against sandbags. It cannot penetrate a double layer but can penetrate up to 10 inches at 600 meters.

(2) The penetration of the 7.62-mm round is best at 600 meters. Most urban targets are closer. The longest effective range is usually 200 meters or less. Table 7-3 explains the penetration capabilities of a single 7.62-mm (ball) round at closer ranges.

RANGE (meters)	PENETRATION (inches)			
	PINE BOARD	DRY LOOSE SAND	CINDER BLOCK	CONCRETE
25	13	5	8	2
100	18	4.5	10	2
200	41	7	8	2

Table 7-3. Penetration capabilities of a single7.62-mm (ball) round.

(3) The caliber .50 round is also optimized for penetration at long ranges (about 800 meters). For hard targets, obliquity and range affect caliber .50 penetration. Both armor-piercing and ball ammunition penetrate 14 inches of sand or 28 inches of packed

earth at 200 meters, if the rounds impact perpendicular to the flat face of the target. Table 7-4 explains the effect of a 25-degree obliquity on a caliber .50 penetration.

THICKNESS (feet)	100 METER (rounds)	200 METERS (rounds)
2	300	1,200
3	450	1,800
4	600	2,400

Table 7-4. Number of rounds needed to penetrate a reinforced concrete wall at a 25-degree obliquity.

c. **Protection.** Barriers that offer protection against 5.56-mm rounds are also effective against 7.62-mm rounds with some exceptions. The 7.62-mm round can penetrate a windowpane at a 45-degree obliquity, a hollow cinder block, or both sides of a car body. It can also easily penetrate wooden frame buildings. The caliber .50 round can penetrate all the commonly found urban barriers except a sand-filled 55-gallon drum.

d. **Wall Penetration.** Continued and concentrated machine gun fire can breach most typical urban walls. Such fire cannot breach thick reinforced concrete structures or dense natural stone walls. Internal walls, partitions, plaster, floors, ceilings, common office furniture, home appliances, and bedding can be easily penetrated by both 7.62-mm and caliber .50 rounds (Tables 7-5 and 7-6).

TYPE	THICKNESS (inches)	HOLE DIAMETER (inches)	ROUNDS REQUIRED
Reinforced concrete	8	7	100
Triple brick wall	14	7	170
Concrete block with single brick veneer	12	6 and 24	30 and 200
Cinder block (filled)	12	*	18
Double brick wall	9	*	45
Double sandbag wall	24	*	110
Log wall	16	*	1
Mild steel door	3/8	*	1
* Penetration only, no loophole.			

Table 7-5. Structure penetrating capabilities of 7.62-mm round (NATO ball) against typical urban targets (range 25 meters).

(1) The medium machine gun can be difficult to hold steady enough to repeatedly hit the same point on a wall. The dust created by the bullet strikes also makes precise aiming difficult. Firing from a tripod is usually more effective than without, especially if sandbags are used to steady the weapon. Short bursts of three to five rounds fired in a U-type pattern are best.

(2) Breaching a brick veneer presents a special problem for the medium machine gun. Rounds penetrate the cinder block but leave a net-like structure of unbroken block.

Excessive ammunition is required to destroy a net since most rounds only pass through a previously eroded hole. One or two minutes work with an E-tool, crowbar, or axe can remove this web and allow entry through the breach hole.

(3) The caliber .50 machine gun can be fired accurately from the tripod using the single-shot mode. This is the most efficient method for producing a loophole. Automatic fire in three- to five-round bursts, in a U-type pattern, is more effective in producing a breach.

TYPE	THICKNESS (inches)	HOLE DIAMETER (inches)	ROUNDS REQUIRED
Reinforced concrete	10	12 24	50 100
	18	7	140
Triple brick wall	12	8 26	15 50
Concrete block with single brick veneer	12	10 33	25 45
Armor plate	1	*	1
Double sandbag wall	24	*	5
Log wall	16	*	1
* Penetration only, no loophole.			

Table 7-6. Structure penetrating capabilities of caliber .50 ball against typical urban targets (range 35 meters).

7-4. GRENADE LAUNCHERS, 40-MM (M203 AND MK 19)

Both the M203 dual-purpose weapon and the MK 19 grenade machine gun fire 40-mm high-explosive (HE) and high-explosive dual-purpose (HEDP) ammunition. Ammunition for these weapons is not interchangeable, but the grenade and fuze assembly hitting the target is identical. Both weapons provide point and area destructive fires as well as suppression. The MK 19 has a much higher rate of fire and a longer range; the M203 is much lighter and more maneuverable.

a. **Employment.** The main consideration affecting the employment of 40-mm grenades within urban areas is the typically short engagement range. The 40-mm grenade has a minimum arming range of 14 to 28 meters. If the round strikes an object before it is armed, it will not detonate. Both the HE and HEDP rounds have 5-meter burst radii against exposed troops, which means the minimum safe firing range for combat is 31 meters. The 40-mm grenades can be used to suppress the enemy in a building, or inflict casualties by firing through apertures or windows. The MK 19 can use its high rate of fire to concentrate rounds against light structures. This concentrated fire can create extensive damage. The 40-mm HEDP round can penetrate the armor on the flank, rear, and top of Soviet-made BMPs and BTRs. Troops can use the M203 from upper stories to deliver accurate fire against the top decks of armored vehicles. Multiple hits are normally required to achieve a kill.

b. **Weapon Penetration.** The 40-mm HEDP grenade has a small shaped charge that penetrates better than the HE round. It also has a thin wire wrapping that bursts into a dense fragmentation pattern, creating casualties out to 5 meters. Because they explode on contact, 40-mm rounds achieve the same penetration regardless of range. Table 7-7 explains the penetration capabilities of the HEDP round.

TARGET	PENETRATION (inches)
Sandbags	20 (double layer)
Sand-filled cinder block	16
Pine logs	12
Armor plate	2

Table 7-7. Penetration capabilities of the HEDP round.

(1) If projected into an interior room, the 40-mm HEDP can penetrate all interior partition-type walls. It splinters plywood and plaster walls, making a hole large enough to fire a rifle through. It is better to have HEDP rounds pass into a room and explode on a far wall, even though much of the round's energy is wasted penetrating the back wall (Figure 7-1). The fragmentation produced in the room causes more casualties than the HE jet formed by the shaped charge.

(2) The fragments from the HEDP round do not reliably penetrate interior walls. Office furniture sandbags, helmets, and protective vests (flak jackets) also stop them. The M203 dual-purpose weapon has the inherent accuracy to place grenades into windows at 125 meters and bunker apertures at 50 meters. These ranges are significantly reduced as the angle of obliquity increases. Combat experience shows that M203 gunners cannot consistently hit windows at 50 meters when forced to aim and fire quickly.

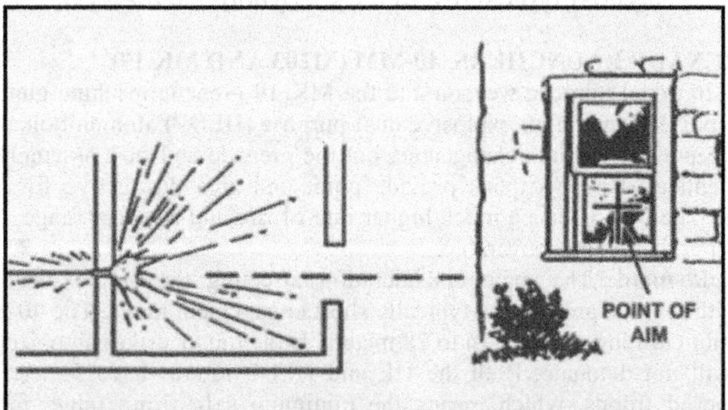

Figure 7-1. Aiming point for 40-mm HEDP.

c. **Wall Penetration.** The M203 cannot reasonably deliver the rounds needed to breach a typical exterior wall. The MK 19 can concentrate its fire and achieve wall penetration. Firing from a tripod, using a locked down traversing and elevating mechanism is best for this role. Brick, cinder block, and concrete can be breached using the MK 19 individual HEDP rounds, which can penetrate 6 to 8 inches of brick. The only material that has proven resistant to concentrated 40-mm fire is dense stone such as that used in some European building construction. No precise data exist as to the number of rounds required to produce loopholes or breach holes with the MK 19; however, the rounds' explosive effects are dramatic and should exceed the performance of the caliber .50 machine gun.

7-5. LIGHT AND MEDIUM RECOILLESS WEAPONS

Light and medium recoilless weapons are used to attack enemy personnel, field fortifications, and light armored vehicles. They have limited capability against main battle tanks, especially those equipped with reactive armor (except when attacking from the top, flanks, or rear). The light category of recoilless weapons includes the AT4 M136 series; the 84-mm M3 Carl Gustaf recoilless rifle; and the shoulder-launched, multipurpose, assault weapon—disposable (SMAW-D) also known as the bunker defeat munitions (BDM). The medium recoilless weapons are the Javelin and Dragon.

a. **Employment.** Other than defeating light armored vehicles, the most common task for light recoilless weapons is to neutralize fortified firing positions. Due to the design of the warhead and the narrow blast effect, these weapons are not as effective in this role as heavier weapons such as a tank main gun round. They are lightweight, allowing soldiers to carry several AT4 rounds. Light recoilless weapons can be fired from the tops of buildings or from areas with proper ventilation.

(1) Light and medium recoilless weapons, with the exception of the SMAW-D, employ shaped-charge warheads. As a result, the hole they punch into walls is often too small to use as a loophole. The fragmentation and spall these weapons produce are limited. Normally, shaped-charge warheads do not neutralize enemy soldiers behind walls unless they are located directly in line with the point of impact.

(2) Against structures, shaped-charge weapons should be aimed about 6 inches below or to the side of a firing aperture (Figure 7-2), which enhances the probability of killing the enemy behind the wall. A round passing through a window wastes much of its energy on the back wall. Since these shaped-charge rounds lack the wire wrapping of the 40-mm HEDP, they burst into few fragments and are often ineffective casualty producers.

Figure 7-2. Point of aim for a shaped-charge weapon against a masonry structure.

(3) Sandbagged emplacements present a different problem (Figure 7-3). These positions may be encountered in urban areas that are adjacent to or contain natural terrain. Because sandbags absorb much of the energy from a shaped-charge, the rounds should be aimed at the center of the firing aperture. Even if the round misses the aperture, the bunker wall area near it is usually easier to penetrate.

Figure 7-3. Point of aim for sandbagged emplacement.

(4) Light and medium recoilless weapons obtain their most effective short-range antiarmor shots by firing from upper stories, or from the flanks and rear. When firing at main battle tanks, these weapons should always be employed against weaker areas in volley or paired firing. They normally require multiple hits to achieve a kill on a tank. Flanks, top, and rear shots hit the most vulnerable parts of armored vehicles. Firing from upper stories protects the shooter from tank main gun and coaxial machine gun fire since tanks cannot sharply elevate their cannons. The BMP-2 can elevate its 30-mm cannon to engage targets in upper stories. The BTR-series armored vehicles can also fire into upper stories with their heavy machine gun.

(5) Modern threat infantry fighting vehicles, such as the BMP-2 and the BTR-80, have significantly improved frontal protection against shaped-charge weapons. Many main battle tanks have some form of reactive armor in addition to their thick armor plate. Head-on, ground-level shots against these vehicles have little probability of obtaining a kill. Even without reactive armor, modern main battle tanks are hard to destroy with a light antiarmor weapon.

(6) The most effective method of engagement for hitting and killing an armored vehicle is to fire from an elevated position. A 45-degree downward firing angle doubles the probability of a first-round hit as compared to a ground-level shot (Figure 7-4).

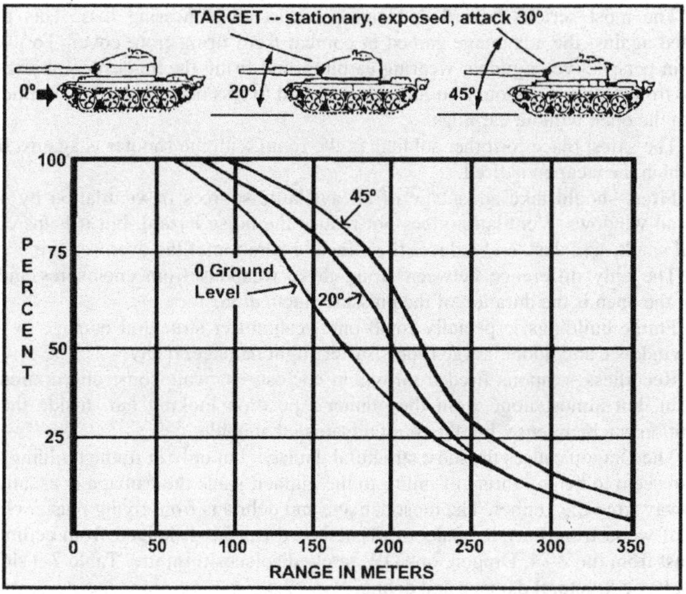

Figure 7-4. Probability of achieving a hit at different angles

b. **Backblast.** Backblast effects must be considered when employing recoilless weapons. During combat in urban areas, the backblast area in the open is more hazardous due to loose rubble and the channeling effect of the narrow streets and alleys.

(1) When firing recoilless weapons in the open, soldiers should protect themselves from blast and burn injuries caused by the backblast. All personnel should be out of the danger zone. Anyone not able to vacate the caution zone should be behind cover. Soldiers in the caution zone should wear helmets, protective vests, and eye protection. The shooter and all soldiers in the area should wear earplugs.

(2) Since the end of World War II, the US Army has conducted extensive testing on the effects of firing recoilless weapons from within enclosures. Beginning as early as 1948, tests have been conducted on every type of recoilless weapon available. In 1975, the US Army Human Engineering Laboratory at Aberdeen Proving Grounds, Maryland, conducted extensive firing of the LAW, Dragon, and TOW from masonry and frame buildings, and from sandbag bunkers. These tests showed the following:

(a) Firing these weapons from enclosures presented no serious hazards, even when the overpressure was enough to produce structural damage to the building.

(b) Little hazard exists to the gunnery or crew from any type of flying debris. Loose items were not hurled around the room.

(c) No substantial degradation occurs to the operator's tracking performance as a result of obscuration or blast overpressure.

(d) The most serious hazard that can be expected is hearing loss. This must be evaluated against the advantage gained in combat from firing from cover. To place this hazard in perspective, a gunner wearing earplugs and firing the loudest combination (the Dragon from within a masonry building) is exposed to less noise hazard than if he fired a LAW in the open without earplugs.

(e) The safest place for other soldiers in the room with the shooter is against the wall from which the weapon is fired.

(f) Firers should take advantage of all available sources of ventilation by opening doors and windows. Ventilation does not reduce the noise hazard, but it helps clear the room of smoke and dust, and reduces the effective duration of the overpressure.

(g) The only difference between firing these weapons from enclosures and firing them in the open is the duration of the pressure fluctuation.

(h) Frame buildings, especially small ones, can suffer structural damage to the rear walls, windows, and doors. Large rooms suffer slight damage, if any.

(3) Recoilless weapons fired from within enclosures create some obscuration inside the room, but almost none from the gunner's position looking out. Inside the room, obscuration can be intense, but the room remains inhabitable.

(4) The Dragon causes the most structural damage, but only in frame buildings. There does not seem to be any threat of injury to the gunner, since the damage is usually to the walls away from the gunner. The most damage and debris is from flying plaster chips and pieces of wood trim. Large chunks of plasterboard can be dislodged from ceilings. The backblast from the AT4, Dragon, or TOW rarely displaces furniture. Table 7-9 shows the test results of structural damage and debris.

NOTE: While the results of the tests may have shown that the threat of injury from debris is rare, commanders must ensure that proper safety precautions are followed prior to firing weapons inside a room.

BUILDING	WEAPON	STRUCTURE DAMAGE	WALL DAMAGE	DEBRIS MOVEMENT
Masonry	LAW Dragon	None None	Slight Slight	Slight Slight
Bunker	Dragon TOW	None None	None None	None Leaves & dust disturbed
Small Frame	LAW Dragon	None Severe	Slight Severe	None None
Medium Frame	LAW Dragon	None Slight	None Slight	Slight Lamps and chairs overturned
Large Frame	LAW Dragon TOW	None Slight Slight	Slight Moderate Severe	Slight None None

Table 7-9. Structural damage and debris movement.

(5) To fire an 84-mm Carl Gustaf recoilless rifle, the AT4, or SMAW-D from inside a room, the following safety precautions must be taken (Figure 7-5).

(a) The building should be of a sturdy construction.

(b) The ceiling should be at least 7 feet high with loose plaster or ceiling boards removed.

(c) The floor size should be at least 15 feet by 12 feet. (The larger the room, the better.)

(d) At least 20 square feet of ventilation (room openings) should exist to the rear or side of the weapon. An open 7- by 3-foot door would provide minimum ventilation.

(e) All glass should be removed from windows and small, loose objects removed from the room.

(f) Floors should be wet to prevent dust and dirt from blowing around and obscuring the gunner's vision.

(g) All personnel in the room should be forward of the rear of the weapon.

(h) All personnel in the room should wear helmets, body armor, ballistic eye protection, and earplugs.

(i) If the gunner is firing from the prone position, his lower body must be perpendicular to the bore of the weapon or the blast could cause injury to his legs.

Figure 7-5. Firing an 84-mm Carl Gustaf recoilless rifle, an AT4, or SMAW-D from inside a building.

c. **Weapon Penetration.** The most important tasks to be performed against structures are the neutralization of fortified fighting positions, personnel, and weapons behind barriers. Recoilless weapons can be used in this role, but none of them are as effective as heavy direct-fire weapons or standard demolitions. Each recoilless weapon has different penetrating ability against various targets. Penetration does not always mean the destruction of the integrity of a position. Usually, only those enemy soldiers directly in the path of the spall from a HEAT round become casualties. Other soldiers inside a fortification could be deafened, dazed, or shocked but eventually return to action. (See Table 7-10, page 7-14.)

TARGET	EFFECT WHEN FIRED AT TARGET	RECOMMENDED AIMING POINT
Firing port or aperture	Rounds fired into firing ports or apertures may be wasted; rounds detonate inside on the rear of the position, causing little or no damage to the position or equipment and personnel unless hit directly.	Coordinate fire: fire light antiarmor weapons at a point 6 to 12 inches from the edge of the aperture or berm.
Berm	Firing at the berm causes the round to detonate outside the position or in the berm, producing only a small hole in the berm, but no damage to the position or equipment and personnel unless hit directly.	Coordinate fire: fire light antiarmor weapons at a point 6 to 12 inches from the edge of the aperture or berm.
Windows	The round may travel completely through the structure before detonating; if not, it causes dust, minor damage to the rear wall, but no damage to the position or equipment and personnel unless they are hit directly.	Fire 6 to 12 inches from the sides or bottom of a window. Light antiarmor rounds explode on contact with brick and concrete, creating an opening whose size is determining by the type of round used.
Wall	The round detonates on contact, creating dust, a small hole, and minor structural damage, but little or no damage to the position or equipment and personnel unless hit directly.	Fire 6 to 12 inches from the sides or bottom of a window. Light antiarmor rounds explode on contact with brick and concrete, creating an opening whose size is determining by the type of round used.
Corners	Corners are reinforced and thus harder to penetrate than other parts of the wall. Any light antiarmor round will detonate sooner on a corner than on less dense surfaces. Detonation should occur in the targeted room, creating dust and overpressure. The overpressure can temporarily incapacitate personnel inside the structure near the point of detonation.	Fire 6 to 12 inches from the sides or bottom of a window. Light antiarmor rounds explode on contact with brick and concrete, creating an opening whose size is determining by the type of round used.

Table 7-10. Light antiarmor weapons effects on urban targets.

(1) ***M136 84-mm Launcher (AT4).*** The AT4 is a lightweight, disposable, direct fire antiarmor weapon. The round has a diameter of 84 millimeters, which gives the warhead much greater penetration. The AT4 can penetrate more than 17.5 inches (450 millimeters) of armor plate. Its warhead produces highly destructive results behind the armor. The AT4 has a minimum arming distance of 10 meters, which allows it to be fired successfully against close targets. Firers should be well covered by protective equipment when firing at close targets.

(2) ***M3 84-mm Carl Gustav Recoilless Rifle.*** The 84-mm, M3 Carl Gustav recoilless rifle is lightweight and maneuverable with great penetrating power, making it a very

useful weapon during urban combat. The Carl Gustav can fire a variety of rounds against a variety of targets.

(a) The FFV HEAT 551 round, for use against armored targets, will penetrate more than 400 millimeters of armor. The HEAT round arms at 5 to 8 meters and may throw fragments back as far as 50 meters.

(b) The FFV HEDP 502 round with a dual mode fuze can be set to detonate on impact against nonreinforced structures, or delayed to detonate after penetrating 1 meter into an earthen bunker. It will penetrate more than 150 millimeters of armor. The HEDP round is probably the most useful during urban combat. It is effective against light-armored vehicles, thick concrete and brick walls, thin wood walls and field fortifications, and unprotected troops. The HEDP round arms at 15 to 40 meters and produces only slight fragmentation out to 50 meters.

(c) The FFV HE 441B is used primarily against personnel and light-skinned vehicles. The HE round can be set for either airburst or impact burst. It contains 800 steel balls that are distributed in a lethal pattern upon detonation. The HE round arms at 20 to 70 meters and may throw its steel balls back as far as 250 meters.

(d) The FFV Illumination 545 round produces 650,000 candlepower, illuminating a 400- to 500-meter area for 30 seconds.

(e) The FFV Smoke 469B round provides a screening and blinding smoke cloud. It is useful to cover friendly units crossing small open areas.

(3) ***Shoulder-Launched, Multipurpose, Assault Weapon—Disposable (SMAW-D).*** The SMAW-D is a lightweight, man-portable, assault weapon easily carried and placed into action by one man. It is used against fortified positions, but is also effective against light-armored vehicles. The SMAW-D has a rifle type sighting system with a three-post front and a peep rear sight. It fires an 83-mm HEDP rocket that is effective against walls, bunkers, and light-armored vehicles. The SMAW-D can destroy most bunkers with a single hit while multiple shots create breach holes even in reinforced concrete; it will not cut reinforcing steel bars.

(4) ***Javelin.*** The Javelin is a dual-mode (top attack or direct fire), man-portable antitank missile with an increased capability to engage and defeat tanks and other armored vehicles. The Javelin has a missile contained in a disposable launch tube-container and reusable tracker. The Javelin is a fire-and-forget weapon system, which significantly increases the gunner's survivability because the gunner is no longer required to track the target for the duration of the missile's flight. Additionally, compared to the Dragon, the Javelin has a soft launch that significantly reduces the visual and acoustical signature the missile makes. The minimum engagement range is 75 meters, and the Javelin can penetrate all urban targets. Penetration, however, does not mean destruction of the structural integrity of a position. Firing ATGMs is the least efficient means to defeat structural walls.

d. **Wall Breaching.** Wall breaching is a common combat task in urban areas for which light recoilless weapons can be used. Breaching operations improve mobility by providing access to building interiors without using existing doors or windows. Breaching techniques can also be used to create loopholes for weapons positions or to allow hand grenades to be thrown into defended structures. Breach holes for troop mobility should be about 50 inches high by 30 inches wide. Loopholes should be about 8 inches in diameter (Figure 7-6, page 7-16). None of the light recoilless weapons organic to maneuver battalions (with the possible exception of the SMAW-D) provide a one-shot wall-breaching capability. To breach walls, a number of shots should be planned.

(1) Of all the common building materials, heavy stone is the most difficult to penetrate. The AT4 or the Carl Gustav usually will not penetrate a heavy European-style stone wall. Surface cratering is usually the only effect.

(2) Layered brick walls are also difficult to breach with light recoilless weapons. Some brick walls can be penetrated by multiple firings, especially if they are less than three bricks thick. Weapons such as the AT4 and the Carl Gustav may require 3 to 5 rounds in order to penetrate brick walls. The SMAW-D produces a hole in brick walls that is often large enough to be a breach hole.

(3) Wooden structural walls offer little resistance to light recoilless weapons. Even heavy timbered walls are penetrated and splintered. The AT8 and SMAW have a devastating effect against a wood-frame wall. A single round produces a breach hole as well as significant spall.

(4) Because of its high velocity, the AT4 may penetrate a soft target, such as a car body or frame building, before exploding.

(5) None of the light recoilless weapons are as effective against structural walls as demolitions or heavier weapons such as a tank main gun, or field artillery. Of all the light recoilless weapons, the SMAW-D is the most effective.

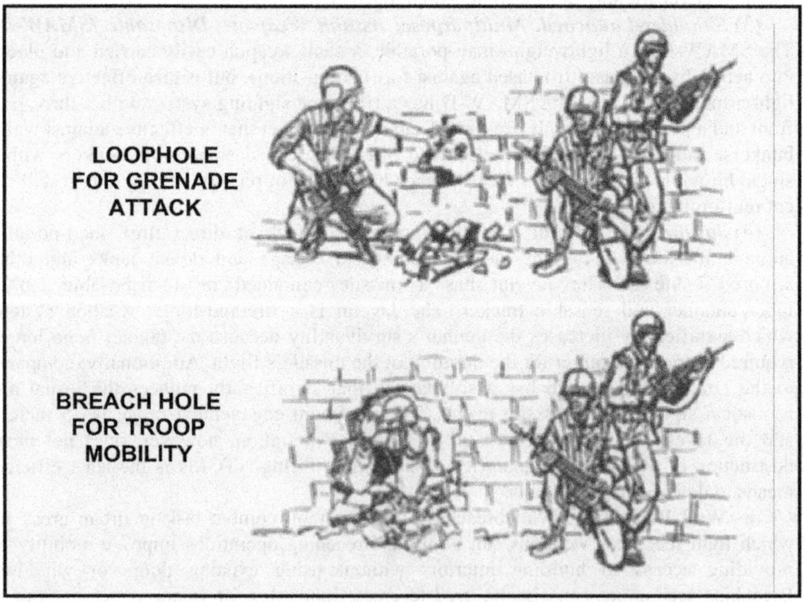

Figure 7-6. Tactical use of holes in masonry walls, after ballistic breaching.

7-6. ANTITANK GUIDED MISSILES

Antitank guided missiles (ATGMs) are used mainly to defeat main battle tanks and other armored combat vehicles. They have a moderate capability against bunkers, buildings, and other fortified targets commonly found during combat in urban areas. This category of weapons includes the TOW and Dragon missiles.

a. **Employment.** TOWs and Dragons provide overwatch antitank fires during the attack of an urban area and extended range capability for engaging armor during the defense. Within urban areas, they are best employed along major thoroughfares and from the upper stories of buildings to attain long-range fields of fire. Their minimum firing range of 65 meters could limit firing opportunities in the confines of densely urban areas.

(1) ***Obstacles.*** When fired from street level, rubble or other obstacles could interfere with missile flight. At least 3.5 feet (1 meter) of vertical clearance over such obstacles must be maintained. Figure 7-7 shows the most common obstacles to ATGM flights in urban areas. Power lines are a special obstacle that presents a unique threat to ATGM gunners. If the power in the lines has not been interrupted, the ATGM guidance wires could create a short circuit. This would allow extremely high voltage to pass to the gunner in the brief period before the guidance wires melted. This voltage could either damage the sight and guidance system, or injure the gunner. Before any ATGM is fired over a power line, an attempt must be made to determine whether or not the power has been interrupted.

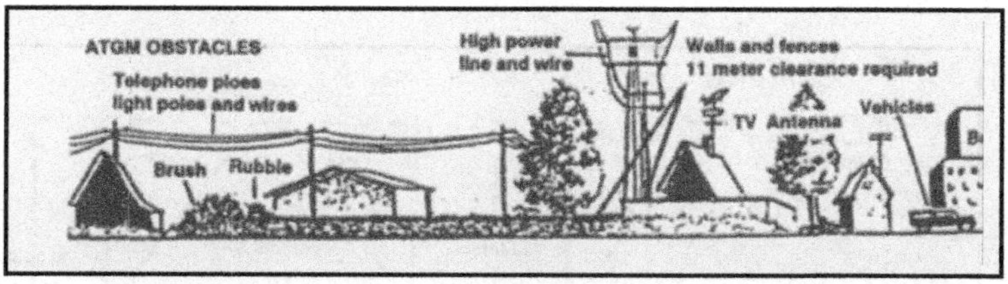

Figure 7-7. Common obstacles to ATGM flights.

(2) ***Dead Space.*** Three aspects of dead space that affect ATGM fires are arming distance, maximum depression, and maximum elevation.

(a) Both the Dragon and TOW missiles have a minimum arming distance of 65 meters, which severely limits their use in urban areas. Few areas in the inner city permit fires much beyond the minimum arming distance—ground-level long-range fires down streets or rail lines and across parks or plazas are possible. ATGMs may be used effectively from upper stories or roofs of buildings to fire into other buildings.

(b) The TOW is limited much more than the Dragon by its maximum depression and elevation. The maximum depression and elevation limits of the TOW mount could result in dead space and preclude the engagements of close targets (Figure 7-8, page 7-18). A target located at the minimum arming range (65 meters) cannot be engaged by a TOW crew located any higher then the sixth floor of a building due to maximum depression limits. At 100 meters the TOW crew can be located as high as the ninth floor and still engage the target.

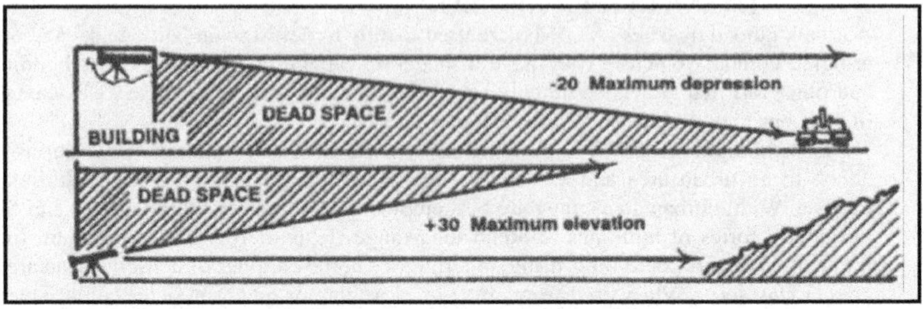

Figure 7-8. TOW maximum elevation and depression limitations.

(3) ***Backblast***. Backblast for ATGMs is more of a concern during combat in urban areas than in open country. Any loose rubble in the caution zone could be picked up and thrown by the backblast. The channeling effect of walls and narrow streets is even more pronounced due to the greater backblast. If the ATGM backblast strikes a wall at an angle, it can pick up debris, or be deflected and cause injury to unprotected personnel (Figure 7-9). Both ATGMs can be fired from inside some buildings. In addition to the helmet and body armor, all personnel in the room should wear eye protection and earplugs.

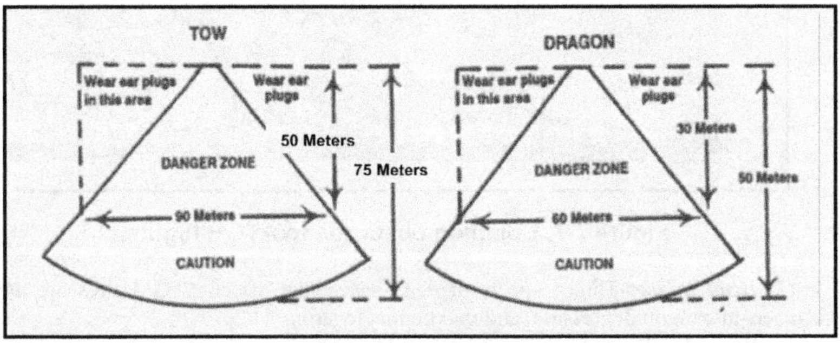

Figure 7-9. ATGM backblast in an open street.

(a) To fire a TOW from inside a room, the following safety precautions must be taken (Figure 7-10).

- The building must be of sturdy construction.
- The ceiling should be at least 7 feet high.
- The floor size of the room should be at least 15 by 15 feet; larger, if possible.
- At least 20 square feet of room ventilation should exist, preferably to the rear of the weapon. An open 7- by 3-foot door is sufficient. Removing sections of interior partitions can create additional ventilation.

- All glass must be removed from the windows, and all small loose objects removed from the room, the room should be cleaned.
- All personnel in the room should be forward of the rear of the TOW.
- All personnel in the room should wear ballistic eye protection and earplugs.
- A clearance of 9 inches (23 centimeters) must be between the launch tube and aperture from which it is fired. (See AR 385-62 and AR 385-63 for more detailed safety information.)

Figure 7-10. TOW fired from inside a room.

(b) To fire a Dragon from inside a room, the following safety precautions must be taken.

- The building must be of sturdy construction.
- The ceiling should be at least 7 feet high.
- The floor size should be at last 15 by 15 feet; larger, if possible.
- At least 20 square feet of ventilation should exist (room openings), preferably to the rear of the weapon. An open 7- by 3-foot door would provide minimum ventilation.
- All glass should be removed from windows, and small loose objects removed from the room.
- The room should be clean or the floors must be wet to prevent dust and dirt (kicked up by the backblast) from obscuring the vision of other soldiers in the room.
- All personnel in the room must be forward of the rear of the weapon.
- All personnel in the room must wear ballistic eye protection and earplugs.
- At least a 6-inch clearance must exist between the launch tube and aperture from which it is fired.

b. **Weapon Penetration.** ATGMs can penetrate and destroy heavily armored tanks. They have large warheads employing the shaped-charge principle. Because of their size, these warheads can achieve significant penetration against typical urban targets. Penetration does not mean concurrent destruction of the structural integrity of a position. The shaped-charge warhead produces relatively little spall. Enemy personnel not standing directly behind or near the point of impact of an ATGM may escape injury.

(1) ***Standard TOW missiles.*** The basic TOW missile can penetrate 8 feet of packed earth, 4 feet of reinforced concrete, or 16 inches of steel plate. The improved TOW (ITOW), the TOW 2, and the TOW 2A have been modified to improve their penetration and they penetrate better than the basic TOW. All TOW missiles can defeat triple sandbag walls, double layers of earth filled 55-gallon drums, and 18-inch log walls.

(2) ***TOW 2B***. The TOW 2B uses a different method of defeating enemy armor. It flies over the target and fires an explosively formed penetrator down onto the top of an armor vehicle, where the armor is thinner. Because of this design feature, the TOW 2B missile cannot be used to attack nonmetallic structural targets. When using the TOW 2B missile against enemy armor, gunners must avoid firing directly over other friendly vehicles, disabled vehicles, or large metal objects such as water or oil tanks.

(3) ***Dragon Missile***. The Dragon missile can penetrate 8 feet of packed earth, 4 feet of concrete, or 13 inches of steel plate. It can attain effective short-range fire from upper stories, or from the rear or flanks of a vehicle. These engagements are targeted against the most vulnerable parts of tanks, and can entrap tanks in situations where they are unable to counterfire. Elevated firing positions increase the first-round hit probability. Firing down at an angle of 20 degrees increases the chance of a hit by 67 percent at 200 meters. A 45-degree down angle doubles the first round hit probability, compared to a ground-level shot.

c. **Breaching Structural Walls**. Firing ATGMs is the least efficient means to breach structures. Because of their small basic load and high cost, ATGMs are better used against enemy tanks or fortified fighting positions. They can be effective against bunkers or other identified enemy fighting positions.

7-7. FLAME WEAPONS

The use of flame weapons, such as Fougasse, the M202A1 Flash, white phosphorous, thermobaric, and other incendiary agents, against targets is not a violation of current international law. They should not, however, be employed to just cause unnecessary suffering to individuals. The use of flame weapons should be addressed in the ROE. Flame weapons are characterized by both physical (flame and overpressure) and psychological casualty-producing abilities. Flame does not normally need to be applied with pinpoint accuracy to accomplish its mission. Efforts must be made to ensure that certain types of flame munitions effects do not spread to structures needed by friendly forces. Large fires in urban areas are catastrophic, and these fires can create an impenetrable barrier for hours. The most common US flame weapons currently employed are the M202A1 Flash, flame field expedients (Fougasse), and the M14 TH3 incendiary hand grenade. In the future, the M202A1 Flash may be replaced by a new shoulder-fired, thermobaric warhead, soft-launched rocket to support soldiers. (See Chapter 3, paragraphs 3-31 and 3-32 for more details.)

a. **Employment.** Flame weapons can be used against fortified positions, interior buildings, tunnels (to include subways and sewers), and open areas. They can also be used to control avenues of approach for personal and lightly armored vehicles. When employed properly, even if the round or burst misses, enough flaming material and overpressure enters the position or area to cause casualties and disrupt operations. Thermobaric munitions will provide a more effective and selective flame capability that is easier and safer to employ at all levels of tactical operations without the side effect of large area destruction due to uncontrolled fires.

b. **Capabilities.** Flame weapons have different effects against typical urban targets and complex terrain.

(1) ***M202A1 Flash Rocket Launcher.*** The M202A1 Flash is a lightweight, individual rocket launcher aimed and fired from the right shoulder using either the standing, the kneeling, or the prone positions. The launcher is loaded with a clip (M74), which contains four 66-mm rockets. It can fire one to four rockets semi-automatically at a rate of one rocket per second and can be reloaded with a new clip (Figure 7-11). The M202A1 can deliver area fire out to 500 meters. During urban combat, the range to targets is normally much less. Point targets, such as an alleyway or bunker, can usually be hit from 200 meters. Precision fire against a bunker aperture is possible at 50 meters.

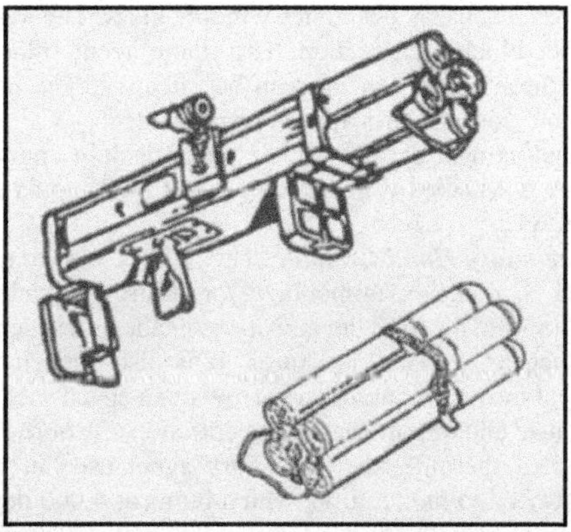

Figure 7-11. M202A1 Flash, 66-mm rocket launcher with M74 rocket clip.

(a) The rocket's warhead contains tri-ethyl aluminum (TEA), which ignites when exposed to air. The minimum safe combat range is 20 meters, which is the bursting radius of the rocket warhead due to splash back. If the projectile strikes a hard object along its flight path and breaks open, it will burst into flames even if the fuse has not armed. M202A1 rocket packs must be protected from small-arms fire and shell fragments that could ignite them. The M202A1 has a backblast that must be considered before firing (Figure 7-12, page 7-22).

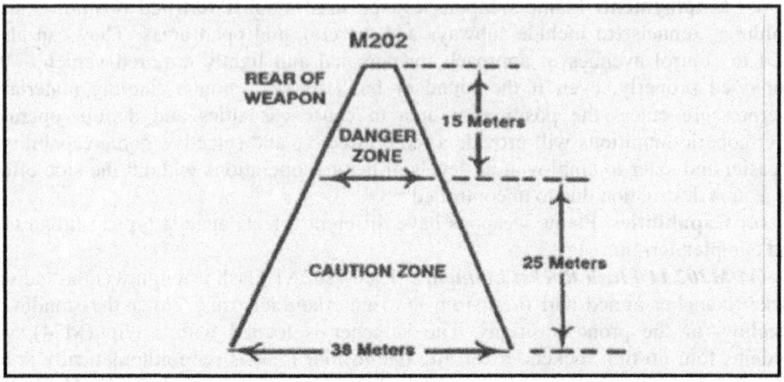

Figure 7-12. Backblast area of an M202A1 Flash.

(b) The M202A1 Flash is not effective in penetrating typical urban targets. It can penetrate up to 1 inch of plywood at 200 meters, and at close range it can penetrate some wooden doors. The rocket reliably penetrates window glass. The M202A1 is not effective against brick or cinder block construction. The flame agent splattered against the top, flanks, and rear of light armored vehicles can be effective. The psychological effect of hits by flame rockets on closed-in crewmen is significant.

(c) A round detonating near or on a vehicle's rear deck or engine compartment could set the vehicle on fire. A wheeled vehicle, such as the BTR, could have its tires severely damaged by the M202A1.

(2) ***M14 TH3 Incendiary Hand Grenade.*** The M14 is used to destroy equipment and start fires. It is used to damage, immobilize, or destroy vehicles, weapons systems, shelters, and ammunition. The M14 incendiary grenade is especially effective against flammable objects such as wooden structures. It is also used to create an immediate smoke cloud to conceal movement across a narrow open space such as a street. Its smoke is not toxic but can cause choking in heavy concentrations. A portion of thermate mixture (an improved version of thermite, the incendiary agent used in hand grenades during World War II) is converted to molten iron, which burns at 4,000 degrees Fahrenheit. The mixture fuses together the metallic parts of any object that it contacts. The thermate filler can burn through a 1/2-inch homogenous steel plate. It produces its own oxygen and burns under water.

(a) The grenade's intense light is hazardous to the retina and can cause permanent eye damage. The brilliant light, smoke, and molten iron particles all combine to make the M14 a very effective psychological weapon. Because it weighs 32 ounces, most infantrymen can throw this grenade only 25 meters.

(b) The M14 incendiary grenade is an effective weapon against enemy armored vehicles when used in the close confines of combat in urban areas. It can be thrown or dropped from upper stories onto enemy vehicles. The M14 can be combined with flammable liquids, detonating cords, blasting caps, and fuse igniters to create the eagle fireball, a field-expedient antiarmor device. (See FM 21-75, Appendix H for more information.)

(3) ***Flame Field Expedients.*** Flame field expedients (Fougasse) are used chiefly in defensive operations; however, they may also be used in offensive operations to—

- Warn of enemy approach when used in a defilade area or during periods of limited visibility.
- Produce casualties by radiant heat or contact with the flaming materiel.
- Deter the enemy by psychological impact.
- Produce limited battlefield illumination to silhouette the opposing force and defeat their IR sensors.
- Restrict terrain to adversaries.
- Defeat underground tunnels and structures.
- Flame minefields that are controll-detonated.

(a) Exploding flame devices consist of a container, thickened fuel, and a firing system to scatter and ignite fuel. The size area to be covered depends on the size of the container and type of firing system. Normal containers will range from 1 to 55 gallons of thickened fuel.

(b) Flame illuminators give an initial flash and then burn for several hours. The normal size container is from 5 to 55 gallons and is filled with earth and then thickened fuel is poured in. The illuminator can be stored for long periods of time if sealed.

(4) ***Thermobarics.*** This type of munition has been used by many nations of the world and their proliferation is an indication of how effectively these weapons can be used in urban and complex terrain. The ability of thermobaric weapons to provide massed heat and pressure effects at a single point in time cannot be reproduced by conventional weapons without massive collateral destruction. Thermobaric weapon technologies provide the ground commander a new choice in protecting the force, and a new offensive weapon that can be used in a mounted or dismounted mode against complex environments. Currently, there are no thermobaric weapons in the US inventory.

7-8. HAND GRENADES

Hand grenades are used extensively during combat in urban areas. Smoke grenades are used for obscuration and signaling. Riot control grenades are used to control civil disturbances. Fragmentation, concussion and stun grenades are used to clear the enemy from rooms and basements. Hand grenades are the most used explosive munition during intense combat in urban areas. In World War II, it was common for a battalion fighting in a city to use over 500 fragmentation grenades each day. Stun grenades are used primarily during precision clearing of an urban structure when the presence of noncombatants is likely.

a. **Employment.** Smoke and riot control grenades have similar employment techniques. Fragmentation and concussion grenades are used to produce enemy casualties. The stun grenade is used as a distraction device.

(1) ***AN-M8 HC smoke grenade.*** The AN-M8 HC smoke grenade produces a dense white or gray smoke. It burns intensely and cannot be easily extinguished once it ignites. The smoke can be dangerous in heavy concentrations because it makes breathing difficult and causes choking. The M8 grenade is normally used for screening. It produces a slowly

building screen of longer duration than the obsolete M34 WP grenade without the problem of collateral damage caused by scattered burning particles.

(2) ***M18-Series Smoke Grenades.*** The M18-series smoke grenades produce several different colors of smoke, which are used for signaling. Yellow smoke is sometimes difficult to see in urban areas. Newer versions of yellow smoke grenades are more visible than the old type.

(3) ***Riot Control Grenade.*** The M7A3 CS riot control grenade can be used to control riots or disperse personnel. (See Appendix F.) Urban areas often create variable and shifting wind patterns. When using CS grenades, soldiers must prevent the irritating smoke from affecting friendly troops. The CS grenade burns intensely and can ignite flammable structures. Enemy troops wearing even rudimentary chemical protective masks can withstand intense concentrations of CS gas.

NOTE: National Command Authority (NCA) approval is required before using riot control agents (RCAs).

(4) ***Concussion Grenade.*** The MK3A2 offensive hand grenade, commonly referred to as the concussion grenade, produces casualties during close combat while minimizing the danger to friendly personnel. The grenade produces severe concussion effects in enclosed areas. For this reason, it is the preferred hand grenade during offensive operations in a MOUT environment. It can be used for light blasting and demolitions, and for creating breach holes in interior walls. The concussion produced by the MK3A2 is much greater than that of the fragmentation grenade. It is very effective against enemy soldiers in bunkers, buildings, and underground passages.

(5) ***Fragmentation Grenade.*** The M67 fragmentation grenade is the most commonly available grenade during combat in urban areas. It provides suppression during room-to-room or house-to-house fighting, and is used while clearing rooms of enemy personnel. When used at close ranges, it can be cooked off for two seconds to deny the enemy time to throw it back. The fragmentation grenade can be rolled, bounced, or ricocheted into areas that cannot be reached by 40-mm grenade launchers. Soldiers must be cautious when throwing grenades up stairs. This is not the most desired method of employment.

(6) ***Stun Grenade.*** The M84 stun hand grenade is the most recent addition to the Army inventory of grenades. Stun hand grenades are used as diversionary or distraction devices during building and room clearing operations when the presence of noncombatants is likely or expected and the assaulting element is attempting to achieve surprise. The following is a description of the M84 stun hand grenade and its components.

(a) *Body.* The body is a steel hexagon tube with holes along the sides to allow for the emission of intense light and sound when the grenade is ignited.

(b) *Fuze.* The fuze is the M201A1 modified with a secondary safety pin installed with a triangular pull ring attached.

(c) *Weight.* The grenade weighs 8.33 ounces.

(d) *Safety Clip.* The fuze has a secondary safety pin installed with a triangular pull ring attached.

(e) *Field-Expedient Use*. In combat, you may need to use the M84 stun hand grenade as an early warning device. Use the following procedures *in combat only*:

- Attach grenade to a secure object such as a tree, post, or picket.
- Attach tripwire to a secured object, extend across path, and attach wire to the pull ring of the grenade.
- Bend the end of the pull pin flat to allow for easy pulling.
- Remove the secondary safety pin.

b. **Effects.** Each type of hand grenade has its own specific effect during urban operations.

(1) The effects of smoke grenades in urban areas are nominal. Smoke grenades produce dense clouds of colored or white smoke that remain stationary in the surrounding area. They can cause fires if used indiscriminately. If trapped and concentrated within a small space, their smoke can suffocate soldiers.

(2) The fragmentation grenade has more varied effects during urban combat. It produces a large amount of small high-velocity fragments, which can penetrate plasterboard partitions and are lethal at short ranges (15 to 20 meters). Fragments lose their velocity quickly and are less effective beyond 25 meters. The fragments from a fragmentation grenade cannot penetrate a single layer of sandbags, a cinder block, or a brick building, but they can perforate wood frame and tin buildings if exploded close to their walls.

(3) Fragmentation barriers consisting of common office furniture, mattresses, doors, or books can be effective against the fragmentation grenade inside rooms. For this reason, a room should never be considered safe just because one or two grenades have been detonated inside. Fragmentation grenades detonated on the floor not only throw fragments laterally but also send fragments and spall downward to lower floors. Predicting how much spall will occur is difficult since flooring material varies, but wooden floors are usually affected the most.

(4) Some foreign grenades throw fragments much larger than those of the US-made M67. Light barriers and interior walls would probably be less effective against these grenades than against the M67. A major problem with the US-made fragmentation grenade is its tendency to bounce back off hard targets. Grenades are often directed at window openings on the ground floor or second floor. At ranges as close as 20 meters, a thrower's chances of missing a standard 1-meter by 1-meter window are high. The fragmentation grenade normally breaks through standard window glass and enters a room. If the grenade strikes at a sharp angle or the glass is thick plate, the grenade could be deflected without penetrating.

(5) The M84 stun hand grenade is designed to be thrown into a room (through an open door, a standard glass window, or other opening) to deliver a loud bang and bright flash sufficient enough to temporarily disorient personnel in the room.

(6) Hand grenades are difficult weapons to use. They involve a high risk of fratricide. Commanders should conduct precombat training with hand grenades as part of normal preparations. Soldiers must be very careful when throwing hand grenades.

(7) The pull pins of all these hand grenades can be replaced if the thrower decides not use the weapon. This pin replacement must be done carefully (see FM 23-30).

(8) The METT-TC and ROE dictates what type of grenade the soldier uses to clear each room. Because of the high expenditure of grenades, units should carry additional grenades of all types. Additional grenades can be carried in empty ammunition or canteen pouches.

7-9. MORTARS

The urban environment greatly restricts low-angle indirect fires because of overhead masking. While all indirect fire weapons are subject to overhead masking, mortars are less affected than field artillery weapons due to the mortar's higher trajectory. For low-angle artillery fire, dead space is about five times the height of the building behind which the target sits. For mortar fire, dead space is only about one-half the height of the building. Because of these advantages, mortars are even more important to the infantry during urban combat.

a. **Employment.** Not only can mortars fire into the deep defilade created by tall buildings, but they can also fire out of it. Mortars emplaced behind buildings are difficult for the enemy to locate accurately and even harder for him to hit with counterfire. Because of their lightweight, even heavy mortars can be hand carried to firing positions that may not be accessible to vehicles.

(1) Mortars can be fired through the roof of a ruined building if the ground-level flooring is solid enough to withstand the recoil. If there is only concrete in the mortar platoon's area, mortars can be fired using sandbags as a buffer under the baseplate and curbs as anchors and braces. (This is recommended only when time is not available to prepare better firing area.) Aiming posts can be placed in dirt-filled cans.

(2) The 60-mm and 81-mm mortars of the US Army have limited effect on structural targets. Even with delay fuzes they seldom penetrate more than the upper stories of light buildings. However, their wide area coverage and multioption fuzes make them useful against an enemy force advancing through streets, through other open areas, or over rubble. The 120-mm mortar is moderately effective against structural targets. With a delay fuze setting, it can penetrate deep into a building and create great destruction.

(3) Mortar platoons often operate as separate firing sections during urban combat. The lack of large open areas can preclude establishing a platoon firing position. Figure 7-13 shows how two mortar sections, which are separated by only one street, can be effective in massing fires and be protected from countermortar fire by employing defilade and dispersion.

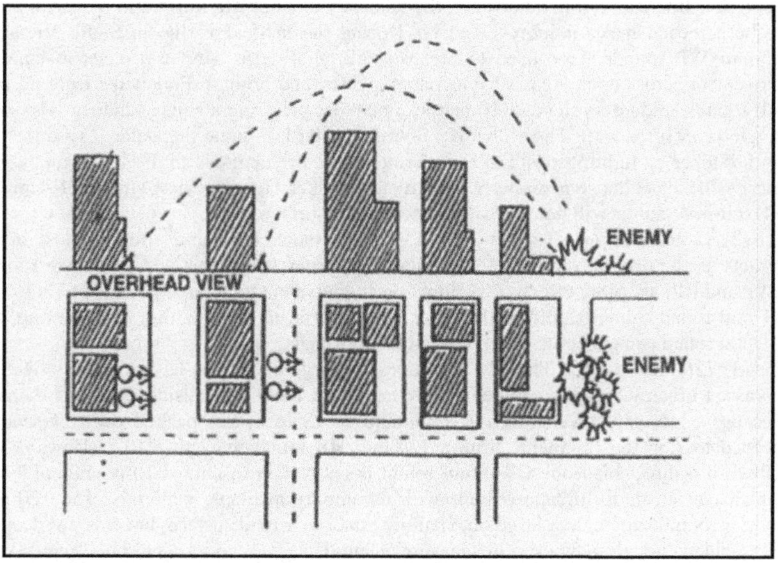

Figure 7-13. Split-section mortar operations on adjacent streets.

(4) All three of the standard mortar projectiles are useful during combat in urban areas. High-explosive fragmentation is the most commonly used round. WP is effective in starting fires in buildings and forcing the enemy out of cellars and light-frame buildings, and is the most effective mortar round against dug-in enemy tanks. Even near misses blind and suppress the tank crew, forcing them to button up.

(5) The artificial relief of urban terrain reduces wind speed and increases atmosphere mixing, so mortar smoke tends to persist longer and give greater coverage in urban areas than in open terrain.

(6) Urban masking impacts the use of illumination. In urban areas, it is often necessary to plan illumination behind friendly positions placing friendly troops in shadows and enemy troops in the light. Illumination rounds are difficult to adjust and are often of limited use because of the deep canyon nature of the urban area. Rapidly shifting wind currents in urban areas also affect mortar illumination, making it less effective.

b. **Effects of Mortar Fire.** The multioption fuze on newer US mortar rounds makes them effective weapons on urban terrain. Delay settings can increase penetration slightly, while proximity bursts can increase the lethal area covered by fragments. Tall buildings can cause proximity fuzed mortar rounds to detonate prematurely if they pass too closely.

(1) ***60-mm Mortar.*** The 60-mm mortar round cannot penetrate most rooftops, even with a delay setting. Small explosive rounds are effective, however, in suppressing snipers on rooftops and preventing roofs from being used by enemy observers. The 60-mm WP round is not normally a good screening round due to its small area of

coverage. In urban combat, however, the tendency of smoke to linger and the small areas to be screened make it more effective. During the battle for Hue in South Vietnam, 60-mm WP rounds were used to create small, short-term, smoke screens to conceal movement across open areas such as parks, plazas, and bridges. Fragments from 60-mm HE rounds landing as close as 10 feet away cannot penetrate a single sandbag layer or a single-layer brick wall. The effect of a 60-mm mortar HE round that achieves a direct hit on a bunker or fighting position is equivalent to 1 or 2 pounds of TNT. Normally, the blast will not collapse a properly constructed bunker but can cause structural damage. The 60-mm mortar will not normally crater a hard-surfaced road.

(2) ***81-mm Mortar***. The 81-mm mortar has much the same effect against urban targets as the 60-mm mortar. It has a slightly greater lethal area and its smoke rounds (WP and RP) are more effective. A direct hit is equivalent to about 2 pounds of TNT. The 81-mm round cannot significantly crater a hard-surfaced road. With a delay setting, the 81-mm round can penetrate the roofs of light buildings.

(3) ***120-mm Mortar***. The 120-mm mortar is large enough to have a major effect on common urban targets. It can penetrate deep into a building, causing extensive damage because of its explosive power. A minimum of 18 inches of packed earth or sand is needed to stop the fragments from a 120-mm HE round impacting 10 feet away. The effect of a direct hit from a 120-mm round is equivalent to almost 10 pounds of TNT, which can crush fortifications built with commonly available materials. The 120-mm mortar round can create a large but shallow crater in a road surface, but it is not deep or steep-sided enough to block vehicular movement. However, craters could be deep enough to damage or destroy storm drain systems, water and gas pipes, and electrical or phone cables.

7-10. 25-MM AUTOMATIC GUN

The 25-mm automatic gun mounted on the M2/M3 Bradley fighting vehicle offers infantrymen an effective weapon for urban combat. The primary role of the BFVs during combat in urban areas is to provide suppressive fire and to breach exterior walls and fortifications. (See paragraph 7-3 for the suppression effects and penetration of the 7.62-mm coaxial machine gun.) The wall and fortification breaching effects of the 25-mm automatic gun are major assets to infantrymen fighting in urban areas.

a. **Obliquity.** The 25-mm gun produces its best urban target results when fired perpendicular to the hard surface (zero obliquity). During urban combat, however, finding a covered firing position that permits low obliquity firing is unlikely, unless the streets and gaps between buildings are wide. Most shots impact the target at an angle, which normally reduces penetration. With the APDS-T round, an angle of obliquity of up to 20 degrees can actually improve breaching. The rounds tend to dislodge more wall material for each shot but do not penetrate as deeply into the structure.

b. **Target Types.** The 25-mm gun has different effects when fired against different urban targets.

(1) ***Reinforced Concrete***. Reinforced concrete walls, which are 12 to 20 inches thick, present problems for the 25-mm gun when trying to create breach holes. It is relatively easy to penetrate, fracture, and clear away the concrete, but the reinforcing rods remain in place. These create a "jail window" effect by preventing entry but allowing grenades or rifle fire to be placed behind the wall. Steel reinforcing rods are normally 3/4 inch thick

and 6 to 8 inches apart—there is no quick way of cutting these rods. They can be cut with demolition charges, cutting torches, or special power saws. Firing with either APDS-T or HEI-T rounds from the 25-mm gun will not always cut these rods.

(2) ***Brick Walls***. The 25-mm gun more easily defeats brick walls, regardless of their thickness, and the rounds produce the most spall.

(3) ***Bunker Walls***. The 25-mm gun is devastating when fired against sandbag bunker walls. Obliquity has the least affect on the penetration of bunker walls. Bunkers with earth walls up to 36 inches thick are easily penetrated. At short ranges typical of combat in urban areas, defeating a bunker should be easy, especially if the 25-mm gun can fire at an aperture.

c. **Burst Fire.** The 25-mm gun's impact on typical urban targets seems magnified if the firing is in short bursts. At close ranges, the gunner might need to shift his point of aim in a spiral pattern to ensure that the second and third bursts enlarge the hole. Even without burst fire, sustained 25-mm gunfire can defeat almost all urban targets.

d. **Weapon Penetration.** Although the penetration achieved by the two combat rounds—armor-piercing, discarding sabot with tracer (APDS-T) and high explosive, incendiary with tracer (HEI-T)—differ slightly, both are eventually effective. However, the best target results are not achieved with either of the combat rounds. At close range against structural targets, the training round (TP-T) is significantly more effective. The TP-T round, however, has little utility when used against enemy armored vehicles.

(1) ***APDS-T***. The APDS-T round penetrates urban targets by retaining its kinetic energy and blasting a small hole deep into the target. The APDS-T round gives the best effects behind the wall, and the armor-piercing core often breaks into two or three fragments, which can create multiple enemy casualties. The APDS-T needs as few as four rounds to achieve lethal results behind walls. Table 7-11 explains the number of APDS-T rounds needed to create different size holes in common urban walls.

TARGET	LOOPHOLE	BREACH HOLE
3-inch brick wall at 0-degree obliquity.	22 rounds	75 rounds
3-inch brick wall at 45-degree obliquity.	22 rounds	35* rounds
5-inch brick wall at 0-degree obliquity.	32 rounds	50* rounds
8-inch reinforced concrete at 0-degree obliquity.	22 rounds	75 rounds(NOTE: Reinforcing rods still in place)
8-inch reinforced concrete at 45-degree obliquity.	22 rounds	40* rounds (NOTE: Reinforcing rods still in place)
*Obliquity and depth tend to increase the amount of wall material removed.		

Table 7-11. Breaching effects of APDS-T rounds.

(a) When firing single rounds, the APDS-T round provides the greatest capability for behind-the-wall incapacitation. The APDS-T round can penetrate over 16 inches of reinforced concrete with enough energy left to cause enemy casualties. It penetrates through both sides of a wood frame or brick veneer building. Field fortifications are easily penetrated by APDS-T rounds. Table 7-12 explains the number of APDS-T rounds needed to create different-size holes in commonly found bunkers.

TYPE BUNKER	OBLIQUITY	PENETRATION	LOOPHOLE	SMALL BREACH HOLE
36-inch sand/timber	0 degree	1 round	25 rounds	40 rounds
36-inch sand/ 6-inch concrete	0 degree	6 rounds	6 rounds	20 rounds

Table 7-12. Number of APDS-T rounds needed to create different size holes in bunkers.

(b) The APDS-T round creates a hazardous situation for exposed personnel because of the pieces of sabot that are thrown off the round. Personnel not under cover forward of the 25-mm gun's muzzle and within the danger zone could be injured or killed by these sabots, even if the penetrator passes overhead to hit the target. The danger zone extends at an angle of about 10 degrees below the muzzle level, out to at least 100 meters and about 17 degrees left and right of the muzzle. Figure 7-14 shows the hazard area of the APDS-T round.

Figure 7-14. APDS-T danger zone.

(2) ***HEI-T.*** The HEI-T round penetrates urban targets by blasting away chunks of material.

(a) The HEI-T round does not penetrate an urban target as well as the APDS-T, but it creates the effect of stripping away a greater amount of material for each round. The HEI-T does more damage to an urban target when fired in multiple short bursts because the accumulative impact of multiple rounds is greater than the sum of individual rounds. Table 7-13 explains the number of HEI-T rounds needed to create different-size holes.

TARGET	LOOPHOLE	BREACH HOLE
3-inch brick wall at 0-degree obliquity.	10 rounds	20 rounds
3-inch brick wall at 45-degree obliquity.	20 rounds	25 rounds
5-inch brick wall at 0-degree obliquity.	30 rounds	60 rounds
8-inch reinforced concrete at 0-degree obliquity.	15 rounds	25 rounds
8-inch reinforced concrete at 45-degree obliquity.	15 rounds	30 rounds

Table 7-13. Number of HEI-T rounds needed to create different-size holes.

(b) The HEI-T round does not provide single-round perforation or incapacitating fragments on any external masonry structural wall. It can create first-round fragments behind wood frame and brick veneer walls. HEI-T rounds cannot penetrate a bunker as quickly as APDS-T, but they can create more damage inside the bunker once the external earth has been stripped away. Against a heavy bunker, about 40 rounds of HEI-T are needed to strip away the external earth shielding and breach the inner lining of concrete or timber. The HEI-T round is also used for suppression against known or suspected firing ports such as doors, windows, and loopholes.

7-11. TANK CANNON

The powerful, high-velocity cannon mounted on the M1-series tanks provides the infantryman heavy direct-fire support. Although the infantry assumes the lead role during combat in urban areas, tanks and infantry work as a close team. Tanks move down streets, after the infantry has cleared them of any suspected ATGM positions, and, in turn, support the infantry with fire. The tank is one of the most effective weapons for heavy fire against structures. The primary role of the tank cannon during urban combat is to provide heavy direct-fire against buildings and strongpoints that are identified as targets by the infantry. The wall and fortification breaching effects of the 105-mm and 120-mm tank cannon are major assets to infantrymen fighting in urban areas.

a. **Obliquity.** Tank cannons produce their best urban target effects when fired perpendicular to the hard surface (zero obliquity). During urban combat, however, finding a covered firing position that permits low-obliquity firing is unlikely. Most shots strike the target at an angle that would normally reduce penetration. With a tank cannon, a HEAT multipurpose round is the item of choice against urban targets; the size of the hole is reduced by approximately 1/3 when the firing angle is 45 degrees.

b. **Ammunition.** Armor-piercing, fin-stabilized, discarding sabot (APFSDS) rounds work best against armored vehicles. Other types of ammunition are carried that are more effective against masonry targets and other urban structures. The 105-mm cannon has HEAT, HEP, and WP rounds in addition to APFSDS. The 120-mm cannon has an effective high-explosive, antitank, multipurpose (HEAT-MP) round, which also has capability against helicopters. The 120-mm tank can also carry a high-explosive, concrete-obstacle reduction cartridge that has rubbling capability.

c. **Characteristics.** Both 105-mm and 120-mm tank cannons have two specific characteristics that affect their employment in urban areas: limited elevation and depression, and short arming ranges. In addition, the M1 and M1A1/M1A2 tanks have another characteristic not involved with its cannon but affecting infantrymen working with it—extremely hot turbine exhaust.

(1) The M1 and M1A1/M1A2 tanks can elevate their cannon +20 degrees and depress it -10 degrees. The lower depression limit creates a 35-foot (10.8-meter) dead space around a tank. On a 16-meter-wide street (common in Europe) this dead space extends to the buildings on each side (Figure 7-15). Similarly, there is a zone overhead in which the tank cannot fire (Figure 7-16). This dead space offers ideal locations for short-range antiarmor weapons and allows hidden enemy gunners to fire at the tank when the tank cannot fire back. It also exposes the tank's most vulnerable areas: the flanks, rear, and top. Infantrymen must move ahead, alongside, and to the rear of tanks to provide close protection. The extreme heat produced immediately to the rear of the M1-series tanks prevents dismounted infantry from following closely, but protection from small-arms fire and fragments is still provided by the tank's bulk and armor. The M1-series tanks also have a blind spot caused by the 0-degree of depression available over part of the back deck. To engage any target in this area, the tank must pivot to convert the rear target to a flank target.

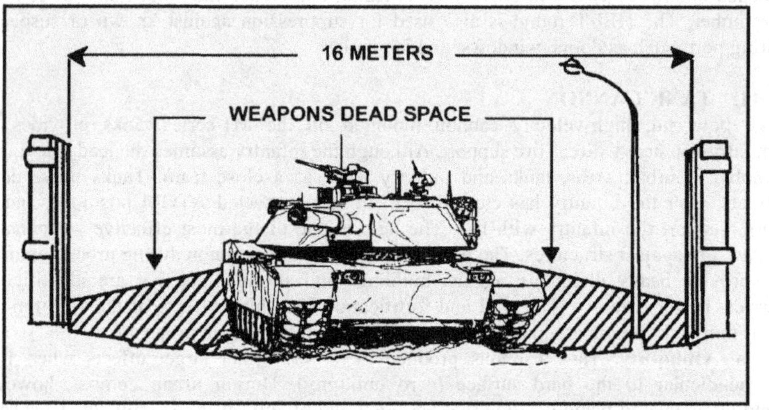

Figure 7-15. Tank cannon dead space at street level.

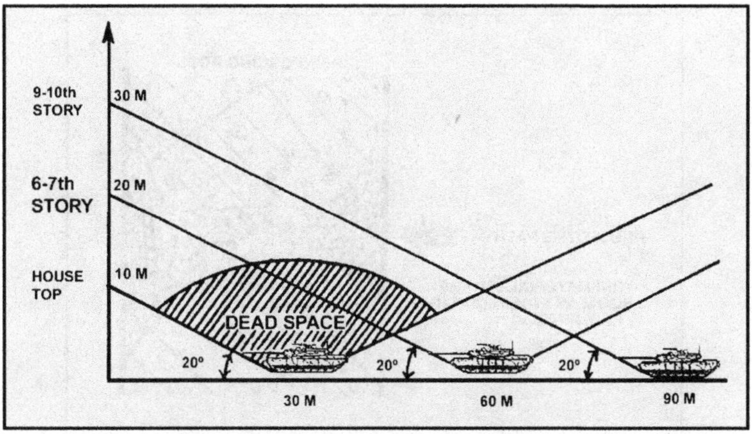

Figure 7-16. Tank cannon dead space above street level.

(2) HEAT type rounds arm within 15 to 30 meters from the gun muzzle. On a 16-meter-wide street, HEAT type ammunition does not arm quickly enough to engage a structure directly perpendicular to the direction of travel. HEAT type rounds fired at structures less than 30 meters from the muzzle will provide some of the desired effects, particularly if the desired effect is casualties inside the building. However, the effectiveness of unarmed HEAT type rounds will be unpredictable and highly variable. These arming distances allow the tank to engage targets from short ranges. The armor of the tank protects the crew from both the blowback effects of the round and enemy return fire. The APFSDS round does not need to arm and, therefore, can be fired at almost any range. The discarding portions of tank rounds can be lethal to exposed infantry forward and to the side of the tank. Additionally, HEAT rounds have an infrequent early burst occurrence. Therefore, exposed infantry should not be forward of a firing tank (60-degree frontal arc).

d. **Target Effects.** High-explosive, antitank rounds are most effective against masonry walls. The APFSDS round can penetrate deeply into a structure but does not create as large a hole or displace as much spall behind the target. In contrast to lighter HEAT rounds, tank HEAT rounds are large enough to displace enough spall to inflict casualties inside a building. One HEAT round normally creates a breach hole in all but the thickest masonry construction—a single round demolishes brick veneer and wood frame construction. Even the 120-mm HEAT round cannot cut all the reinforcing rods, which are usually left in place, often hindering entry through the breach hole (Figure 7-17, page 7-34). The 105-mm HEP round cuts the reinforcing rods and leaves a 20-inch hole.

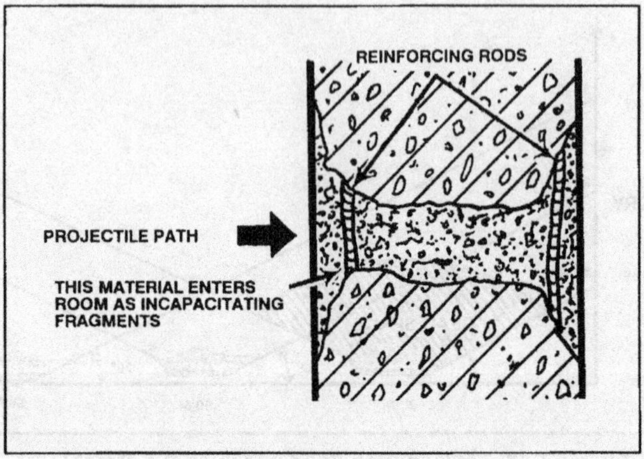

Figure 7-17. Tank HEAT round effects on reinforced concrete walls.

e. **Employment.** Tank-heavy forces could be at a severe disadvantage during urban combat, but a few tanks working with the infantry can be very effective, especially if they work well together at the small-unit level. Tank, infantry, and engineer task forces are normally formed to attack a fortified area. Individual tanks or pairs of tanks can work together with rifle squads or platoons.

(1) Tanks need infantry on the ground to provide security in urban areas and to designate targets. Against targets protected by structures, tanks should be escorted forward to the most covered location that provides a clear shot. On-the-spot instructions by the infantry unit leader ensure the tank's fire is accurate and its exposure is limited.

(2) When the tank main gun fires, it creates a large fireball and smoke cloud. In the confines of an urban area, dirt and masonry dust are also picked up and add to this cloud. The smoke and dust of the explosion further obscure the target. Depending on local conditions, this obscuration could last as long as two or three minutes. Infantry can use this period to reposition or advance unseen by the enemy. Caution must be exercised because the enemy might also move.

(3) Tank cannons create an overpressure and noise hazard to exposed infantrymen. All dismounted troops working near tanks should wear their Kevlar helmet and protective vest, as well as ballistic eye protection. If possible, they should also wear earplugs and avoid the tank's frontal 60-degree arc during firing.

(4) Tanks are equipped with powerful thermal sights that can be used to detect enemy personnel and weapons hidden in shadows and behind openings. Dust, fires, and thick smoke significantly degrade these sights.

(5) Tanks have turret-mounted grenade launchers that project screening smoke grenades. The grenades use a bursting charge and burning red phosphorous particles to create this screen. Burning particles can easily start uncontrolled fires and are hazardous to dismounted infantry near the tank. The tank commander and the infantry small-unit

leader must coordinate when and under what conditions these launchers can be used. Grenade launchers are a useful feature to protect the tank but can cause significant problems if unwisely employed.

(6) The tank's size and armor can provide dismounted infantry cover from direct-fire weapons and fragments. With coordination, tanks can provide moving cover for infantrymen as they advance across small open areas. However, enemy fire striking a tank but not penetrating is a major threat to nearby infantry. Fragmentation that is generated by antitank rounds and ricochets off tank armor have historically been a prime cause of infantry casualties while working with tanks in urban areas.

(7) Some tanks are equipped with dozer blades that can be used to remove rubble barriers, breach obstacles, or seal exits under fire.

f. **Multipurpose Antitank (MPAT) Round Target Effects.** The MPAT round can be very effective during UO. Different MPAT effects are described below.

(1) ***Heavy Armor.*** Because of a relatively small explosive warhead, MPAT effectiveness against heavy armor (tanks) is limited to attacks from the side and rear. Mobility kills of heavy armor can be achieved when fired at from these orientations (especially if tracks and or road wheels are struck); however, the vehicle armament is likely to remain operational.

(2) ***Light Armored Vehicles (LAVs).*** The heavy nose of the MPAT projectile makes it extremely effective against LAVs, such as the BMP. Vehicle kills can be achieved with an impact on varying locations on the hull or (if so equipped) the turret. Mobility kills can be achieved if the wheels or tracks are struck, and it is likely that a road wheel or track impact will also produce penetration of the hull structure. MPAT is effective when it impacts targets from perpendicular to highly oblique, but will function with a reduced reliability when striking excessively oblique surfaces (nearing that of a graze impact).

(3) ***Bunkers.*** The heavy nose of the MPAT projectile makes it extremely effective against earthen, timber, and or sandbag bunkers with the projectile "burying" itself into the bunker structure before warhead detonation. When this occurs, the projectile detonation produces not only lethal effects to personnel within, but a highly-destructive effect to the bunker structure itself.

(4) ***Buildings.*** MPAT is effective against buildings with wooden walls over 1 inch thick. Impact against a thinner wall structure (plywood sheathing without striking supporting members) may produce only a small hole as the projectile passes through the wall without detonating. Impact against a supporting structure (roof rafter, wall stud) causes detonation of the warhead and a subsequent hole and lethal fragmentation effects to personnel located inside. Impact against concrete walls yield holes of about 24 inches in diameter, but reinforcing bars embedded within the concrete are not likely to be cleared from the hole, unless struck directly.

(5) ***Helicopters.*** MPAT, when switched to the "A" or "air" mode, is effective against attack helicopters because of its proximity switch, which can produce mission abort kills without actually impacting the aircraft. The design of the proximity switch is such that if the projectile (set in the "A" mode) is fired against a helicopter, and is on a direct impact flight path, the projectile warhead will not function in the proximity mode, but will be detonated when the projectile strikes the target. If the projectile, however, strikes lightly armored parts of the structure (such as windows or the aluminum skin of the aircraft), it is likely to pass directly through the aircraft without detonating. Impact with heavier

structures, such as the engine or transmission components, will cause detonation of the warhead.

(6) ***Concrete Obstacles (XM908 OR-T Projectile).*** The OR-T projectile, because of its steel nose, is effective against large concrete obstacles. This effectiveness comes from the projectile's striking the face of the obstacle and penetrating several inches before the warhead is detonated. This penetration fractures the concrete obstacle from within, breaking it into smaller blocks, which can be cleared with an ACE. A concrete block 6 feet in diameter and 6 feet long is broken up into rubble, which can be cleared by a tank equipped with a bulldozer blade.

7-12. ARTILLERY AND NAVAL GUNFIRE

Field artillery and naval gunfire can both provide support to the infantry fighting in urban areas. The infantry must understand their capabilities and limitations to employ them effectively.

a. **Indirect Fire.** Indirect artillery fire is not effective for attacking targets within walls and masonry structures. It tends to impact on roofs or upper stories rather than structurally critical wall areas or pillars.

(1) Weapons of at least 155-mm are necessary against thick reinforced concrete, stone, or brick walls. Even with heavy artillery, large expenditures of ammunition are required to knock down buildings of any size. Tall buildings also create areas of indirect-fire dead space due to a combination of building height and angle of fall of the projectile (Figure 7-18). Usually the dead space for low-angle indirect fire is about five times the height of the highest building over which the rounds must pass.

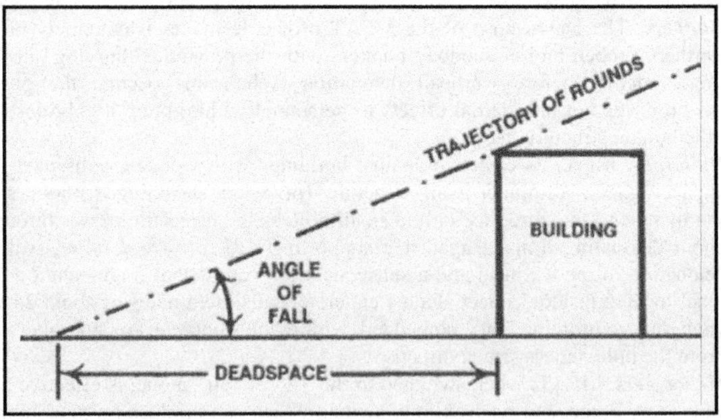

Figure 7-18. Indirect-fire dead space (low angle).

(2) Even when it is theoretically possible to hit a target in a street over a tall building, another problem arises because of range probable error (PE). Only 50 percent of the rounds fired on the same data can be expected to fall within one range PE of the target.

This means when firing indirect fire into urban areas with tall buildings, it is necessary to double the normal ammunition expenditure to overcome a reduced target area and range PE. Also, up to 25 percent of all HE rounds fail to detonate because they glance off hard surfaces.

(3) Naval gunfire, because of its flat trajectory, is even more affected by terrain masking. It is usually difficult to adjust onto the target, because the gun-target line is constantly changing.

b. **Direct Fire.** Self-propelled artillery pieces are not as heavily armored as tanks, but they can still be used during urban combat if adequately secured by infantry. The most likely use of US artillery in an urban direct-fire role is to reinforce tank fires against tough or important urban targets. Because of their availability and habitual relationship with infantry, tanks remain a more common direct-fire support means than self-propelled artillery. Self-propelled artillery should be used in this role only after an analysis of the need for heavy direct fire and the tradeoff involved in the extreme decentralization of artillery firepower. It has the same need for close security and target designation as tanks.

c. **Target Effects.** 155-mm direct fire has a devastating effect against masonry construction and field fortifications. Smaller artillery pieces (105-mm) are normally towed and, therefore, are difficult to employ in the direct-fire mode. Their target effects are much less destructive than the larger caliber weapons.

(1) ***155-mm howitzers***. The 155-mm self-propelled howitzer offers its crew mobility and limited protection in urban areas. It is effective due to its rate of fire and penetration. HE rounds can penetrate up to 38 inches of brick and nonreinforced concrete. Projectiles can penetrate up to 28 inches of reinforced concrete with considerable damage beyond the wall. HE rounds fuzed with concrete-piercing fuzes provide an excellent means of penetrating strong reinforced concrete structures. One round can penetrate up to 46 inches. Five rounds are needed to create a 1.5-meter breach in a 1-meter thick wall. About 10 rounds are needed to create such a breach in a wall 1.5 meters thick. Superquick fuzing causes the rubble to be blown into the building, whereas delay fuzing tends to blow the rubble outward into the street.

(2) ***Naval Cannon.*** The most common naval cannon used to support ground troops is the 5-inch 54-caliber gun. In either single or double mounts, this weapon has a high rate of fire and is roughly equivalent to the 155-mm howitzer in target effect.

7-13. AERIAL WEAPONS

Both rotary- and fixed-wing aircraft can quickly deliver large volumes of firepower over large urban areas. Specific targets are often hard to distinguish from the air. Good ground-to-air communications are vital to successfully employ aerial firepower.

a. **Rotary-Winged Aircraft.** Armed attack helicopters can be used to engage targets in urban areas. Enemy armored vehicles in small parks, boulevards, or other open areas are good targets for attack helicopters.

NOTE: The target effects of TOW missiles and 40-mm grenades carried by attack helicopters have already been discussed

(1) The HELLFIRE missile has a larger warhead and greater range than the TOW. It has a shaped-charge warhead and is not specifically designed for use against masonry

targets. Laser target designation for the HELLFIRE sometimes may not be possible due to laser reflections off glass and shiny metal surfaces. The use of attack helicopters to deliver ATGMs against targets in the upper stories of high buildings is sometimes desirable.

(2) The 2.75-inch folding fin aerial rocket and the 20-mm cannon common to some attack helicopters are good area weapons to use against enemy forces in the open or under light cover. They are usually ineffective against a large masonry target. The 20-mm cannon produces many ricochets, especially if AP ammunition is fired into urban areas.

(3) The 30-mm cannon carried by the Apache helicopter is an accurate weapon. It penetrates masonry better than the 20-mm cannon.

b. **Fixed-Wing Aircraft.** Close air support to ground forces fighting in urban areas is a difficult mission for fixed-wing aircraft. Targets are hard to locate and identify, enemy and friendly forces may be intermingled, and enemy short-range air defense weapons are hard to suppress.

(1) Because only one building can separate enemy and friendly forces, accurate delivery of ordnance is required. Marking panels, lights, electronic beacons, smoke, or some other positive identification of friendly forces is needed.

(2) General-purpose bombs from 500 to 2,000 pounds are moderately effective in creating casualties among enemy troops located in large buildings. High-dive angle bomb runs increase accuracy and penetration but also increase the aircraft's exposure to antiaircraft weapons. Low-dive angle bomb runs using high drag (retarded) bombs can be used to get bombs into upper stories. Penetration is not good with high-drag bombs. Sometimes aerial bombs pass completely through light-clad buildings and explode on the outside.

(3) Aerial rockets and 20-mm cannons are only moderately effective against enemy soldiers in urban areas since rockets lack the accuracy to concentrate their effects. The 20-mm cannon rounds penetrate only slightly better than the caliber .50 round; 20-mm AP rounds can ricochet badly; and tracers can start fires.

(4) The 30-mm cannon fired from the A-10 aircraft is an accurate weapon. It is moderately effective against targets in urban areas, penetrating masonry better than the 20-mm cannon.

(5) The AC-130 aircraft has weapons that can be most effective during combat in urban areas. This aircraft can deliver accurate fire from a 20-mm Vulcan cannon, 40-mm rapid-fire cannon, and 105-mm howitzer. The 105-mm howitzer round is effective against the roof and upper floors of buildings. The AC-130 is accurate enough to concentrate its 40-mm cannon and 105-mm howitzer fire onto a single spot to create a rooftop breach, which allows fire to be directed deep into the building.

(6) Laser and optically-guided munitions can be effective against high-value targets. The US Air Force has developed special, heavy, laser-guided bombs to penetrate hardened weapons emplacements. Problems associated with dense smoke and dust clouds hanging over the urban area and laser scatter can restrict their use. If the launching aircraft can achieve a successful laser designation and lock-on, these weapons have devastating effects, penetrating deep into reinforced concrete before exploding with great force. If launched without a lock-on, or if the laser spot is lost, these weapons are unpredictable.

7-14. DEMOLITIONS

Combat in urban areas requires the extensive use of demolitions, which requires all soldiers, not only engineers, to be trained in demolition employment. (See FM 5-25 for specific information on the safe use of demolitions.)

a. **Bulk Demolitions.** Bulk demolitions come in two types: TNT and C4 (see Chapter 8).

(1) TNT comes in 1/4-, 1/2-, and 1-pound blocks. About 5 pounds of TNT are needed to breach a nonreinforced concrete wall 12 inches thick if the explosives are laid next to the wall and are not tamped. If the explosives are tamped, about 2 pounds are sufficient.

(2) C4 comes in many different sized blocks. About 10 pounds of C4 placed between waist and chest high will blow a hole in the average masonry wall large enough for a man to walk through.

b. **Shaped Charges.** The two sizes of US Army shaped charges are a 15-pound M2A3 and a 40-pound M3A3. The M3A3 is the most likely shaped charge to be used in urban areas. It can penetrate 5 feet of reinforced concrete. The hole tapers from 5 inches down to 2 inches. The amount of spall thrown behind the target wall is considerable. There is also a large safety hazard area for friendly soldiers.

c. **Satchel Charges.** Satchel charges are very powerful. The two standard US Army satchel charges are the M183 and the M37. Both come in their own carrying satchel with detonators and blasting cords. Each weighs 20 pounds. The M183 has 16 individual 1 1/4-pound blocks that can be used separately. When used untamped, a satchel breaches a 3-foot thick concrete wall. Debris is thrown great distances. Friendly troops must move away and take cover before detonation.

d. **Cratering Charges.** The standard US Army cratering charge is a 43-pound cylinder of ammonium nitrate. This explosive does not have the shattering effect of bulk TNT or C4, and it is more useful in deliberate demolitions than in hasty ones.

7-15. COMMON EFFECTS OF URBAN COMBAT

Regardless of what weapons are used, there are several common effects during urban combat. Leaders must take them into account and use them or avoid them, as the situation demands.

a. **Penetration and Damage to Structures.** Most tactical situations call for penetration of buildings and walls. No one can be sure of how much penetration a round will achieve against a specific target until it has been tried. Some situations require units to limit penetration to the minimum obtainable. Generally, the smaller the round, the less penetration. High-explosive rounds normally penetrate less than armor-piercing rounds. High-explosive rounds and aerial bombs can cause extensive damage to buildings surrounding the target area (Figure 7-19, page 7-40).

Figure 7-19. Examples of penetration and damage to buildings.

b. **Rubble.** Combat experience has shown that after an urban area is reduced to rubble by weapons fire, it often becomes more of an obstacle to advancing troops, and a stronger position for defending troops, than it was before. (See Figure 7-20.)

Figure 7-20. Examples of rubble.

c. **Fire.** The risk of fire during urban combat is very high. Once a large fire starts, it is nearly impossible to put out. Damage to gas lines and water mains, the scarcity of fire fighting equipment and trained firemen, the general lack of access caused by rubble blocking the streets, and the danger posed by the combat itself make containing fires very difficult. Fires that rage out of control can cause more damage to the urban area than any other factor (Figure 7-21).

Figure 7-21. Examples of urban fires.

d. **Smoke and Haze.** Limited visibility is a common factor during urban combat. Fires produce large clouds of often toxic or irritating, choking smoke (Figure 7-22). Explosions add significant amounts of dust to the atmosphere. Even the effort to rescue personnel trapped within collapsed buildings creates dust.

Figure 7-22. Example of smoke as a result of fires.

e. **Trapped and Injured Survivors.** Intense urban combat will inevitably result in large-scale destruction to buildings. Survivors, both military and civilian, may be trapped in the rubble and must be extracted (Figure 7-23, page 7-42). This extraction effort may be impossible without heavy construction equipment. Unless specially trained personnel are available, the rescue effort itself can result in more casualties as the rubble shifts and collapses on would-be rescuers. Once they are located, casualties must be evacuated quickly and safely. This is often difficult to do without causing additional injury.

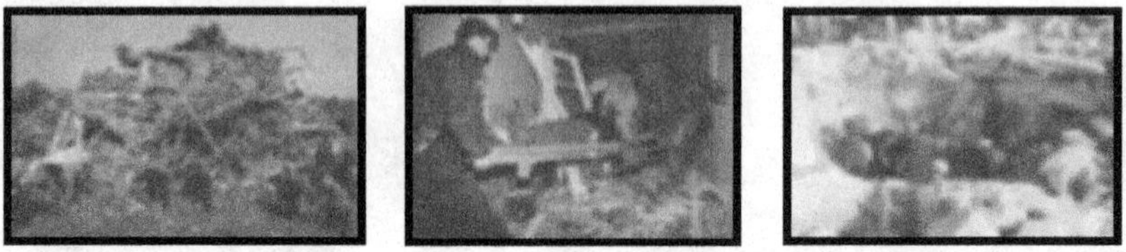

Figure 7-23. Examples of personnel trapped under rubble.

f. **Damaged and Destroyed Transportation Systems.** Urban areas are transportation hubs. Road, rail, barge, air, and ship traffic rely on functioning systems for the movement, loading, unloading, and distribution of supplies and goods. Battles in urban areas disrupt the normal flow of traffic destroying or damaging roads, ports, bridges, and rail lines (Figure 7-24). Large numbers of civilian inhabitants can be fed and cared for only if these are intact.

Figure 7-24. Example of a damaged rail transportation system.

g. **Displaced Civilian Occupants.** Although many civilian inhabitants of a town will flee the fighting, experience has shown that many others will remain behind. They may be trying to protect their property, or they may feel that they have no other place to go. For whatever reason, they will be in the immediate area of the fighting and in danger. These civilians must be considered in all military planning, and commanders must make provisions for their protection and evacuation. (See Figure 7-25.)

Figure 7-25. Example of displaced persons.

CHAPTER 8
OBSTACLES, MINES, AND DEMOLITIONS

*In urban combat, obstacles and mines are used extensively by the defender to canalize the enemy, impede his movement, and disrupt his attack. The national policy of the United States severely restricts the use of antipersonnel land mines, beginning with those that do not self-destruct but eventually including all types. This policy, currently in effect, applies to all units either engaged in, or training for, operations worldwide. US national policy forbids US forces from using standard or improvised explosive devices as booby traps. This policy does **not** affect the standard use of antivehicular mines. It does **not** affect use of the M18 Claymore mine in the command-detonated mode. For the immediate future, units may still use self-destructing antipersonnel mines, such as the ADAM, when authorized by the appropriate commander. Under proper command authority, units may still emplace mixed minefields containing self-destructing antipersonnel land mines used to protect antivehicular land mines; for example, MOPMS or Volcano. Consider all references to antipersonnel mines and the employment of minefields in the light of the national policy limiting the use of nonself-destructing antipersonnel land mines. Readers should not construe any uses of the term mines, antipersonnel obstacle, protective minefield, or minefield contained in this manual to mean a US-emplaced obstacle that contains nonself-destructing antipersonnel land mines or booby traps.*

Section I. OBSTACLES

Obstacles are designed to slow or prevent movement by personnel, to separate infantry from tanks, and to slow or stop vehicles.

8-1. TYPES OF OBSTACLES

Command-detonated mines, barbed wire, and exploding flame devices are used to construct antipersonnel obstacles (Figures 8-1 through 8-5, pages 8-2 through 8-4). (See FM 5-25 for more detailed information.) These obstacles are used to block the following infantry approaches:

- Streets.
- Buildings.
- Roofs.
- Open spaces.
- Dead space.
- Underground systems.

a. The three types of obstacles used in defensive operations are protective, tactical, and supplementary.

(1) Protective obstacles are usually located beyond hand-grenade range (40 to 100 meters) from the defensive position.

(2) Tactical obstacles are positioned to increase the effectiveness of friendly weapons fire. Tactical wire is usually positioned on the friendly side of the machine gun's final protective line (FPL).

(3) Supplementary obstacles are used to break up the pattern of tactical obstacles to prevent the enemy from locating friendly weapons.

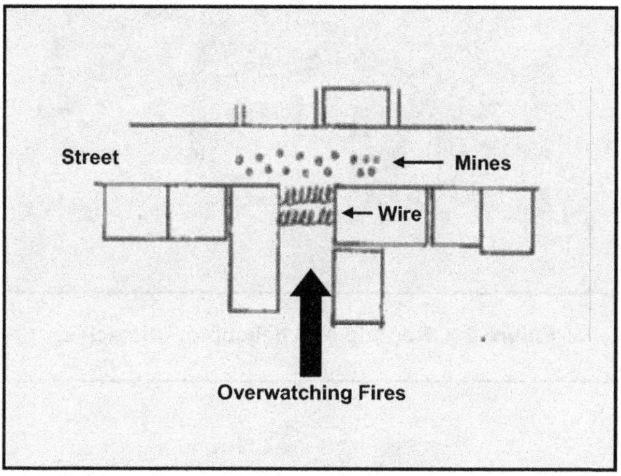

Figure 8-1. Mines and wire.

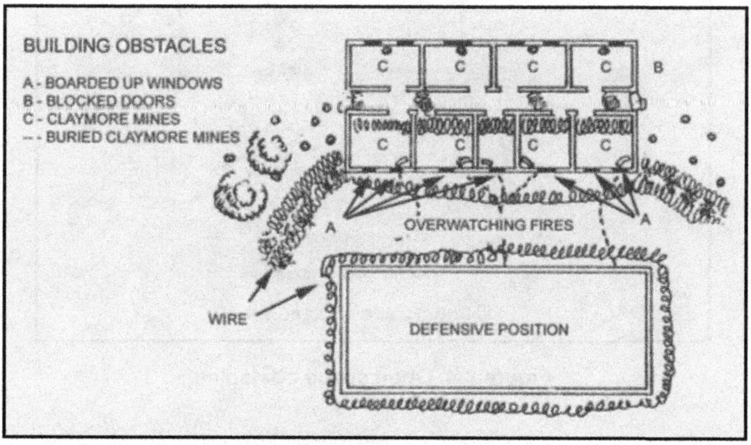

Figure 8-2. Building obstacles.

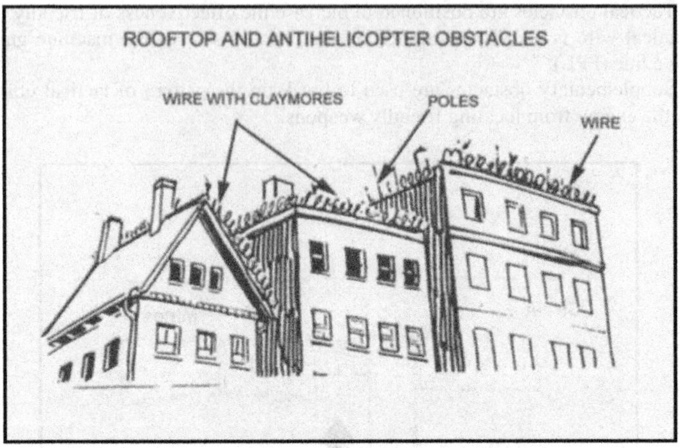

Figure 8-3. Rooftop and helicopter obstacles.

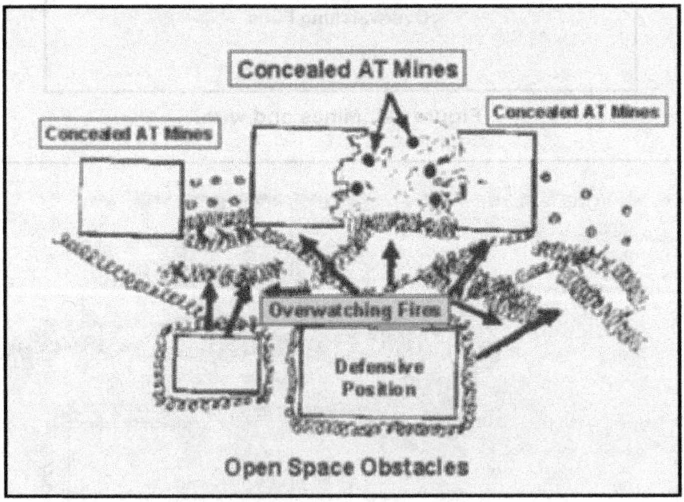

Figure 8-4. Open space obstacles.

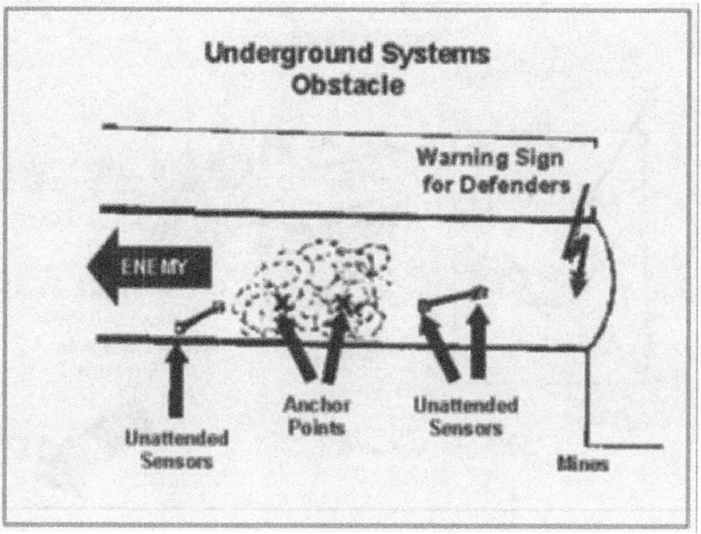

Figure 8-5. Underground systems obstacle.

b. Dead space obstacles are designed and built to restrict infantry movement in areas that cannot be observed and are protected from direct fires.

c. Antiarmor obstacles are restricted to streets and open areas (Figures 8-6 through 8-11, pages 8-5 through 8-7).

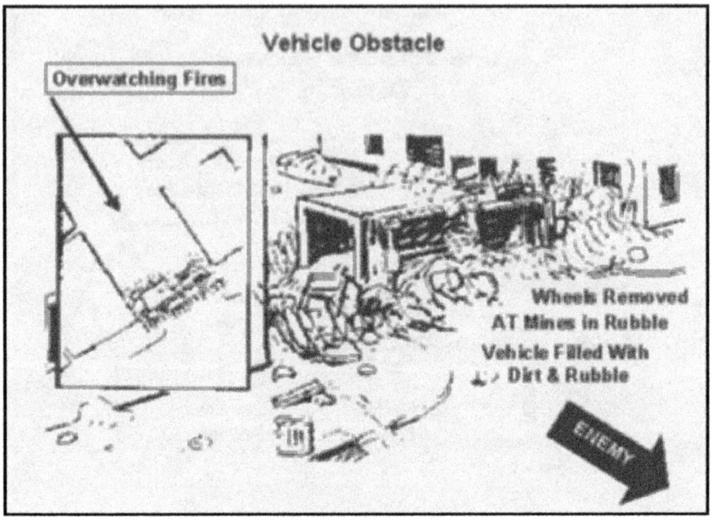

Figure 8-6. Vehicle obstacle.

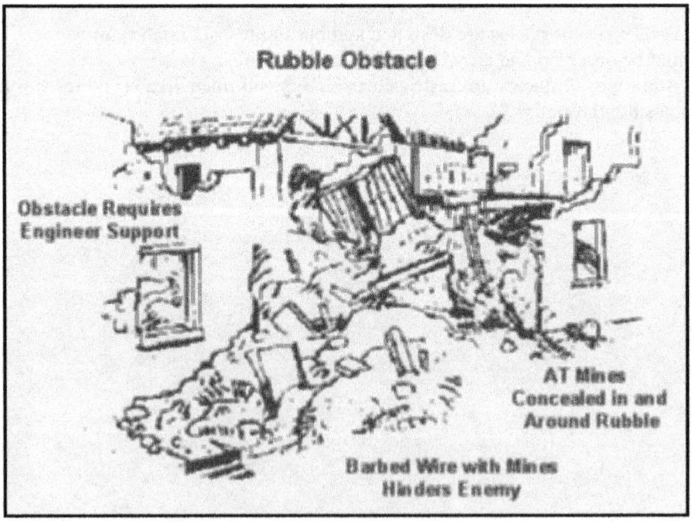

Figure 8-7. Rubble obstacle.

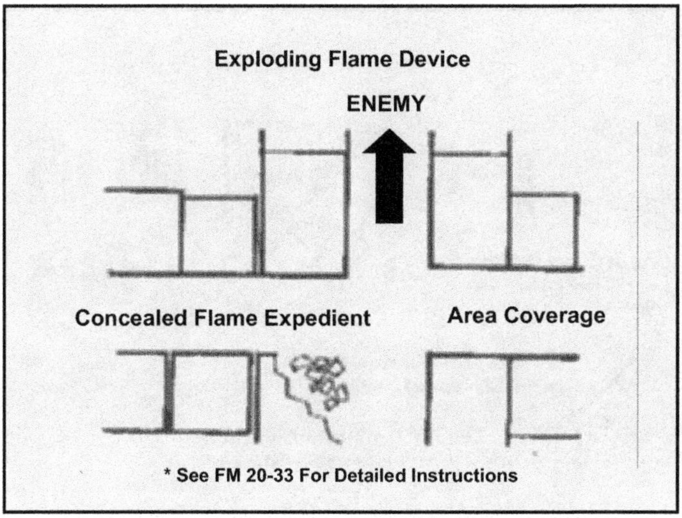

Figure 8-8. Exploding flame device.

Figure 8-9. Steel hedgehog.

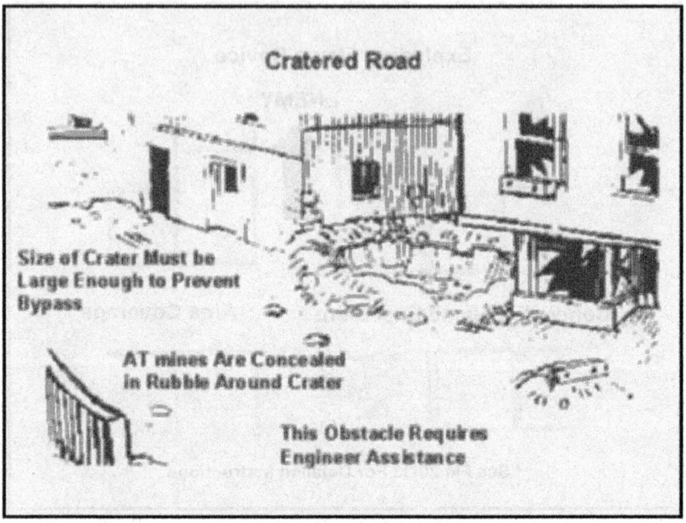

Figure 8-10. Road craters.

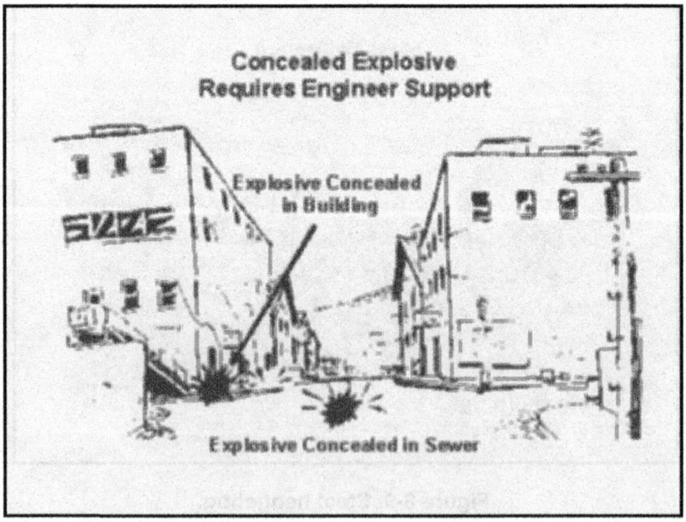

Figure 8-11. Concealed explosive.

8-2. CONSTRUCTION OF OBSTACLES

Obstacles are constructed in buildings to deny enemy infantry covered routes and weapons positions close to friendly defensive positions. They can be constructed by rubbling with explosives or flame, or by using wire. The building can be prepared as an explosive or flame trap for execution after enemy occupation.

Section II. MINES

On 16 May 1996, Public Law 104-295 (Leahy Amendment) took effect. US policy forbids the use of booby traps and nonself-destructing antipersonnel land mines. Mines found in built-up areas should be recorded on a building sketch. The sketch should include the number of the building (taken from a city map) and floor plans. It should also include the type of mine and firing device, if known. When identified, mined buildings should be marked on the friendly side. Clearing areas or buildings that have been mined is extremely difficult. Therefore, they should be considered "NO GO" areas. This factor must be carefully considered when planning and authorizing the placement of mines. (See Table 8-1 for the approving authority for minefields.)

TYPE MINEFIELD	APPROVING AUTHORITY
Protective hasty	Brigade commander (may be delegated down to battalion level or company level on a mission basis
Deliberate	Division or installation commander
Tactical	Division commander (may be delegated to brigade level).
Point	Brigade commander (may be delegated to battalion. level).
Interdiction	Corps commander (may be delegated to division level).
Phony	Corps commander (may be delegated to division level).
Scatterable long duration (24 hours or more).	Corps commander (may be delegated to division level).
Short duration (less than 24 hours)	Corps commander (may be delegated to division, brigade, or battalion level).

Table 8-1. Minefield employment authority.

8-3. TYPES OF MINES AND EMPLOYMENT TECHNIQUES

Several types of mines can be employed in built-up areas.

a. The M18A1 Claymore mine can be employed during the reorganization and consolidation phase on likely enemy avenues of approach. It does not have to be installed in the street but can be employed on the sides of buildings or any other sturdy structure.

(1) Claymore mines can be used for demolition against thin-skinned buildings and walls, or the 1 1/2 pounds of composition C4 can be removed from the mine and used as an explosive, if authorized.

(2) Claymore mines can be mixed with antitank mines in nuisance minefields. They can fill the dead space in the final protective fires of automatic weapons (Figure 8-12).

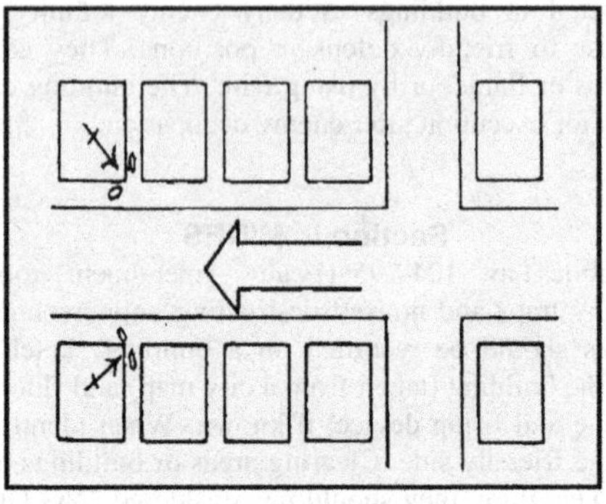

Figure 8-12. Claymore mines used to cover dead space of automatic weapons.

(3) Claymore mines can be used several ways in the offense. For example, if friendly troops are advancing on a city, Claymore mines can be used in conjunction with blocking positions to cut off enemy escape routes (Figure 8-13).

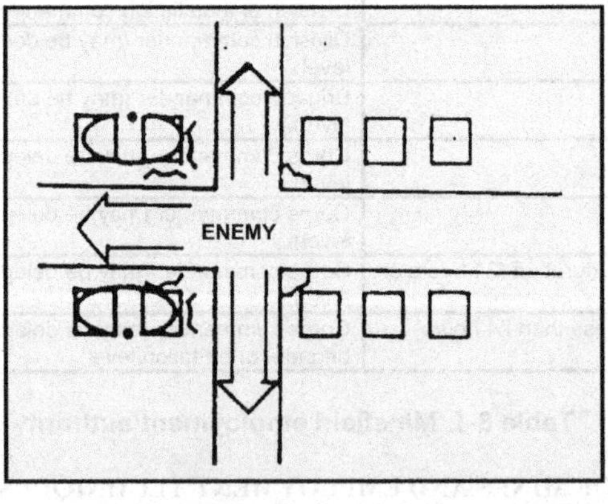

Figure 8-13. Claymore mines used to block enemy escape routes.

b. The M15, M19, and M21 antitank mines are employed (Figure 8-14, page 8-10)—

- In conjunction with other man-made obstacles and covered with fire.
- As standard minefields in large open areas with the aid of the M57 dispenser.
- In streets or alleys to block routes of advance in narrow defiles.
- As command detonated mines with other demolitions.

Figure 8-14. Emplacement of antitank mines.

8-4. ENEMY MINES AND BOOBY TRAPS

Buildings contain many areas and items that are potential hiding places for booby traps such as doors, windows, telephones, stairs, books, canteens, and so on. When moving through a building, soldiers must not pick up anything—food, valuables, enemy weapons, and so on. Such items could be rigged with pressure devices that explode when moved. Soldiers must be well dispersed so that if a booby trap explodes, the number of casualties will be few. Many different types of mines and booby traps could be encountered during urban combat (Figure 8-15).

a. Equipment used in clearing operations:

- Mine detectors.
- Probes.
- Grappling hooks.
- Ropes.
- Bulk explosives and firing devices.
- Protective vests (at least PASGT). (EOD vests are best, but heavy).
- Eye protection.
- Engineer tape or other marking devices such as florescent spray paint.

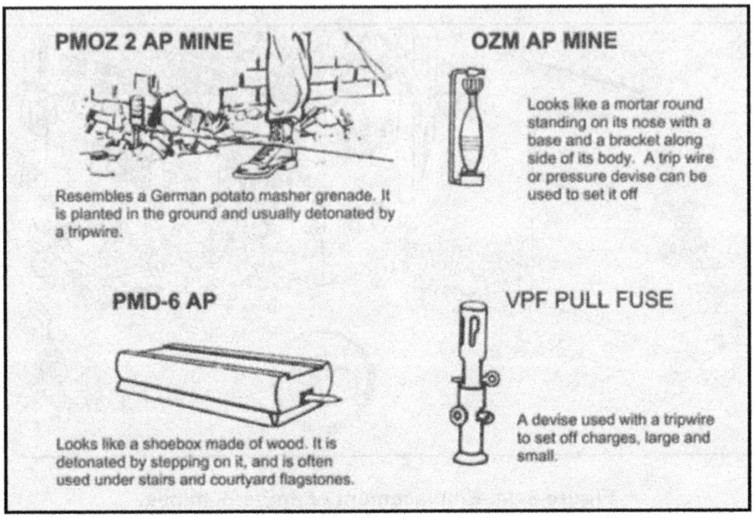

Figure 8-15. Threat mines and booby traps.

b. If available, scout dogs should be used to *alert* soldiers to trip wires or mines.

c. To detect tripwires, soldiers can use a 10-foot pole with 5 feet of string tied on one end. He attaches a weight to the loose end of the string, which snaps on the trip wire. This allows the lead man to easily detect a trip wire (Figure 8-16, page 8-12).

d. Many standard antipersonnel mines are packed in boxes and crates. If a soldier discovers explosive storage boxes, he should sketch them and turn the sketch over to the platoon leader or S2.

e. Explosive ordnance disposal (EOD) personnel should neutralize most booby traps. If EOD teams are not available, booby traps can be blown in place. Personnel should move to adequate cover. If the booby trap is in a building, all personnel should go outside before the booby trap is destroyed. Engineer tape placed around the danger area can be used to mark booby traps. If tape is not available, strips ripped from bedsheets can be used.

f. If possible, a guide should lead personnel through known booby-trapped areas. Prisoners and civilians can be a good source of information on where and how booby traps are employed.

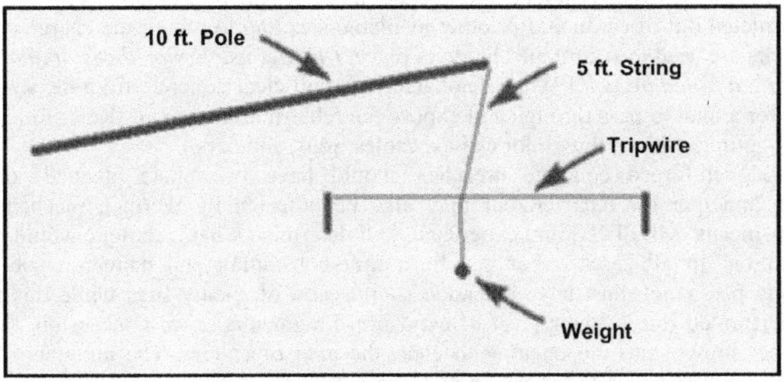

Figure 8-16. Trip wire detection.

Section III. DEMOLITIONS

Demolitions are used more often during urban combat than during operations in open terrain. The engineers that support the brigade, battalion task force, and company team should enforce demolition operations. However, if engineers are involved in preparing and executing the barrier plan, infantrymen can prepare mouseholes, breach walls, and rubble buildings themselves, assisted and advised by the brigade, task force, or team engineer.

8-5. OFFENSIVE USE

When assaulting or clearing an urban area, demolitions enable the maneuver commander to create an avenue of approach through buildings. As discussed earlier, the infantry commander forms his personnel into assault and support elements to seize and clear buildings.

a. Prefabricate expedient charges and initiation systems. Cross load charges and explosives so as not to overburden any one individual with unnecessary explosives, and to keep the charges dispersed if compromised. Ensure that the caps and charges are separate, but within the breach element. As METT-TC dictates, cross loading explosives within elements of the assault may be necessary. Once a foothold is established, redistribute the explosives for the next COA during consolidation and reorganization.

b. One of the most difficult breaching operations faced by the assault element is the breaching of masonry and reinforced concrete walls. Always look for an alternate entry point, including safe firing positions and movement routes, before committing the element. Also consider the effects of blast waves, overpressure, and secondary missile projectiles in direct relationship to the charge and charge placement.

(1) Normally, building walls are 15 inches thick or less and will vary depending on the theatre of operations. C4 is an ideal charge to use when assuming all outer walls are constructed of reinforced concrete. When breaching external walls, place six to eight horizontal blocks of C4 10 to 12 inches apart in two columns (three or four blocks in each column). Prime at the outer edge of each block using a ring main and sliding uli knots.

The supplied adhesive may not be strong enough to hold the blocks in place. A frame can be constructed out of cardboard or other available sheeting to mount the charge on, then prop sticks are used to secure the charge in place. *(Always use two methods of attachment precluding a failed breach.)* When detonated, this will clear concrete from the wall large enough for a man to pass through and expose the rebar reinforcement. (Refer to FM 5-34 for steel-cutting rules of thumb for chains, cables, rods, and bars.)

(2) All reinforced concrete breaches should have two shots planned—one for concrete and one for rebar. Rebar may also be defeated by thermal, mechanical or ballistics means. METT-TC and experience will determine what technique would best fit the situation. In all cases, rebar can be a time-consuming and dangerous objective. Breaching personnel must have extended suppression of enemy fires while this task is being performed due to their level of exposure. Fragmentation or concussion grenades should be thrown into the opening to clear the area of enemy. The amounts of TNT required to breach concrete are shown in Table 8-2.

NOTE: Prop sticks become secondary projectiles when used.

REINFORCED CONCRETE		
THICKNESS OF MATERIAL	TNT	SIZE OF OPENING
Up to 10 CM (4 inches)	5 KG (11 LBS)	10 to 15 CM (4 to 6 inches)
10 to 15 CM (4 to 8 inches)	10 KG (22 LBS)	15 to 25 CM(6 to 10 inches)
15 to 20 CM (6 to 8 inches)	20 KG (44 LBS)	20 to 30 CM (8 to 12 inches)
NONREINFORCED CONCERT MASONARY		
THICKNESS OF MATERIAL	TNT	SIZE OF OPENING
Up to 35 CM (14 inches)	1 KG (2.2LBS)	35 CM (14 inches)
35 to 45 CM (14 to 18 inches)	2 KG (4.4 LBS)	45 CM (18 inches)
45 to 50 CM (18 to 20 inches)	3 KG (6.6 LBS)	50 CM (20 inches)

Table 8-2. TNT required to breach concrete.

c. Mouseholes provide the safest method of moving between rooms and floors. Although they can be created with explosives, all mechanical means should be used first. When assaulting a unit in the defense, mouseholes may be provided.

d. When enemy fire prevents an approach to the wall, the breaching charge may be attached to a pole and slid into position for detonation at the base of the wall (Figure 8-17, page 8-14). Small-arms fire will not detonate C4 or TNT. The charge must be primed with detonating cord. Soldiers must take cover before detonating the charge.

Figure 8-17. Charge placement when small-arms fire cannot be suppressed.

8-6. DEFENSIVE USE

The use of demolitions in defensive operations is the same as in offensive operations. When defending an urban area, demolitions are used to create covered and concealed routes through walls and buildings that can be used for withdrawals, reinforcements, or counterattacks. Demolitions are also used to create obstacles and clear fields of fire.

a. Infantrymen use demolitions for creating mouseholes and constructing command-detonated mines. Expedient C4 satchel charges can be concealed in areas that are likely enemy weapons positions, in individual firing positions, or on movement routes. Expedient-shaped charges (effective in equipment destruction and against lightly armored vehicles) can also be placed on routes of mounted movement and integrated into antiarmor ambushes.

b. Engineers must furnish technical assistance for selective rubbling. Normally, buildings can be rubbled using shaped charges or C4 on the supports and major beams.

c. Charges should be placed directly against the surface to be breached unless a shaped charge is used. Depending on the desired effect and target material, charges may be tamped, untamped, or buffed depending on the situation. Tamping materials can be sandbags, rubble, or even water blivits when filled (Figure 8-18).

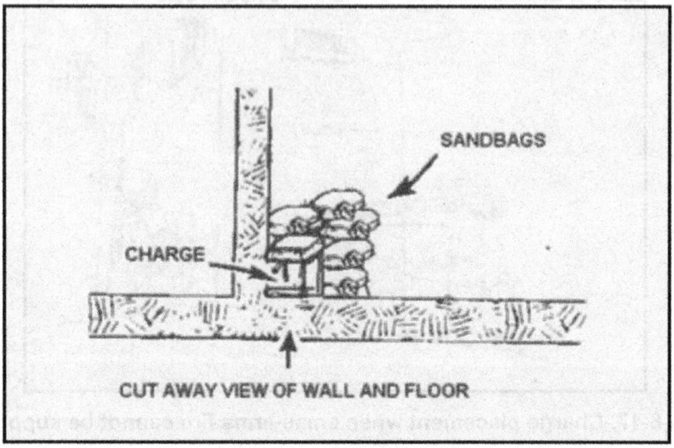

Figure 8-18. Sandbags used to tamp breaching charge.

d. For most exterior walls, tamping of breaching charges could be impossible due to enemy fire. Thus, the untamped ground level charge requires twice the explosive charge to produce the same effect as an elevated charge (Table 8-3).

METHODS OF PLACEMENT				
THICKNESS OF CONCRETE (FEET)	ELEVATED CHARGE		GROUND-LEVEL CHARGE	
	POUNDS OF TNT	POUNDS OF C4	POUNDS OF TNT	POUNDS OF C4
2	14	11	28	21
2 ½	27	21	54	41
3	39	30	78	59
3 ½	62	47	124	93
4	93	70	185	138
4 ½	132	99	263	196
5	147	106	284	211
5 ½	189	141	376	282
6	245	186	490	366

Table 8-3. Breaching reinforced concrete.

e. The internal walls of most buildings function as partitions rather than load-bearing members. Therefore, smaller explosive charges can be used to breach them. In the absence of C4 or other military explosives, internal walls can be breached using one or more fragmentation grenades primed with modernized demolition initiator (MDI), or a

Claymore mine (Figure 8-19). These devices should be tamped to increase their effectiveness and to reduce the amount of explosive force directed to the rear.

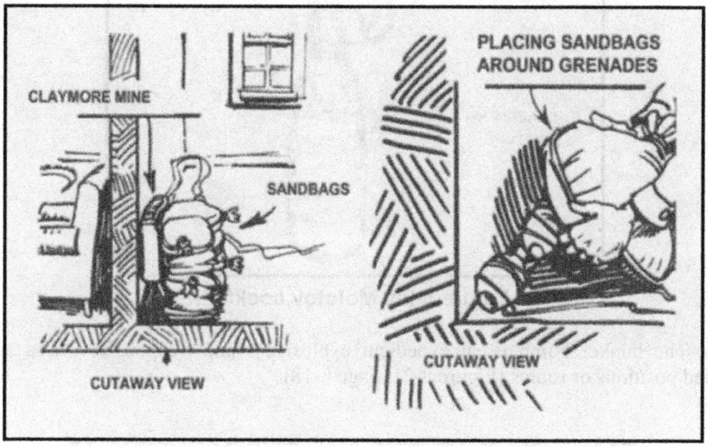

Figure 8-19. Tamping of a Claymore mine and fragmentation grenades to breach internal walls.

f. The Molotov cocktail (Figure 8-20) is an expedient device for disabling both wheeled and tracked vehicles. It is easy to make since most materials are readily available. Results can be very effective because of the close nature of engagements in urban areas. The objective is to ignite a flammable portion of the vehicle or its contents, such as the fuel or ammunition it is transporting.

> **WARNING**
>
> **Ensure that a safe distance is maintained when throwing the Molotov cocktail. Caution troops against dropping the device. Throw it in the opposite direction of personnel and flammable materials. Do not smoke while making this device.**

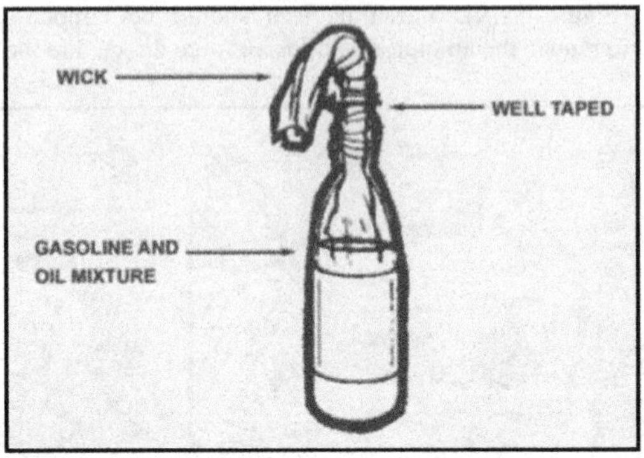

Figure 8-20. Molotov cocktail.

g. The bunker bomb is an expedient explosive flame weapon best used against fortified positions or rooms (Figure 8-21, page 8-18).

> **WARNING**
> **Never carry the device by the handle or igniter. Remove the igniter safety pin only when it is time to use the device. Use extreme care when handling or carrying MDI systems. Protect blasting caps from shock and extreme heat. Do not allow the time fuse to kink or become knotted. Doing so may disrupt the powder train and may cause a misfire. Prime detonating cord and remove the MDI igniter safety pin only when it is time to use the device.**

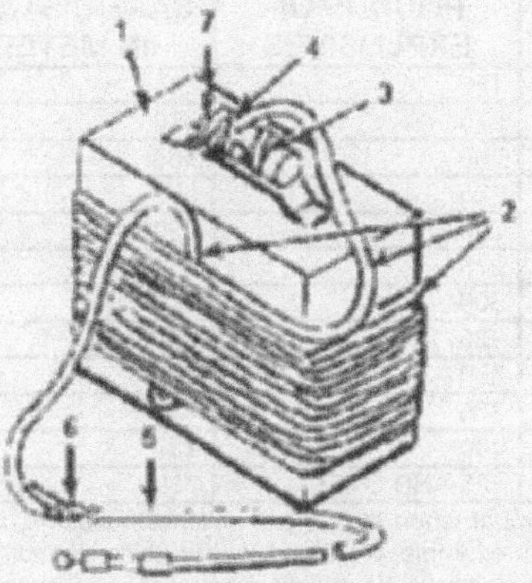

Figure 8-21. Bunker bomb made from ammunition can.

8-7. SAFETY

The greatest dangers to friendly personnel from demolitions are the debris thrown by the explosion and blast overpressure. Leaders must ensure protective measures are enforced, and have personnel trained in procedures for determining overpressure, be it indoors or outdoors, in respect to net explosive weights and room size. The minimum safe distances listed in Table 8-4 indicate the danger of demolition effects.

a. Rules for using demolitions:

- Team/squad leaders and the platoon engineer(s) supervise the employment of demolitions.
- Wear helmets, body armor, ear and eye protection when firing explosives.
- Handle misfires with extreme care.
- Clear the room and protect personnel when blowing interior walls.

b. Some charges should be prepared, minus initiators, beforehand to save time; for example, 10- or 20-pound breaching charges of C4 and expedient-shaped charges in No. 10 cans.

- Use C4 to breach hard targets (masonry construction).
- Do not take chances.
- Do not divide responsibility for explosive work.
- Do not carry explosives and blasting caps together.

POUNDS OF EXPLOSIVES	SAFE DISTANCE IN METERS	POUNDS OF EXPLOSIVES	SAFE DISTANCE IN METERS
1 to 27	300	150	514
30	311	175	560
35	327	200	585
40	342	225	605
45	356	250	630
50	369	275	651
60	392	300	670
70	413	325	688
80	431	350	705
90	449	375	722
100	465	400	737
125	500	425 AND OVER	750
NOTE: These distances will be modified in combat when troops are in other buildings, around corners, or behind intervening walls. For example, a platoon leader using demolitions in an urban area with heavy-clad, mass construction buildings available to protect his soldiers, may use this information in conjunction with making a risk assessment and reduce the MSD to 50 meters if he is using a 20- to 25-pound charge.			

Table 8-4. Minimum safe distances (MSD) for personnel in the open.

Section IV. FIELD-EXPEDIENT BREACHING OF COMMON URBAN BARRIERS

Urban combat requires access to compounds, buildings, and rooms. Mechanical breaching of doors or windows using sledgehammers, bolt cutters, saws, or crowbars; or ballistic breaching using weapons fire are options. However, mechanical or ballistic breaching are sometimes too slow or exposes soldiers to enemy fire. Explosive breaching is often the fastest and most combat-effective method. With a little time to prepare, units can use slightly modified standard Army demolitions to breach all common urban barriers.

NOTE: The techniques described in this section should be employed by soldiers who have been trained in their use.

8-8. FORCE PROTECTION

Soldiers must take care when fabricating, carrying, and using field-expedient explosive devices. Leaders must ensure all standard procedures and precautions outlined in doctrinal and training material for Army demolitions activities are followed. This is dictated by more than just the commander's concern for the safety and welfare of his individual soldiers. Accidental or premature detonation of demolitions during combat not only can injure or kill friendly soldiers but can jeopardize the unit's mission. During combat, soldiers often need to position themselves close to breach points to enter quickly and overcome enemy resistance before the effects of the explosion subside. However, a soldier who is too close to an explosion and injured by flying debris becomes a casualty. Fire and extreme dust volumes may be encountered, preventing fluid movements, and must be prepared for. The unit must accomplish its mission without sustaining friendly casualties from its own demolitions.

8-9. BREACHING REINFORCED AND NONREINFORCED EXTERIOR WALLS

The Army issues both bulk explosives (TNT or C4) and prepackaged satchel charges that are powerful enough to breach all but the most heavily reinforced exterior walls. In some situations, satchel charges may be unavailable or may prove too powerful for the breach required. In high-intensity urban combat, the situation may call for large amounts of bulk explosive, but in many precision conditions the commander may want to create a smaller-size hole than the M37 or M183 satchel charges normally produce. Smaller satchel charges can be improvised. Research and development can determine the correct size of these improvised satchels, depending on the types of walls found in the battle area.

a. General-purpose satchel charges can be assembled using empty machine gun ammunition bandoleers filled with various amounts of C4 explosive.

(1) Connect a short length of detonation cord firmly to the explosive and leave it dangling. Tape the explosive securely into the bandoleer.

(2) Hang the charge on a wall by the bandoleer strap or prop it against the wall using a stick or other object. Satchel charges detonated while firmly secured against the target wall at about shoulder height produce the best effects.

(3) Prime the charge with an MDI firing system to the short length of detonation cord left dangling. When used against a nonreinforced concrete wall, a satchel charge containing 2 pounds of C4 usually produces a mousehole; 5 pounds creates a hole large enough for a man to move through; 7 pounds creates a hole large enough for two men to move through simultaneously; and 10 pounds of C4 can blow a hole large enough to drive a vehicle through. The 10-pound charge may also destroy the entire building if it is not of sturdy construction.

b. A useful breaching charge improvised by light engineer sappers during combat operations in Somalia consisted of a 3-foot length of engineer picket (U-shaped engineer stake) packed with 4 to 8 pounds of C4. The explosive was primed with detonation cord and taped securely to the picket. When needed, the picket was placed upright with its flat side against the wall, held to the wall by another stake, and then detonated. This charge could be rapidly fabricated, was sturdy, and could be easily and quickly emplaced. According to reports from the field, this device would blow a hole about 4 feet wide and 8 feet high in a nonreinforced concrete wall (common in the third world). The charge would throw fragments from the picket straight back for long distances (from 50 to 100 meters) but was fairly safe to either side. In combat, infantrymen could stand about 20 meters from the picket, crouched tightly against the wall with their backs turned to the explosive, without undue risk. This allowed them to follow up on the explosion with a rapid assault into the compound or building before the occupants could recover.

8-10. BREACHING INTERIOR WALLS AND PARTITIONS

Interior walls generally require much less explosive to create a satisfactory breach than do exterior, load-bearing walls. An easily fabricated silhouette charge can further reduce the amount of explosive needed to breach plywood, Sheetrock, or light plaster walls.

It can also be used to breach wooden or metal doors. This charge can be emplaced quickly and creates a hole large enough for a man to move through.

a. Tape two E-type silhouette targets, or similar stiff cardboard, together. To make the charge easier to carry, it can be built to fold in the middle (Figure 8-22). Rounding the corners makes the charge easier to handle.

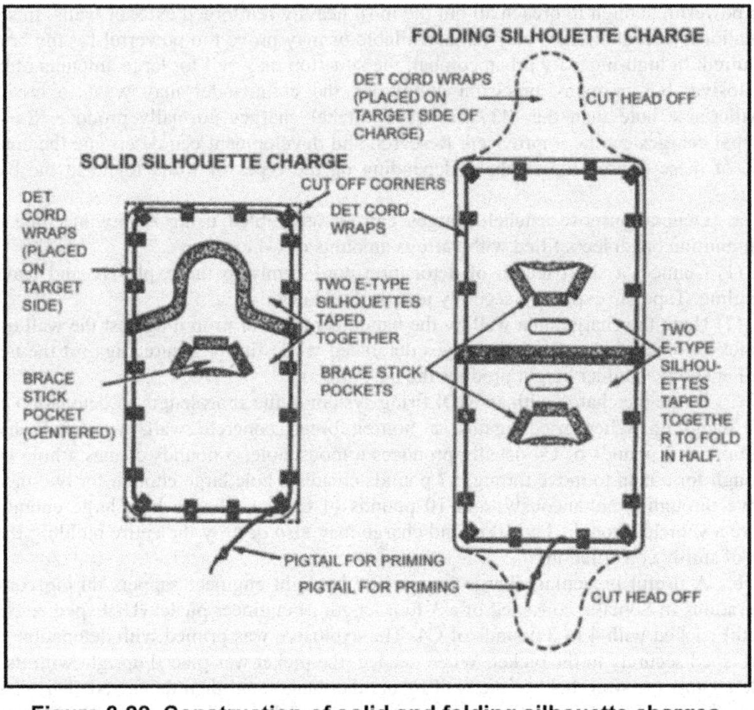

Figure 8-22. Construction of solid and folding silhouette charges.

b. Place detonation cord or flexible linear-shaped charge (FLSC) around the edges of the silhouettes, leaving a 6-inch tail for priming. Secure the cord to the silhouette using sturdy tape (for example, "100-mph tape"). (See Table 8-5 on page 8-22 for the appropriate number of wraps of detonation cord or FLSC to breach various barriers.) Tape several small dowels or other materials at various places around the silhouette if using FLSC. This provides the necessary standoff distance to ensure the maximum shaped charge effect. (See Table 8-6 on page 8-22 for the required standoff distance for various sizes of FLSC.)

NOTE: Always consider the silhouette material (about 1/8 inch) when determining standoff distance.

Type of Obstacle	Detonation Cord Needed	FLSC Needed
Hollow-core door	1 wrap	75 grain/foot
Particle-filled door (1 inch)	2 wraps	75 grain/foot
Solid wood door (2 inches)	3 wraps	75 grain/foot
High-quality solid door	4 wraps	225 grain/foot
1/4-inch plywood	1 wrap	75 grain foot
1/2-inch plywood	2 wraps	75 grain/foot
3/4-inch plywood	3 wraps	75 grain/foot
Light metal door	NA	225 grain/foot
Medium steel door	NA	300 grain/foot
Heavy steel door	NA	300 grain/foot

Table 8-5. Silhouette charge explosive loads.

NOTE: The information, in Table 8-5, is based on US manufactured materials. Building materials of other countries may be of inferior quality; however, some European companies have standards that surpass that of the US.

Standoff Required for FLSC	Standoff
75 grain	0 - 1/16 inch
225 grain	1/8 inch
300 grain	1/8 inch - 3/16 inch
NOTE: FLSC that is 300 grains or higher cannot be molded.	

Table 8-6. Standoff required for flexible linear shaped charges.

c. Place three or four strips of heavy-duty, double-sided contact tape on the front of the silhouette from top to bottom. Construct a sturdy pocket for a brace stick in the appropriate position on the back of the silhouette.

d. Pull the covering off the double-sided tape and place the charge against the wall at knee height, bracing it if necessary. Prime the charge, take cover, and detonate.

8-11. DOOR-BREACHING CHARGES

Several different field-expedient charges can be used to breach interior or exterior doors and chain link fence. Among these are the general-purpose charge, the rubber band charge, flexible linear charge, doorknob charge, rubber strip charge, and the chain link charge. All can be made ahead of time and are simple, compact, lightweight, and easy to emplace.

CAUTION
Any time explosive charges are used to breach doors, the knobs, locks and hinges made of steel and metal can become lethal projectiles.

a. **General-Purpose Charge.** The general-purpose charge is the most useful preassembled charge for breaching a door or other barrier. As its name implies, it is useful not only for door breaching, but it can also cut mild steel chain and destroy captured enemy equipment.

(1) Start building the general-purpose charge with a length of detonation cord about 2 feet long. Using another length of detonation cord, tie two uli knots (Figure 8-23) around the 2-foot long cord. The uli knots must have a minimum of six wraps and be loose enough for them to slide along the main line, referred to as a uli slider. Trim excess cord from the uli knots and secure them with tape, if necessary.

(2) Cut a block of C4 explosive to a 2-inch square. Tape one slider knot to each side of the C4 block, leaving the length of detonation cord free to slide through the knots (Figure 8-24).

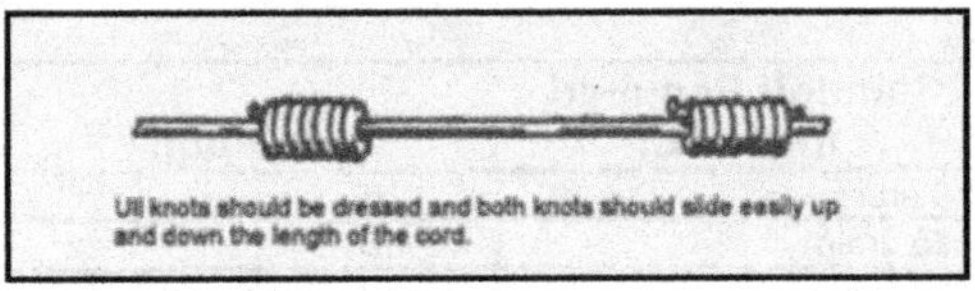

Figure 8-23. Sliding uli knots.

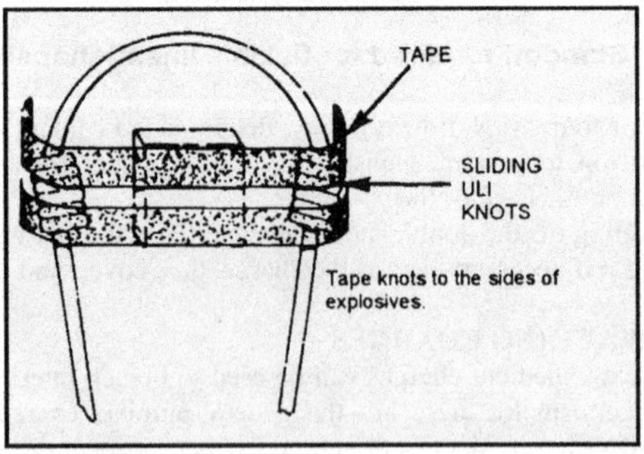

Figure 8-24. Completed general-purpose charge.

(3) To breach a standard door, place the top loop of the charge over the doorknob. Slide the uli knots taped to the C4 so that the charge is tight against the knob. Prime the loose ends of the detonation cord with a MDI firing system and detonate (Figure 8-25).

To cut mild steel chain, place the loop completely around the chain link to form a girth hitch. Tighten the loop against the link by sliding the uli knots.

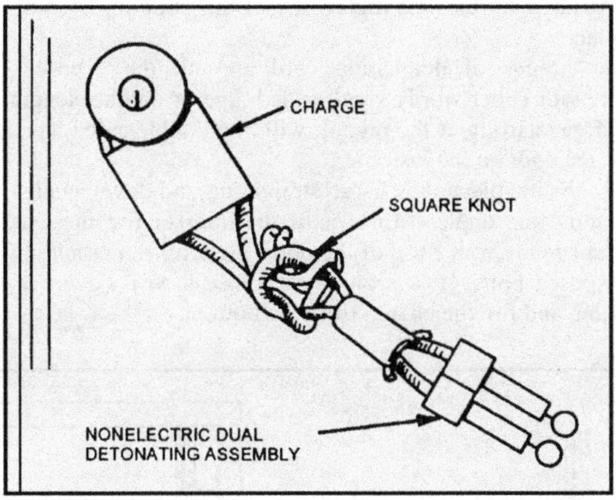

Figure 8-25. Charge placement against doorknob.

b. **Rubber Band Charge.** The rubber band charge is another easily fabricated lightweight device that can be used to remove the locking mechanism or doorknob from wooden or light metal doors, or to break a standard size padlock at the shackle.

(1) Cut a 10-inch piece of detonation cord and tie an overhand knot in one end. Using another piece of detonation cord, tie a uli knot with at least eight wraps around the first length of cord. Slide the uli knot tightly up against the overhand knot. Secure it in place with either tape or string. Loop a strong rubber band around the base of the uli knot tied around the detonation cord. Tie an overhand knot in the other end of the cord to form a pigtail for priming the charge.

(2) Attach the charge to the doorknob (or locking mechanism) by putting the loose end of the rubber band around the knob. The charge should be placed between the knob and the doorframe. This places the explosive over the bolt that secures the door to the frame.

c. **Flexible Linear Charge.** One of the simplest field-expedient charges for breaching wooden doors is the flexible linear charge. It can be made in almost any length, and it can be rolled up and carried until needed. It is effective against hollow-core, particle-filled, and solid wood doors. When detonated, the flexible linear charge cuts through the door near the hinges (Figure 8-26).

(1) Lay out a length of double-sided contact tape with the topside adhesive exposed. Place the necessary number of strands of detonation cord down the center of the double-sided tape, pressing them firmly in place. Military detonation cord has 50 grains of explosives per foot and there are 7,000 grains in a pound. Most residential doors are 80 inches tall and commercial doors are 84 inches tall. This must be considered when calculating the quantities of explosives, overpressure and MSDs. For hollow-core doors,

use a single strand. For particle-filled doors, use two strands, and for solid wood doors use three. If the type doors encountered are unknown, use three strands. One of the strands must be cut about a foot longer than the others and should extend past the end of the double-sided tape. This forms a pigtail where the initiating system is attached once the charge is in place.

(2) Cover the strands of detonation cord and all the exposed portions of the double-sided tape with either sturdy single-sided tape or another length of double-sided tape. Roll the charge, starting at the pigtail, with the double-sided tape surface that is to be placed against the door on the inside.

(3) At the breach site, place the charge straight up and down against the door tightly. If the charge is too long, angle it to best fit the door or use the excess to defeat the possibility of a door return at the top of the door. Sometimes but not always visible from the outside by exposed bolts. If it is too short, place it so it covers at least half of the door's height. Prime and fire the charge from the bottom.

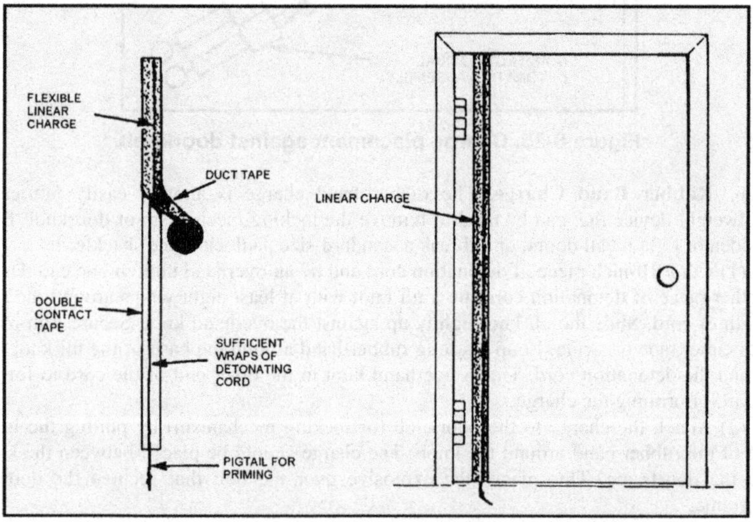

Figure 8-26. Placement of the flexible linear charge.

d. **Doorknob Charge.** A doorknob charge is easy to make and highly effective against wooden or light metal doors. Charges for use against wooden doors can be made with detonation cord. If the charge is to breach a light metal door, either detonation cord (three lengths) or 225 grain/foot flexible linear shaped charge (FLSC) should be used.

(1) Cut the appropriate amount of detonation cord for the charge. Use a 30-inch length for a hollow-core door. For a particle-filled door, use one 30-inch length and one 18-inch length. For a solid-core wooden door or a light metal door, use one 30-inch length and two 18-inch lengths.

(2) Cut the charge holder from a piece of stiff cardboard.

(3) Place double-sided tape on the face of the charge holder in the shape of a large "C". Place the detonation cord on top of the double-sided tape, also in the shape of a large "C" along the edge of the charge holder. Leave a 12-inch pigtail for priming (Figure 8-27).

(4) If using FLSC, cut a length 21 inches long. Tape the FLSC to the outside of the charge holder, leaving a 3-inch tail for priming. Bend the tail upward. Tie a Sliding Uli knot to a 12-inch length of detonation cord and tie an overhand knot on each end of the knot. Tape the slider and detonation cord combination to the tail end of the FLSC and on the inside of the "V" shape to insure detonation.

(5) Hang the charge on the doorknob or locking mechanism. Secure the FLSC charge in place with the double-sided tape, and the "Det cord" charge with "100 mph" tape. The detonation cord must be held firmly against the door's surface.

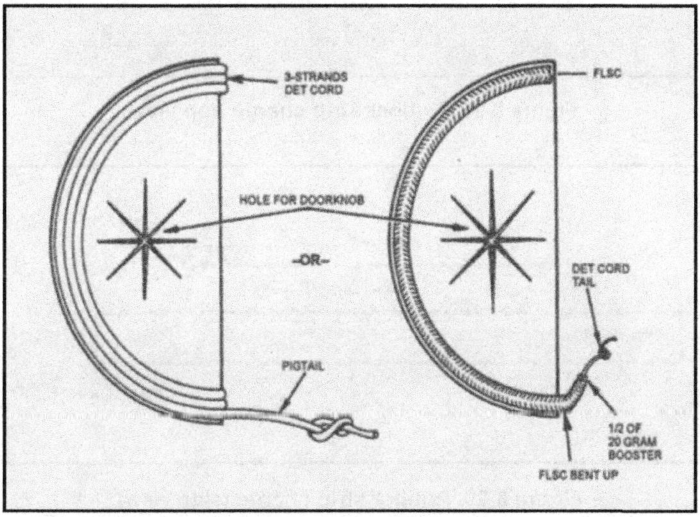

Figure 8-27. Doorknob charge.

e. **Rubber Strip Charge.** The rubber strip charge (Figures 8-28 and 8-29) can be used to open a solid wood door with multiple locking devices or a metal door. It defeats the locking mechanism and dislodges the door from the frame. It can also be used to defeat windows with a physical security system. Place the charge on the target between locking devices and doorjamb. Ensure that the rubber strip covers the area where the locking bolts are located. For a metal door with standard locking devices, place the charge in the center of the door, parallel with the locking mechanism. When detonated it will buckle and or bend the door, pulling the locking mechanisms from their catches.

WARNING
Net explosive weight should not exceed 8 ounces for a complete charge.

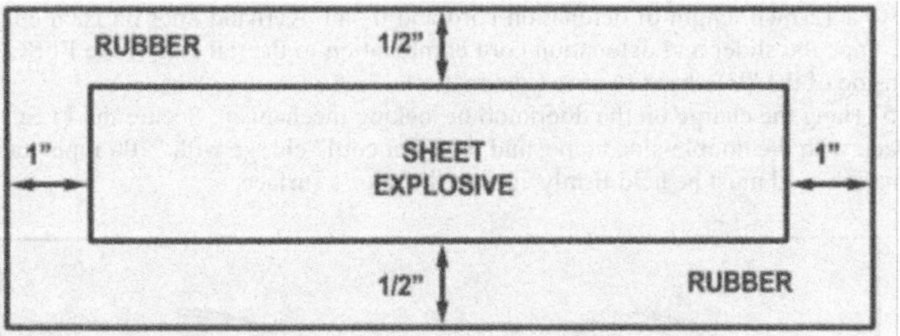

Figure 8-28. Rubber strip charge (top view).

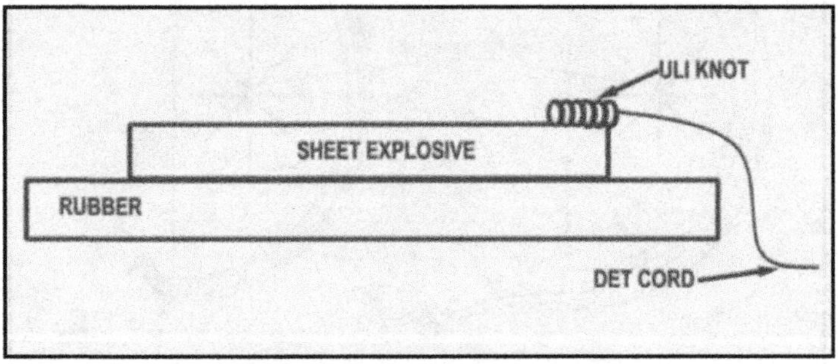

Figure 8-29. Rubber strip charge (side view).

WARNING
Friendly troops must be clear of the area 90 degrees from the target. The doorknob will be blown away from the door with considerable force.

f. **Chain-Link Ladder Charge.** The chain-link ladder charge (Figure 8-30, page 8-28) is designed to create a man-sized hole in a chain-link fence. The charges run lengthwise along the detonation cord on sliding uli knots. Once in the desired location, secure in place with tape. Determine where charge is to be emplaced on the fence. To ensure that the full impact is received by the fence, place the charge next to a fence post so that the fence will not "give." Secure the top of the charge to the breach location on the fence. Simply hang the large hook into a link of the fence. Secure the charge at the bottom of the fence using surgical tubing and a small hook.

NOTES: 1. Use six-wrap uli knots constructed of detonation cord.

2. Only slight pressure is required to secure the charge; too much tension may cause the fence to buckle and result in poor cutting of the fence.

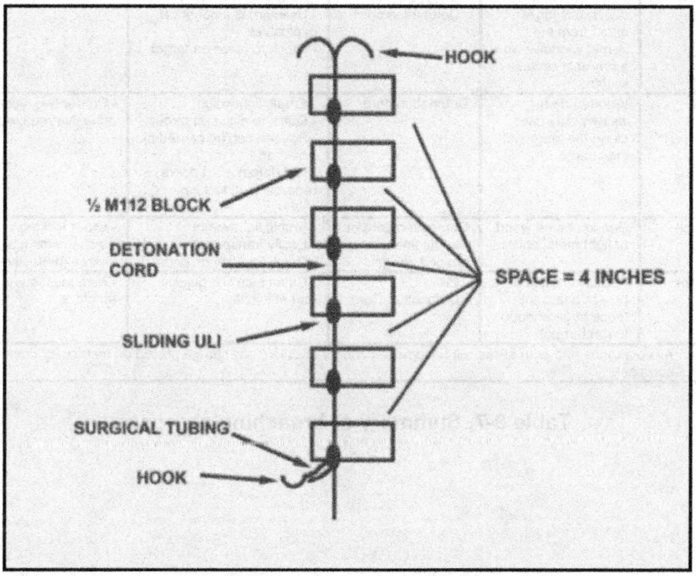

Figure 8-30. Chain-link ladder charge.

NOTE: Table 8-7 provides a summary of the different charges discussed in this chapter.

CHARGE	OBSTACLE	EXPLOSIVES NEEDED	ADVANTAGES	DISADVANTAGES
Wall Breach Charge (Satchel Charge or U-Shaped Charge)	Breaches through wood, masonry, or brick, and reinf concrete walls	- Detonation cord - C4 or TNT	- Easy and quick to make - Quick to place on target	- Does not destroy rebar - High overpressure - Appropriate attachment methods needed - Fragmentation
Silhouette Charge	Wooden doors; creates man-sized hole. Selected walls (plywood, Sheet-rock, CMU)	- Detonation cord	- Minimal shrapnel - Easy to make - Makes entry hole to exact specifications	- Bulky; not easily carried
General Purpose Charge	Door knobs, mild steel chain, locks, or equipment	- C4 - Detonation cord	- Small, lightweight - Easy to make - Very versatile	- Other locking mechanisms may make charge ineffective
Rubber Strip Charge	Wood or metal doors; dislodges doors from the frame, windows with a physical security system	- Sheet Explosive - Detonation cord	- Small, easy to carry - Uses small amount of explosives - Quick to place on target	
Flexible Linear Charge	Wooden doors; widows cuts door along the length off the charge.	Detonation cord	- Small, lightweight - Quick to place on target - Several can be carried by one man - Will defeat most doors regardless of locking systems	- Proper two-sided adhesive required
Doorknob Charge	Doorknobs on wood or light metal doors	Detonation cord or flexible linear shaped charge	- Small, lightweight - Easily transported - Quick to place on door	- Other locking mechanisms may make charge ineffective
Chain-link Ladder Charge	Rapidly creates a hole in chain-link fence large enough to run through	- C4 - Detonation Cord	- Cuts chain link quickly and efficiently	- Man must stand to emplace it
NOTE: All doorknobs and prop-sticks will become secondary missiles; any charge placed on metal may create shrapnel.				

Table 8-7. Summary of breaching charges.

CHAPTER 9

EMPLOYMENT OF ATTACK AND ASSAULT/CARGO HELICOPTERS

Ground maneuver commanders must understand that aviation forces can provide a significant advantage during UO. In addition, ground maneuver planners must understand that the unique capabilities of Army aviation also require unique planning and coordination. Army aviation forces must be fully integrated in the military decision-making process (MDMP) to ensure effective combined arms employment. Effective combined arms employment also requires that aviation and ground maneuver forces synchronize their operations by operating from a common perspective. This chapter highlights some possible procedures that will aid in creating a common air-ground perspective.

9-1. SUPPORT FOR GROUND MANEUVER UNITS

Ground units may receive support from a variety of attack helicopters including (but not limited to) the AH-64, AH-1, and OH-58D. Attack helicopters can provide area fire to suppress targets, and precision fire to destroy specific targets or breach structures. Attack helicopters can also assist with ISR and communications using their advanced suite of sensors and radios. Other supporting (lift) helicopters, such as the UH-60 and CH-47, may also have weapon systems (7.62-mm machine gun, caliber .50 machine gun, 7.62-mm minigun) that aid in the suppression of enemy forces when operating in urban terrain. However, their primary role is to transport personnel, equipment, and supplies to those critical urban areas. Lift helicopters can provide a distinct advantage by placing personnel and weapon systems at critical locations at critical times to surprise and overwhelm the enemy. Lift helicopters can also transport needed supplies to urban areas that may be inaccessible to ground transportation

9-2. ROLE DURING URBAN OPERATIONS

Army aviation's primary role during UO is to support the ground maneuver force's operations. Army aviation is normally most effective conducting shaping operations. Aviation forces operating on the urban periphery effectively enhance isolation, reconnaissance, resupply, troop movement, evacuation, and support by fire for ground forces. Army aviation also enhances the combined arms team's ability to quickly and efficiently transition to new missions. Aviation forces normally avoid over-flight of built up terrain due to the high risk of being engaged by enemy forces in close proximity. When aviation forces cannot avoid the built up areas, special measures and thorough risk analysis may minimize the associated dangers. The following missions are commonly performed during UO.

a. **Assess**. Identify the portions of the urban area essential to mission success. Aviation forces provide reconnaissance capability, security to ground forces, movement of troops and supplies, and augmentation of communication and surveillance capabilities (Table 9-1, page 9-2).

b. **Shape**. Isolate those areas essential to mission success or avoid isolation while in the defense. In the offense, aviation forces attack to isolate the objective, move troops and supplies, enhance C2, conduct reconnaissance, and augment ground forces. In the defense, aviation forces act as a maneuver element to set the conditions for the main battle and prevent isolation (Table 9-1).

c. **Dominate**. Precisely mass the effects of combat power to rapidly dominate the area. Army aviation supports the ground maneuver commander's intent and scheme of maneuver by providing maneuver and support assets. Aviation supports the combined arms effort by providing support by fire, movement of troops and supplies, enhanced C2, air assaults, reconnaissance, and continued isolation of the objective (Table 9-2).

d. **Transition**. Transition the urban area to the control of another agency and prepare for follow-on operations. Aviation forces conduct combat, combat support and combat service support missions that facilitate the combined arms transition to follow-on operations (Table 9-3). Stability and support missions are shown in Table 9-4, page 9-4.

ASSESS AND SHAPE
(Reconnaissance, Movement, Isolation of Objective)

Lift (Utility/Cargo) Helicopter Units

- Conduct air assaults to the flanks and rear of urban areas to deny LOCs from enemy.
- Provide CASEVAC.
- Perform CSAR.
- Perform DART.
- Conduct air movement of troops and supplies.
- Air move/assault ground R&S elements.
- Emplace logistical resupply points and FARPs.
- Conduct C2 operations.
- Conduct EW operations.
- Provide NEO support.
- Conduct countermobility operations/emplace Volcano mines.

Attack/Cavalry Helicopter Units

- Perform reconnaissance of urban peripheral area to establish enemy strength and disposition.
- Conduct route and area reconnaissance for ground maneuver forces.
- Establish initial security of urban flanks and rear until relieved by ground forces.
- Augment ground forces for isolation of urban area.
- Employ indirect fires and CAS, AF CAS, urban CAS, and JAAT to enforce isolation.
- Perform air assault security.
- Provide suppressive fires in support of ground maneuver and security elements.
- Employ direct fires to destroy enemy elements attempting to escape, resupply, or reinforce the urban area.
- Destroy key targets with direct fires.

Table 9-1. Assessing and shaping missions.

DOMINATE
(Isolate, Secure Foothold, Clear Objective)

Lift (Utility/Cargo) Helicopter Units

(In addition to the missions listed under assess and shape, *utility and cargo helicopters may conduct these missions.)*
- Perform air assault in proximity to urban area to insert infantry elements.
- Support CA/PSYOPS operations.

Attack/Cavalry Helicopter Units

(In addition to the missions listed under assess and shape, *attack and cavalry helicopters may conduct these missions.)*
- Provide security to flanks of advancing ground forces.
- Provide suppressive fires in support of attacking ground forces.
- Engage high priority targets influencing point of penetration with precision direct fires.

Table 9-2. Dominating missions.

TRANSITION
(Consolidate, Reorganize, Prepare for Future Operations)

Lift (Utility/Cargo) Helicopter Units

- Provide CASEVAC.
- Perform CSAR.
- Perform DART missions.
- Conduct air movement of troops and supplies.
- Emplace logistical resupply points and FARPs.
- Conduct C2 operations.
- Augment TCF.

Attack/Cavalry Helicopter Units.

- Provide screen or area security.
- Conduct route and area reconnaissance for ground maneuver forces.
- Serve as TCF or reserve.
- Conduct deep operations to set conditions for follow-on missions.

Table 9-3. Transition missions.

Lift (Utility/Cargo) Helicopter Units	Attack/Cavalry Helicopter Units
• Perform show of force. • Conduct air assaults with QRF. • Provide medical evacuation. • Conduct DART. • Conduct SAR/CSAR. • Transport troops and equipment. • Emplace FARPs. • Perform C2. • Conduct aerial search. • Conduct movement of dignitaries and or civilians. • Conduct offensive operations. • Conduct defensive operations. • Conduct aerial rescue and evacuation.	• Perform reconnaissance. • Conduct aerial surveillance. • Conduct security operations. • Provide air assault security. • Provide over-watch for ground forces. • Conduct offensive operations. • Conduct defensive operations.

Table 9-4. Stability and support missions.

9-3. COMMAND AND CONTROL

Army aviation forces may be employed organic to a division or higher level of command to conduct maneuver or provide support (CS/CSS). Aviation forces may also be attached or under operational control (OPCON) of another command. Army aviation units normally will not be OPCON to echelons below battalion level; however, attack helicopters may conduct direct air-to-ground coordination with companies and platoons during combat operations.

9-4. MANEUVER GRAPHIC AIDS

The greatest strength of aviation is the ability to maneuver in the third dimension. In an urban environment, this strength can be a detriment due to associated challenges. One associated challenge is that aircrews have different visual cues and perspectives than do ground forces. Common graphics and sketches can help alleviate these differences.

a. A network route structure of air control points (ACP) and routes (preferably surveyed) may be used to facilitate route planning, navigation, and command, control, and communications (C3).

b. Sketches help correlate air and ground control measures with predominate urban features. The area sketch offers the ground commander and the aircrew a means of identifying friendly and enemy locations for planning and coordination. It is best used for smaller towns and villages but can be applied to a certain engagement area or specific area of operations in a larger city. The area sketch captures the natural terrain features, man-made features, and key terrain in an area and designates a letter or numeral code to each. Buildings are coded and each corner of the building is coded. This gives the aircrews an accurate way to identify specific buildings as requested by the ground unit commander or to identify friendly locations.

c. Inclusion of maneuver graphic, fire support control measures (FSCM), and airspace control measures (ACM) further allow aircrews and maneuver elements to better visualize the urban portion of the battlespace. It is the responsibility of both the aviation unit and the ground maneuver unit to ensure they use the same area sketch for accurate coordination (Figure 9-1).

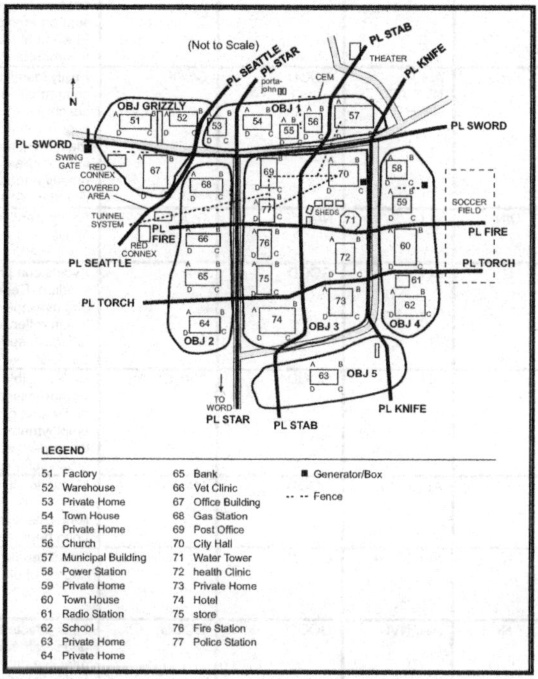

Figure 9-1. Simplified area sketch.

9-5. IDENTIFYING FRIENDLY POSITIONS, MARKING LOCATIONS, AND ACQUIRING TARGETS

In the urban environment, friendly and enemy forces, along with noncombatants, may operate in close vicinity to one another. Furthermore, structures and debris can cause problems with the identification of precise locations. Reliable communication is essential to ensure aircrews know the locations of all participants in UO. To further enhance air-ground coordination, methods must be established to allow aircrews to visually identify key locations.

a. **Marking Methods.** Table 9-5, on pages 9-6 and 9-7, describes different marking methods.

METHOD	DAY/ NIGHT	ASSETS	FRIENDLY MARKS	TARGET MARKS	REMARKS
SMOKE	D	ALL	GOOD	GOOD	Easily identifiable, may compromise friendly position, obscure target, or warn of fire support employment. Placement may be difficult due to structures.
SMOKE (IR)	D/N	ALL/ NVD AT NIGHT	GOOD	GOOD	Easily identifiable, may compromise friendly position, obscure target, or warn of fire support employment. Placement may be difficult due to structures. Night marking is greatly enhanced by the use of IR reflective smoke
ILLUM, GROUND BURST	D/N	ALL	N/A	GOOD	Easily identified, may wash out NVDs.
SIGNAL MIRROR	D	ALL	GOOD	N/A	Avoids compromise of friendly location. Dependent on weather and available light and may be lost in reflections from other reflective surfaces (windshields, windows, water, etc.)
SPOT LIGHT	N	ALL	GOOD	MARGINAL	Highly visible to all. Compromises friendly position and warns of fire support employment. Effectiveness is dependent upon degree of urban lighting.
IR SPOT LIGHT	N	ALL NVD	GOOD	MARGINAL	Visible to all with NVGs. Less likely to compromise than overt light. Effectiveness dependent upon degree of urban lighting.
IR LASER POINTER (below .4 watts)	N	ALL NVG	GOOD	MARGINAL	Effectiveness dependent upon degree of urban lighting.
IR LASER POINTER (above .4 watts)	N	ALL NVD	GOOD	GOOD	Less affected by ambient light and weather conditions. Highly effective under all but the most highly lit or worst weather conditions. IZLID-2 is the current example.
VISUAL LASER	N	ALL	GOOD	MARGINAL	Highly visible to all. Risk of compromise is high Effectiveness dependant upon degree of urban lighting.
LASER DESIGNATOR	D/N	PGM OR LST EQUIPPED	N/A	GOOD	Highly effective with PGM. Very restrictive laser acquisition cone and requires line of sight to target. May require pre-coordination of laser codes
TRACERS	D/N	ALL	N/A	MARGINAL	May compromise position. May be difficult to distinguish mark from other gunfire. During daytime use, may be more effective to kick up dust surrounding target.

Table 9-5. Marking methods.

METHOD	DAY/ NIGHT	ASSETS	FRIENDLY MARKS	TARGET MARKS	REMARKS
ELECTRONIC BEACON	D/N	SEE REMARKS	EXCELLENT	GOOD	Ideal friendly marking device for AC-130 and some USAF fixed wing (not compatible with Navy or Marine aircraft). Least impeded by urban terrain. Can be used as a TRP for target identification. Coordination with aircrews essential to ensure equipment and training compatibility.
STROBE (OVERT)	N	ALL	MARGINAL	N/A	Visible by all. Effectiveness dependent upon degree of urban lighting.
STROBE (IR)	N	ALL NVD	GOOD	N/A	Visible to all NVDs. Effectiveness dependent upon degree of urban lighting. Coded strobes aid in acquisition
FLARE (OVERT)	D/N	ALL	GOOD	N/A	Visible by all. Easily identified by aircrew.
FLARE (IR)	N	ALL NVD	GOOD	N/A	Visible to all NVDs. Easily identified by aircrew.
GLINT/IR PANEL	N	ALL NVD	GOOD	N/A	Not readily detectable by enemy. Very effective except in highly lit areas.
COMBAT IDENTIFICATI ON PANEL	D/N	ALL FLIR	GOOD	N/A	Provides temperature contrast on vehicles or building. May be obscured by urban terrain.
VS-17 PANEL	D	ALL	MARGINAL	N/A	Only visible during daylight. Easily obscured by structures.
CHEMICAL HEAT SOURCES	D/N	ALL FLIR	POOR	N/A	Easily masked by urban structures and lost in thermal clutter. Difficult to acquire, can be effective when used to contrast cold background or when a/c knows general location.
SPINNING CHEM-LIGHT (OVERT)	N	ALL	MARGINAL	N/A	Provides unique signature. May be obscured by structures. Provides a distinct signature easily recognized. Effectiveness dependent upon degree of urban lighting.
SPINNING CHEM-LIGHT (IR)	N	ALL NVD	MARGINAL	N/A	Provides unique signature. May be obscured by structures. Effectiveness dependent upon degree of urban lighting.

Table 9-5. Marking methods (continued).

b. **Targeting Grids and Reference Techniques** (Figures 9-2 through 9-5 on pages 9-8 and 9-9). Ground maneuver elements generally use a terrain-based reference system during urban operations. MGRS coordinates have little meaning at street level. To facilitate combined arms operations, aviation and ground maneuver forces must use common control methods. Possible techniques include urban grid, checkpoint targeting, objective area reference grid, and TRPs. These techniques are based on the street and structure pattern present, without regard to the MGRS grid pattern. Using common techniques allows aircrews to transition to the system in use by the ground element upon arrival in the objective area. For example, references to the objective or target may include local landmarks such as, "The third floor of the Hotel Caviar, south-east corner." This transition should be facilitated by using a "big to small" acquisition technique.

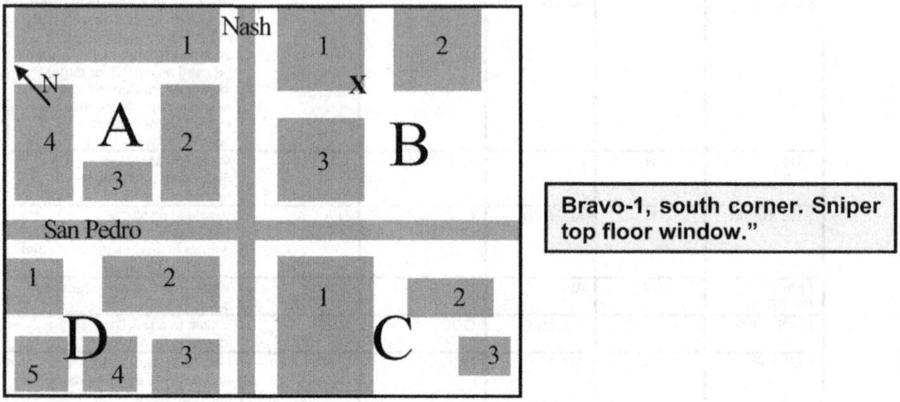

Figure 9-2. Urban grid technique.

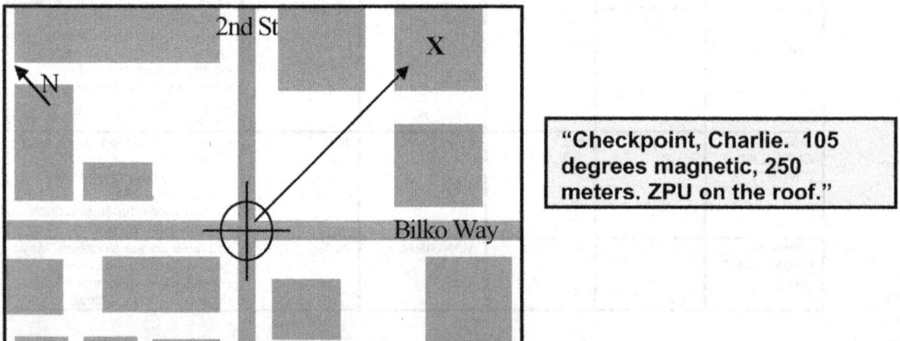

Figure 9-3. Checkpoint technique.

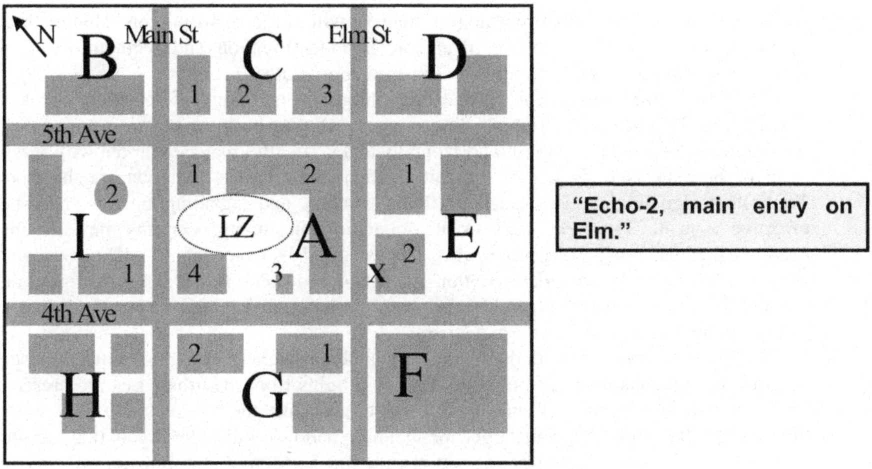

Figure 9-4. Objective area reference grid technique.

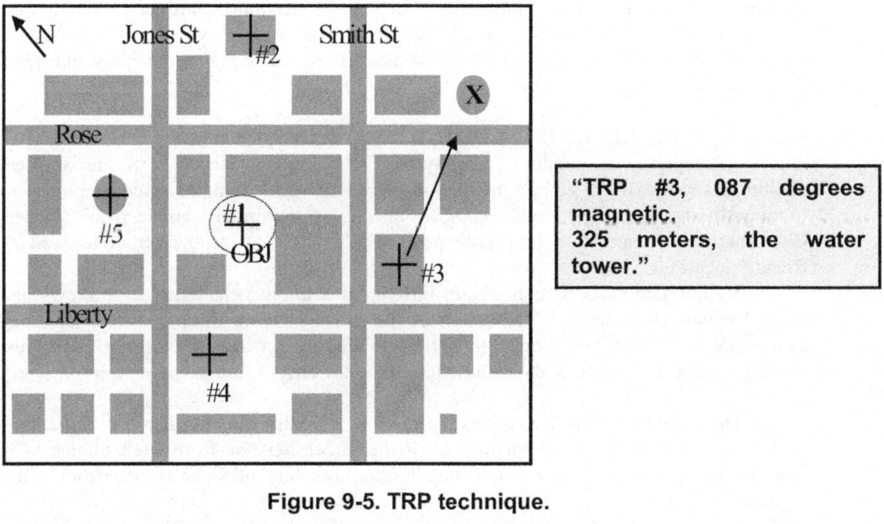

Figure 9-5. TRP technique.

c. **Additional Cues.** Aviation forces also use these additional cues.

(1) ***Roof Characteristics.*** Flat roofs, pitched roofs, domed roofs, and roofs with towers or air conditioning units on top will aid in visual and thermal acquisition.

Additional structural features revealed in imagery will aid in confirmation. This method of terrain association may prove invaluable for identification and recognition since structures are often too close to rely on mere grid coordinates.

(2) ***Visual Markings.*** The visual signaling or marking of positions makes determining the location of friendly forces easier. During building clearing operations, the progress of friendly units (both horizontally and vertically) may be marked with spray paint or bed sheets hung out of windows. Often, the simplest methods are the best. Traditional signaling devices, such as flares, strobes, and signaling mirrors, may be effective as well. Target marking or an orientation on enemy positions may also be accomplished using signaling procedures. The use of GLINT tape, combat ID panels, and IR beacons assist in the identification of friendly ground forces on urban terrain. Standardized usage of ground lighting, thermal contrast, and interposition of structures influence the effectiveness of these devices.

(3) ***Shadows.*** During both high and low ambient light conditions, expect to see significant urban shadowing from buildings when lights from the urban area are present. Shadows will hide personnel and or vehicular targets, like the shadows that hide small hills against the background of larger mountains. Shadows will hide targets that are not thermally significant, but thermal targets should still be seen. A combination of sensors must be used to acquire and identify the target; therefore, a sensor hand-off plan must be thoroughly briefed.

(4) ***GPS.*** The use of aircraft with integrated GPS reduces the amount of time spent finding the target area. If ground forces can provide accurate coordinates, inputting a target grid into the GPS or inertial navigation system (INS) provides fire control cues (range, heading, time) to the target, which aids in quicker target acquisition and helps distinguish friendly forces from enemy forces.

9-6. ATTACK HELICOPTER TARGET ENGAGEMENT

Attack helicopters will conduct a variety of TTP to engage targets in the urban area. Techniques range from support by fire/attack by fire at maximum standoff ranges to running/diving fire and close combat attack at minimum engagement ranges. Coordination is imperative to ensure positive identification of the target as well as friendly locations.

a. **Urban Canyons.** Urban terrain introduces a unique challenge to aircrews and ground personnel alike with the notion of the urban canyon. Simply stated, an urban canyon exists when a target or target set is shielded by vertical structures. Unlike most natural terrain, the vertical characteristics of urban terrain can greatly affect delivery options.

(1) Urban terrain typically creates corridors of visibility running between structures. Street-level targets are only visible along the street axis or from high angles. The interposition of structures around a target interrupts line of sight (LOS) from many directions. The presence of buildings and other structures in urban terrain creates corridors of visibility along streets, rivers, and railways. LOS must be maintained for enough time to acquire the target, achieve a weapons delivery solution, and fly to those parameters.

(2) This timeline is reduced during the employment of the AH-64D. Its precise navigation system enables the aircraft to slave its sensors and weapons to a stored target,

thereby significantly reducing target acquisition times. In some cases, the AH-64D may employ the gun or FFARs in an “area supression” mode and never have to expose the aircraft to the target area.

(3) Visibility limitations on marking devices in the urban environment are geometric in nature. The use of any pointer or laser requires LOS. In addition, the aircraft must have LOS with the target to see the mark. Urban terrain severely limits LOS opportunities. Due to the close proximity of structures to one another, there may be very narrow fields of view and limited axes of approach.

NOTE: Ground forces should make every attempt to pass along accurate 8-digit grid coordinates. The AH-64D can easily and accurately engage targets using this method.

b. **Reflective Surfaces.** The high number of reflective surfaces in an urban area presents an additional challenge. Laser energy can be reflected and present multiple false returns. For these reasons, fire support can be expected to be more time consuming and be much more dependent on good communications. Combinations of marking devices and clear talk-on procedures will be essential to safe and effective fire support. Ground forces should consider using buddy team or remote laser marking tactics for laser guided munitions when urban effects preclude the attacking aircraft from maintaining LOS with the target until ordnance impact. However, when designating a target with a ground based laser along a narrow street bounded by tall buildings, LOS geometry may allow the weapon to receive reflected laser energy. Aircrews must also consider the potential miss distances for “precision” munitions when their guidance source is interrupted or removed.

c. **Weapons Mix.** Armed helicopters can carry a mix of weapons. Commanders must choose the weapons to use on a specific mission based on the effects on the target, employment techniques, and the target's proximity to ground forces. Planners must consider proportionality, collateral damage, and noncombatant casualties. Planners and aircrew must consider the following when choosing weapons.

(1) Hard, smooth, flat surfaces with 90-degree angles are characteristic of man-made targets. Due to aviation delivery parameters, munitions will normally strike a target at an angle less than 90 degrees. This may reduce the effect of munitions and increase the chance of ricochets. The tendency of rounds to strike glancing blows against hard surfaces means that up to 25 percent of impact-fused rounds may not detonate when fired onto rubbled areas.

(2) Identification and engagement times are short.

(3) Depression and elevation limits create dead space. Target engagement from oblique angles, both horizontal and vertical, must be considered.

(4) Smoke, dust, and shadows mask targets. Additionally, rubble and man-made structures can mask fires. Targets, even those at close range, tend to be indistinct.

(5) Urban fighting often involves units attacking on converging routes. The risks from friendly fires, ricochets, and fratricide must be considered during the planning of operations.

(6) The effect of the weapon and the position of friendly and enemy personnel with relation to structures must be considered. Chose weapons for employment based on their effects against the building's material composition rather than against enemy personnel.

(7) Munitions can produce secondary effects, such as fires.

9-7. AIR GROUND INTEGRATION IN THE HASTY ATTACK/CLOSE FIGHT

Attack helicopter employment in urban operations will typically involve the close fight and often, the hasty attack. The hasty attack in the close fight, historically, lacks proper coordination between air and ground elements to ensure mission success. The key to success for enhancing air-ground coordination, and the subsequent execution of the tasks involved, begins with standardizing techniques and procedures. The end-state is a detailed SOP between air and ground maneuver units that addresses the attack in a close combat situation.

Effective integration of air and ground assets begins with the ground maneuver brigade. When the aviation brigade or task force receives a mission to provide assistance to a ground unit engaged in close combat and planning time is minimal, the initial information provided by the brigade in contact should be sufficient to get the aviation attack team out of the aviation tactical assembly area to a holding area to conduct direct coordination with the engaged maneuver unit. The attack teams utilized in this procedure are under aviation brigade control. This procedure contains five major steps:

- Maneuver brigade planning requirements.
- Battalion close fight SITREP.
- Attack team check-in.
- Coordination for aviation close fires (ACF).
- Battle damage assessment/reattack.

a. **STEP 1, Maneuver Brigade Planning Requirements.** The maneuver brigade, through their aviation liaison officer, provides the aviation brigade headquarters the necessary information to meet planning requirements (Figure 9-6). The initial planning and information to be passed to the aviation brigade headquarters includes the location of the holding area, the air axis, and the route or corridor for entry and exit through the brigade and battalion sector.

(1) The holding area should be in the sector of the ground maneuver battalion involved in close combat. The holding area may be a concealed position or an aerial holding area that allows for final coordination between the attack team leader and the ground unit leader. It must be located within frequency modulated (FM) radio range of all units involved. Alternate holding areas, along with ingress and egress routes, must be designated if occupation is expected to last longer than 15 minutes.

(2) The brigade also provides the call signs and frequencies of single-channel ground and air radio system (SINCGARS), Hopsets, and communications security (COMSEC) information regarding the battalion in contact. If the unit is SINCGARS-equipped, the attack team must also have the common "time," which may be taken from global positioning systems (GPSs).

(3) In addition, the brigade provides a current situation update for their AO and specifically for the supported battalions AO. This will include a recommended engagement area, which will allow for initial planning for battle positions (BPs) or attack by fire/support by fire positions (ABF/SBF) and possibly prevent unintentional overflight of enemy positions.

MINIMUM BRIGADE PLANNING REQUIREMENTS
1. **Current situation** should include friendly forces location and situation, enemy situation highlighting known ADA threat in the AO, and tentative engagement area coordinates. 2. **Brigade/Battalion level graphics update** this can be via MCS-P, or via radio communications, updating critical items such as LOA, fire control measures, base maneuver graphics so as to better integrate into the friendly scheme of maneuver. 3. **Fire support** coordination information; i.e. location of DS artillery, and organic mortars, call signs and frequencies 4. **Ingress/egress routes** into their AO, this includes PP into sector or zone and air routes to the holding area. 5. **Holding Area** for face-to-face coordination between the attack team and the battalion in contact. A holding area equates to an assault position. It must be adequate in size to accommodate the number of aircraft assigned the mission, be out of range of enemy direct fire systems, should be out of enemy mortar range. 6. **Call signs/frequencies** of the battalion in contact down to the company in contact. Air ground coordination must be done on command frequencies to provide situational awareness for all elements involved. 7. **SINCGARS** time hack.

Figure 9-6. Maneuver brigade planning requirements.

b. **STEP 2, Battalion Close Fight SITREP.** En route to the holding area, the attack team leader contacts the ground maneuver battalion on its FM command net to receive a close fight situation report (SITREP) (Figure 9-7). This SITREP verifies the location of the holding area and a means to conduct additional coordination. The attack team leader receives an update from the ground maneuver battalion on the enemy and friendly situations. The battalion also verifies frequencies and callsigns of the unit in contact. By this time, the ground maneuver battalion has contacted the ground maneuver unit leader in contact to inform him that attack aviation is en route to conduct a hasty attack. Figure 9-8 on page 9-14 shows an example of radio traffic and what may occur.

BATTALION CLOSE FIGHT SITREP
1. **Enemy situation** focusing on ADA in the AO, type of enemy vehicles/equipment position (center mass) and direction of movement if dispersed provide front line trace. 2. **Friendly situation** location of company in contact, mission assigned to them, method of marking their position. 3. **Call sign/frequency verification** 4. **Holding area verification** if intended to be used for face-to-face coordination, a sign counter sign must be agreed upon; i.e. using a light/heat source to provide a recognizable signature, answered by either aircraft IR lights or visible light flashes to signify which aircraft to approach.

Figure 9-7. Battalion close fight SITREP.

Attack Team	Ground Maneuver Battalion
"Bulldog 06 this is Blackjack 26, over."	"Blackjack 26 this is Bulldog 06, L/C, over."
"Bulldog 06, Blackjack 26 enroute to HA at grid VQ 98454287, request SITREP, over."	"Blackjack 26 this is Bulldog 06, enemy situation follows, Hardrock 06 is taking direct fire from a platoon size armor element at grid VQ 96204362, (or reference grid, Echo-2, main entry on Elm) Hardrock 06 elements are established on phase line Nevada center mass VQ 96000050, holding area VQ 94004000 expect radio coordination only, contact Hardrock 06 on FH 478, over."

Figure 9-8. Example radio conversation.

(1) Upon receiving the required information from the ground maneuver battalion, the attack team leader changes frequency to the ground company's FM command net to conduct final coordination before progressing on attack routes to BPs or ABF/SBF positions. Coordination begins with the ground maneuver company commander and ends with the leader of the lowest-level unit in contact.

(2) Regardless of which key leader the attack team leader conducts coordination with, the ground command net is the most suitable net on which both air and ground elements can conduct the operation. It allows all key leaders on the ground, including the fire support team (FIST) chief and the attack team leader and his attack crews, to communicate on one common net throughout the operation. Operating on the command net also allows the attack team to request responsive mortar fire for either suppression or immediate suppression of the enemy. The AH-64 Apache and the AH-1 Cobra are limited to only one FM radio due to aircraft configuration. However, the OH-58 is dual-FM capable, which gives the attack team leader the capability to maintain communications with the ground maneuver company, as well as its higher headquarters, or a fire support element (Figure 9-9).

Attack Team	Ground Maneuver Battalion
"Hardrock 06 this is Blackjack 26 on FH 478, over."	"Blackjack 26 this is Hardrock 06, L/C over."

Figure 9-9. Attack team/maneuver company communications check.

c. **STEP 3, Attack Team Check-In.** Upon making initial radio contact with the ground maneuver unit in contact, the attack team leader executes a succinct check-in (Figures 9-10 and 9-11).

(1) This check-in includes the attack team's present location, which is normally the ground or aerial holding area; the composition of the attack team; the armament load and weapons configuration; total station time; and the night-vision device capability of the attack team. In the event a ground holding area is not used because of METT-TC considerations, the attack team will select and occupy an aerial holding area within FM communications range until all required coordination is complete.

(2) The attack team leader and ground unit's key leaders must consider the effects on friendly forces of the various weapons carried by the attack aircraft prior to target selection and engagement (see figure TBD, SDZ for 30 mm, 2.75, and Hellfire). Weapon systems and munition selection for a given engagement is METT-TC dependent. Point target weapon systems, such as Hellfire or TOW, are the preferred system for armor or hardened targets when engaging targets in the close fight. The gun systems and the 2.75-inch rockets are the preferred system/munition for engaging troops in the open and for soft targets such as trucks and trenchworks. These area fire weapon systems pose a danger to friendly soldiers who may be in the lethality zone of the rounds or rockets. In this case, the leader on the ground must be very precise in describing the target he wants the aircraft to engage.

ATTACK TEAM CHECK-IN

1. Aircraft present location
2. Team composition
3. Munitions available
4. Station time
5. Night vision device capable and type

Figure 9-10. Attack team check in.

Attack Team	Ground Maneuver Battalion
"Hardrock 06, Blackjack 26 is currently holding at grid VQ 98454287, 2 Kiowa Warriors with 450 rounds of .50 cal, 2 Hellfires each, half hour station time, all aircraft are NVG and FLIR capable, over."	
"Blackjack 26, roger."	
	"Blackjack 26, Hardrock 06, stand by, over."

Figure 9-11. Example attack team check in radio conversation.

d. **STEP 4, Coordination for Aviation Close Fires (ACF).** Time is the primary constraining factor for coordinating ACF in the hasty attack. When possible, ACF should be coordinated face to face using the ACF coordination checklist (Figure 9-12, page 9-16), but if time is not available to accomplish face-to-face coordination, then radio only communications will be the means for coordination using the request for immediate ACF (Figure 9-13). Additionally, during an ACF attack, when face-to-face coordination has been conducted, targets of opportunity may require engagement through a target handoff between the ground and aviation elements using the request for immediate ACF. Although face-to-face coordination is preferred, METT-TC will dictate how the coordination between the commander in contact and the attack team leader is accomplished. A major benefit of face-to-face coordination is the ability to talk to the ground commander with a map available and integrate into the ground scheme of maneuver. This also provides an opportunity for the attack team to update their maps with the maneuver battalion's latest graphics.

AVIATION CLOSE FIRES (ACF) COORDINATION CHECKLIST

1. **Enemy situation specific target ID.**
2. **Friendly situation location and method of marking friendly positions.**
3. **Ground maneuver mission/scheme of maneuver.**
4. **Attack aircraft scheme of maneuver.**
5. **Planned engagement area and BP/SBF.**
6. **Method of target marking.**
7. **Fire coordination and restrictions.**
8. **Map graphics update.**
9. **Request for immediate ACF should be used for targets of opportunity or for ground to air target handoff.**

Figure 9-12. ACF coordination checklist.

NOTE: To employ immediate ACF, at a minimum, essential elements from the coordination checklist, **in bold letters**, should be briefed by way of radio as a SITREP by the ground commander.

Attack Team	Ground Maneuver Battalion
	"Blackjack 26, Hardrock 06, stand by for update, friendly platoon in contact located at VQ 96000050, marked by IR strobes, enemy platoon size armor element is 800 meters due north, there has been sporadic heavy machine gun fire and main tank gun fire into our position, fire appears to be coming from road intersection vic VQ 96204362, negative knowledge on disposition of enemy ADA, I'll be handing you down to Hardrock 16 for the ACF request, over."
"Hardrock 06, Blackjack 26, good copy, standing by at HA for ACF request, over."	
	"Roger Blackjack 26, Hardrock 16 request follows, friendly location VQ96000050, 360 degrees to target, 800 meters, 2 T-80s at the road intersection, target location VQ96000850, PAQ-4 spot on, no friendlies north of the 00 grid line, low wires directly over our position, over."
"Hardrock 16, Blackjack elements will attack from the southeast, turn on IR strobes at this time, we will establish a BP to the west of your position 100 meters, over."	
	"Blackjack 26, Hardrock 16, strobes on at this time, over."
"Roger Hardrock, Blackjack has your position, enroute for attack 30 seconds, over."	
	"Hardrock 16, roger."
"Hardrock 16, Blackjack 26, engagement complete, 2 T-80s destroyed, over."	
	"Blackjack 26, Hardrock 16, roger 2 T-80s destroyed, end of mission, out."

Figure 9-13. Example request for immediate ACF.

(1) After receipt of a request for immediate ACF, the attack team leader informs the ground unit leader of the battle position (BP), support-by-fire position (SBF), or the series of positions his team will occupy that provide the best observation and fields of fire into the engagement or target area.

(a) The BP or SBF is a position from which the attack aircraft will engage the enemy with direct fire. It includes a number of individual aircraft firing positions and may be preplanned or established as the situation dictates. Size will vary depending on the number of aircraft using the position, the size of the engagement area, and the type of terrain.

(b) The BP or SBF is normally offset from the flank of the friendly ground position, but close to the position of the requesting unit to facilitate efficient target handoffs. This also ensures that rotor wash, ammunition casing expenditure and the general signature of the aircraft does not interfere with operations on the ground. The offset position also allows the aircraft to engage the enemy on its flanks rather than its front, and lessens the risk of fratricide along the helicopter gun target line.

(2) The attack team leader then provides the ground maneuver unit leader with his concept for the team's attack on the objective. This may be as simple as relaying the direction the aircraft will be coming from or attack route, time required to move forward from their current position, and the location of the BP. Only on completion of coordination with the lowest unit in contact does the flight depart the holding area for the battle position. As the attack team moves out of the holding area, it uses NOE flight along attack routes to mask itself from ground enemy observation and enemy direct fire systems. The attack team leader maintains FM communications with the ground unit leader while he maintains internal communications on either his very high frequency (VHF) or ultra high frequency (UHF) net.

NOTE: Grid locations may be difficult for the ground maneuver, depending on the intensity of the ongoing engagement, and actual FM communications between the ground and air may not work this well.

e. **STEP 5, Battle Damage Assessment and Reattack.** After completing the requested ACF, the attack team leader provides a battle damage assessment (BDA) to the ground maneuver commander. Based on his intent, the ground maneuver commander will determine if a reattack is required to achieve his desired endstate. Requests for ACF can be continued until all munitions or fuel is expended. Upon request for a reattack, the attack team leader must consider the effects on duration and strength of coverage he can provide the ground maneuver commander. The attack team may be required to devise a rearming and refueling plan, maintaining some of his aircraft on station with the unit in contact, while the remainder returns to the forward arming and refueling point (FARP). In addition to coordinating with the ground maneuver unit in contact, the attack team leader is required to coordinate this effort with his higher headquarters.

9-8. EMPLOYMENT OF ASSAULT/CARGO HELICOPTERS

Assault and cargo helicopters play a crucial role in UO. Rapid mobility of ground forces and equipment is essential in all urban combat operations. In battle in an urban area, troop movement may become a major requirement. Units engaged in house to house fighting normally suffer more casualties than units fighting in open terrain. The casualties must be evacuated and replaced quickly with new troops. At the same time, roads are likely to be crowded with resupply and evacuation vehicles, and may also be blocked with craters or rubble. Helicopters provide a responsive means to move troops by flying nap of the earth (NOE) flight techniques down selected streets already secured and cleared of obstacles. Aircraft deliver the troops at the last covered position short of the fighting and then return without exposure to enemy direct fire. Similar flight techniques can be used for air movement of supplies and medical evacuation missions. High risk-high payoff missions, like the air assault, may similarly be conducted. Air assaults into enemy-held urban territory are extremely difficult. One technique is to fly NOE down a broad street or commercial ribbon while attack helicopters and door gunners from lift helicopters suppress buildings on either side of the street. Scheduled artillery preparations can be incorporated into the air assault plan through the H-hour sequence. Feints and demonstrations in the form of false insertions can confuse the enemy as to the real assault landings.

a. **Small-Scale Assaults.** Small units may have to be landed onto the rooftop of a key building. Success depends on minimum exposure and the suppression of all enemy positions that could fire on the helicopter. Inspections should be made of rooftops to ensure that no obstacles exist, such as electrical wires, telephone poles, antennas, or mines and wire emplaced by the enemy, that could damage helicopters or troops. In many modern cities, office buildings often have helipads on their roofs for landing helicopters. Other buildings, such as parking garages, are usually strong enough to support the weight of a helicopter. Insertion/extraction techniques to rooftops include:

- Remaining light on the landing gear after touchdown.
- Hovering with a single skid or landing gear touching the structure.
- Rappelling.
- Jumping from the aircraft.
- Using the fast rope.
- Using the special purpose insertion/extraction system (SPIES).
- Hoisting.

b. **Large-Scale Assaults.** For large-scale air assaults, rooftop landings are not practical. Therefore, open spaces (parks, parking lots, sports arenas) within the urban area must be used. Several spaces large enough for helicopter operations normally can be found within 2 kilometers of a city's center. Most major cities have urban parks near the central business district. Other potential LZs include athletic stadiums and parking lots. These areas are often suitable LZs, if secure.

(1) ***Technical Considerations for LZ/PZ Selection.*** The technical considerations for urban LZ/PZ selection are similar to those in natural terrain. The employment of Pathfinders or ground security elements is invaluable in the preparation of the LZ. Air traffic services (ATS) units can provide tactical air control team (TACT) support for LZs/PZs and they can also provide critical information such as LZ condition, enemy situation, and landing data. Communications should be established with ground elements at the LZ as early as possible during ingress.

(2) ***Tactical Considerations for LZ/PZ Selection.*** The tactical considerations for LZ/PZ selection, mission, location, and security are exaggerated in urbanized terrain. Missions may dictate the number and type of aircraft required and, therefore, the size of LZ selected. The opposite may also be true—the physical limitations of available LZs dictate the number of aircraft used. If there are more aircraft than a single LZ can accommodate, multiple LZs near the objective should be selected. Control measures must be adequate to safely deconflict the movement of all elements. Formations of helicopters should not be larger than the ability to simultaneously land on the LZ. All secondary or alternate LZs should be the same size (or larger) to prevent unnecessary stack-ups of exposed aircraft waiting to land. LZs should be located as close to the objective as

possible to enable swift accomplishment of the ground mission. Generally, the more difficult the LZ and approach/departure paths are to observe from the ground or surrounding structures, the more easily the area can be secured.

9-9. AVIATION URBAN OPERATIONS RISK ASSESSMENT

Risk assessment, as a step in the military decision-making process (MDMP), must identify and assess unique hazards associated with aviation UO. The following list is not all-inclusive, but it provides a good starting point in identifying possible UO hazards.

a. **Fires.**

(1) ***Weapon Utilization.*** Minimum arming range and minimum slant ranges within urban areas limit the usage of some weapons.

(2) ***Coordination.*** Heavy concentration of precision weapon systems along a narrow front may cause coordination problems.

(3) ***Laser Designation.*** Multiple flat, polished surfaces in an urban area may degrade laser use, thereby rendering some weapon systems useless.

(4) ***Limited Close Air Support.*** Heavily developed urban centers will limit the close air support. This is due to vulnerabilities at higher altitudes, thus placing a greater dependence on rotary-wing aircraft support.

(5) ***Need for Direct and Indirect Suppressive Ground Fires.*** Direct and indirect suppressive ground fire should augment the escort suppressive fires as helicopters approach intended landing zones.

(6) ***ROE.*** Operations could be in areas with a high concentration of civilians or cause significant collateral damage of property.

(7) ***Weapons Effects.*** Certain weapons may produce less or more damage on structures. Specific knowledge of weapons effects is critical to target destruction, the reduction of unnecessary collateral damage, and overall mission accomplishment. In addition, collateral damage may introduce the risk of exposure to toxic industrial materials (TIM).

b. **Threat.**

(1) ***Small Arms.*** There is a high risk to aircraft from close-range, small arms fires that is complicated by the close proximity friendly forces and non-combatants.

(2) ***Exposure to Direct Fire.*** Small-arms fire may increase risk, as the urban area forces concentration of units and provides excellent cover and concealment.

(3) ***Individual Soldiers.*** The individual soldiers may be the greatest threat to rotary-wing aircraft in an urban environment.

(4) ***Detection of ADA.*** Portable surface to air missile systems will be difficult to detect in and among the buildings.

(5) ***Predictable Landing Zones.*** As landing zones may be scarce and therefore predictable, air assault operations in mass may be vulnerable to enemy fires.

c. **Navigation.**

(1) Most maps do not show the vertical development of the urban terrain.

(2) Accurate navigation in the urban environment requires familiarization with current military city maps which pilots have limited experience.

(3) The planning of missions will depend heavily on the upgraded photo-imagery products to make accurate assessments of key features.

(4) Newly developed areas and buildings may not appear on a navigational map.

(5) The numerous buildings and streets and few map references complicate navigation over built-up areas. Flight routes over urban terrain may increase employment time and fuel consumption.

d. **Weather.**

(1) Continuous smoke and fire within the buildings cause obscuration.

(2) Urban areas directly affect weather, especially wind patterns.

(3) Night-vision systems are degraded due to city lights and thermal inversion.

e. **Terrain.** Obstacles may be more numerous and dangerous than in any other environment. These obstacles will include, but are not limited to, the following:

(1) Buildings limit maneuverability and engagement ranges.

(2) Urbanized terrain masks intelligence and electronic warfare acquisition capabilities.

(3) Landing and pickup zones may be severely limited; operations from rooftops may be required.

(4) Vertical development blocking line-of-sight radio communication will severely affect air-to-ground and low-level air-to-air communication.

CHAPTER 10
FIRES

Coordinating lethal and nonlethal effects are an extremely important function for fire support officers (FSOs) at all levels. This chapter provides planning considerations and practical TTP for FSOs at the brigade, battalion, and company levels.

10-1. BRIGADE FIRE SUPPORT FOR URBAN OPERATIONS

Fire support (FS) is the use of fires delivered from various means (armed aircraft, land-based or sea-based indirect fire systems, and electronic warfare systems) against ground targets to support land combat operations at both the operational and tactical levels. The direct support (DS) field artillery (FA) battalion commander is the fire support coordinator (FSCOORD) for the supported brigade. As such, he is specifically responsible for fire support planning and coordination.

a. The fire support system is made up of five distinct components that function together to give the brigade commander the fire support he needs to accomplish the mission. The components of the fire support system are:

- Command, control, communications, computers and intelligence systems.
- Target acquisition systems.
- Both lethal and non-lethal FS means to support the concept of operations.
- Combat service support.
- Trained personnel in the above areas.

The key to effective FS is the brigade commander's ability to bring the five components of the system to bear throughout his battle space to strike the enemy in an integrated and coordinated manner that is synchronized with the concept of operations. This is the current doctrine of fire support. (Refer to Chapter 2 of FM 6-20-40.)

b. In UO, however, effects must be used as the standard to ensure the five components of fire support adequately supports each phase or mission and defeats the challenges and restrictions presented by urban missions. Effects are determined by munitions, and munitions are limited by delivery platforms. As munitions advance, smart and brilliant munitions with increased range and lethality provide precision and terminal effects on the target. UO argues for allocation of delivered effects rather than allocation of delivery platforms. The selection of a delivery system should reflect, based solely the nature and priority of the target, the effects to be achieved and the rapidity with which they must be produced.

10-2. COMMAND, CONTROL, COMMUNICATIONS, COMPUTERS AND INTELLIGENCE (C4I)

The first component of the fire support system deals with effects. The mix of platforms and munitions brings a wide range of desired effects to the fight. The critical function of automation in the effects management process simplifies and accelerates the management decision. The real time status visibility over all relevant lethal and nonlethal delivery capabilities, together with automated target filtration, target/effects matching, and information presentation mechanisms permitting rapid tactical comparison of available effects against competing requirements, falls on the FSO. The implication for the FSO is that he must provide the commander robust support from available fires platforms, which are not necessarily *tucked in their back pocket*, but are positioned to best deliver the required effects. Fires platforms can be positioned to optimize the ability of the total system to apply quick and decisive overmatching effects, when and where needed. One of the key organizational enablers that allows the fires community to migrate towards an "effects-based" fires construct is the effects coordination cell (ECC). The ECC is tailored to the mission and controls sensors and delivery systems organic to the force, and coordinates the joint and combined systems resourced to the force.

10-3. MISSION SUPPORT OF OFFENSIVE AND DEFENSIVE OPERATIONS

Fire support planning considerations for offensive and defensive UO are discussed in this paragraph.

a. **Offensive Considerations.** The mission flow for a brigade combat team (BCT) conducting offensive military operations in UO generally includes moving some distance from a line of departure to an urban area. This mission includes:

- Breaching obstacles to enter the urban area.
- Gaining a foothold.
- Defeating enemy forces and seizing a designated area.
- Conducting a follow-on mission.

For the FSO this means integrating fires into a scheme of maneuver involving a battalion task force or larger BCT. This integration includes:

- Supporting movement to contact or air assault (or combination of the two).
- Supporting breaching operations.
- Supporting a deliberate attack to seize objectives in a city or town.
- Providing fires for a follow-on mission.

For the fire support system, the fight begins with fires setting the conditions for interdiction fires 24 hours prior to disrupting enemy forces preparing their defense. This fight continues when units cross the line of departure (LD) rather than at the breach site or in the city or town. Fire support planning for missions involving a deliberate attack on urban terrain objectives must include synchronization of fires during the fight from the LD to the breach site.

b. **Offensive Support.** In the offense, fire support plans should include fires to isolate the objective area, support the assault, and support the clearing action. Fires are delivered to isolate and fix the enemy and deny him the use of avenues of approach into and out of the urban area. Fires are employed to rupture the enemy's established defenses and screen friendly maneuver, in order to maintain the momentum of the attack. Fire support is also allocated to units involved in clearing operations. Fire support plans should incorporate the employment of aerial observers and unmanned aerial vehicles (UAVs) to compensate for restrictions to observation and to assist in the delivery of deep fires. Procedures for designating the forward line of own troops (FLOT), marking targets, shifting fires, and communicating in the urban environment should also be considered.

c. **Defensive Considerations.** The mission flow for a BCT conducting defensive UO generally includes fires to disrupt and slow the enemy attack. This mission includes:

- Delaying and disrupt armor and mechanized infantry columns.
- Canalization of armor and mechanized infantry.

- Destruction of infrastructure, communications, power, transportation, water, sewer, and so on.

For the FSO this means integrating fires into a defensive plan. The objective of the defense could be to delay and give the force time to execute a rearward movement, or it could be to save the infrastructure of the built up area for future operations. In either case the integration considerations include:

- Fires delivered at maximum ranges along avenues of approach.
- Fires to separate armor and infantry forces.
- Fires to canalize the enemy into killing zones.
- Fires to deny the enemy use of key facilities in the city, such as communications and transportation facilities, or other infrastructure of an urban area.

For the fire support system, the fight relies on external or available assets more than internal and organic assets, though not to their exclusion. Depending on the situation and availability, use of Naval and Air Force assets will provide extended ranges necessary, and prevent any unintended or undesired effects on infrastructure from overpressures or maneuvering artillery in the urban area. Additionally use of available assets does not reveal capabilities the commander has organically, adding surprise and depth to the urban fight from the viewpoint of the aggressor.

d. **Defensive Support.** In the defense, fire support plans address fires to disrupt and slow the enemy attack. Fires are delivered at maximum ranges along avenues of approach to separate armor and infantry forces, to canalize the enemy into killing zones, and to deny the enemy use of key facilities in the city, such as communications and transportation facilities. Defensive fires planned at lower echelons support fighting within the urban area. Defensive fire support plans identify fire support coordination procedures necessary to execute the delivery of fire support.

e. **Assets and Urban Fire Support.** Battle calculus must determine effects (Figure 10-1, page 10-4), which in turn determine systems, ammunition requirements, and coordination measures for fires to support missions like suppress, obscure, secure, reduce obstacles, deny, delay and damage. During urban combat, fire support planning must address unique challenges created by urban operations. They include effects affected by:

- Buildings and structures of varying heights.
- Rules of engagement (ROE) restricting use of indirect fires.
- Noncombatants.
- Observer inability to locate and observe enemy targets.
- Future actions or need to preserve infrastructure.

Step 1	Translate commander's guidance into quantifiable effects required to achieve target defeat criteria. The FSO should consult with the S2 for assistance in translating the commander's guidance into a quantifiable number of enemy soldiers and equipment that should be in the target area.
Step 2	Equate the required effects to the required ammunition. Determine from a munitions effects database the number of rounds required to accomplish the essential FS task (suppress, neutralize, or destroy).
Step 3	Determine the available time to conduct attack. Based on the rate of fire of the selected system(s), determine how long it will require to execute the attack.
Step 4	Determine if required ammunition can be delivered in time available.
Step 5	Determine max volleys that can be fired at one target location. Based on the facts and assumptions determined in Step 1, determine the time required for a moving target to pass through a specific point. Determine the number of volleys than can be delivered in that timeframe.
Step 6	Determine the number of attacks needed to deliver required ammunition.
Step 7	Determine if time and space are available to execute attacks.

Figure 10-1. Battle effects calculus.

f. **Planning.** Fire support should be planned through the depth and breadth of the zone of attack for each branch and sequel. For fire support planners (primarily the fire support coordinator [FSCOORD]), brigade and battalion FSOs can provide valuable assistance during the planning process. Additional coordination with the FA battalion S3 and or XO, aviation, ALO, and joint may be required depending on circumstances.

(1) Planning and coordinating fire support for a complex scheme of maneuver must be completed before units cross the line of departure (LD) to conduct air assaults or a movement to contact or approach march en route to the final objective. Planned and synchronized fire support during the movement to contact/approach march toward a town are as important as the fires provided during the attack into the town because they enable the commander to arrive at the objective with maximum maneuver combat power.

(2) During the fight in the city, positioning of FA units, counterfire radar, and observers becomes critical. Ammunition resupply for special munitions (Copperhead) and sustained fires could possibly exceed the FA unit transportation capacity. Planning should also include actions necessary to rapidly transition to the follow-on phase or mission.

(3) The checklists in Figures 10-2 through 10-5, on pages 10-6 through 10-15, provide urban planning considerations broken down by echelon—brigade, battalion, company, and FA battalion S3. Within each echelon, the calculus considerations follow an offensive sequence but could be tailored for defensive sequences.

g. **Effects—Ammunition Considerations.** Planners must understand ammunition effects to achieve desired results and avoid undesired results. Special consideration must be given to shell and fuze combinations when effects of munitions are limited by buildings or ROE. Special considerations must be given to the delivery platform because of angle of fall, canalization of rounds, and blockage of one target by another building. Illuminating or obscuring an enemy position degrades the ability to see him more than his ability to employ his weapons. Other ammunition considerations are:

(1) Mortar smoke is white phosphorous (WP) incendiary.

(2) Variable time (VT) fuzes help clean off building tops, but varying heights of surrounding buildings may cause premature detonation. Observer-adjusted time fuzes may be better against targets among buildings of varying heights.

(3) The MK 399 Mod 1 PD/delay fuze for UO is available for all artillery units. It provides improved penetration through wood frame, reinforced concrete, and multilayered brick structures for high-explosive artillery projectiles. In the UO battle, one role of the artillery is to supply indirect fire, which will typically be at high angles. The objective using the MK 399 Mod 1 fuze would be to penetrate a building or structure and achieve blast effects inside of the target using its delay mode, or to rubble the structure using the PD mode. If a projectile fired with an MK 399 Mod 1 fuze impacts a substantial object (either intentionally or unintentionally), a high-order explosive function may result even when the object is located inside the 400-caliber minimum arming distance from the weapon (138 feet for 105-mm and 203 feet for 155-mm weapon systems). FM 6-50, Chapter 6 clearly outlines a very deliberate and necessary process required to determine the minimum quadrant elevation to safely fire the howitzers. Field artillery units must follow those procedures to ensure targets are not engaged (either in the indirect or direct fire mode) before the round has reached the 400-caliber minimum arming distance from the weapon. Point detonating (PD) fuzes on delay only allow penetration of the first wall or roof.

(4) Calculated minimum safe distances (MSDs) are no longer accurate. Buildings provide cover that reduces MSDs to a few meters. Inaccurate or stray fires can be attributed to projectiles careening or skipping off tall buildings, towers, cables, and so forth.

(5) In the offense, illuminating rounds are planned to burst above the objective to put enemy troops in the light. However expended canisters may fall and cause injuries or hazards undesired and unintended. If the illumination occurs behind the objective, the enemy troops would be in the concealment of shadows rather than in the light. In the defense, illumination is planned to burst behind friendly troops to put them in the shadows and place the enemy troops in the light. Buildings reduce the effectiveness of the illumination by creating uncontrolled shadows and inject the undesirable hazard of fire, which must be put out or left to burn uncontrolled. Use of illumination and obscuring fires on UO objectives favor the defender. Also, the effects of using WP or Illumination may create unwanted smoke screens or limited visibility conditions that could interfere with the tactical plan. WP may also have significant choking effects on personnel, and may cause urban fires.

(6) During suppression, obscuration, securing, and reduction (SOSR) or preparation fires, accurately adjusted, concentrated artillery fire (HE fuzed with quick and delay) at breach sites is effective in obstacle reduction. These fires significantly weaken wire obstacles with mines and booby traps. They will not significantly affect metal tetrahedrons or concrete dragon's teeth.

(7) Careful use of VT is required to avoid premature arming. VT, TI, and ICM are effective for clearing enemy positions, observers, and antennas off rooftops. Fuze delay should be used to penetrate fortifications. Indirect fires may create unwanted rubble. The close proximity of enemy and unfriendly troops requires careful coordination. SCATMINE may be used to impede enemy movements. SCATMINE effectiveness is

reduced when delivered on a hard surface. When using HE ammunition in urban fighting, only point detonating fuzes should be used. The use of proximity fuzes should be carefully considered, because the nature of urban areas may cause proximity fuzes to function prematurely. Proximity fuzes, however, are useful in attacking targets such as OPs on tops of buildings.

MISSION ANALYSIS/BATTLE CALCULUS

_____What is the mission flow for maneuver units' movement to contact, air assault, breaching operations, deliberate attack into city for each COA?
_____What is the FA organization for combat?
_____What other assets are available (mortars, attack aviation, CAS, NGF, AC-130)?
_____Prepare asset matrix listing all artillery and other assets, ranges of each, ammunition available, time available, controlling HQ.
_____Essential fire support tasks (EFSTs) by mission for each COA? Refine "method" of EFSTs--highlight special ammunition requirements: SCATMINE, dimensions/duration of smoke, preps, destruction/reduction fires.
_____HPTs? According to S2 collection plan, what will find each HPT?
_____Calculus--are more assets needed to accomplish all EFSTs and attack each HPTs? (By mission for each COA)
_____By mission for each COA, draft fire support plan = fires paragraph___, FSEM___, target list scheme of fires and priority of fires worksheet___.
_____Does the S2 R&S plan cover each HPT by mission for each COA?
_____Plan critical friendly zones (CFZs).
_____Brief draft fire support plan for each COA during war-gaming and COA selection.
_____Upon COA selection, send fire support WARNO (fires paragraph, FSEM, target list, TSM) to subordinate FSOs and DS battalion S3.
_____Finalize plan/clean up products/add to order.
_____Rehearse.

FIGHT FROM LD TO BREACH SITE

_____Disseminate the friendly scheme of maneuver and EFSTs for this phase to FS and FA leaders.
_____Identify all fire support assets available for this phase.
_____What are probable locations and azimuths of fire of enemy indirect fire systems?
_____Identify counterfire radar positions that give the best aspect to detect enemy indirect fire.
_____What FA unit has the counterfire mission?
_____Are maneuver forces tasked to locate and destroy enemy mortars?
_____By shell/fuze type, how much FA ammunition is needed to fire scheduled/ preplanned fires? How much is available for emergency missions?
_____What is the communications link to each asset tasked to assess effects on each HPT attacked by FA or another fire support asset?
_____Will FA units displace during this phase? What is the trigger?
_____Positioning of DFSCOORD/Bde FSE/ALO/COLTS during this phase.
_____Other information needed by the battalion FSOs and DS FA, S3 from the brigade FSE?

Figure 10-2. Example brigade fire support officer planning checklist.

BREACHING OPERATIONS

_____What are the indirect fire rules of engagement?

_____Disseminate the friendly scheme of maneuver and EFSTs for this phase to FS and FA leaders.

_____What fire support assets are available?

_____By shell/fuze type, how much FA ammunition is available for SOSR fires? How much is needed to fire all scheduled/preplanned fires?

_____Specify who is controlling SOSR fires. Specifically at the main breach point, have a primary and alternate observer.

_____Where are COLTs positioned? TACPs?

_____ Is there a deception breach?

THE URBAN FIGHT

_____What are the indirect fire rules of engagement? What is on the restricted target list?

_____Disseminate the friendly scheme of maneuver and EFSTs for this phase to FS and FA leaders.

_____Determine FS assets available for this phase. Who controls each?

_____Specify who positions COLTs.

_____Where will FA units and counterfire radar be positioned?

_____Determine radar zones and cueing agents needed in the objective city.

_____Identify the locations of underground fuel and industrial storage tanks, gas distribution lines, storage tanks, and gas lines above ground (locations needed for friendly unit warning since below-MSD fires may produce secondary explosions).

_____Determine how the enemy is reinforcing buildings--sandbagging rooftops and upper floors, adding internal bracing/structural support, sandbagging walls.

_____Determine which maps will be distributed to FS and FA personnel. Map references must be the same as numbers assigned to specific buildings.

_____ Determine how fire support personnel determine 8-digit grid coordinates with altitudes to targets in urban areas.

_____Identify the general construction or composition of buildings, road surfaces, and barrier obstacles that require breaching. Identify buildings that have basements.

_____Identify buildings or structures requiring large-caliber weapon/howitzer direct fire before assaulting.

_____Locate the dead space areas where tall building masking prevents indirect fire from engaging targets. Locate "urban canyon" areas where aircraft cannot engage targets between tall buildings.

_____Identify buildings that provide the best OPs for enemy and friendly observers. Identify buildings providing vantage points for employment of laser designators.

_____Locate possible firing points for 81-/82-/107-/120-mm mortars, for towed howitzers, for SP howitzers. Which positions permit 6400-mil firing?

_____Identify enemy mortar capability. Does enemy have a large number of 60-mm of smaller "knee" mortars?

_____Identify areas of the city that are likely to be affected by the incendiary effects of detonating artillery and mortar rounds.

_____Determine the best positions outside the objective city for employing G/VLLDs and other ground designators.

_____Identify targets and trigger points for blocking fires outside the city.

_____Have the effects of certain weapon systems and munitions available (Hellfire, Copperhead, Maverick, 155 DPICM, VT, CP, and so forth).

_____ Integrate TF mortars into the scheme of fires.

_____ Plan and refine CFZs.

Figure 10-2. Example brigade fire support officer planning checklist (continued).

THE FOLLOW-ON MISSION
_____Disseminate the friendly scheme of maneuver and EFSTs for the follow-on mission (or sustained combat and occupation in the objective city) to FS and FA leaders. _____Identify fire support assets available for follow-on missions. _____Identify ammunition requirements for follow-on missions. _____Determine optimal FA and radar position areas to support follow-on missions.

Figure 10-2. Example brigade fire support officer planning checklist (continued).

MISSION ANALYSIS/BATTLE CALCULUS
_____Review the brigade fire support WARNO. _____What is the mission flow for the battalion--movement to contact, air assault, breaching operations, deliberate attack into city? _____What fire support assets are available to support the battalion (FA, mortars, attack aviation, CAS , NGF, AC-130)? _____Prepare a fire support asset matrix (nondoctrinal) listing all artillery and other systems, ranges of each, ammunition available, time available, and controlling HQ. _____Identify essential fire support tasks (EFSTs) for battalion fire support personnel for each phase of the mission. _____What are the HPTs? What asset is tasked to find each of the HPTs? What asset is tasked to assess effects when an HPT is attacked? _____Determine how much ammunition by shell/fuze type is needed to accomplish all scheduled or preplanned fires. How much is available for emergency missions? _____Determine if enough fire support assets are available to attack all HPTs and provide on-call fire support during each phase of the mission. What additional assets are needed? _____Identify special ammunition requirements (SCATMINE, Copperhead, dimensions/duration of obscuration fires, DPICM, concrete piercing fuzes, preparations, reduction/destruction fires). _____Develop a communication plan to defeat range and compatibility problems between fire support personnel and FA units or other assets during each phase. _____Identify special equipment needs, especially for breaching operations and the fight in the city--COLT or other laser designator, climbing rope, wire gloves, axes or sledge hammers, kneepads, goggles. _____Determine what types of maps fire support personnel will use. (During the fight in the city, fire support personnel must be able to locate targets by 8-digit grid coordinates.) _____Develop observer plan for each phase--observer positioning and observer/target link-up--which should include primary, backup observer and trigger. _____Develop and disseminate products (fires paragraph, FSEM, target list, TSM) to subordinate FSOs, battalion mortars, DS FA battalion S3, and other supporting fire support elements. _____Conduct fire support rehearsal, and participate in FA technical rehearsal.

Figure 10-3. Example battalion fire support officer planning checklist.

FIGHT FROM LD TO BREACH SITE
_____Disseminate battalion scheme of maneuver and EFSTs to FS and FA leaders.
_____Identify all fire support assets available and method of control of each.
_____Does the battalion have priority of fires? Allocate priority targets and FPFs to companies.
_____Plan targets on known and suspected enemy positions and obstacles along the route (SEAD if conducting an air assault).
_____Ensure all reconnaissance elements are included in the fire support plan.
_____By shell/fuze type, how much FA ammunition is needed to fire scheduled/pre-planned fires? How much is available for emergency missions?
_____What is the communication link to each fire support asset supporting the battalion?
_____What is the communication link to each asset tasked to assess effects on each HPT attacked by FA or another fire support asset?
_____Will FA units displace during this phase? What is the trigger?
_____What is the battalion mortar employment plan?
_____Positioning of the battalion FSE/ALO.
_____Other information needed by the company FSOs, DS FA battalion S3, FSE, and brigade FSE.

THE BREACHING OPERATION
_____What are the indirect fire rules of engagement?
_____Disseminate the battalion scheme of maneuver and EFSTs for this phase to FS and FA leaders. Is the battalion mission inside (breach, clear, and secure in city) or outside (isolate) the objective city?
_____What fire support assets are available?
_____What type of breaching operation is being conducted? Time required?
_____Specify who is initiating and controlling SOSR fires.
_____By shell/fuze type, how much FA ammunition is available for SOSR fires? How much is needed to fire all scheduled/preplanned fires? What are the dimensions/duration of obscuration fires?
_____Develop the air/ground observer plan to adjust obscurants.
_____Positioning of the battalion FSE/ALO or TACP/COLT, if under battalion control.
_____What is the communication link between all FA units or fire support assets and the observers controlling the assets?
_____Is the control of any fire support asset being handed over from one observer to another? What is the trigger point to initiate the handover?

THE URBAN FIGHT
_____What are the indirect fire rules of engagement?
_____Disseminate the battalion scheme of maneuver and EFSTs for this phase to the FS and FA leaders.
_____Determine who controls each fire support asset.
_____Exchange fire plan and observer plan with adjacent battalions.
_____Determine exact locations for battalion mortars; submit to brigade FSE.
_____Develop observer plan; identify special requirements (laser designator positions, observer positions to overwatch trigger points, observer positions in tall buildings).
_____Disseminate maneuver graphics to FS and FA leaders to preclude fratricide.
_____Identify uses of obscurants in city.
_____Identify targets and trigger points for interdiction fires against counterattack force.

Figure 10-3. Example battalion fire support officer planning checklist (continued).

_____Identify the locations of underground fuel and industrial storage tanks, gas distribution lines, storage tanks, and gas lines above ground (locations needed for friendly unit warning because below-MSD fires may produce secondary explosions).
_____Determine how the enemy is reinforcing buildings--sandbagging rooftops and upper floors, adding internal bracing/structural support, sandbagging walls.
_____What maps are battalion fire support personnel using? How is the maneuver building numbering system going to be translated into 8-digit grid coordinates for building locations?
_____Identify the general construction or composition of buildings, road surfaces, and barrier obstacles that require breaching. Identify buildings that have basements.
_____Identify buildings or structures requiring large-caliber weapon/howitzer direct fire before assaulting. Will an escalating response matrix be used?
_____Locate dead space areas where tall building masking prevents indirect fire from engaging targets. Locate "urban canyon" areas where aircraft cannot engage targets between tall buildings.
_____Identify buildings providing the best OPs for friendly and enemy observers. Identify buildings providing vantage points for employment of laser designators.
_____Locate firing points for battalion mortars and supporting howitzers. Which positions provide 6400-mil firing capability?
_____Identify areas of the city most likely to be affected by the incendiary effects of detonating artillery and mortar rounds.
_____Identify routes/roads in the objective city that permit/do not permit artillery convoy (prime mover, howitzer, ammunition carrier) travel.
_____Identify buildings/structures capable of hiding artillery prime movers, howitzers, and ammunition carriers.
_____Do enemy forces in the city use or have access to laser designators, pointers, spotlights, or other light sources that may be used to incapacitate observation devices and NVGs?
_____Where are radio communications dead spaces? Is a communications visibility plot available?
_____Determine where use of obscurants will favor friendly forces, and where it will favor the enemy.
_____Determine where building masking, overhead power lines, structures or towers will degrade GPS accuracy.
_____Will electrical lines in the objective city be "hot?"
_____Will dense/congested structures containing metal and electrical lines affect compasses and gyro-based directional equipment?
_____Determine weather effects in and around the objective city--low industrial fog and smoke; updrafting winds caused by tall, congested buildings; temperature increase caused by buildings/pavement/industrial activity.
_____Will there be a need for artillery illumination?
_____Determine likely enemy azimuths of fire for indirect systems.
_____Will friendly local or US/allied personnel with in-depth knowledge of the objective city layout be available to accompany/assist fire support personnel?
_____If required, could observers conduct howitzer/mortar registration?
_____What is the sniper threat against fire support personnel, especially those occupying OPs in tall buildings? What is the mine/booby trap threat?
_____Will buildings or structures require fire support personnel to carry/use equipment not normally carried--field expedient antennas, climbing rope, wire gloves, axes or sledge hammers, kneepads, goggles, or B/LPS?
_____Will enemy forces attempt to limit friendly use of indirect fires by using civilians as "human shields?"

Figure 10-3. Example battalion fire support officer planning checklist (continued).

THE FOLLOW-ON MISSION
_____Disseminate the battalion scheme of maneuver and EFSTs for the follow-on mission (or sustained combat and occupation in the objective city) to FS and FA leaders.
_____Identify fire support assets available and ammunition requirements for follow-on missions.

Figure 10-3. Example battalion fire support officer planning checklist (continued).

MISSION ANALYSIS/BATTLE CALCULUS
_____What is the mission flow for the company-movement to contact, air assault, breaching operations, deliberate attack into city?
_____What fire support assets are available to support the company (FA, battalion and company mortars, attack aviation)?
_____Prepare a fire support asset matrix listing all artillery and other systems available, ranges of each, ammunition available, time available, and controlling HQ.
_____Refine essential fire support tasks (EFSTs) for company fire support personnel for each phase of the mission.
_____Where are the HPTs in the company sector? What asset will find each HPT? What asset will assess effects when an HPT is attacked?
_____Determine if enough fire support assets are available to attack all HPTs and provide on-call fire support during each phase of the mission. What additional assets are needed?
_____Determine how much ammunition by shell/fuze type is needed to accomplish all scheduled or preplanned fires. How much is available for emergency missions?
_____Identify special ammunition requirements (SCATMINE, Copperhead, dimensions/duration of obscuration fires, DPICM, concrete piercing fuzes, preps, reduction/destruction fires).
_____Develop radio plan to talk to platoon FO parties, the battalion FSE, supporting FA units, mortars, and other assets. This plan must defeat range and compatibility problems. Is planning digital and execution voice?
_____Identify special equipment needs for fire support personnel, especially for breaching operations and the fight in the city--COLT or other laser designator, climbing rope, wire gloves, axes or sledge hammers, goggles or B/LPS.
_____Determine how observers will locate targets by 8-digit grid coordinates during the fight in the city.
_____Develop observer plan for each phase--observer positioning and observer/target link-up.
_____Participate in fire support and FA technical rehearsals.

Figure 10-4. Example company fire support officer planning checklist.

FIGHT FROM LD TO BREACH SITE
_____Disseminate company scheme of maneuver and EFSTs to FS and FA leaders.
_____Identify fire support assets available and method of control.
_____How many priority targets and FPFs does the company have?
_____Plan targets on known and suspected enemy positions and obstacles along the route.
_____By shell/fuze type, how much FA and mortar ammunition is needed to fire scheduled/preplanned fires? How much FA and mortar ammunition is available for emergency missions?
_____What is the communication link to each fire support asset supporting the company?
_____What is the communication link to each asset assessing effects on each HPT in the company sector?
_____What are the battalion and company mortar employment plans?
_____Positioning of the company FIST.
_____Other information needed by platoon FOs, company and battalion mortars, DS FA battalion S3, and FSE.

THE BREACHING OPERATION
_____What are the indirect fire rules of engagement?
_____Disseminate company scheme of maneuver and EFSTs for this phase to the FS and FA leaders.
_____What fire support assets are available?
_____What type of breaching operation is being conducted? Time required?
_____How are SOSR fires initiated and controlled?
_____By shell/fuze type, how much FA and mortar ammunition is needed for SOSR fires? For all scheduled/pre-planned fires? What are the dimensions/duration of obscuration fires?
_____What is the communication link between FA units and mortars and the observers controlling the assets?
_____Is the control of any fire support asset being handed over from one observer to another? What is the trigger point to initiate the handover?

THE URBAN FIGHT
What are the indirect fire rules of engagement?
_____Disseminate the company scheme of maneuver and offsets for this phase to the FS and FA leaders.
_____Determine who controls each fire support asset.
_____Exchange fire plan and observer plan with adjacent companies.
_____Determine how company mortars will be employed (direct lay or deliberate emplacement), firing points, and azimuths of fire. Pass to battalion FSE for consideration during development of the radar deployment order.
_____Develop observer plan--OPs in specific buildings, location of laser designators, overwatch of trigger points, and so forth.
_____Identify locations of hazardous sites--below and above-ground fuel and industrial storage tanks, gas distribution lines, and so forth., that may produce secondary explosions caused by detonating mortar or artillery rounds.
_____Identify which buildings or structures the enemy is fortifying--sandbagging the rooftop or upper floors, adding internal bracing/structural support, sandbagging walls.

Figure 10-4. Example company fire support officer planning checklist (continued).

_____Identify method fire support will use to identify targets using 8-digit grid coordinates (city map of maneuver building diagram versus military tactical map with UTM grid coordinates). 8-digit grid coordinate accuracy is needed for engaging targets in a city.
_____Identify the general construction or composition of buildings, road surfaces, and barrier obstacles that require breaching. Identify buildings with basements.
_____Identify buildings or structures requiring large-caliber weapon/howitzer direct fire before assaulting. Will an escalating response matrix be used?
_____Locate dead space and "urban canyon" areas where tall-building masking prevents indirect fire and aircraft from engaging targets.
_____Identify buildings providing the best OPs for friendly and enemy observers and employment of laser designators.
_____Locate firing point for company mortars and howitzers. Do they allow 6400-mil firing capability?
_____Identify areas of the city where incendiary effects of detonating artillery and mortar rounds will start fires.
_____Identify routes artillery convoy (prime mover, howitzer, and ammunition carrier) travel, and buildings capable of hiding this equipment.
_____Does the enemy posses laser designators, pointers, spotlights, or other light sources capable of incapacitating observation devices and NVGs?
_____Where are radio communication dead spaces?
_____Where does building masking, overhead power lines, structures, or towers degrade GPS, gyro-based directional devices, and compass functioning?
_____Will use of obscurants and artillery or mortar illumination favor friendly units or the enemy?
_____Will friendly local or US/allied personnel with in-depth knowledge of the objective city layout accompany or assist fire support personnel?
_____If required, could observers observe howitzer/mortar registrations?
_____What is the sniper threat against fire support personnel occupying OPs in tall buildings? What is the mine/booby trap threat?
_____Will buildings or structures require fire support personnel to carry/use equipment not normally carried-field expedient antennas, climbing rope, wire gloves, axes or sledge hammers, kneepads, goggles?
_____Will enemy forces attempt to limit friendly use of indirect fires by using civilians as "human shields?"

THE FOLLOW-ON MISSION

_____Disseminate company scheme of maneuver and EFSTs for the follow-on mission (or sustained combat and occupation in the objective city) to FS and FA leaders.
_____Identify fire support assets and ammunition available for follow-on missions.

Figure 10-4. Example company fire support officer planning checklist (continued).

MISSION ANALYSIS/BATTLE CALCULUS

_____Review the brigade fire support WARNO.

_____What is the mission flow for maneuver units--movement to contact, air assault, breaching operations, deliberate attack into city - for each COA?

_____What is FA organization for combat?

_____Essential field artillery tasks (EFATs) by phase for each COA.

_____Calculus--how much ammunition by shell/fuze type is needed for each EFST?

_____Determine special ammunition requirements--SCATMINE, Copperhead, DPICM, concrete piercing fuzes, smoke and WP, RAP, Charge 8, or Red Bag. Coordinate ammunition resupply as early as possible. Ammunition resupply is an EFAT!

_____What are the radar zone and cueing requirements by phase?

_____Determine artillery position areas required by phase. Will they be secure? Coordinate for position areas, movement times, and security support (if needed).

_____Determine radar position areas required by phase. Will they be secure? Coordinate for position areas, movement times, and security support (if needed).

_____Artillery and radar positioning should facilitate rapid transition from one phase to the next or to a follow-on mission. Targets for the next phase or mission should be within range of firing units according to ammunition available.

_____Develop communication, MET and survey plan. How will extended ranges and communications dead spaces in the objective city be defeated?

_____Develop and disseminate FASP.

_____Participate in fire support rehearsal (maneuver rehearsal if conducted), and conduct FA technical rehearsal.

_____Are missions planned and executed digitally, planned digitally, and executed by voice communications, or a mix?

FIGHT FROM LD TO BREACH SITE

_____EFATs for this phase.

_____Ensure firing units in position ready to fire in support of scheme of maneuver.

_____Rehearse all scheduled/preplanned fires. Ammunition for SOSR fires (suppression, reduction or obscuration fires during obstacle breaching) must be available and readied for sustained fires.

_____Ensure inactive firing units follow active missions, or are laid on priority targets that support the scheme of maneuver.

_____Determine triggers for ammunition resupply and repositioning of firing units during this phase.

_____Positioning of battalion TOC/TAC during this phase of the operation.

_____Identify communications links to all supported unit observers during this phase.

_____What unit is the counterfire HQ?

_____Determine radar zones and cueing schedule for this phase? Who are cueing agents? What are their triggers?

_____Determine other information required from DFSCOORD, brigade and battalion FSEs.

Figure 10-5. Example direct support field artillery battalion S3 planning checklist.

THE BREACHING OPERATION

_____EFATs for this phase.
_____Ensure firing units in position ready to fire in support of breach.
_____Ensure required ammunition for SOSR fires is ready to sustain scheduled suppression or smoke fires.
_____Are registrations required to assure accuracy of SOSR fires?
_____Determine if amount of ammunition is available for emergency fires.
_____Identify triggers for ammunition resupply or repositioning of firing units during this phase.
_____Is counterfire radar positioned at the optimum aspect angle to detect enemy indirect fire trajectories?
_____Determine and identify triggers for activation/deactivation of radar zones. Are radar zones activated to protect the breaching forces and prevent fratricide of friendly mortars?
_____Are observer/designators at proper observer target angle (Angle T) to designate for Copperhead?
_____Location of TOC/TAC during this phase.
_____Other information needed from the DFSCOORD, brigade and battalion FSEs.

THE URBAN FIGHT

_____EFATs for this phase.
_____Is the brigade system to clear fires in place and functioning?
_____What are the communications links to supported unit observers?
_____Position areas must adequately cover the objective city and blocking targets outside the city to interdict reinforcement/escape.
_____Identify routes for artillery convoy travel in and around objective city.
_____Determine triggers for ammunition resupply and repositioning of firing units.
_____What unit has the counterfire mission?
_____Determine and identify triggers for activation/deactivation of radar zones.
_____Determine the sniper/mine/booby trap threats to firing units, if position in or very near the objective city. Are firing unit howitzer sections identified to fire "Killer Junior" or direct fire, self-defense missions?
_____Are all inactive firing units laid on priority targets to support the scheme of maneuver?
_____Are registrations required to ensure accurate fires into the objective city?
_____Does MET data collection account for atmospheric conditions in and around the city--updrafting winds around tall buildings, temperature increases caused by smog, buildings and pavement, industrial activity?
_____Is survey available to give accurate firing unit positions, and, when possible, accurate building/landmark locations in the objective city?
_____Location of battalion TOC/TAC during this phase.
_____Other information needed from the DFSCOORD, brigade and battalion FSEs.

THE FOLLOW-ON MISSION

_____EFATs for the follow-on mission (or sustained combat and occupation in the objective).
_____Determine triggers for ammunition resupply and repositioning of firing units.
_____Determine trigger for repositioning of radar and for activation/deactivation of radar zones.
_____Location of battalion TOC.

Figure 10-5. Example direct support field artillery battalion S3 planning checklist (continued).

10-4. ACQUISITION PLATFORMS

If ground observation is limited, consider the use of aerial observers. FOs may be placed on upper floors of buildings to improve visibility. They are vulnerable if positioned on rooftops. Adjustment of fires is difficult. FOs should identify size and location of dead space (area in which indirect fires cannot fall). Dead space is generally five times the height of buildings for low-angle fire and one-half the height of buildings for high-angle fire. Radars may become more effective because of the increased use of high-angle fires. Radars lose effectiveness if sited too close behind tall buildings.

a. If Copperhead and other laser-guided munitions are used, OH-58Ds, AH-64s, and Combat Observation Lasing Teams (COLTs) need a series of well-defined numbered aerial attack-by-fire (ABF) positions or observation posts. These aerial attack positions must meet the angle-T requirements (800 mils for Copperhead, 1065 mils for Hellfire, and so on) for key buildings and terrain features (bridges, parks, military installations) throughout the city (Figure 10-6). Considerations for use of laser designators in urban terrain include the following:

- Tall structures may degrade the effectiveness of the designator.
- Maintaining a continuous laser track on moving targets is difficult.
- The presence of highly reflective surfaces such as windows may refract laser energy and or pose a hazard to friendly troops.
- The presence of highly absorptive surfaces such as open windows or tunnels may degrade designator effectiveness.
- Because of a fluid FLOT, designators may have to reposition more often.

b. Structures and urban infrastructure will reduce radio ranges. Use of wire, messenger, and visual signals should be increased. Antennas should be remoted on upper floors to increase their range. They are vulnerable if positioned on rooftops. Existing civilian telephone systems should be used for unsecured communication. Wire should be routed through sewers and buildings for protection. Generators should be placed near existing walls outside occupied buildings.

(1) Observers with maneuver elements will encounter ground obstacles broken glass, rubble piles, burning buildings, smoke, downed electrical lines, mines, snipers, to name a few—that will impede movement. They cannot rapidly reposition and will have limited visibility. Observers maneuvering outside the city can help fill gaps.

(2) Observers will locate targets by the maneuver unit building numbering system for a particular city. Building numbers must be translated into grid coordinates for FA units and mortar fire direction centers (FDCs). City tourist maps, inaccurate 1:12.5K, and smaller scale maps may be used. This increases the difficulty of determining accurate target grid coordinates.

(3) Global positioning system (GPS) functioning is greatly degraded in cities with tall buildings (since these buildings mask satellite coverage).

(4) Observation Posts (OPs) should be positioned to observe these fires, and trigger points must be identified.

(5) Tactical air control party (TACP) and enlisted tactical air controller (ETAC) positions require visibility, not just on the target, but also of the surrounding terrain and sky to allow for terminal control of close air support (CAS)/ground attack aircraft.

(6) Air and naval gunfire liaison companies (ANGLICOs) have been inactivated. Elements designated to observe and adjust naval gunfire and control Navy and Marine CAS/ground attack aircraft have the same positioning requirements as TACPs and ETACs.

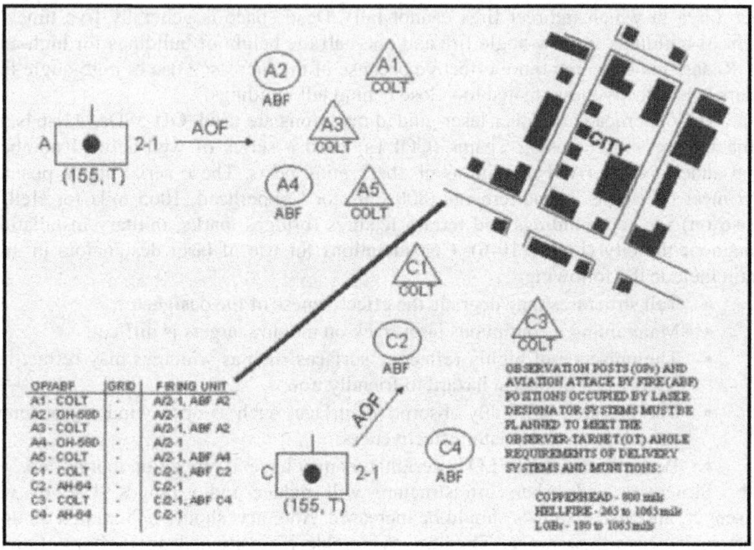

Figure 10-6. Observer positioning for laser-guided munitions.

10-5. METEOROLOGICAL AND SURVEY REQUIREMENTS

Accurate meteorological (MET) information and surveys are required, as most targets are point targets.

a. Conventional surveying is hampered by decreased line of sight. Map spotting is difficult in large cities. Establishment of multiple survey control points (SCPs) should be anticipated. Artillery may be used in the direct fire role for precision operations. Survey datum from geodetic markers around cities, especially in developing countries, is considered unreliable (different datum, different calculation techniques, geodetic markers that have been moved or tampered with). If SCPs cannot be extended from known, reliable surveys, use hasty techniques before using datum found around the cities.

b. MET conditions in cities are different than surrounding terrain (ambient heat radiated from buildings, industrial smog conditions common to cities in developing countries, and deviation in winds to extremely high altitudes caused by large urban areas). The precision for indirect fires during urban combat may increase the need or frequency of MET measurements.

10-6. DELIVERY ASSETS

Appendix A of FM 6-20-40 has extensive and current information on delivery systems and munitions available for them. The key piece of information that must be remembered when viewing this data from UO perspective is that the risk estimates were based on open field environment and must be adapted for the urban environment. As mentioned at the beginning of the chapter, effects will drive the fire support effort, which will be focused on munitions that provide the desired effect and then limited by the delivery systems available to deliver those munitions without incurring undesired effects.

10-7. TACTICAL AIR

A unit may be supported by USAF, USN, USMC, or allied fighters and attack aircraft while fighting in urban areas.

a. The employment of CAS depends on the following.

(1) ***Shock and Concussion.*** Heavy air bombardment provides tactical advantages to an attacker. The shock and concussion of the bombardment reduce the efficiency of defending troops and destroy defensive positions.

(2) ***Rubble and Debris.*** The rubble and debris resulting from air attacks may increase the defender's cover while creating major obstacles to the movement of attacking forces.

(3) ***Proximity of Friendly Troops.*** The proximity of opposing forces to friendly troops may require the use of precision-guided munitions and may require the temporary disengagement of friendly forces in contact. The AC-130 is the air weapons platform of choice for precision UO as the proximity of friendly troops precludes other tactical air use.

(4) ***Indigenous Civilians or Key Facilities.*** The use of air weapons maybe restricted by the presence of civilians or the requirement to preserve key facilities within a city.

(5) ***Limited Ground Observation.*** Limited ground observation may require the use of airborne FAC.

b. CAS maybe employed during offensive operations to support the isolation of the city by interdicting entry and exit routes. It can be used to support attacking units by reducing enemy strong points with precision guided munitions. Or it may conduct tactical air reconnaissance and to provide detailed intelligence of enemy dispositions, equipment, and strengths.

c. CAS maybe employed during defensive operations to strike enemy attack formations and concentrations outside the urban area. TACP concerns, such as identification of IPs, communications difficulties, and how to provide terminal control of aircraft when TACP cannot observe aircraft due to buildings, may require setting up numerous communication relays or retransmission sites (with assistance of signal officer) or splitting CP operations with the forward CP controlling one area of the fight, while the main CP controls the other area of the fight.

10-8. NONLETHAL MEANS

The divisional MI battalion directly supports the division commander and G2 by providing dedicated multidiscipline battlefield IEW support to the division and its subordinate maneuver brigades. At the brigade level the focus is on the intelligence products and services needed by commanders to plan, fight, and win battles at the tactical level. In addition to the GS intelligence support provided by the MI battalion, the brigade will normally receive a DS MI company from the MI battalion. The DS MI company capabilities include:

- Automated multidiscipline intelligence and information processing.
- Analytical control team.
- UAV control.
- Interrogation of prisoners of war and limited document exploitation.
- CI support.
- C2 of organic and reinforcing IEW assets.
- JSTARS coverage and product dissemination.

IEW is integrated into unit operations regardless of the type of unit, level of war, or the scope of the mission. IEW complements other destructive systems in the context of overall strategy. When IEW is synchronized with lethal fires, the friendly commander gains agility by slowing the reaction time of his adversary.

10-9. ARTILLERY USED IN DIRECT FIRE

> *"Having discovered on the first day that some apartment buildings and air-raid shelters could withstand the fire of tanks and tank destroyers, Colonel Corley [Commander, 3rd Bn, 26th Inf Regt] called for a 155mm self-propelled rifle. Early the next morning the big weapon proved its worth in the first test when one shot leveled one of the sturdy buildings. Impressed, the regimental commander, Colonel Seitz, sent one of the big rifles to support his other battalion as well."*
>
> Charles B. MacDonald
> The Siegfried Line Campaign,
> US Army in World War II

Division or joint task force commanders in coordination with their FSCOORD and the force artillery commander will determine the parameters under which field artillery assets may be used as direct fire platforms for support to ground maneuver units. Normally, field artillery will only be employed in direct fire under high-intensity conditions of urban combat where tanks, BFVs, and other direct fire systems may not be available or able to have an effect on the target. Subordinate maneuver commanders in coordination with their respective fire support officers (FSOs) will then direct the employment of individual howitzer sections to support the unit in contact. This paragraph will provide TTP for the use of howitzer sections in direct fire at the company level and below.

a. **General.** Large-caliber artillery rounds provided by direct fire are effective for destroying targets in buildings. If available, self-propelled 155-mm howitzers can use direct fire to destroy or neutralize bunkers, heavy fortifications, or enemy positions in reinforced concrete buildings (Figure 10-7, page 10-20). The self-propelled artillery can be used to clear or create avenues of approach. Towed artillery can be used in a direct fire role but is not the preferred artillery for UO, because it does not have ballistic crew protection. This should not preclude its use if the situation calls for it and the commander is willing to accept increased risk for crew casualties. When artillery is used in the direct fire role, infantry must be provided to the howitzer for security.

Figure 10-7. Artillery in direct-fire role.

b. **Commander's Intent.** The maneuver commander's intent must be clearly understood to ensure proper application of artillery support. The FSCOORD will use the commander's guidance to develop the ammunition requirements and restrictions; for example, "Killer Junior" for maximum fragmentation, PD fuzes set on delay or CP fuzes to penetrate structures, WP for incendiary effects, and so forth. Subordinate artillery commanders should identify their most proficient direct fire sections.

c. **Command and Control.** Artillery may be placed in tactical control (TACON) to maneuver commanders, for example a platoon of three guns may be TACON to a company, a battery TACON to a battalion. As systems, howitzers should be controlled through the FSO since he knows the capabilities and limitations of the system. The company team commander then has the option to control the system as a company support by fire asset or to place the system in support of the platoon leader requiring the direct fire support. For example, the company team commander may designate a platoon to isolate the objective through a support by fire mission. The howitzer system would receive its fire commands from the infantry platoon leader that is given this mission. Control of the howitzer system would not normally go below platoon level.

d. **Other Considerations.** Other considerations for using artillery in direct fire are:

(1) ***Communications.*** A dedicated FM radio frequency between the controlling maneuver headquarters speaking directly to the howitzer section will be necessary to control fires and prevent fratricide.

(2) ***Protection.*** Each howitzer must be protected by infantry; for example, a BFV and a squad of dismounted mechanized infantry, or a squad of light infantry with antiarmor weapons. The infantry must provide local security and prevent enemy ground assault, sniper fire, and antitank fire (such as RPG).

(3) ***Positions.*** The infantry will need to reconnoiter and occupy positions where the howitzer can provide fire support. These positions must be free from enemy direct fire but still allow direct fire by the howitzer on the target. Self-propelled artillery systems provide minimal ballistic protection from fragmentation for the crew. Although these systems seem formidable, they provide less crew protection than a BFV and contain large amounts of onboard ammunition and propellant. They are susceptible to catastrophic destruction by heavy automatic weapons, light cannon, and antitank fire. Towed artillery

systems provide no protection for crew members. The howitzer should provide the necessary support and then move to an alternate position to avoid being identified by the enemy. Overwatching infantry should always move ahead of and with the howitzer to prevent enemy ambushes.

CHAPTER 11

MOBILITY, COUNTERMOBILITY, SURVIVABILITY

"A squad of engineers from a platoon of the 1st Engineer Battalion was to accompany each assault company. The squad would be equipped with flame-throwers and dynamite charges for the reduction of pillboxes."

LTC Darrel M. Daniel
Commander, 2nd Bn, 26th Inf Regt
October, 1944, Battle of Aachen

This chapter provides considerations for engineer planners and leaders to employ when battalions and brigades conduct UO. While the considerations in this chapter apply specifically to offensive UO, they can be tailored for defensive, stability, and support operations.

11-1. GENERAL

While the process for planning engineer support for UO follows existing decision-making steps, engineer planners must understand how this diverse terrain impacts engineer operations. Terrain enhances the enemy's countermobility and survivability efforts and increases the friendly force's mobility requirements. Critical points include:

- Structures become key terrain.
- Below ground and multi-layered above-ground dimensions are added.
- Decentralized execution—while staying collectively synchronized—is required.
- Urban specific pre-combat checks (PCCs), precombat inspections (PCIs), and rehearsals must be conducted.

Engineer planners must account for these factors to provide effective engineer support to maneuver forces.

11 2. MISSION ANALYSIS

Mission analysis sets the conditions for planning and ultimate success of UO. All planners must identify specified, implied, and essential tasks as well as constraints and limitations. Well-prepared engineer battlefield assessments (EBA) and terrain analysis products are essential to successful UO planning. Answering the following questions will help engineer planners, in conjunction with the principal battle staff, develop an effective UO mission analysis.

a. **S2, S3, Engineer, FSO.**

(1) ***Where is the key/decisive terrain?*** Identify this terrain for the approach march and for seizing buildings. Conduct a line-of-sight analysis along the route and compare it to the enemy template. Identify the most likely sites for enemy sniper and observer positions. Target these positions for deliberate reconnaissance to confirm or deny enemy presence. Plan obscuration and suppression to facilitate friendly movement.

(2) ***Where are the best obstacle reduction sites and support-by-fire positions for securing a foothold?*** Consider the terrain, the enemy force template, and massing fires. Determine the minimum engineer force required to seize a foothold, seize essential facilities, and provide mobility support to mounted forces, such as sequencing of engineer

tasks and changing the engineer task organization to accomplish essential tasks. Identify the key leaders required to facilitate command and control of critical events and task organization changes. Decide how to best integrate cannon-delivered smoke, hand-emplaced smoke, and smoke generators to conduct breaching operations.

b. **S3, Engineer, S4.** ***How should subordinate units execute in-stride versus deliberate breaching operations based on the enemy template and results of reconnaissance and surveillance (R&S) efforts?*** Decide where to use the mine-clearing line charge (MICLIC), tank-mounted countermine equipment (CME), and manual breach techniques. Balance exposure of the breach force to enemy fires with the probability that a system may be destroyed before it can be employed. Determine acceptable collateral damage when employing the MICLIC. Plan for resupply of Class V (explosives, smoke, machine-gun ammunition) items after the initial foothold is seized.

c. **S3.** Decide how reconnaissance forces link up, guide, or mark obstacles for bypass/breaching operations.

d. **S2, Engineer, FSO.**

(1) ***What are the counterattack routes of the enemy force?*** Consider the terrain and weather. Determine if enemy counterattack routes can be used to move friendly combat service support assets based on the enemy event template and time phasing of the counterattack. Determine what situational obstacles (rapid mining, scatterable mining) the enemy counterattack force has available.

(2) ***What is the safety zone and trigger for using scatterable mines?*** Ensure that this information is disseminated at all rehearsals.

e. **Engineer.** ***What is the composition of the buildings to be attacked?*** Determine the effects weapons will have on these structures (this drives the selection of fuze/shell combinations and aircraft attack munitions).

f. **S2, Engineer.** ***What is the "layout" of the town both above and below ground?*** Determine the protected areas, such as churches, hospitals, and museums. Sources for this information are imagery from the division, gun camera tapes from OH-58/AH-64 helicopters, road maps, and tour books.

11-3. SUPPORT PRODUCTS

The engineer staff planner uses the following products developed to support the military decision-making process (MDMP). All of these products must be developed in conjunction with the S2. These products are updated based on the results of reconnaissance and surveillance.

a. **Engineer Battlefield Assessment (EBA).** The EBA feeds many of the subsequent products. Clearly articulate the enemy engineer capability based on the most likely and most dangerous courses of action. Consider past experience with this enemy, his current strength, anticipated barrier material basic loads, expected resupply rates, and locally available materials he can use to prepare his defense. This information will support development of the situation template (SITEMP).

(1) Identify friendly engineer capabilities for mobility, countermobility, and survivability operations. Explicitly state the number and types of breaches each engineer unit is capable of executing based on its personnel, equipment, and logistical status. Leader proficiency and audacity impact on this estimate, so plan two levels down based

on the particular unit. Use this information to develop the task organization later in the MDMP.

(2) Estimate the impact of terrain and weather on both friendly and enemy capabilities. Line-of-sight, hydrology, cross-country movement, and line-of-communication overlays are helpful and can be provided by the division terrain detachment or quickly approximated from maps.

b. **Situation Template.** Know the enemy capability based on an estimated unit basic load of Class IV and V materials and anticipated resupply. The time available to prepare the defense is essential. Reconnaissance assets should observe the delivery and emplacement of barrier materials. The S2 and the engineer will template enemy obstacles and counterattack routes based on terrain and weather conditions. Determine what resources are available in the urban area (ammonium nitrate, acetylene, propane, lumber yards, jersey barriers, vehicles, and construction equipment) that can contribute to enemy defensive preparation.

(1) Based on this analysis, the engineer and S2 will jointly template the enemy engineer countermobility/survivability capability on the SITEMP. It should include minefields, tactical and protective wire obstacles, and vehicles and other barriers in roads. This overlay is used to plan the engineer task organization, because this and the friendly scheme of maneuver determine the number of sapper squads needed and where mobility assets are placed in the movement.

(2) Time and materials will impact enemy defensive capability. The force array in the security zone and main defensive belt impacts the amount of defensive preparation. Indirect-fire systems can only service one priority target and must shift to cover other targets, which may help with refining the obstacle template. Locations and movement of mounted weapons may indicate usable lanes for friendly infiltration of vehicles.

c. **Event Template.** Determine what triggers the commitment of enemy counterattack forces. The engineer planner can assist the S2 in determining what situational obstacle capabilities he has, where and for what purpose the capabilities will be committed, and what the triggers are. Determine the structures likely to be set for destruction (such as petroleum and natural gas storage facilities).

d. **Friendly Forces Survivability Time Line.** The engineer and the S4 plan to construct positions to support the forward displacement of combat support and combat service support assets and limited command and control nodes. The survivability effort should be an essential part of the maneuver deception plan.

e. **Breach Execution Matrix.** This matrix helps the task force allocate engineer assets and determine when in-stride and deliberate breach techniques are required. Specify where to use MICLIC, hand-emplaced explosives, armored combat earthmover (ACE), armored vehicle-launched bridge (AVLB), and tank-mounted counter-mine equipment to reduce enemy obstacles. It is important to keep in mind that rubble can be a more significant obstacle than conventional mines and wire obstacles.

f. **Decision Support Template/Decision Support Matrix.** Help the S3 identify and plan viable branches and sequels to the plan. It is essential to know where engineers will culminate and how rapidly engineer platoons/squads can be consolidated, reorganized, and put back into the fight.

g. **Execution Checklist/Operations Schedule.** Develop with the S3 the operations schedule (OPSKED), which is a combination of key events from the synchronization matrix and associated code words. This product supports the decision support template and helps the battle captain and maneuver commander track the battle and make decisions. Prepare a rough execution checklist after receiving the warning order and continue to refine it during mission analysis. Finalize the checklist during war-gaming and provide advance copies to task force engineers and squad leaders.

h. **Troop-Leading Procedures Timeline.** Ensure that adequate time is available for engineers to both prepare the task force rehearsal site and conduct their own internal rehearsals.

11-4. ENGINEER STAFF PLANNING CHECKLIST (BRIGADE AND BELOW)

a. **General.**

— Identify and resource all mobility/survivability essential tasks.

— Address all the breach tenets during planning and rehearsals.

— Request terrain products, urban layout diagrams, and data on building composition from higher headquarters.

— Study available terrain products to determine which sub-surface routes to use and how to defend against enemy use of these systems.

— Study available maps and photos to determine the best routes to use when approaching the city and within the city, as well as identifying tentative locations for casualty collection points, aid stations, and ammunition and water resupply points.

— Use scatterable mines to support engagement areas that block mounted counterattack routes. Disseminate this plan to critical maneuver and combat service support leaders.

— Establish essential engineer friendly forces' information requirements and no-later than (NLT) report times.

— Nominate engineer-specific PIR and associated named areas of interest (NAIs) to support the reconnaissance plan. Ensure that the latest time information is of value (LTIOV) is clearly understood. Decide what actions to take if the PIR are not answered before LTIOV.

— Disseminate the enemy obstacle template to all engineer leaders.

— Task-organize engineers to support essential mobility/survivability reconnaissance missions.

— Determine how much and what types of obscuration smoke are available. Determine the wind direction and speed, which will impact the effects of smoke. Coordinate with the fire support officer for recommended uses of white phosphorus (both mortar and artillery delivered) and handheld smoke. Coordinate with the smoke platoon leader for duration of smoke and level of obscuration.

— Designate and clear routes for mounted forces and reserve forces.

— Identify the *conditions* and a decision point for initiating deliberate breaching operations during each critical event of the operation.

b. **Approach March.**

— Designate routes for ground convoys and allocate engineers to clear them.

— Determine the clearance method and acceptable risk.

— Ensure that all vehicles have lane and by-pass marking materials on board.
— Designate ground evacuation routes.
— Determine the decision point for using alternate routes.
— Determine when to establish TCPs/guides at critical obstacles on the route.
— Establish NAIs along the ground route to confirm or deny the enemy obstacle template.

c. **Secure a Foothold.**

— Designate the best reduction site and technique based on enemy force array, terrain, and trafficability.
— Nominate NAIs for breaching operations.
— Designate one lane for each simultaneously assaulting platoon and the engineers needed to reduce it.
— Explain the lane-marking system.
— Establish a traffic-control plan for dismounted and mounted traffic.
— Establish a vehicle route and a dismounted route of evacuation from the foothold to the helicopter landing zone.
— Designate locations for blocking positions to keep counterattacks from interfering with breaching operations.
— Resource blocking positions with MOPMS, conventional mines, and expedient barrier capability (such as abatis). Depict the planned locations of scatterable mines (include the safety zone) on maneuver and combat service support graphics to reduce fratricide.

d. **Seize Objective.**

— Designate buildings to enter and a reduction site that will support maneuver to the point of penetration.
— Designate where the support force will enter buildings.
— Resource battalions and their engineers with sufficient explosives and hand-emplaced and artillery smoke.
— Explain the cleared-building and cleared-lane marking systems.

e. **Prepare/Execute.**

— Construct appropriate rehearsal sites to support maneuver and CSS operations.
— Provide enough detail in the troop-leading procedure timeline to encourage both engineer and combined arms rehearsals.
— Issue sketch maps and terrain products to engineers.
— Construct a lane marking system and by-pass marking system that all vehicle drivers must go through en route to the objective area.
— Provide enough detail in the maneuver and engineer execution checklists to effectively use the Decision Support Matrix.
— Specify times for engineer-specific pre-combat inspections conducted by platoon leaders, company commanders, and first sergeants.

11-5. RECONNAISSANCE AND SURVEILLANCE PLANNING CONSIDERATIONS

Consider the following during reconnaissance and surveillance planning:

a. **Integrate Engineer Reconnaissance Teams.** Consider the integration of engineer reconnaissance teams into the brigade R&S plan. Focus these teams on engineer targets such as landing zone denial, obstacles in the reduction area, enemy survivability on the objective, and obstacles on approach routes. The NAI assigned to engineers should have priority intelligence requirements (PIR) that determine the best reduction sites in the urban area and confirm or deny enemy fortification of key sites.

b. **Precombat Inspections.** After conducting precombat checks (PCCs), inspect materials used to mark obstacle by-pass lanes. Conduct FM radio communications exercises using the OPSKED and reports specific to the current operation. Inspect all maps for operations security considerations. Sterile maps are not required, but information provided on overlays should not compromise the attack plan. Overlays should portray only NAIs. Targets, pickup and landing zones, and link-up locations should not be on overlays taken into the objective area. All soldiers must clearly understand the NAI priority and associated PIR, casualty evacuation (CASEVAC) plan, abort criteria, compromise plan, exfiltration and linkup plan, and communications windows.

11-6. MOBILITY PLANNING CONSIDERATIONS

Providing mobility support to a maneuver force during UO normally will require engineers to support multiple combined arms breaching operations. The reverse planning process (discussed in FM 90-13-1) applies to all terrain situations. The considerations discussed in this paragraph complement this process.

a. **Conduct the Approach March.** The S3 and the battle staff plan a primary route and an alternate route to support the movement of each maneuver battalion's combat forces. The engineer makes recommendations based on trafficability of the terrain and the ability to clear these routes using standard tactics, techniques, and procedures (TTP). Control of movement routes is critical, particularly when ground evacuation is the primary method of removing casualties. The S4, S3, and XO coordinate one-way, two-way, and alternating-direction traffic on task force routes and identify decision criteria for switching to alternate routes. Maximize aerial reconnaissance of routes to identify possible obstacles, combat outposts, and ambushes. The engineer planner ensures that the task force has enough engineer assets dedicated to accomplish the implied tasks and ensures that enough Class IV and V are available to support the movement.

(1) ***Precombat Inspections.*** The engineer ensures that subordinate engineer squads conduct standard route-clearance PCCs and PCIs, which should be listed in the unit SOP. As a minimum, the task force engineer should check initiation systems, demolition charges, reduction equipment, marking materials, and mine detectors, and have a basic understanding of the concept of engineer operations.

(2) ***Rehearsals.*** The engineer, with the S3, ensures that all of the breach tenets and control measures are understood by key leaders at the task force rehearsal.

b. **Secure a Foothold.** Create lanes through obstacles using one sapper squad per lane, with a minimum of one lane per simultaneously assaulting platoon. (This does not necessarily mean nine lanes per Infantry battalion; this requirement should be carefully analyzed.) Use adequate marking materials, guides for assault and follow-on forces, and lane hand-over procedures. It takes at least 30 minutes to *cycle* this engineer squad back into the fight. A squad cannot support breaching operations continuously. A decision point or trigger must support any changes in task organization and missions for engineers. Establish decision points for changing approach routes, reduction sites, and initiation of SOSRA (suppress, obscure, secure, reduce, assault).

(1) ***Precombat Inspections.*** Equip the unit with bolt cutters (two per engineer squad), grappling hooks (three per engineer squad), a lane-marking kit, hand-emplaced explosives (10 per squad, per lane), mine detectors, and probes. Ensure that handheld smoke is available for each Infantry soldier and that vehicles or utility helicopters carry smoke pots. Mass this smoke with the breach force at the breach point. Ballast load marking system upgrade materials on gun trucks. Use expedient reduction tools, such as SKEDCO litters, for wire reduction.

(2) ***Rehearsals.*** No matter what rehearsal type or technique is used, perform basic rehearsals IAW SOSR factors.

(a) *Suppress.* Ensure that all personnel understand the location of support-by-fire positions and the pyrotechnic and radio signals to initiate obstacle reduction and indicate when the lanes are open (proofed and marked). The rehearsal site should have a full-scale lane-marking system visible to every soldier. All key leaders should understand the commitment criteria for the breach force.

(b) *Obscure.* Rehearse triggers for artillery-delivered, hand-emplaced, and vehicle-generated smoke. Consider the position of the moon relative to the support-by-fire position, the percent of illumination, and the night-vision goggle window.

(c) *Secure.* Hold a combined arms rehearsal of the breach force using the full-dress rehearsal technique. This rehearsal includes engineers and attached maneuver elements dedicated to suppressing direct fires and destroying local counterattacks.

(d) *Reduce.* The combined arms rehearsal should include handing over lanes from engineers to maneuver soldiers. The rehearsal should occur at the "NCO to NCO" level and discuss details of linkup and handover. Consider the need to back-haul casualties when planning the number of lanes.

c. **Seize Key Facilities.** Plan procedures for dynamic entries into buildings and vertical envelopment, which require prepared special demolition charges (see Chapter 8) and expedient assault ladders. Rehearse the TTP for getting into windows on second and third floors. Have cutting tools available to prepare climbing poles at the breach point. Plan for sub-surface entry, if necessary. Consider the use of reducing wire in stairwells and hallways.

(1) ***Precombat Inspections.*** Inspect special breaching charges (see Chapter 8.). Ensure that charges are properly constructed and that they will "stick" when placed on walls and doors. Use double-sided foam tape when placing vertical breaching charges during warm, dry conditions. Use spikes, braces, or Ramset-type power-actuated fasteners during rain or when temperatures are below freezing. Ensure that sufficient handheld and hand-emplaced smoke is available. Assaulting soldiers can carry smoke pots and additional explosives. Where METT-TC factors permit, consider using mechanical breaching tools to enter doors. Conserve explosives by bringing one or two 24-inch crowbars to lift manhole covers and pry open entryways in buildings and sewers. Provide night-vision goggles to soldiers who reduce obstacles, because Infantry leaders

use infrared "tactical pointers" extensively, and reduction element soldiers must be able to see these signals. Use all available infrared lights. Mount and zero all AN/PAQ-4s and AN/PVS-4s during the preparation phase of the mission. Engineers must bring handheld infrared and visible light sources to help move and reduce obstacles inside buildings and sub-surface structures. (See Appendix B.) Ambient light inside hallways and underground is virtually zero, so plan for additional light sources. Mark cleared buildings so the marking is visible from rotary-wing aircraft, armored vehicles, and by dismounted soldiers (Appendix I).

(2) ***Rehearsals.*** Focus on the location and control of support forces and signals for committing the breach force. Ensure that soldiers understand the minimum safe distance and the best reduction site based on the building structure. Clearly identify routes between buildings and the marking method for "safe routes." Deconflict building clearance markings from collection points for casualties, displaced civilians, and enemy prisoners of war. Rehearse precision clearing techniques with the Infantry for interior building clearing. Basic SOSR rehearsals from "secure the foothold" apply to dynamic entry into buildings, but these rehearsals usually focus on the Infantry platoon and an engineer squad.

NOTE: The company, battalion, and brigade engineers must be knowledgeable of demolitions effects and recommend minimum safe distances based on the amount of and type of explosives which will be used and the construction of buildings to be breached.

(3) ***Noncombatants and Enemy Prisoners of War.*** Establish "protected areas" for noncombatants, and clearly mark routes for displaced civilians. Consider an expedient countermobility effort to restrict access to noncombatants and enemy prisoners of war (EPWs). Liaison officers from psychological operations, civil affairs, and the military police should address this topic in the brigade maneuver rehearsal. Although there are no specific engineer requirements, engineers should be prepared to provide technical assistance during planning and execution phases.

(4) ***Subsurface Considerations.*** Important points are:

- Entering the tunnel or sewer complex using hand tools or explosives
- Identifying and neutralizing mines and booby traps.
- Marking cleared areas.

Subterranean navigation inside sewers and radio communications from underground to soldiers above ground are challenging. There is no ambient light inside tunnels, so plan and rehearse using infrared and visible light signals. (See Appendix D.)

d. **Movement Within the Urban Area.** Plan one vehicle lane per mounted platoon entering each section of the urban area. The lane through tactical and perimeter protective obstacles will become an axis of advance within the urban area. These lanes initially will support one-way traffic. Plan and rehearse traffic control when lanes become alternating traffic lanes to allow for MEDEVAC/CASEVAC. Improve at least one lane to two-way traffic and designate this as the primary MEDEVAC/CASEVAC route. Designate, clear, and mark a route from the casualty collection point to the MEDEVAC/CASEVAC primary and alternate helicopter landing zones. Use combat route-clearance techniques to clear the ground MEDEVAC/CASEVAC route. Reduce or bypass obstacles created by disabled vehicles, rubble, and so forth. If by-passing is part of the plan, make it a branch to the plan and include decision points and conditions.

(1) ***Precombat Inspections***. Inspect mine clearing line charge (MICLIC) and tank-mounted CME. Ensure that designated dismounted sappers have at least 20 blocks of TNT or C4 and 500 feet of detonating cord to reduce a 100-meter deep "lane" for vehicles. Inspect mine detectors carried by engineers designated to execute this mission. Sandbag one vehicle to use for proofing vehicle lanes, and dismount all passengers when proofing the lane. Ballast load additional lane marking material on vehicles. To assist the maneuver force in locating the correct lane to support their tactical plan, ensure that markings for multiple lanes are easily distinguished by day and at night. MEDEVAC/CASEVAC lanes must have a dedicated traffic control post (TCP). Integrate a tank-mounted plow or properly prepared heavy vehicle (dozer, loader, or 5-ton truck with winch) into the plan to reduce rubble or obstacles caused by disabled vehicles.

(2) ***Rehearsals.*** A combined arms breaching rehearsal is required (according to FM 90-13-1). This rehearsal will serve as the final check for mission-essential equipment and final adjustments to the plan based on PCIs. Synchronize the establishment of support-by-fire positions to isolate reduction sites and trigger conditions for initiating reduction operations (the conditions and who makes the decision). Determine who shifts obscuration and suppressive fires and when they are shifted. Leaders must rehearse handing over lanes to follow-on forces. Rehearse time-phasing the ground MEDEVAC/CASEVAC route clearance to helicopter landing zones and ambulance exchange points. Construct the unit's standard lane-marking system and route signs at the rehearsal site.

11-7. COUNTERMOBMITY PLANNING CONSIDERATIONS

These issues should be addressed in brigade, battalion, and company-level rehearsals. Plan to issue a scatterable mine warning (SCATMINWARN) to prevent fratricide.

a. **Tactical Employment of Scatterable Mines.** The S3, engineer, and FSO should plan, in detail, the employment of artillery-delivered antipersonnel mines/remote antiarmor mines (ADAM/RAAM) and multiple-delivery mine systems (VOLCANO). Specify the target to be attacked, a tentative location, its effect (disrupt, turn, fix, or block), the delivery system, the observer, and the trigger. To reduce the risk of fratricide, the scatterable mine execution plan must be clearly understood by leaders of mounted elements.

b. **Protective Employment of Scatterable Mines.** Ballast load the modular pack mine system (MOPMS) on vehicles moving into objective area blocking positions. Consider sling loading the MOPMS, conventional mines, and limited barrier materials to support transitioning to the defense and blocking enemy counterattacks.

c. **Engagement Area Development.** The S3 should specify the engagement area to interdict the enemy counterattack force. Ensure that battalion and brigade reserve forces have specified routes to move to the engagement area. The engineer must ensure that these movement routes are obstacle restricted zones. Engineers may not be available to emplace obstacles, so specify the engagement area development tasks, including obstacle emplacement and fire integration, to maneuver units.

11-8. SURVIVABILITY PLANNING CONSIDERATIONS

Perform this work concurrently with initial reconnaissance and the effort to shape the area of operations (setting conditions) by the brigade, to support the brigade and division deception plans. Specific considerations include:

- **Field Artillery.** Determine positioning areas and plan counterfire radars and ammunition.
- **Forward Area Refuel Point.** Establish locations for stocking fuel and ammunition. Plan for multiple refueling sites to support the attack and lift aviation simultaneously.
- **Battalion Aid Station.** Locate forward treatment facilities and ingress/egress routes. The implied task is to establish helicopter landing zones for these sites.

CHAPTER 12

COMBAT SUPPORT

"The third battalion with a platoon of medium tanks, a platoon of tank destroyers, a platoon of engineers, and a 155 mm self-propelled rifle arrived in the Farwick Park section...on 13 October 1944...On 14 October a section of 4.2 chemical mortars was attached to Company 'M'. A general counterattack all along the battalion front was stopped...on 15 October."

Charles B. MacDonald
The Siegfried Line Campaign
U.S. Army in World War II

12-1. MORTARS

Mortars are the most responsive indirect fires available to Infantry commanders and leaders. Their mission is to provide close and immediate fire support to the maneuver units. Mortars are well suited for combat in urban areas because of their high rate of fire, steep angle of fall, and short minimum range. Company commanders must plan mortar support with the FIST Chief as part of the total fire support system. (See FM 7-90 for detailed information on the tactical employment of mortars.)

a. **Role of Mortar Units.** The role of mortar units is to deliver suppressive fires to support maneuver, especially against dismounted Infantry. Mortars can be used to obscure, neutralize, suppress or illuminate during urban combat. Mortar fires inhibit enemy fires and movement, allowing friendly forces to maneuver to a position of advantage. The most common and valuable use for mortars is often harassment and interdiction fires. One of their greatest contributions is interdicting supplies, evacuation efforts, and reinforcement in the enemy rear just behind his forward defensive positions. During World War II and the recent Middle East conflicts, light mortar HE fires have been used extensively during urban combat to deny the use of streets, parks, and plazas to enemy personnel. Finally, mortars can be used, with some limitations, against light armor and structures. Effectively integrating mortar fires with dismounted maneuver is key to the successful combat in an urban area.

b. **Position Selection.** The selection of mortar positions depends on the size of buildings, the size of the urban area, and the mission. Rubble can be used to construct a parapet for firing positions. Positions are also selected to lessen counterbattery fire.

(1) ***Existing Structures and Masking.*** The use of existing structures (for example, garages, office buildings, or highway overpasses) for positions is recommended to afford the best protection and lessen the camouflage effort. Proper masking enhances survivability. If the mortar is fired in excess of 885 mils to clear a frontal mask, the enemy counterbattery threat is reduced. These principles are used in both the offense and the defense.

(2) ***Use of Sandbags.*** Do not mount mortars directly on concrete but use sandbags as a buffer. Sandbags should consist of two or three layers; butt the sandbags against a curb or wall; and extend at least one sandbag width beyond the baseplate.

(3) ***Placement.*** Mortars are usually not placed on top of buildings because lack of cover and mask makes them vulnerable. They should not be placed inside buildings with

damaged roofs unless the structure's stability has been checked. Overpressure can injure personnel, and the shock on the floor can weaken or collapse the structure.

c. **Communications.** Initially, radio is the primary means of communication during urban combat. An increased use of wire, messenger, and visual signals are required. However, wire is usually the primary means of communication between the forward observers, fire support team, fire direction center, and mortars since elements are close to each other. Also, FM radio transmissions in urban areas are likely to be erratic. Structures reduce radio ranges; however, remoting the antennas to upper floors or roofs may improve communications and enhance operator survivability. Another technique that applies is the use of radio retransmissions. A practical solution is to use existing civilian systems to supplement the unit's capability, understanding that this is an unsecure method of communication.

d. **Magnetic Interference.** In an urban environment, all magnetic instruments are affected by surrounding structural steel, electrical cables, and automobiles. Minimum distance guidelines for the use of the M2 aiming circle (FM 23-90) is difficult to apply. To overcome this problem, an azimuth is obtained to a distant aiming point. From this azimuth, the back azimuth of the direction of fire is subtracted. The difference is indexed on the red scale and the gun manipulated until the vertical cross hair of the sight is on the aiming point. Such features as the direction of a street may be used instead of a distant aiming point.

e. **High-Explosive Ammunition**. During urban combat mortar high-explosive (HE) fires are used more than any other type of indirect fire weapon. Although mortar fires are often targeted against roads and other open areas, the natural dispersion of indirect fires will result in many hits on buildings. Leaders must use care when planning mortar fires during UO to minimize collateral damage.

(1) HE ammunition, especially the 120-mm projectile, gives good results when used against lightly built structures within cities. However, it does not perform well against reinforced concrete found in larger urban areas.

(2) When using HE ammunition in urban fighting, only point detonating fuzes should be used. The use of proximity fuzes should normally be avoided, because the nature of urban areas causes proximity fuzes to function prematurely. Proximity fuzes, however, are useful in attacking some targets such as OPs on tops of buildings.

f. **Illumination.** In the offense, illuminating rounds are planned to burst above the objective. If the illumination were behind the objective, the enemy troops would be in the shadows rather than in the light. In the defense, illumination is planned to burst behind friendly troops to put them in the shadows and place the enemy troops in the light. Buildings reduce the effectiveness of the illumination by creating shadows. Continuous illumination requires close coordination between the FO and FDC to produce the proper effect by bringing the illumination over the defensive positions as the enemy troops approach the buildings (Figure 12-1).

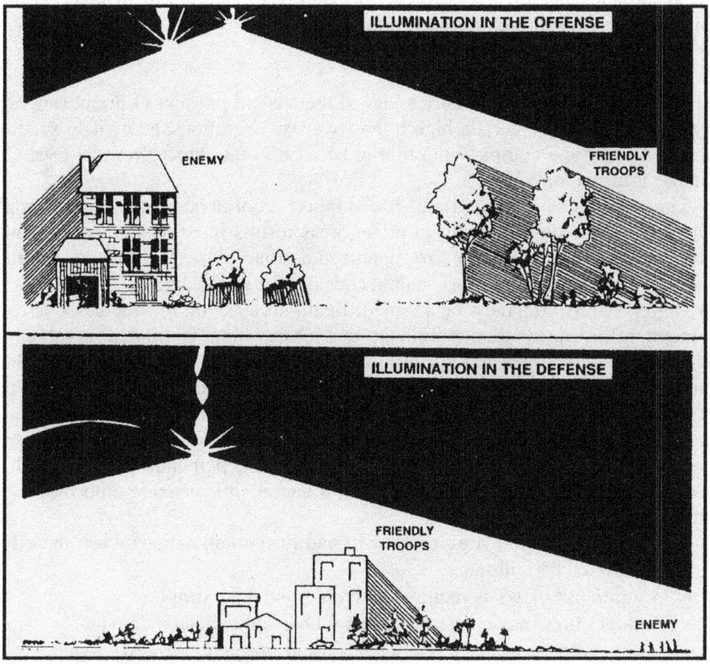

Figure 12-1. Illumination during urban operations.

g. **Special Considerations.** When planning the use of mortars, commanders must consider the following: (See Appendix K for additional TTP.)

(1) FOs should be positioned where they can get the maximum observation so target acquisition and adjustments in fire can best be accomplished. This is not necessarily on tops of buildings.

(2) Commanders must understand ammunition effects to correctly estimate the number of volleys needed for the specific target coverage. Also, the effects of using WP or RP may create unwanted smoke screens or limited visibility conditions that could interfere with the tactical plan.

(3) FOs must be able to determine dead space in urban terrain. Dead space is the area in which indirect fires cannot reach the street level because of buildings. This area is a safe haven for the enemy. For mortars, the dead space is about one-half the height of the building.

(4) Mortar crews should plan to provide their own security.

(5) Commanders must give special consideration to where and when mortars are to displace while providing immediate indirect fires to support the overall tactical plan.

Combat in urban areas adversely affects the ability of mortars to displace because of rubbling and the close nature of urban combat.

12-2. FIELD ARTILLERY

A field artillery battalion is normally assigned the tactical mission of direct support (DS) to a maneuver brigade. In certain high-intensity urban operations, a howitzer section may be placed TACON to a company in order to be used in the direct fire role. (See Chapter 10 for more information.)

a. **Considerations.** Appropriate fire support coordination measures should be carefully considered since fighting in urban areas results in opposing forces fighting in close combat. When planning for fire support in a urban area, the company commander, in coordination with his FIST chief, should consider the following:

(1) Target acquisition may be more difficult because of the increased cover and concealment afforded by the terrain. Ground observation is limited in urban areas, therefore FOs should be placed high. Adjusting fires is difficult since buildings block the view of adjusting rounds; therefore, the lateral method of adjustment may be most useful.

(2) Initial rounds are adjusted laterally until a round impacts on the street perpendicular to the FEBA. Airburst rounds are best for this adjustment. The adjustments must be made by sound. When rounds impact on the perpendicular street, they are adjusted for range. When the range is correct, a lateral shift is made onto the target and the gunner fires for effect.

(3) Special consideration is given to shell and fuze combinations when the effects of munitions are limited by buildings.

- Careful use of VT is required to avoid premature arming.
- Indirect fires may create unwanted rubble and collateral damage.
- The close proximity of enemy and friendly troops requires careful coordination.
- WP may create unwanted fires and smoke.
- Fuze delay should be used to penetrate fortifications.
- Illumination rounds can be effective; however, friendly positions should remain in shadows and enemy positions should be highlighted. Tall buildings may mask the effects of illumination rounds.
- VT, TI, and improved conventional munitions (ICM) are effective for clearing enemy positions, observers, and antennas off rooftops.
- Swirling winds may degrade smoke operations.
- FASCAM may be used to impede enemy movements. FASCAM effectiveness is reduced when delivered on a hard surface.

(4) Target acquisition is difficult in urban terrain because the enemy has many covered and concealed positions and movement lanes. The enemy may be on rooftops and in buildings, and may use sewer and subway systems. Aerial observers are extremely valuable for targeting because they can see deep to detect movements, positions on rooftops, and fortifications. Targets should be planned on rooftops to clear away enemy FOs as well as communications and radar equipment. Targets should also be planned on major roads, at road intersections, and on known or likely enemy positions. Employing artillery in the direct fire mode to destroy fortifications should be considered, especially when assaulting well prepared enemy positions. Also, restrictive fire support coordination

measures, such as a restrictive fire area or no-fire area may be imposed to protect civilians and critical installations.

(5) 155-mm self-propelled howitzers are effective in neutralizing concrete targets with direct fire. Concrete-piercing 155-mm rounds can penetrate 36 inches of concrete at ranges up to 2,200 meters. The mounted .50 caliber machine gun is also used as direct fire support. The Infantry closely protects the howitzer when it is used in the direct-fire mode, but the howitzer does not have any significant protection for the crew.

(6) Forward observers must be able to determine where and how large the dead spaces are. This area is a safe haven for the enemy because he is protected from indirect fires. For low-angle artillery, the dead space is about five times the height of the building. For high-angle artillery, the dead space is about one-half the height of the building.

(7) Aerial observers are effective for seeing behind buildings immediately to the front of friendly forces. They are extremely helpful when using the ladder method of adjustment because they may actually see the adjusting rounds impact behind buildings. Aerial observers can also relay calls for fire when communications are degraded due to power lines or masking by buildings.

(8) Radar can locate many artillery and mortar targets in an urban environment because of the high percentage of high-angle fires. If radars are sited too close behind tall buildings, some effectiveness will be lost.

(9) The use of airburst fires is an effective means of clearing snipers from rooftops. HE shells with delay fuzes may be effective against enemy troops in the upper floors of buildings, but, due to the overhead cover provided by the building, such shells have little effect on the enemy in the lower floors. The planning and use of field artillery in offensive, defensive, and other operations are also addressed in Chapters 4, 5, and 10.

b. **Direct Fire Control.** See Chapter 4 for information on direct fire control.

12-3. AIR DEFENSE ARTILLERY

This paragraph discusses the role of short-range air defense (SHORAD) artillery in support of UO. The mission of ADA at the tactical level is to protect the maneuver force and critical assets from fixed-wing and rotary-wing aircraft, unmanned aerial vehicles, cruise missiles, and surveillance platforms.

a. **Air Defense Artillery Employment Principles.** Commanders apply four principles when planning active air and missile defense operations. These principles are mass, mix, mobility, and integration.

(1) ***Mass.*** Mass is the concentration of air and missile defense combat power. It is achieved by assigning enough firepower to successfully defend the force or the asset against air and missile attack or surveillance. To mass air and missile defense combat power, commanders may have to accept risks in other areas of the battlefield.

(2) ***Mix.*** Mix is the employment of a combination of weapon and sensor systems to protect the force and assets from the threat. Mix offsets the limitations of one system with the capabilities of another and complicates the situation for the attacker. All joint and multinational arms resources are considered when applying this principle. Proper mix causes the enemy to adjust their tactics. Enemy tactics designed to defeat one system may make the enemy vulnerable to another friendly system.

(3) ***Mobility.*** Mobility is the capability to move from place to place while retaining the ability to perform the air defense mission. The mobility of air and missile defense

resources must be equivalent to the mobility of the supported force. First priority for mobility should be planning moves that support accomplishment of the mission. Tactical situations may dictate additional moves to enhance survivability. Strategic mobility is essential to support force-projection operations.

(4) ***Integration.*** Integration is the close coordination of effort and unity of action, which maximizes operational effectiveness. It is applicable, regardless of command relationships established. Active air and missile defense operations must be integrated into the supported commander's concept of the operation. The AD plan describes vertical and horizontal integration of air defense systems across the width and depth of the battlefield and includes integration with joint and multinational forces.

b. **Air Defense Artillery Planning.** SHORAD weapon systems provide low-altitude air defense coverage to ensure the force has the freedom to maneuver during combat operations. The basic air defense planning process does not change when the units operate in urbanized terrain. When determining ADA priorities in an urban environment, the air defense officer and ground commander must consider the entire spectrum of air threat, and when and how it is incorporated into the fight. The air portion of the IPB process provides a good picture of what, when, and how of threat air employment. Some basic considerations are:

(1) Rotary-wing aircraft can be employed in a variety of roles to include air assault, fire support, and CSS. They can also conduct the same missions conducted by FW aircraft. UAVs can provide vital intelligence and target acquisition data to threat forces throughout urban operations. Therefore, it is likely that they will be employed throughout the entire operation.

(2) Based on limited maneuverability and difficulty in targeting within a urban area, FW aircraft will normally target key logistics facilities, C2 nodes, and troop concentrations on the outskirts of the city, and infrastructure (key road networks, communications facilities, bridges, and rail networks) both in and out of the urban area.

(3) Air attacks normally will happen just after beginning morning nautical twilight when visibility is greatly reduced. During AD planning, the ADO weighs the risk of placing ADA assets within the urban area. He considers the following:

- Mutual support of weapon systems will be drastically reduced.
- Radar masking and degraded communications reduce air defense warning time for all units. Air defense control measures must be adjusted to permit responsive air defense within this reduced warning environment.
- Moving weapon systems will be difficult.
- Providing security for isolated firing positions will be difficult.
- Communications between elements may be difficult.
- Digital communications will require an enhanced position location reporting system (EPLRS) via SINCGARS to be effective.
- High-altitude ADA systems can protect the force from positions on or outside of the edge of the city.

c. **SHORAD Employment.** Linebackers, Bradley Stinger fighting vehicles (BSFVs), and Avengers are most effective on the outskirts of the urban area in a weighted coverage type defense where they can maximize the effects of their weapon systems, to cover the most likely enemy air avenues of approach. They are employed within the urban area based on METT-TC, and only when absolutely necessary.

Man-portable air defense systems (MANPADS) teams may be deployed outside the urban area in firing positions that are high and clear enough to provide a clear primary target line (PTL). Special cases will put Stinger gunners on rooftops within the urban area only after the area is secured.

d. **Early Warning.** The planning process, the IPB, METT-TC, and the commander's intent will dictate where sensors should be deployed adjacent or forward of the urban area. Sentinel sensor employment should not be thought of as either-or, but as two techniques that may be combined. In any case, the sensor employment design should be the one that best supports the commander's intent. Early warning dissemination during UO will be digital from the Sentinel to the SHORAD elements, and voice from the SHORAD elements to the maneuver elements. The SHORAD elements will broadcast EW IAW TSOP on the appropriate command voice net.

12-4. ENGINEERS

Normally, an engineer *platoon* will be attached to a light infantry company; an engineer *company* will be attached to a mechanized infantry company. Most engineer manual labor tasks (for example, preparing fighting positions) will have to be completed by Infantry units, with reinforcing engineer heavy-equipment support and technical supervision. (See Chapter 11 for discussion of mobility, countermobility, and survivability considerations.)

a. **Offensive Missions.** Engineers may perform the following missions during offensive operations.

- Conduct a technical reconnaissance to determine the location and type of enemy obstacles and minefields, and to make breaching recommendations.
- Clear barricades and heavy rubble with earth-moving equipment or explosives to assist forward movement.
- Use explosives to destroy fortifications and strongpoints that cannot be reduced with the maneuver unit's organic assets.
- Use the ACE, if available, to destroy structures or to clear rubble.
- Lay mines to protect flanks and rear areas.
- Conduct mobility operations (gap crossing).
- Locate and remove mines that may hamper the unit's movement.
- Conduct breaching operations.
- Conduct route reconnaissance.

b. **Defensive Missions.** Engineers may perform the following missions during the defense of a urban area.

- Construct complex obstacle systems.
- Rubble buildings.
- Lay mines.
- Assist in the preparation of defensive positions and strongpoints.
- Maintain counterattack, communications, and resupply routes.
- Enhance movement between buildings, catwalks, bridges, and so on.
- Crater roads.
- Clear fields of fire.
- Fight as Infantry, when needed.

c. **Defense Against Armor.** In defensive situations, when opposed by an armor-heavy enemy, priority should be given to the construction of antiarmor obstacles throughout the urban area. Use of local materials, where possible, makes obstacle construction easier and reduces logistics requirements. Streets should be barricaded in front of defensive positions at the effective range of antitank weapons. These weapons are used to increase the destruction by antiarmor fires, to separate enemy Infantry from their supporting tanks, and to assist in the delay and destruction of the attacker. Antitank mines with anti-handling devices, integrated with antipersonnel mines in and around obstacles and covered by fires, help synchronize a defensive fire plan.

12-5. MILITARY POLICE

Military police operations play a significant role by assisting the Infantry commanders in meeting the challenges associated with conducting UO. MPs provide a wide range of diverse support in urban terrain, to include area damage control, area security, and EPW operations and non-combatant operations. MP operations require continuous coordination with host nation civilian police to maintain control of the civilian population and to enforce law and order. These MP assets may be attached or OPCON to infantry units for the duration of a specific mission and then will be released to the control of their parent unit. Their training in urban operations can be of great assistance for help in crowd control, roadblocks/checkpoints, marking and controlling routes, and EPW control.

a. **Area Damage Control.** MP units take measures to support area damage control operations that are frequently found in urban areas. With the increased possibility of rubbling, MP units report, block off affected areas, and re-route movement to alternate road networks.

b. **Area Security.** MP units also secure critical areas, such as communications centers and water and electrical supply sources. These MP assets can assist an Infantry unit that is assigned a mission like this. (See Chapter 4 for more information.)

c. **EPW/Noncombatant Operations.** MP units are tasked with EPW operations and civilian operations and to perform them as far forward as possible. MPs operate collecting points and holding areas to briefly retain EPWs and civilians. EPW and civilian operations are of great importance in urban areas because the rate of capture of EPWs and the presence of civilians can be higher than normal. Infantry units can use MP assets to assist them in conducting these types of operations when assigned these types of missions.

d. **Other Urban Operations Considerations.** Other considerations include support for:

(1) ***Maneuver and Mobility.*** MPs can support breaching operations, passage of lines, and operate straggler control points.

(2) ***Area Security.*** MPs conduct area damage control, secure critical sites, and conduct response force operations.

(3) ***Internment and Resettlement.*** MPs collect EPWs/CIs from combat units; safeguard and maintain accountability; protect and provide humane treatment for all personnel under their control including the management of dislocated civilians.

(4) ***Law and Order.*** MPs can conduct investigations during UO from minor crimes to war crimes.

(5) ***Police Intelligence.*** MPs gather criminal intelligence, which can affect the operational and tactical environment during UO.

12-6. COMMUNICATIONS

See Appendix L for information on communications.

CHAPTER 13
COMBAT SERVICE SUPPORT

During urban operations (UO), the terrain and the nature of the operations create unique demands on the battalion combat service support (CSS) system. Increased ammunition consumption, casualties, transportation difficulties resulting from rubble, and the decentralized nature of operations all challenge the battalion CSS operators and planners. Solutions to these problems require innovative techniques and in-depth planning. This chapter focuses primarily on battalion-level CSS, but brigade- and company-level tactics, techniques, and procedures (TTP) have been inserted where applicable. All types of Infantry forces can use the TTP in this chapter, with modifications.

Section I. GENERAL

Although UO present a different set of problems, the supply and movement operations of the support platoon change little. The guidelines and principal functions of CSS are explained in this section.

13-1. GUIDELINES

Regardless of the conditions under which UO are conducted, there are some general guidelines for CSS.

- Preconfigure resupply loads and push them forward at every opportunity.
- Provide supplies to using units in required quantities as close as possible to the location where those supplies are needed.
- Protect supplies and CSS elements from the effects of enemy fire both by seeking cover and avoiding detection.
- Disperse and decentralize CSS elements with proper emphasis on communication, command and control, security, and proximity of main supply route (MSR) for resupply.
- Plan for carrying parties and litter bearers.
- Plan for and use host country support and civil resources when practical.
- Position support units as far forward as the tactical situation permits.
- Plan for requesting and distributing special equipment such as body armor, toggle ropes with grappling hooks, ladders, and hand tools.
- Position support units near drop or landing zones (DZ/LZ) for resupply from corps to forward units to reduce surface movement.

13-2. PRINCIPAL FUNCTIONS

The principal functions of CSS in urban areas are to arm, fuel, fix, and man the combat systems.

a. **Arm.** Urban combat is characterized by extremely high ammunition expenditure rates. Not only do individual soldiers fire more small arms ammunition, they use more munitions such as smoke, stun, concussion, and fragmentation grenades; LAWs; AT4s; Claymore mines; and explosives. The ammunition consumption rate for the first day of combat in an urban area can be up to four times the normal rate. Even though it decreases during succeeding days, consumption remains high. Commanders and S4s must plan to meet these high consumption rates. The plan must include how ammunition and explosives are to be moved forward to the companies. Bradley fighting vehicles (BFVs) or armored personnel carriers (APCs) may have to be allocated for the movement of ammunition if rubble or glass prevents wheeled-vehicle traffic. Carrying parties may be used if streets are blocked by rubble.

b. **Fuel.** The amount of bulk fuel needed by a battalion or task force during combat in urban areas is usually reduced, due to the density of the terrain. Combat vehicles normally use less fuel in urban areas, because they travel shorter distances and perform less cross-country traveling. Engineer equipment and power generation equipment may use more fuel but requirements are small. The exception to this rule is a force equipped with M1 tanks. Because of the engine design, these tanks use fuel at a fairly steady rate whether they drive long distances or not. A unit may not use much fuel daily, but when it does need fuel, a problem exists in delivering bulk fuel to vehicles. In open terrain, a vehicle that has run out of fuel can be recovered later. But in urban areas, the same vehicle may be difficult to recover and could be lost quickly. Commanders and S4s must plan and provide the means of moving bulk fuel forward to combat units. They must pay special attention to any M1 tank units they have attached. These vehicles are not easily refueled by hand.

c. **Fix.** Maintenance teams must operate well forward to support units fighting in urban areas. Some maintenance operations may be consolidated in civilian facilities and many vehicles may have to be fixed near the fighting positions. Battle damage assessment and repair (BDAR) procedures allow mechanics to be inventive and make maximum use of battlefield damage, analysis, and repair techniques to return damaged vehicles to a serviceable condition (see applicable TMs). Other considerations include:

(1) Combat in urban areas generates a high demand for tires.

(2) The dust and rough handling characteristic of combat in urban areas also places great strains on communication and night observation devices.

(3) The unit armorer and their small-arms repair kits provide only limited maintenance. S4s should plan for increased weapon maintenance demands and coordinate maintenance support from higher headquarters. Based on recommendations from the staff (S3, S4, and motor officer), the commander may choose to consolidate and cross level major items of equipment and weapons.

NOTE: During recent experimentation, a higher number of M16A2 and M4 rifles were rendered unserviceable due to rounds impacting on the weapons as a result of close combat during precision and high intensity engagements. CSS planners must be prepared to replace a higher number of unserviceable weapons during urban combat.

d. **Man.** Units conducting combat in urban areas frequently experience higher casualty rates. Casualty feeder reports must be prepared and forwarded to the battalion personnel and administration center (PAC). Battalion S1s must be prepared to request replacements based on the OPTEMPO, realizing that assaulting units will probably sustain higher casualties than supporting units.

(1) ***MEDEVAC.*** The battalion surgeon (medical platoon leader), in coordination with the S1 and company first sergeants, must plan to expedite the evacuation of the wounded out of the urban area. Forward aid station locations and evacuation routes must be planned and disseminated to the lowest level. Higher casualty rates should be expected and may require stockpiling medical supplies and augmenting medical personnel and nonmedical personnel to serve as litter bearers.

(2) ***Replacements.*** The battalion PAC should process replacements quickly and transport them to their new unit. The battalion PAC is responsible for reviewing assignment orders, welcoming soldiers to the battalion, assigning soldiers IAW commanders priorities, obtaining personal information, and collecting medical records and forwarding them to the aid station. It is also responsible for adding names to the battle roster, preparing SIDPERS input, and processing the names into the servicing postal activity. The S1 and PAC should brief the new soldiers on the tactical situation, provide mess and medical support as needed, inspect for combat critical equipment shortages, and coordinate transportation to units. Company replacements should be brought forward from the field trains with the LOGPAC and linked up with their new unit's first sergeant. If replacements are brought forward at unscheduled times, the logistic release point (LRP) should still be used as the linkup point. The S1 must coordinate with the S3 or S4 to transport replacements over long distances and to issue missing individual combat equipment. At night, replacements may need to be sent to their new unit with guides. These groups may be used to carry critical supplies and ammunition forward.

(3) ***Personnel Accountability.*** Proper accountability of platoon personnel and accurate strength reporting are essential to support decision making by platoon leaders, company commanders, and the battalion commander. Using battle rosters, leaders in the platoon maintain accurate, up-to-date records of their personnel. At periodic intervals, they provide strength figures to the company CP. During combat, they provide hasty strength reports on request or when significant changes in strength occur.

(4) ***Casualty Reporting.*** By-name casualty information is reported by secure means to company headquarters during lulls in the tactical situation. Soldiers having direct knowledge of an incident complete a DA Form 1155 (Witness Statement on Individual) to report missing or captured soldiers, or casualties no longer under US control. DA Form 1156 (Casualty Feeder Report) is used to report soldiers who are killed or wounded. (See AR 600-8-1 for instructions on how to complete these forms.) After being collected and reviewed for accuracy by the platoon leader or platoon sergeant, these forms are forwarded to the company headquarters. These forms provide important casualty information and are also used to determine the platoon's replacement requirements.

(5) ***Stress Management.*** The surgeon, in coordination with the S1, must develop a flexible and comprehensive combat health support (CHS) plan to treat and or evacuate those with battle fatigue or combat stress. Medical personnel can provide training to unit leaders on recognizing stress and actions, which can be taken to mitigate its effect. (Refer to FM 22-51 for information on combat stress control for unit leaders.)

(a) Prolonged combat in urban areas generates incredible stress. Some soldiers show signs of inability to cope with such stress. Stress management is the responsibility of commanders at all levels. The surgeon coordinates for trained personnel, such as medical personnel and unit ministry team personnel, to support units when the situation dictates.

(b) The more intense the combat, the higher the casualties. The more extreme the weather, the longer the battle lasts, the more combat exhaustion and stress, the more casualties.

(c) The S1 should plan to provide the soldier with a short rest period along with warm food and hot liquids in a protected section of the battalion rear area. He should take this opportunity to give the soldier command information products (obtained through public affairs channels). These inform the soldier about the larger picture of the battle, the theater of operations, the Army, and the welfare of the nation as a whole. As a result of treating stress problems in the battalion area, a higher percentage of stress casualties can be returned to duty than if they had been evacuated farther to the rear. When recovered, they should be returned to their original units.

13-3. SUPPLY AND MOVEMENT FUNCTIONS

The S4, support platoon leader, and battalion motor officer share responsibility for coordinating supply and movement functions within the battalion. The use of preconfigured LOGPACs pushed forward to the elements in contact is the key to a successful resupply operations. The support platoon contains trucks and trained drivers to move supplies forward. Some classes of supply, and how they are moved, may assume greater importance during UO.

a. **Class I (Rations and Water).** The process of ordering and moving rations to the battalion's forward positions is complicated by the dispersed nature of UO and their increased caloric demands on soldiers. The battalion mess section must try to provide hot meals, whenever possible.

(1) Urban combat causes great stress on soldiers and requires great physical exertion. This combination of stress and exertion causes rapid dehydration. Unless potable water is continuously provided, soldiers will seek local sources, which are usually contaminated by POL runoff, sewage, bacteria, or unburied corpses. Soldiers who are not provided sufficient quantities of potable water become casualties due to drinking from contaminated sources or from dehydration. Waterborne contamination can quickly render entire units combat-ineffective.

(2) Water and other liquid supplements such as coffee, tea, or soup that must be brought forward to exposed positions may need to be backpacked there at night.

b. **Class II (General Supplies).** UO also places a great strain on uniforms and footgear. The battalion S4 should increase his on-hand stocks of uniforms, boots, and individual combat equipment such as protective masks and armored vests. Chemical protective overgarments (CPOGs) either tear or wear out quickly when worn in the rubble typical of combat in urban areas. Limited amounts of other Class II and IV items may be available locally. These should be gathered and used if authorized and practical. Local shops may provide such items as hand tools, nails, bolts, chains, and light construction equipment, which are useful in preparing a defense or reducing enemy-held positions. The unit's organic wire communications net may be augmented with locally obtained telephone wire and electrical wire.

c. **Class III (POL).** Bulk fuel may have to be brought forward from fuel tankers by 5-gallon cans. One man can carry a fuel can long distances, even over rubble, if it is lashed to a pack frame. Supplies of bulk Class III items and some prepackaged POL may be available at local gas stations and garages. These may be contaminated or poor quality.

The S4 should coordinate with the brigade S4 to have a fuel test performed by a qualified member of the supporting forward support battalion (FSB) or the forward support company (FSC).

d. **Class IV (Barrier Materials).** Barriers can be built from abandoned cars and buses, which are dragged into position, turned on their sides, and chained together through the axles. A unit defending an urban area may require less Class IV materials than in other areas because of local availability. After coordinating the effort with higher headquarters, the S4, support platoon leader, and supporting engineer officer can gather materials for use in strengthening a defense. Cargo trucks from the support platoon, wreckers or recovery vehicles from the maintenance platoon, and engineer construction equipment can be used to load and move barrier materials. Normally, division or corps-level assets bring Class IV materials forward. Defense of an urban area may require concertina wire and or barbed wire to restrict the enemy infantry's movements.

e. **Class V (Munitions).** Urban combat causes ammunition to be expended at extremely high rates. Commanders should plan for early resupply of explosives, grenades, and ammunition for small arms and direct/indirect fire weapons.

(1) In the defense, the S4 should prestock as much ammunition as practical in dispersed storage areas. These storage areas should be protected and have easy access from the forward defensive positions. In the offense, attacking troops should not be overburdened with excessive ammunition. Mobile distribution points may be set up as low as company level.

(2) Commanders and S4s must plan to deliver ammunition continuously to the leading elements as they advance. Armored vehicles may carry this close behind the advancing troops or by designated carrying parties. Modern ammunition is characterized by extensive amounts of packing material. The S4 must plan to have an element remove the ammunition depot's overpack before it is transported forward. Resupply by helicopter (prepackaged slingloads) may be feasible.

(3) Removing the outer packaging from large amounts of ammunition can be time-consuming. It may require the efforts of the entire support platoon, augmented by available soldiers. If carrying parties are used to move ammunition forward, an individual can carry about 75 to 90 pounds using a pack frame or rucksack. Carry the bulky and heavier loads by lashing them to litters and using carrying teams (two to four men). Loads up to 400 pounds can be carried moderate distances using four-man teams.

NOTE: Do not use medics to carry ammunition forward; it is a violation of the Geneva Accords. Consider using aid and litter teams to bring ammunition forward and then to back haul casualties.

f. **Class VIII (Medical Supplies).** Due to the decentralized nature of combat in urban areas, medical supplies should be dispersed throughout the battalion and not consolidated only with the aid station and the individual combat medic. Individual soldiers, especially trained combat lifesavers, should carry additional bandages, cravats, and intravenous sets. Companies should request additional splints and litters.

13-4. COMPANY RESUPPLY OPERATIONS

The company XO, first sergeant, and supply sergeant normally share the responsibility for coordinating all resupply and transportation requirements for the company. As stated previously, preconfigured LOGPACs are normally pushed forward to the companies in contact as the normal method of resupply operations. The support platoon contains the trucks and trained drivers needed to move supplies forward. The XO normally coordinates transportation requirements with the battalion S4. Generally, the priorities of resupply for infantry companies in urban combat are ammunition, medical supplies, water, and food. These priorities may change based on the factors of METT-TC; however, they will generally remain constant. This paragraph will discuss how infantry companies conduct resupply operations during UO.

a. **LOGPAC.** Resupply operations normally occur once a day. When possible, they should be conducted during periods of limited visibility. Company resupply is primarily a *push* system. This is accomplished through the reception of a LOGPAC from battalion. The contents of the LOGPAC are planned by the S4 based on mission requirements. The supplies are normally organized and assembled in the battalion field trains by the company supply sergeant under the supervision of the HHC commander and the support platoon leader. The LOGPAC should provide all supplies, equipment, and personnel needed to sustain the company for the next 24 hours or until the next scheduled LOGPAC delivery. Adjustments to the LOGPAC are sent to the battalion S4 by either the supply sergeant or the XO. These changes should be specific and based on the company commander's priorities. The company status reports usually serve as the basis for the LOGPAC and provide the necessary troop strength figures to determine resupply needs.

NOTE: Recent experimentation with different UO TTP has seen increased use of the company support team (CST) technique. The CST may consist of a squad or platoon (-) of soldiers that would come under the control of the company supply sergeant, XO, or 1SG. The soldiers come from the company headquarters section and organic platoons/sections. The CST fills magazines and canteens/camelbacks, removes packing material from ammunition, and generally prepares and delivers LOGPACs for immediate use by the rifle platoons. The CST also assists with CASEVAC by administering first aid and moving casualties from the point of injury to patient (casualty) collection points and/or battalion aid station.

b. **Distribution of Supplies.** Supplies are distributed as follows:

(1) ***Movement of LOGPACs.*** Once the LOGPAC is formed in the field trains, it is ready to move forward under the control of the supply sergeant. The support platoon leader normally organizes a convoy for moving the LOGPACs along a supply route to the logistics release point (LRP), where the first sergeant or XO takes control. The LRP is usually a covered and concealed position that offers protection to those distributing supplies (for example, a large enough building that can be secured locally).

(a) The LRP should be positioned close enough to the combat area so that LOGPACs can be off-loaded and carried to the platoons. LOGPACs should be broken down into 50- to 75-pound loads to be carried in rucksacks. Litters can be used to carry heavier loads, when excess are available or evacuating wounded is not required. Companies should plan for carrying supplies and identify soldiers to do this.

(b) The platoon sergeant assumes control of the supplies once they arrive in the platoon area. Carrying parties should use covered and concealed routes through buildings to move from the LRP to the company sector. Carrying parties should be prepared to provide their own security to and from the LRP. When necessary, a security force (for example, a fire team) can accompany the carrying party.

(2) ***Pre-positioned Supplies.*** Based on METT-TC factors, LOGPACs may be pre-positioned in predetermined locations by the battalion S4. The LOGPAC would be placed in a covered and concealed location where the company retrieves the supplies. This system works best when the company is moving from one location within the urban area to another in a relatively secured area. Care should be exercised to ensure supplies are not left unprotected or pilfered through by the enemy or civilians. For example, the company can provide a two- to four-man security team to the S4 to guard the pre-positioned supplies until the company can recover them.

c. **Company Resupply Techniques.** Once the LOGPACS are brought to the company area, the XO or first sergeant has three options for resupplying the platoons: *in position*, *out of position*, and *pre-position*.

(1) ***In Position.*** This is the most common technique that will be used on urban terrain when the company is conducting operations that require platoons to maintain combat power forward (during contact or when contact is imminent). For example, this technique would probably be used during the consolidation and reorganization phase of an offensive operation where a counterattack is expected. Ammunition, medical supplies, and water would be brought forward by the support element and resupplied directly to the platoons in cleared buildings. All the assaulting platoons would remain in position.

(2) ***Out of Position.*** This technique is used when the situation does not require all combat power to be forward (contact is not likely). The XO or first sergeant would establish a resupply point in a covered and concealed position (a cleared building with overhead cover) to the rear of the platoon. Platoons would send selected personnel back to the resupply point, pick up the supplies, and move back to position.

(3) ***Pre-Position.*** This technique is most often used during defensive operations when supplies are often cached (pre-positioned and concealed) in buildings throughout the company sector or subsequent battle positions.

d. **Other Considerations.** Lessons learned from Operation Just Cause in Panama provide additional key points for company commanders to consider:

(1) LOGPAC vehicles must be prepared to back-haul captured enemy equipment and prisoners. Ensure drivers and company supply personnel are proficient in handling EPWs.

(2) Train and qualify supply personnel to configure and rig supplies for external helicopter lift.

(3) Consider deploying with airload slings and nets, and have procedures in place to expedite resupply operations.

(4) Train resupply personnel to receive container delivery system (CDS) bundles from a C-130 or C-141. This is a viable means of resupply in urban terrain when convoys and Army aviation are limited.

(5) Soldiers must deploy with extra sets of clothing. Uniforms deteriorate quickly in urban combat; especially in hot, humid conditions.

(6) Soldiers must deploy with sufficient toilet articles to sustain them for at least 15 days. Sundry packs are not maintained in the depot system and soldiers may not be resupplied with health and hygiene articles for up to 60 days after deployment.

(7) Soldiers may have compassion for disadvantaged people and may give away clothing and rations. Commanders and first sergeants must monitor this situation, and caution soldiers in austere conditions.

13-5. LOAD PLANNING AND MANAGEMENT

The soldier's load (what the soldier carries) is a crucial concern for leaders at all levels. This load is especially true during UO, where the demands of physical and mental stress are combined with the need to carry additional ammunition and water. (FM 7-10, Chapter 8, provides detailed information concerning load planning and management.) This paragraph highlights specific load management concerns for infantry company commanders during UO.

a. **Army Research.** Army research indicates that a soldier can carry an amount equal to 30 percent of his body weight and still retain a high percentage of his agility, stamina, alertness, and mobility (all of these are qualities that directly affect the ability of a soldier to fight on urban terrain). For the average soldier weighing 160 pounds this would be a 48-pound load. For each pound over 30 percent, the soldier loses a proportional amount of his functional ability. When his load exceeds 45 percent of his body weight, or 72 pounds, his functional ability drops rapidly and the chance of his becoming a casualty increases. The company commander must be directly involved in load planning and management. He must weigh the needs of the mission based on METT-TC against the physical reality of what a soldier can carry into the fight.

b. **Loads.** FM 7-10 divides loads into three major types: the combat, sustainment, and contingency loads. The combat load is defined as the minimum mission-essential equipment, as determined by the commander. This includes only what is needed to fight and survive immediate combat operations. The combat load is further broken down into two levels—the fighting load and the approach march load. The fighting load is defined as what the soldier carries once contact has been made with the enemy. It consists of only essential items the soldier needs to accomplish his task during the engagement. The remainder of this discussion will concentrate on describing the manner in which to manage a fighting load in urban combat. Company commanders must recognize that urban combat will place additional physical stress on soldiers. Part of this physical stress will be caused by the additional weight that soldiers will carry. The calculations in Table 13-1 are provided to assist in load planning and management.

COMMON ITEMS	POUNDS
BDUs and boots	8.2
Pistol belt, straps, and first aid kit	1.6
Canteen, cup and cover (with water)	3.3
Canteen, cover (with water)	2.4
Gloves	0.3
Socks	0.3
TOTAL	**16.1**
DUTY LOAD	**POUNDS**
M16A2 w/30-round magazine	8.2
Two ammunition pouches	1.8
Six magazines/180 rounds	6.3
Two fragmentation grenades	2.0
Two stun grenades	1.8
One smoke grenade	2.6
TOTAL	**22.7**
THREAT PROTECTION	**POUNDS**
M40 protective mask with cover	4.2
Knee and elbow pads (average weight)	1.5
Helmet	3.4
Eye protection	.3
Body armor	17.5
TOTAL	**26.9**
VARIABLES	**POUNDS**
Bayonet with scabbard	1.3
Rifle launched entry munition	1.9
TOTAL	**3.2**
TOTAL FIGHTING LOAD	**POUNDS**
Common items	16.1
Duty load	22.7
Threat protection	26.9
Variables	3.2
TOTAL	**68.9**
NOTE: All weights were rounded to the nearest tenth of a pound. Body armor was calculated with the front and back plates included. If a double basic load of ammunition (includes 2 more ammo pouches, 6 additional magazines and 180 additional rounds of ammunition) is added, an additional 8.10 pounds must be calculated, for a total of 77 pounds. If two additional quarts of water (a two-quart canteen with cover) are added, an additional 4.80 pounds must be calculated, for a total weight of 81.80 pounds.	

Table 13-1. Load calculation.

c. **Load Management Techniques.** Based on the calculations, soldiers will probably carry more than the recommended 48-pound fighting load during urban combat, which will tax soldiers and create a greater amount of physical exertion. Commanders must be aware of this fact and manage loads accordingly. The following are key-load management techniques used during UO.

(1) Standardize the way items are carried within the unit. Soldiers should evenly distribute items on their load-carrying equipment (LCE) and the cargo pockets of their BDUs. Do not place anything on the firing side of the LCE that interferes with aiming the weapon.

(2) Distribute loads throughout the unit. Have the assault element only carry the items necessary to accomplish the mission, usually ammunition and water. Designate individuals in the support element to bring additional ammunition, medical supplies, and water forward as the tactical situation permits. Replace ammunition and water carried on the LCE as soon as possible. Medical personnel can only carry medical supplies; carrying ammunition is a violation of the Geneva Conventions.

(3) Designate individuals who will perform breaches and modify ammunition loads accordingly. Rotate these individuals when they tire, or after they have made numerous breaches.

(4) Rotate the assault element after each intermediate objective is secured, as the tactical situation permits. Try to maintain fresh assault troops to the maximum extent possible.

(5) Always consider the use of augmented transportation assets to carry loads. (For example, host nation or allied force vehicles, pack animals, civilian volunteers, wheelbarrows, and bicycles; however, do not procure them without authority.)

(6) Avoid unnecessary movement and displacements. To conserve the soldier's stamina, plan the mission as efficiently as possible.

(7) All leaders must supervise the soldier's load closely, through precombat inspections (PCIs), to ensure that soldiers carry only the items necessary.

13-6. OTHER COMBAT SERVICE SUPPORT TACTICS, TECHNIQUES, AND PROCEDURES

The following TTP are a result of lessons learned from combat training centers (CTCs) and different unit standing operating procedures (SOP).

a. **Soldier Top-Off Point.** This technique is primarily used at battalion or brigade level. A position is established that provides simple services to soldiers in a centralized location. The soldier top-off point can be as detailed as assets and time permit. These services are normally provided in a relatively secure location such as a lodgment area, the brigade support area, or the combat trains. Security considerations and mission requirements will determine the exact location of the soldier top-off point. Examples of activities and services at a soldier top-off point are:

- Distribution of mail and newspapers.
- Serving of hot food, to include fresh fruit and cold/hot drinks.
- Chaplain services/support.
- Showers.
- Combat health support, to include restocking of aid bags and combat lifesaver bags.
- Supply issue points for the distribution of water, MRE, ammunition, and so on.
- Distribution of A-bags.
- Sleep/rest area, which may include heated tents with cots.
- Briefing area.

b. **Resupply Techniques.** The following resupply techniques may be used in UO.

(1) ***Plastic Bags.*** This technique involves using double or triple stacked plastic bags to preconfigure soldier resupply. Resupply could include prefilled magazines, MREs, bottled water, and first aid dressing. Plastic bags are readily available and do not require back haul.

(2) ***Plastic Bladders***. Plastic bladders used in milk dispenser machines in most military dining facilities make excellent water containers. These bladders hold about five gallons of water, have a spout that permits canteens to be filled easily, and will fit into a rucksack or other container. These bladders have survived 60-foot drops from hovering helicopters when placed inside an empty MRE box. The bladder and MRE box are expendable. A box of 100 bladders is inexpensive and readily available through supply channels.

(3) ***Water Bottles.*** In many recent operations bottled water was the standard for supplying water to the soldier for individual consumption. During UO, this type of water distribution can be both a benefit and challenge. Although plastic water bottles are easy to transport and eliminate the need for back haul, they are expensive; usually require a contract with a commercial provider; and, in bulk, come in flimsy cardboard boxes. Regardless, this may be the best solution for water resupply in many situations.

(4) ***Speed Balls.*** This technique uses helicopters and preconfigured loads to resupply units. In urban areas, rooftops or secured drop-off points, such as small parking lots or playing fields, are used (Figure 13-1). CSS personnel prepackage supplies in aviation kit bags, duffle bags, or other suitable containers. Helicopters fly as close to the drop point as possible, reduce speed, drop supplies, and leave the area quickly in order to reduce exposure time. Supplies should be prepackaged in bubble wrap or other shock-absorbing material to minimize damage. This technique can work well where there is a minimal or no air defense threat and where the units receiving supplies are light infantry or other dismounted type forces.

Figure 13-1. Speed ball delivery.

(5) ***Rope.*** A technique similar to *speed balls* is to lower water cans or supply containers from a rope. Water cans or containers are attached to the rope using a snap link and slid down. Once the supplies are received, the same rope can be used to back haul empty water cans or other items. Care must be taken to insure that a counterweight remains at the end of the rope to insure that it does not flap into the wind and catch the helicopter propellers. Heavy items can also quickly slide down the rope and damage the supplies, for example a five gallon water can weighs 40 pounds. Multiple water cans and/or "speed balls" can be linked together by means of a sling rope that has a snap link attached. To slow the rate of descent, a round turn can be applied to the snap link holding the supplies. Additionally, the rope can be belayed from the ground, which will also control the rate of descent.

(6) ***SKEDCO Litters.*** SKEDCO litters can be used effectively during UO at the squad and platoon level to move supplies and equipment. The litter can be used to move heavier items, such as mortar rounds, through difficult areas such as rubble. Additionally, SKEDCO litters can be used in conjunction with ropes and pulleys to haul supplies along the side of a building or through elevator shafts or destroyed stairwells.

(7) ***Body Bags.*** A "human remains bag" (NSN 9930-01-331-6244) Type 2 can be used to move supplies. The word "SUPPLIES" should be clearly stenciled in a bright color on the bag to avoid confusion. These bags are useful because they are durable, waterproof, have carrying handles, can hold a large number of items, and can be folded and carried in a rucksack. They are readily available through supply channels.

13-7. PERSONNEL SERVICES

The S1 plans for all personnel services supporting and sustaining the morale and fighting spirit of the battalion. Among the most important of these services are:

- Religious support.
- Postal services.
- Awards and decorations.
- Rest and recuperation.
- Replacement operations.
- Strength accounting.
- Casualty reporting.
- Finance support.
- Legal support and services.
- Public affairs activities.

A unit may encounter severe problems if it allows civilians to steal or destroy its equipment. Even friendly civilians may steal supplies or furnish intelligence to the enemy. Civilians should be evacuated to prevent pilferage, sabotage, and espionage. Military police and civil affairs units normally provide control of the civilian population. Collection points for noncombatants are established in rear areas. The S1 is the battalion's link to the population control programs of the higher command.

13-8. DECEASED PERSONNEL

The commander is responsible for the evacuation of deceased personnel to the nearest mortuary affairs collection point, whether they are U.S., allied, enemy, or civilian. (See

FMs 10-63 and 10-497 for specific information on the handling of deceased personnel.) Some general considerations for handling deceased personnel include:

- The theater commander is the approval authority for hasty burial.
- The deceased's personal effects must remain with the body to assist in the identification of the body and to facilitate shipment of personal effects to the next of kin. Unauthorized retention of personal items is considered looting and is punishable by UCMJ.
- When operating under NBC conditions, the bodies of deceased personnel should be decontaminated before removal from contaminated areas to prevent further contamination and casualties.
- Care must be exercised when handling deceased personnel. Improper handling can result in a significant decrease in unit and civilian morale.

Section II. COMBAT HEALTH SUPPORT

"During the earlier years of our involvement in the Viet Nam War, it was a rare Medal of Honor list that failed to relate the story of some devoted, selfless medic. Wherever his infantry platoon, artillery battery, or tank troop went, the 'Doc' was always there, ready with skilled, competent hands to do his tasks."

Major Henry J. Waters
MS Medical Implications of Combat in Cities
Unpublished Research Paper, April 1974

13-9. MEDICAL CONSIDERATIONS FOR THE BATTALION STAFF

The battalion surgeon, in coordination with the battalion S1, physician's assistant, and field medical assistant, is responsible for planning and executing medical functions within the battalion. The most critical functions during urban combat include preventive medicine, trauma treatment, and evacuation. In addition, there should be a plan for treating, decontaminating, and evacuating NBC-related casualties that may occur in combat in urban areas.

a. **Disease.** Combat in urban areas exposes soldiers not only to combat wounds but to the diseases endemic to the area of operations. Commanders must enforce prevention measures against the spread of infectious diseases. The medical platoon advises the commander on how best to implement and use preventive measures.

b. **Medical Supplies.** Critical medical supplies should be planned for all operations and resupplied to the company as needed. (See paragraph 13-3f for more information.)

c. **Combat Medics.** Although the combat medic normally attached to each rifle platoon is the soldier best trained in the treatment of traumatic injury, he can quickly become overwhelmed by the number of casualties needing care. The commander must train selected soldiers within the platoons to administer enhanced first aid using the combat lifesaver program. The work of these combat lifesavers, plus the buddy-aid efforts of individual soldiers, eases the burden of the combat medic and allows him to concentrate on the seriously wounded. The medical platoon should plan to care for the mass casualties inherent in combat in urban areas. Combat medics and lifesavers should expect a higher incidence of crushing injuries, eye injuries, burns, and fractures due to falling debris, spall from buildings, rubble, and fire hazards. Additional effects, such as concussive shock and hearing loss due to explosives, should be expected.

d. **Medical/Casualty Evacuation (MEDEVAC/CASEVAC).** The difficulties encountered when evacuating casualties from urban terrain are many and require innovative techniques and procedures (Figure 13-2). The planning for medical evacuation in urban terrain must include special equipment. For example, axes, ropes, pulleys, pitons, and other climbing equipment that will be used by litter teams; use of air ambulances and the rescue hoist, when feasible; use of the litter/ambulance shuttle system; and communications requirements and techniques for locating casualties.

Figure 13-2. Treatment of casualties.

(1) ***Casualty Collection Point (CCP).*** Although litter teams are labor intensive, they are required for evacuation from buildings, where casualties can occur on any level. Also, rubble in the streets, barricades, and demolition of roads impede the use of ground ambulances, requiring a heavy reliance on litter teams. Casualties are brought to the designated CCP and evacuated from there to the battalion aid station. The CCP is placed in a covered and concealed location that has overhead cover (usually a heavy-clad building that has not collapsed). The CCP should be located at a point where the field ambulances can reach them, yet close enough to the combat area so that casualties do not have to be carried great distances. It must be marked well enough that it is easy to find. By predesignating collection points, soldiers who are wounded but still ambulatory can walk to these points, hastening the CASEVAC effort.

(2) ***Air Ambulances/Shuttle System.*** When available, air ambulances equipped with the rescue hoist may be able to evacuate casualties from rooftops or insert medical personnel where they are needed. The vulnerability to enemy fire must be considered and weighed against probable success of the evacuation mission. Pilots must be familiar with over flying urban areas and the atmospheric conditions they may encounter. A litter/ambulance shuttle system with collection points, air exchange points (AXPs), and relay points can be established within the battalion area. The first sergeant or XO can coordinate with the S1 and S4 to determine the location of the AXPs within the urban

area in order to facilitate MEDEVAC for the company. By establishing an ambulance shuttle system, the distance required to carry casualties by litter teams is shortened. Air ambulances can also be used at secured ambulance exchange points to hasten evacuation time.

(3) ***Communications.*** Communications present one of the biggest obstacles to CASEVAC. Due to the terrain, line-of-sight radios probably will not be effective. Also, individual soldiers normally do not have access to radios. Therefore, wounded soldiers may be difficult to find and evacuate. The unit SOP should contain alternate forms of communications such as colored panels or other types of markers that can be displayed to hasten rescue when the battle is over. (See Appendix I for more information.) Also, a systematic search of the area after the battle may be required to recover casualties. Consideration may be given to marking the CCP so that it is visible to individuals trying to find it. Buildings and streets can cause confusion and disorient individuals trying to evacuate wounded.

(5) ***Special Equipment.*** Special equipment requirements include ropes, pulleys, SKEDCO litters and stokes litters, axes, crowbars, and other tools used to break through barriers.

d. **Use of Existing Medical Facilities.** The use of local medical facilities, hospitals, and medical supplies may be available during operations in large urban areas. The commander must adhere to the theater guidelines established for when and how these facilities can be used. A commander cannot confiscate civilian medical supplies unless he makes provisions to provide adequate replacements if civilians are wounded.

13-10. CONSIDERATIONS FOR THE COMBAT MEDIC (TRAUMA SPECIALIST)

The combat medic supporting Infantry companies and battalions during UO will be challenged physically, mentally, and professionally. He must be—

- Proficient in basic soldiering skills to survive on the battlefield and to accomplish his mission.
- Mentally agile to adjust and to adapt medical support as required by the changing battlefield environment.
- Well-versed in emergency medical treatment (EMT) procedures for trauma (especially to the head), burns, and blast, missile, and crush injuries.
- Knowledgeable in extraction techniques and procedures for extricating casualties from damaged aircraft, vehicles, and collapsed structures. The combat medic should request assistance from the supporting engineer unit to assist in freeing soldiers pinned down in collapsed structures. Using the correct techniques for extricating soldiers who are trapped by rubble may prevent doing further injury to the soldier.
- Proficient in locating and evacuating casualties from above, below, and at ground level.

Prior to the operation, the combat medic should ensure that all soldiers participating in the operation are proficient in first aid skills. He should conduct refresher training if any deficiencies are identified. He should also verify that combat lifesavers have all required medical supplies. The command surgeon may determine that it is feasible to train additional first aid skills for all soldiers to enhance their potential for survival, if wounded. The combat medic must determine what specialized equipment will be required to extricate casualties (such as crowbars or extrication devices) and how this equipment will be carried into battle (such as on supporting helicopters).

a. **Acquiring Patients.** The first thing the combat medic must accomplish is to survive on the battlefield. He must be knowledgeable in basic soldier skills such as the techniques for crossing open areas, entering and exiting buildings, and recognizing and avoiding booby traps and mines. He must have the foresight and training not to needlessly expose himself to hostile fire. The combat medic who is wounded or killed cannot help his platoon and cannot accomplish his medical mission. He must also be proficient at and be prepared to defend himself and his patients in sudden, short-range engagements.

(1) If a soldier is wounded in an open area (for example, crossing a street), the first impulse of his fellow soldiers is to rush to his aid. However, since the casualty is located in an open area and probably exposed to sniper fire, soldiers (who are going out to rescue the wounded) are likely to become casualties. The combat medic should approach as closely as possible to the casualty, while still remaining under cover, and try to verbally determine the casualty's status.

(a) If the casualty is responsive to verbal stimulus, the combat medic should try to calm him and encourage him to try to crawl or drag himself out of the line of fire. If that is not possible, the combat medic should try throwing the casualty a rope which he could then hold or attach to his load-bearing equipment (LBE) and be dragged to safety. (The combat medic should carry a length of rope with a D-ring attached to one end for this purpose.)

(b) If the casualty is unresponsive, he may have to remain in the open until visibility can be reduced using smoke to obscure the area, or wait until he can be recovered under the cover of darkness.

NOTE: The use of smoke must be judiciously applied. During the battle for Hue, the enemy realized that U.S. Marines were using smoke on the battlefield and to cross open areas. Therefore, the enemy would blindly fire into the smoke in hopes of wounding U.S. forces crossing the street. The medic can coordinate with the maneuver platoon/company for covering fire and smoke obscuration to aid in rescuing the casualty.

(c) Another technique to provide cover to acquire a patient in an open area is to place a vehicle in between the casualty and the sniper and or hostile force. The rescuers move along the side of the vehicle until they reach the casualty, secure him, and continue to move along with the vehicle until appropriate cover is reached.

(2) The combat medic may be required to move a soldier before he has the opportunity to fully assess the potential for spinal injury. If the casualty is in imminent danger (such as near a burning vehicle or being exposed to gunfire), the combat medic or another soldier may be required to move him before the application of a cervical collar and or spine board can be accomplished. In this case, the combat medic should try to move the wounded soldier as efficiently and smoothly as possible. The rescuer can grasp the shoulders of the patient and use his forearms to temporarily stabilize the neck while the casualty is being dragged to safety.

(3) The combat medic must remain aware of his location when treating casualties as well as being familiar with information about urban construction and weapons effects to knowledgeably select effective cover for himself and his patients. Some building materials will not stop fragments or small arms. For example, casualties may be moved to a relatively safe area behind the wall of a building; however, if the wall is damaged and there are gaping holes, the medic and his patient may still be subjected to sniper fire through the openings. If the area is enclosed, they may be subject to ricochets, or if the wall is made of wood, depending upon its thickness and density (plywood versus oak), small arms fire will probably penetrate it. Another consideration is the type of building construction (for example, framed or massed), and the potential for collapse if hit by artillery or tank fire.

(4) If casualties are located above or below ground level, it may not be feasible to reach them until a lull in or cessation of the battle occurs. Special equipment, such as collapsible ladders or ropes and pulleys, may be required to reach these casualties. The time required and the complexity of erecting a means to reach the casualty may jeopardize the safety of the rescuers and the accomplishment of the mission.

NOTE: In this scenario, if the wounded soldier has medical supplies necessary to administer lifesaving first aid in his possession, such as a tourniquet for a traumatic or incomplete amputation, his chances for survival are improved while he has to wait to be rescued. This is another situation in which the combat medic can provide direction to the casualty who is responsive to verbal stimulus to calm him and reinforce the lifesaving steps the casualty can perform for himself.

CAUTION

The application of a tourniquet may result in the loss of the limb. All other methods of stopping a hemorrhage (direct pressure or pressure bandage), when practical should be tried first.

(5) The combat medic must guard against exhausting or seriously depleting his medical supplies on the initial casualties. The duration of the battle and or the number of patients may increase. Having each soldier carry the medical supplies for his own initial treatment enhances the combat medic's ability to conserve his medical supplies.

b. **Treating Patients.** During UO, the treatment of a patient is complicated by a number factors. As mentioned earlier, the patient may be experiencing some degree of dehydration; this in conjunction with blood loss from wounds or injuries may make the prognosis for certain injuries grim. Urban combat is often conducted within confined or limited space. The noise, smell, and confusion of battle may make it difficult for the combat medic to focus his energies and may eliminate some medical indicators (such as breath sounds) that he would normally rely on in assessing the patient.

(1) While the casualty and combat medic are still subject to hostile fire, there is little medical treatment that the combat medic can provide. If the casualty has a severe

hemorrhage, the combat medic may be able to place a tourniquet on it. The combat medic's first priority, however, is to survive and defend himself and his patient. Once the combat medic can move the casualty to a more secure location, such as under cover or behind a wall, he can more fully assess and treat the casualty. This is the same principle as used in emergency medicine—to remove the patient from the source of injury.

(2) One of hardest and most essential tasks the combat medic must accomplish is to quickly and accurately triage the patient conditions he is faced with. EXPECTANT category patients must be protected from further injury and, if possible, made comfortable, but the focus of the combat medic must be directed towards those patients whose chances of survival are greater.

(3) The combat medic ensures that each patient has an open airway, bleeding is controlled, and circulation maintained. Once this is completed, he must assess the mental status of the wounded soldiers. During UO, the outcome of the battle may hinge on the amount of firepower that can be concentrated at critical times. Patients that are wounded but still mentally alert may be able to contribute to the defense of the position. Further, depending upon the extent of injuries, the wounded soldier may still be able to function after being administered a low dose of pain medication.

CAUTION

If at any time the combat medic feels that a soldier has an altered mental state, the soldier should be disarmed regardless of the cause for that condition (injury/wound, dehydration, stress, medication, or other factors).

(4) The combat medic must effectively use all resources available to him; this includes other injured or wounded soldiers. A patient with an altered mental state may not be able to help defend the position, but he may be capable of applying direct pressure to another soldier's wounds. Another example would be for a relatively stable patient to observe a more critically wounded soldier and to alert the combat medic of any change in the patient's condition (such as the reappearance of bleeding or labored breathing).

(5) The patient conditions presenting and the likelihood of multiple traumas may tax the knowledge base of the combat medic, especially if he must sustain these patients for an extended period of time. If secure communications are available, the combat medic should avail himself of telemedicine support (advice) from the supporting battalion aid station (BAS) and or clearing station (forward support medical company [FSMC]).

(6) After providing initial treatment to all of the patients, the combat medic must continually reevaluate the patients' conditions until they can be evacuated. The length of evacuation to the BAS may not be that great. Treatment teams may be able to move closer than normal to the fight due to the available cover afforded by buildings. Urban structures can provide cover and concealment and facilitate light discipline. Engagement ranges are usually short and a few city blocks may provide a relatively secure area for the provision of advanced trauma management (ATM).

c. **Evacuating Patients.** Medical evacuation (MEDEVAC) during UO is a labor-intensive activity that may have to overcome significant and numerous obstacles. During UO, casualty evacuation (CASEVAC) may entail evacuation by litter, the expedient use of a vehicle of opportunity, or other military vehicle not dedicated for medical evacuation. CASEVAC occurs when medical personnel are not available to provide en route medical care to the casualty while being moved from the point of injury or wounding to a medical treatment facility (MTF). Field Manual 8-10-6 provides an in-depth discussion on the distinct differences between CASEVAC and MEDEVAC operations.

(1) ***Aeromedical Evacuation.*** Aeromedical evacuation is the preferred method; however, it may not be a feasible option. The threat from small arms fire, rocket-propelled grenades (RPG), and other shoulder-fired weapons may prohibit the employment of these resources. Aeromedical evacuation can also be hampered by the lack of landings zones (LZs) due to narrow streets and heavily-damaged buildings.

(2) ***Ground Ambulance.*** Evacuation by ground ambulance may have to surmount obstacles such as rubble, debris, barricades, mines and or booby traps, and wrecked or otherwise inoperable vehicles and or aircraft. If the employment of air ambulances is not feasible, a ground evacuation operation must be executed. Ground ambulances should be-

- Hardened to improve the survivability of the evacuation platform and the safety of the ambulance crew and patients.
- Field-sited as close to the operation as is tactically and physically possible.
- Maneuverable in tight spaces and narrow streets.

(3) ***Litter.*** The same obstacles that interfere with effective aeromedical and ground evacuation of patients will be factors in CASEVAC operations. Nonmedical vehicles and aircraft may not be feasible alternatives in MOUT and the only form of evacuation possible may be by litter carries. Litter evacuation is labor-intensive and the tactical commander may not be able to divert his soldiers to the task of performing litter evacuation until a lull in or a cessation of the battle occurs. Planners should push forward as many standard litters as are feasible.

(a) If litters are not available, the combat medic can improvise a litter from items such as blankets or ponchos and poles, doors, shutters, metal roofing materials, or other flat objects. The item selected to serve as a litter should be sturdy and as light as possible. Adding the weight of a soldier to an already heavy item (such as an oak door) increases the burden of the litter bearers. (Refer to FM 8-10-6 for additional information on improvising litters.)

(b) Depending on the distance the patients must be carried to reach the MTF or an area accessible by ground or air platforms, a litter shuttle system can be established. The litter shuttle system provides brief periods of rest for the litter bearers, and maintains the litter bearers' knowledgeable about the specific urban terrain in that area.

(4) ***Casualty Rates.*** Combat health support planners must take into consideration potentially higher casualty rates. The requirement for close combat by infantry will inevitably produce a higher patient workload in a shorter period of time than will operations where forces have more freedom of maneuver. Plans should address secure routes and marking structures for casualty/patient collection points and aid stations. Means of extracting casualties from damaged structures should be planned for, resourced, and, when possible, coordinated with the supporting engineer unit.

d. **Guidelines and Considerations.** Additional guidelines and considerations for the combat medic are as follows:

(1) ***Overall Health Status of Soldiers.*** In addition to rendering EMT to battlefield casualties, the combat medic is concerned with the overall health status of the soldiers he supports. Disease and nonbattle injuries generate significant combat ineffectiveness. The combat medic should be vigilant for the signs and symptoms of infectious diseases. Early recognition and reporting of diseases coupled with early treatment can avoid epidemic situations.

(2) ***Replacement of Combat Medic.*** The combat medic should designate one of the combat lifesavers to take his place in the event he is injured. Although the combat lifesaver only has advanced first aid skills, he can effectively coordinate for medical evacuation and treatment support.

(3) ***Disease and Nonbattle Injuries (DNBI).*** The prevention of DNBI is a significant responsibility for the combat medic. He must ensure that supported elements are proficient in and practice field hygiene and sanitation procedures. He must also ensure that supported personnel only consume food and drinks from approved sources.

(4) ***Mine and Booby Trap Awareness.*** Mine and booby trap awareness is a crucial consideration when operating in urban areas. The combat medic must be able to recognize mines and booby traps and take precautions as required. Additionally, the combat medic must be proficient in the skills and techniques to extract casualties from minefields. (Refer to FM 8-10-6 for additional information.)

(5) ***Manual Evacuation.*** Because MEDEVAC is a labor-intensive activity during UO, the combat medic must train his supported personnel in the techniques of manual evacuation. It is essential that those soldiers acting as litter bearers correctly handle the litter. Being proficient in using litters decreases the potential for injury to the litter bearers (muscle strains and back problems) and avoids further injury to the wounded soldier; for example, dropping the litter or tilting it so that the patient falls from the litter. (For information on litter-carrying techniques and manual evacuation, refer to FM 8-10-6.)

(6) ***Survivability.*** Soldier survivability is of the utmost importance and cannot be over emphasized. The combat medic should have his fellow Infantrymen train him in these skills.

(7) ***Navigation.*** Urban terrain is often difficult to navigate. The combat medic should have a city map available if at all possible. He should be familiar with possible routes from the battle site to the BAS. The routes do not necessarily have to only be established roads. If crossing gardens and yards or going through basements or battle-damaged buildings (holes in walls) is possible and these avenues provide better cover and concealment than is available on established road, then these routes should be used.

(8) ***Medical Materiel.*** Under the provision of the Geneva Conventions, medical materiel (supplies and equipment) are protected from intentional destruction. Should friendly units come across civilian or enemy medical supplies, medics and leaders must ensure that soldiers do not intentionally destroy these items. (For a discussion of the Geneva Conventions refer to FM 8-10.)

13-11. CONSIDERATIONS FOR THE BATTALION PHYSICIAN'S ASSISTANT AND COMMAND SURGEON

Routine operations in an urban environment performed numerous times without incident can suddenly become an intense battle. Medical personnel must be prepared to sustain casualties for longer periods of time in the event that MEDEVAC support is delayed. This paragraph provides considerations for the battalion physician's assistant (PA) and command surgeon during UO.

a. **Reliance on First-Aid Measures.** Due to decentralized control and the isolating nature of urban combat, each soldier must be proficient in administering first aid to himself and or his buddy. Further, units should have sufficient soldiers trained in advanced first aid (combat lifesaver skills) to assist the combat medic in stabilizing casualties. PAs and surgeons may determine that each individual soldier should carry sufficient medical supplies for his own initial care, if wounded or injured. These medical supply items may include tourniquets, carried in a designated location in the BDUs or rucksack; bandages; intravenous (IV) fluids, and IV starter kits.

b. **Specific Considerations.** Before entering an urban environment, PAs and surgeons must ensure that supporting medical personnel have the skills and knowledge to effectively perform their medical duties in a challenging and, often, adverse environment. They must first assess the skill level of the available medical personnel, provide refresher training to correct deficiencies, and develop treatment protocols/guidance for the care and treatment of patients while under fire and during lulls in the battle. PAs and surgeons must also ensure that they participate in the early planning for UO. To develop a comprehensive, effective, and synchronized combat health support (CHS) plan, the PA and surgeon must know and understand the tactical commander's guidance, intent, and concept of the operation. They can only accomplish this by participating in the planning process. Further, the PA and surgeon must be included in the combined arms rehearsal to ensure the CHS plan is synchronized with the tactical plan. PAs and surgeons should determine if each soldier should carry medical supplies for his own treatment into the battle. This decision is based on patient estimates, anticipated level of hostilities to be encountered, the expected duration of the operation, and any anticipated delays in evacuation. Specific considerations are:

(1) ***Wounds.*** The types and frequencies of wounds and injuries generated by urban combat should be monitored to ensure required specialty teams are available for medical care.

(2) ***Resupply.*** The PA and surgeon must ensure that prepackaged emergency medical resupply bundles are prepared prior to the operation in the event emergency resupply by airdrop is required. This may include complete medical equipment sets (MESs) or specific replacement for high use items, such as bandages.

(3) ***Preventive Medicine***. Preventive medicine (PVNTMED) policies and programs must receive command emphasis. This is particularly important with regards to water discipline. Dehydration is a significant threat in the urban environment. Increased water loss may be due in part to the effects of the weather, additional heat stress of body armor and or MOPP gear, and operating in poorly ventilated and confined spaces. Dehydration can affect mental functions and complicate medical treatment in the event of wounding or injury. Medical surveillance programs are essential for the following reasons:

- To ensure the health of the command during UO.

- To provide a baseline of diseases present in the AO in order to facilitate the identification and confirmation of potential use of biological warfare agents against U.S. forces.
- To identify what immunizations, prophylaxis, pretreatments, and barrier creams will be effective in the AO.

(4) ***Medical Intelligence.*** Current and comprehensive medical intelligence is a critical factor during UO. Medical intelligence can provide the CHS planner with the information needed to protect the fighting force from preventable diseases once engaged within the city confines. It also assists the tactical commander in evaluating the health status of enemy forces, which, if it is poor, may indicate a degradation of morale and will to fight.

(5) ***Remains.*** During UO, it is often difficult to clear the area of remains. Although U.S. forces have always quickly recovered their dead, the enemy may not. The unrecovered remains of enemy dead, civilian casualties, and animals can present serious complications in preventing the spread of infectious diseases. Coordination with supporting civil affairs (CA) units may be required.

(6) ***Disease.*** Food and waterborne diseases can quickly decimate a fighting force. Small-unit leaders must ensure that soldiers only eat and drink from approved water and food sources. Although the water system may be intact in an urban area, the water may still be contaminated. Food procured on the local market can contain infectious organisms or parasites. Food and waterborne diseases are preventable. Field hygiene and sanitation measures must be fully understood and enforced.

(7) ***Triage.*** The PA and surgeon must ensure triage categories and priorities are ingrained within all medical personnel. In mass casualty situations, the focus of health care delivery is to provide the greatest good for the greatest number. Mass casualty situations develop when the patient workload exceeds the capabilities of medical personnel/facilities available. During UO, this type of situation may develop when a single combat medic must treat, sustain, and manage three or four seriously wounded soldiers prior to evacuation. When a casualty has received massive wounds, making survival unlikely even if treated in a major trauma facility, he should be triaged as EXPECTANT. This permits the combat medic to use his medical skills and scarce resources to treat patients with a better chance of survival. (A description of triage categories is provided in FM 8-10-1.)

(8) ***Marking System.*** The PA and surgeon must ensure that a uniform marking system is used to indicate where casualties are located within buildings. This facilitates the quick retrieval and treatment of isolated casualties. The uniform system permits follow-on forces/reinforcements to identify locations where casualties can be found. These procedures should be included in the unit's tactical standing operating procedure (TACSOP). Most combat maneuver units have a system for marking the buildings and rooms as they are cleared. The CHS planner should coordinate with the supported maneuver unit to ensure the location of casualties is also indicated. (See Appendix I for more information on marking systems.)

(9) ***Mobility***. The PA and surgeon must coordinate with supporting engineers to clear medical evacuation routes of rubble, debris, and barricades behind the fighting force. Due to isolated areas of resistance, this may not be possible in all urban scenarios.

(10) ***Communications***. A viable communications capability is essential for successful UO. The PA and surgeon must ensure that medical elements have sufficient

communications resources to remain synchronized with the tactical force, locate casualties, and effect telemedicine activities, if required.

(11) ***Collateral Damage.*** Although every effort is made to limit collateral damage, some civilian casualties will inevitably occur. The PA and surgeon must develop a policy, in conjunction with the staff judge advocate (SJA), on the care and treatment of civilian casualties. Further, CHS planners will be required to coordinate with the local civilian government and health care community for the transfer of these patients to civilian facilities.

13-12. BATTALION AID STATION OPERATIONS

The following paragraphs discuss considerations for establishing and operating a battalion aid station (BAS).

a. **Site Selection.** As in all military operations, site selection is a crucial element in facilitating the accomplishment of the mission. This is particularly true in the urban environment. The BAS must be located close enough to the fight to reduce evacuation times, normally by litter, yet not interfere with ongoing tactical operations. Further, the site selected must be defensible. Progress during urban combat is often measured in feet and yards. Moving and fighting house-to-house and street-by-street is time consuming and often results in higher casualties. Since forward progress during UO may be slow, the BAS has the opportunity to more fully establish the medical treatment facility (MTF) and to fortify its position.

(1) ***Location.*** Whenever possible, the BAS should establish the MTF in a location that is accessible to both ground or air ambulances. The location should permit ambulance turnaround, which facilitates the further evacuation of patients from the immediate combat area.

(2) ***Layout.*** The actual layout of the BAS will depend upon the type of structure and the amount of space available. The anticipated duration of the operation determines the extent to which the BAS is established. If the BAS is expected to move frequently, only essential elements will be established. If an existing building will be used, the structural integrity of the building should be determined prior to establishing the facility for an extended period of time. Heavy-clad mass construction buildings with good roofing are preferred.

(3) ***Fortification.*** The position of the BAS should be fortified when possible. Barricading windows, placing sandbags, or otherwise reducing the vulnerability to attack, provides the physician and PA a more secure location to provide advanced trauma management (ATM). While fortifying their positions, BAS personnel should ensure that observation/firing ports remain to observe the surrounding area and to permit the return of defensive fire, if required. Firing ports can be covered during night operations to facilitate light discipline. If the observation/firing ports are covered with a material that does not blend with the building facade, it must be removed during daylight hours to ensure it does not identify the structure as being in use. Chapter 3 provides information and techniques for fortifying positions; however, in evaluating this information it is important to consider medical requirements. Although it is advisable to move fighting positions away from the wall/window to afford more protection and to reduce the potential that the enemy would recognize the position, it may significantly limit the usable space available to care for patients. Space requirements include a treatment area

and a space to hold soldiers awaiting evacuation. Unless the BAS is reinforced/ augmented with cots and additional medical personnel, the BAS may not have a holding capability. During UO, however, wounded soldiers may have to be sustained for longer periods of time due to limitations and delays in evacuation. Further, the BAS must have sufficient space to establish a patient decontamination area, if required.

b. **Acquiring Patients.** The BAS may acquire patients from ground ambulances or from casualty/patient collection points.

(1) ***Ground Ambulances.*** The employment of ground ambulances forward of the BAS may not be possible due to rubble, debris, and other obstacles. Patients may arrive at the BAS by litter or other means of conveyance or, if ambulatory, they may arrive unassisted.

(2) ***Casualty/Patient Collection Points.*** Casualty/patient collection points are preplanned and included on the CHS overlays. Control devices, such as phase lines and check points, can be used as triggers for the movement of these facilities.

c. **Treating Patients.** The BAS is the first location on the battlefield where the wounded soldier can receive ATM. The BAS has both a physician and a PA assigned. It has the capability of forward-siting a treatment team closer to the fight if required.

(1) ***Triage.*** As with the combat medic, the patient conditions presenting at the BAS must be triaged in order to provide the greatest good for the greatest number of patients. Higher casualty rates in UO can quickly attrit units engaged in combat operations. Those patients who can be treated and quickly returned to duty (minor wounds and illnesses-MINIMAL category) should be expeditiously managed. When evacuation capability is limited, patients in the IMMEDIATE category should be stabilized and further evacuated at the first opportunity. ROUTINE and EXPECTANT category patients should be evacuated after patients in the IMMEDIATE category. (FM 8-10-6 provides more information on categories of triage.)

(2) ***Delays.*** If delays in evacuation are anticipated, the BAS may be reinforced with a limited holding capability (cots and additional medical personnel) and increased stockage of Class VIII materiel.

(3) ***Wounds.*** The BAS must be prepared to treat an increased number of patients with head wounds; missile wounds (shrapnel and shards of brick, concrete, glass, metal, and wood); burns; and inhalation and crush injuries. Medical equipment sets may need to be reconfigured, with the advice of the command surgeon, prior to the mission to ensure adequate supplies to treat trauma injuries are included. When space is a limiting factor, those Class VIII items used for routine sick call may have to be replaced with trauma-related Class VIII items. Routine sick call service may have to be delegated to follow-on forces and coordinated with the supporting medical company.

NOTE: Recent experimentation has shown that most wounds from small arms occur in the front and back of the head and upper torso.

d. **Evacuating Patients.** Patients are evacuated from the BAS as follows:

(1) ***Litter Evacuation.*** If ground ambulances cannot go forward of the BAS due to rubble, obstacles, and barricades, evacuation by litter or manual carries to the BAS may be required. When developing plans for CHS during UO, the requirement for additional litter bearers should be considered and resourced as appropriate. Training in the safe

handling of litters should be provided to reduce the incidence of injuries to the litter bearers and patients.

(2) ***Evacuation Routes.*** Medical evacuation personnel should have the most up-to-date commercial maps available. Evacuation routes should be preplanned, reconnoitered, and, when possible, secured. If commercial maps are not available and strip maps must be used, an ambulance shuttle system should be established. The shuttle system allows the ambulance crews most familiar with the terrain surrounding the BAS to remain in that area, thereby limiting the chances of ambulance crews getting lost and straying into hostile territory. (An in-depth discussion of an ambulance shuttle system is provided in FM 8-10-6.)

(3) ***Aeromedical Evacuation.*** See paragraph 13-9c.

13-13. PRECOMBAT MEDICAL CHECKLISTS.

Tables 13-1 and 13-2 provide examples of precombat checklists for the combat medic during UO and for the BAS.

Map of the AO is available (military, commercial, strip maps, aerial imagery, as required.
Class VIII Supplies are on hand. • Medications and other time-sensitive supplies are current. • Has the command surgeon authorized the carrying of additional and or different Class VIII based upon the tactical scenario?
Is specialized equipment for the extrication of casualties from surface, supersurface, and subterranean levels available? • Ropes and pulleys. • Rope with attached D-ring. • Collapsible light weight ladders. • Picks, axes, crowbars, and shovels. • Mechanical/electrical extrication devices. • Heavy gloves. • Collapsible light-weight litters.
Do combat lifesavers have required Class VIII?
Are soldiers carrying bandages and IV fluids for their own initial care? (Where will they be carried on the uniform?)
How will MEDEVAC/CASEVAC be accomplished? • Are litter teams available? (How will they be requested? Where are the resources from? Supported unit? Medical augmentation?) • Locations of patient collecting points and AXP are preplanned. (What are the triggers for activation? • Will ground and or air ambulances be available? (If yes, how will medical evacuation requests be managed?)

Table 13-2. Example precombat checklist for the combat medic during UO.

Is specialized military equipment available? • Flack jackets and/or other body armor. • Chemical Protective Overgarments (CPOG). • Night vision devices (NVDs). • Personnel locator devices. • Chemical lights. • Flares and colored smoke.
Is a marking system developed to mark buildings where casualties can be located?
Have troops been prescribed/issued pretreatments or chemoprophylaxis based on the anticipated threat?
Have personal protective measures (such as insect repellent) supplies/equipment been issued to unit members?
How will contaminated casualties be managed/evacuated?

Table 13-2. Example precombat checklist for the combat medic during UO (continued).

Map of the AO is available (military, commercial, strip maps, aerial imagery, as required).
Location of patient collecting points and AXP are preplanned. (What are the triggers for activation?)
All authorized shelters are on hand and serviceable.
All authorized collective protective equipment and shelters are on hand and serviceable.
Procedures for the management of medical waste are established.
Provisions for water supply are coordinated.
Patient protective measures are instituted (such as patient bunkers or sandbags placed in areas of patient care).
Ambulance turnaround is planned for and established.
Area for patient decontamination operations is planned for and established, when required.
Camouflage material is available if authorized for use.
Medical unit identification markers are on hand.
Class VIII. • Medical equipment is properly calibrated and serviceable. • Medications and other time-sensitive supplies are current. • All authorized MESs are on hand and complete. • Special medications and equipment authorized by the command surgeon for the operation are on hand and current. • Each MES has a packing list on hand. • Authorized number of days of supply are on hand. • Accountability of controlled substances is maintained. • Medical gases are on hand.

Table 13-3. Example precombat checklist for the BAS.

Ground ambulances. • Authorized MES are on hand and complete. • Medical equipment is calibrated and serviceable. • Authorized medical gases (oxygen) are on hand and serviceable. • Authorized medications are on hand and current. • Packing list is available. • Commercial city maps, strip maps, or road maps are available. • On-vehicle materiel (OVM) is on hand. • Situational awareness equipment (position locator) is on hand and serviceable. • Log book is present and current, if applicable. • Communications equipment is on hand, serviceable, and set to the correct frequency. • Medical unit identification markings (in accordance with the Geneva Conventions) are displayed.
Medical/refresher training for combat medics. • Has battalion surgeon initiated training program for combat medics? • Has refresher training been conducted for combat medics?

Refresher training for BAS personnel.
• Has battalion surgeon initiated a training program for BAS personnel?
• Have BAS personnel received refresher training in the care for—
  • Blast injuries?
  • Crush injuries?
  • Head injuries?
  • Complete/incomplete amputations?
  • Missile injuries?

Table 13-3. Example precombat checklist for the BAS (continued).

NOTE: Markings are a red cross on a white background; camouflaged or subdued markings are not authorized. If the unit is required to camouflage, the Geneva Conventions emblem is removed or covered. In accordance with North Atlantic Treaty Organization (NATO) Standardization Agreement 293-1, the Geneva Conventions emblem can be camouflaged (removed) IAW instructions of a brigade-level commander or higher when the tactical situation dictates.

Section III. LEGAL ASPECTS OF URBAN OPERATIONS

Commanders must be familiar with legal aspects of UO including the control of large groups of civilians, protection of key facilities, and civil affairs operations. While this section primarily addresses considerations for brigade and battalion commanders and staffs, many of the issues discussed will also affect company commanders. A judge advocate (JA) should be involved in the planning process and immediately available to commanders for advice during the conduct of UO.

13-14. CIVILIAN IMPACT IN THE BATTLE AREA

The presence of large concentrations of civilians can greatly impede tactical operations.

a. **Mobility.** Fleeing civilians, attempting to escape over roads, can block military movement. Commanders should plan routes to be used by civilians and should seek the assistance of the civil police in traffic control.

b. **Firepower.** The presence of civilians can restrict the use of potential firepower available. Areas may be designated no-fire areas to prevent civilian casualties. Other areas may be limited to small-arms fire and grenades with prohibitions on air strikes, artillery, mortars, and flame weapons. Target acquisition and the direction of fire missions are complicated by the requirement for positive target identification. Detailed guidance on the use of firepower in the presence of civilians will normally be provided to the battalion commander in the form of ROE. In the absence of guidance, the general rules of the law of land warfare apply. Commanders must ensure that all soldiers receive a copy of the written ROE and that soldiers are thoroughly briefed on its contents.

c. **Security.** Security should be increased to preclude:

- Civilians being used as cover by enemy forces or agents.
- Civilians wandering around defensive areas.
- Pilferage of equipment.
- Sabotage.

d. **Obstacle Employment.** The presence of local civilians and movement of refugees influence the location and type of obstacles employed. Minefields may not be allowed on designated refugee routes or, if allowed, must be guarded until the passage of refugees is completed. Booby traps and flame obstacles cannot be emplaced.

13-15. LIMITS OF AUTHORITY

The limits of authority of commanders at all levels over civilian government officials and the civilian populace must be established and understood. A commander must have that degree of authority necessary to accomplish his mission. However, the host government's responsibility for its populace and territory can affect the commander's authority in civil-military matters. In less secure areas, where the host government may be only partly effective, the commander may be called upon to assume greater responsibility for the safety and well being of the civilian populace. Again, depending on the nature of the operation, the brigade and battalion staff will coordinate concerning the limits of command authority and for any non-organic assets needed to accomplish the mission.

13-16. DIVERSION OF MILITARY RESOURCES

Conducting operations in highly populated areas may require the diversion of men, time, equipment, and supplies from combat tasks to accomplish stability and support tasks. If host government agencies collapse, the impact on military resources could be substantial. Battalion TFs and brigades can be given a wide range of stability and support missions that will tax the logistical capabilities of the unit. Commanders must clarify the limits of their authority concerning the commandeering of civilian supplies or equipment in order to facilitate mission accomplishment.

13-17. HEALTH AND WELFARE

The disruption of civilian health and sanitary services sharply increases the risk of disease among both civilian and military personnel. Commanders should minimize contact with the civilian population, unless contact is specifically required for mission accomplishment.

13-18. LAW AND ORDER

The host government may not be able to control mobs. Brigades and battalion TFs may have to augment civilian forces to protect life and property and to restore order, by controlling civil disturbances. The Provost Marshal/brigade MP has the responsibility of coordinating with host nation police to quell civil disturbances. Brigades and battalion TFs may also have to secure vital government facilities for the host nation (see Chapter 5). (See FM 19-15 for more information on how to control civilians violating civil law.)

13-19. PUBLIC AFFAIRS OFFICER AND MEDIA RELATIONS

It is very likely that there will be media representatives in the brigade and battalion's operational area. Brigade and battalion commanders must be prepared to receive and possibly escort journalists in their respective area of operations. Generally, the best way to deal with the media is to receive them and allow them access to soldiers IAW guidance from higher headquarters. While free access to units in the field is desirable, operational security, existing guidelines, and or ROE considerations will take priority and may limit access. All members of the media visiting the field should have an escort officer. This officer may be detailed from line units due to the shortage of trained public affairs personnel. Ensuring the media follows the established guidelines or ROE will help prevent negative publicity that could jeopardize the mission. Media access will be coordinated through the S1, who may be assisted by a public affairs officer.

13-20. CIVIL AFFAIRS UNITS AND PSYCHOLOGICAL OPERATIONS

Civil affairs (CA) units and psychological operations (PSYOP) have essential roles during UO. They are critical force multipliers that can save lives. The battle in urban terrain is won through effective application of necessary combat power, but CA and PSYOP can help facilitate mission accomplishment. CA and PSYOP offer the possibility of mission accomplishment in urban terrain without the destruction, suffering, and horror of battle. These units may become key factors in shaping the urban battlefield and facilitating movement from shaping directly to transition, thus minimizing the amount of close combat conducted by companies, platoons, and squads (see Chapter 4, Sections IV and V).

a. **Evacuation of Civilians.** If the brigade or battalion TF is tasked to facilitate the evacuation of civilians from the AO, the unit is normally augmented by CA personnel. Infantry units may provide security and command and control for the execution of this operation, which is accomplished in two separate but supporting actions.

(1) CA personnel coordinate with the military police and local police officials for evacuation planning. They plan for establishing evacuation routes and thoroughfare crossing control, and for removing civilians from the military supply routes (MSRs).

(2) CA personnel coordinate with U.S. Army PSYOP assets, local government officials, radio and television stations, newspapers, and so on, to publicize the evacuation plan.

b. **Health and Welfare of Civilians.** CA assets will also conduct coordination for the health and well being of civilians. They can include the reestablishment of water systems; distribution of available food stocks and clothing; and establishment of displaced persons, refugee, and evacuee (DPRE) camps. Again, brigades and battalion

TFs may be tasked to provide security and command and control for some of these missions.

c. **Tactical PSYOP.** Tactical PSYOP in support of UO are planned and conducted to achieve immediate and short-term objectives. PSYOP are an integral and coordinated part of the overall tactical plan. They provide the tactical commander with a system that can weaken the enemy soldier's will to fight, thereby reducing his combat effectiveness. They can also help prevent civilian interference with military operations. PSYOP are designed to exploit individual and group weaknesses. For example, infantry units may be given the mission to clear a specific urban objective where it has been determined that a graduated response will be used. The PSYOP unit would be in support of the unit conducting this mission, and they use loudspeakers to broadcast warnings and or incentives not to resist.

e. **Other PSYOP.** PSYOP units also provide support during UO using television, radio, posters, leaflets, and loudspeakers to disseminate propaganda and information. Television, including videotapes, is one of the most effective media for persuasion. It offers many advantages for PSYOP and is appropriate for use during UO. In areas where television is not common, receivers may be distributed to public facilities and selected individuals.

NOTE: See FM 41-10 for further discussion on civil affairs.

13-21. PROVOST MARSHAL

The provost marshal has the responsibility of coordination with the host nation's law enforcement officials. The provost marshal recommends measures to control civilians and direct MP activities in support of refugee control operations. The provost marshal coordinates his activities with staff sections and supporting units in the area. Refugee control operations are the responsibility of the G5 or S5, host nation authorities, or both. MPs assist, direct, or deny the movement of civilians whose location, direction of movement, or actions may hinder operations. The host nation government is responsible for identifying routes for the safe movement of refugees out of an area of operations.

NOTE: Other military police responsibilities, regarding civil affairs and civilian control, are contained in FM 3-19.1(FM 19-1).

13-22. COMMANDERS' LEGAL AUTHORITY AND RESPONSIBILITIES

Commanders and leaders at all levels are responsible for protecting civilians and their property to the maximum extent allowed by military operations. Looting, vandalism, and brutal treatment of civilians are strictly prohibited, and individuals who commit such acts can be severely punished under the UCMJ. Civilians, along with their religions and customs, must be treated with respect. Women must be especially protected against any form of abuse. In urban combat, however, some situations are not always clear. Discussed below are those civilian-military situations most common during UO, and how an infantry commander might manage them to legally accomplish his mission.

a. **Control Measures.** Commanders may enforce control measures to conduct operations, maintain security, or ensure the safety and well being of the civilians.

(1) ***Curfew.*** A commander may need to establish a curfew to maintain security or to aid in control of military traffic. (Curfews are not imposed as punishment. They are normally established to reduce noncombatant casualties and provide a measure of force protection.)

(2) ***Evacuation.*** A commander can require civilians to evacuate towns or buildings if the purpose of the evacuation is to use the town or building for imperative military purposes, to enhance security, or to safeguard those civilians being evacuated. If the commander takes this action, he must specify and safeguard the evacuation routes. Food, clothing, medical, and sanitary facilities should be provided or available at the destination until the evacuees can care for themselves. Evacuated civilians must be transferred back to their homes as soon as hostilities in the area have ceased. The staff must plan for and coordinate the movement and evacuation of civilians to ensure their actions do not interfere with the military operation. The S5, brigade JA, S3, S2, and supporting civil affairs units working with local officials coordinate the movements of civilians.

(3) ***Forced Labor.*** The Geneva Accords prohibit the use of civilians in combat. However, they may be used before the battle reaches the city. Commanders will be given guidelines for use of civilian labor. The brigade or battalion TF may force civilians over 18 years of age to work if the work does not oblige them to take part in military operations. Permitted jobs include maintenance of public utilities as long as those utilities are not used in the general conduct of the war. Jobs can also include services to local population such as care of the wounded and burials. Civilians can also be forced to help evacuate and care for military wounded, as long as doing so does not involve any physical danger. Prohibited jobs include digging entrenchments, constructing fortifications, transporting supplies or ammunition, or acting as guards. Volunteer civilians can be employed in such work. Brigade and battalion TF commanders normally will not order forced civilian labor. However, they may find themselves supervising such labor in their area of operations.

b. **Civilian Resistance Groups.** Units may encounter civilian resistance groups whose actions may range from lending supplies, services, and noncombat support to the enemy to actively fighting against friendly forces. Members of such resistance groups should be dealt with in accordance with applicable provisions of the law of war. Commanders should seek guidance from the JAG concerning the detention and disposition of persons participating in acts harmful to friendly forces. The S2, brigade JA, PSYOP, MP, and civil affairs units must work together to identify these threats and recommend, within the ROE, the appropriate preemptive action or response, when required. The activities of resistance groups may also benefit friendly forces. They may provide HUMINT; act as guides, liaisons or translators; and provide subject matter expertise on local public facilities such as refineries, power plants, and water works. They may also provide active resistance against the threat. Another situation that brigade and battalion commanders might encounter is combat with a civilian resistance group.

(1) ***Accompanying Civilians.*** Civilians who are accompanying their armed forces with an identity card authorizing them to do so are treated as enemy prisoners of war (EPWs). For example, when captured; civilian members of military aircraft crews, war correspondents, supply contractors, and members of labor units or service organizations responsible for the welfare of the armed forces.

(2) ***EPWs.*** Civilians are treated as EPWs when captured if they—

- Are from a nonoccupying territory and have taken up arms against an invading enemy without time to form regular armed forces.
- Wear a fixed, distinctive insignia that can be seen at a distance and carry their weapons openly.
- Operate according to the rules and customs of war.

Other civilians who provide assistance to such groups may not be entitled to status as combatants, depending upon whether they are actually members of the resistance group. Any person whose status is in question should be treated as an EPW and be accorded all corresponding protections, rights, and privileges. Consult the JA.

(3) ***Armed Civilian Groups.*** Armed civilian groups that do not meet the criteria of a legal resistance (civilians accompanying their armed forces and mass conscription) or individuals caught in the act of sabotage, terrorism, or espionage are not legal combatants. If captured, they must be considered criminals under the provisions of the law of land warfare. They should be detained in a facility separate from EPWs and should be quickly transferred to the military police. Reprisals, mass punishments, taking of hostages, corporal punishment, pillage, and destruction of property are prohibited punishments. However, any person whose status is in question should be treated as an EPW and be accorded all corresponding protections, rights, and privileges. Consult the JA.

(4) ***Civilian Control.*** The law of land warfare lets a commander control the civil population under the conditions already described using his own resources. However, because of language and cultural differences between U.S. and foreign personnel, it is advised to use native authorities, such as the police, for such purposes. Use of the police does not relieve a commander of his responsibility to safeguard civilians in his area. Infantry units may be required to engage in these type of operations as part of a joint task force.

c. **Protection of Property.** Like civilian personnel, buildings and towns normally have a protected status—specifically, they are not legal targets. Buildings and towns lose their protected status if authorities determine that the enemy is using them for military purposes. If doubt exists as to whether a town or building is defended, that doubt should be settled by reconnaissance, not by fire.

(1) ***Legal Targets.*** If the enemy is using a building or a portion of the town for military purposes—for example, as a supply point or a strongpoint—that building or that portion of the town is a legal target. Before engaging the target, the commander must decide if fire on the target is necessary. Only such destruction as is required for military purposes is justified.

(2) ***Restricted Facilities/Objects.*** Normally, religious, historical, and cultural facilities/objects and buildings are not legal targets. They are sometimes marked with symbols to signify their cultural status. Medical facilities, personnel, and equipment are protected under provisions listed in Article 39 (GWS) and shall bear one of the following emblems on a white background: the internationally recognized Red Cross, Red Crescent, Red Lion, or Red Star of David. The fact that such symbols are absent does not relieve a commander of his responsibility to protect objects he recognizes as having religious, cultural, medical, or historical value.

(3) ***Misuse of Restricted Facilities/Objects.*** The misuse of such facilities/objects by the enemy is grounds to disregard their protected status. Whenever possible, a demand should be made for the enemy to stop his misuse of the protected object within a reasonable time. If an enemy forward observer uses a church for an OP, for example, a commander would be justified in destroying it immediately because a delay would allow the enemy to continue the misuse of the church. If a religious shrine were used as a telephone switchboard, a warning would be appropriate, since it would take some time to dismantle the wires. Once the decision to order fires on those objects is reached, destruction should be limited to the least necessary in order to neutralize the enemy installations.

(4) ***Destruction or Damage to Property.*** The destruction, demolition, or military use of other buildings is permitted under the law of land warfare, if required by clear military necessity. Thus, destroying a house to obtain a better field of fire would be a legal act—destroying it as a reprisal would not be. Likewise, firing on any houses that are occupied or defended by an enemy force is legal.

CHAPTER 14
STABILITY OPERATIONS AND SUPPORT OPERATIONS

Units may have to conduct operations in environments that do not involve traditional combat. A unit may be called upon to conduct a stability or support contingency operation and then have to quickly transition into offensive or defensive missions. A unit may also be utilized in a stability or support operation at the successful conclusion of a combat mission. When assigned a stability or support mission, a well-trained unit must be able to rapidly shift its focus from war fighting to stability and support and also from stability and support to war fighting. While stability operations and support operations can occur anywhere, they will most likely occur in an urban environment. During a stability operation or support operation, units perform many activities not necessarily contained in its mission-essential task list (METL). While this chapter specifically addresses companies and company teams, many of the planning factors and TTP are applicable to levels above and below the company, with modifications. (See TC 7-98-1 for additional considerations and TTP.)

Section I. STABILITY OPERATIONS

Conducting stability operations is fundamentally identical to conducting combat operations. While each stability operation is different, the military decision-making process (MDMP) and troop-leading procedures (TLP) methodologies apply. This section will discuss planning considerations and specific TTP for the execution of stability missions.

14-1. PURPOSES AND TYPES OF STABILITY OPERATIONS

Various types of stability operations are conducted for many reasons.

a. **Purposes.** The purposes of stability operations are to—

- Protect national interests.
- Promote peace or deter aggression.
- Satisfy treaty obligations or enforce agreements and policies.
- Reassure allies, friendly governments, and agencies.
- Encourage a weak or faltering government.
- Maintain or restore order.
- Protect life and property.
- Demonstrate resolve.
- Deter or respond to terrorism.
- Reduce the threat of conventional arms and weapons of mass destruction (WMD) to regional security.
- Eliminate or contain subversion, lawlessness, and insurgency.

b. **Types of Stability Operations.** Table 14-1, on page 14-2, depicts the types of stability operations that a battalion TF may be called upon to conduct and the missions it will issue its subordinate companies/company teams in order to execute the stability operations.

TYPE	MISSIONS
Peace Operations	***Peacekeeping***: employ patrols, establish checkpoints, roadblocks, buffer zones, supervise truce, EPW exchange, reporting and monitoring, negotiation and mediation, liaison, investigation of complaints and violations, civil disturbance missions, act as quick reaction force (QRF), and offensive and defensive missions. ***Peace Enforcement:*** separation of belligerents; establishment and supervision of protected zones, sanction enforcement, movement denial and guarantee, restoration and maintenance of order, area security, humanitarian assistance, civil disturbance missions, act as QRF, and offensive and defensive missions. ***Operations in Support of Diplomatic Efforts:*** military to military contacts, exercises, security assistance, restore civil authority, rebuild physical infrastructure, provide structures and training for schools and hospitals, and reestablish commerce.
Foreign Internal Defense	***Indirect Support:*** military to military contacts, exercises, area security. ***Direct Support***: civil-military operations, intelligence and communications sharing, and logistical support. ***Combat Operations:*** offensive and defensive missions.
Support to Insurgencies	Show of force, defensive missions, raids, area security, employ patrols, and provide CSS.
Support to Counterdrug Operations	Liaison and advisor duty, civic action, intelligence support, surveillance support, reconnaissance, logistical support, and information support.
Combating Terrorism	Conduct force protection, offensive and defensive missions.
Noncombatant Evacuation Operations	Attack to seize terrain that secures evacuees or departure area, guard, convoy security, act as QRF, delay, and defend. See FM 90-29.
Arms Control	Seize and destroy weapons, convoy escort, assist and monitor inspection of arms, and conduct surveillance.
Show of Force	Perform tactical movement, demonstration, defensive operations, and perform training exercises.

Table 14-1. Types of stability operations, missions.

14-2. PLANNING CONSIDERATIONS

Planning considerations for stability operations include the following:

a. **Rules of Engagement**. The ROE are directives that explain the circumstances and limitations under which US forces initiate and or continue combat engagement with hostile forces. These rules reflect the requirements of the laws of war, operational concerns, and political considerations when the operational environment shifts from peace to conflict and back to peace. They should be established for, disseminated to, and thoroughly understood by every soldier in the unit. Another important consideration in development and employment of ROE is that commanders must assume that the belligerents they encounter will also understand the ROE; these unfriendly elements will attempt to use the ROE to their own advantage (and to the disadvantage of the friendly force). (See Appendix A for a more detailed discussion of ROE.)

b. **Rules of Interaction.** These directives, known as ROI, embody the human dimension of stability operations; they lay the foundation for successful relationships with the myriad of factions and individuals that play critical roles in these operations. ROI encompass an array of interpersonal communication skills such as persuasion and negotiation. These are tools the individual soldier will need to deal with the nontraditional threats that are prevalent in stability operations, including political friction, unfamiliar cultures, and conflicting ideologies. In turn, ROI enhance the soldier's survivability in such situations. ROI are based on the applicable ROE for a particular operation; they must be tailored to the specific regions, cultures, and or populations affected by the operation. Like ROE, ROI can be effective only if they are thoroughly rehearsed and understood by every soldier in the unit.

c. **Force Protection.** Commanders must implement appropriate security measures to protect the force. Establishment of checkpoints, effective base camp security procedures, and aggressive patrolling are examples of force protection measures.

d. **Task Organization.** Because of the unique requirements of stability operations, the company team may be task-organized to operate with a variety of units. This includes some elements with which the team does not normally work such as linguists, counterintelligence teams, and civil affairs teams.

e. **CSS Considerations.** The operational environment the company team faces during stability operations may be very austere, creating special CSS considerations. These factors include, but are not limited to, the following:

- Reliance on local procurement of certain items.
- Shortages of various critical items, including repair parts, Class IV supply materials, and lubricants.
- Special Class V supply requirements, such as pepper spray.
- Reliance on bottled water.

f. **Media Considerations.** The presence of the media is a reality that confronts every soldier involved in stability operations. All leaders and soldiers must know how to deal effectively with broadcast and print reporters and photographers. This should include an understanding of which subjects they are authorized to discuss and which ones they must refer to the PAO.

g. **Operations with Non-Army Agencies.** US Army units may conduct certain stability operations in coordination with a variety of outside organizations. These include other US armed services or government agencies as well as international organizations (including private volunteer organizations, nongovernmental organizations [NGO], and UN military forces or agencies).

14-3. ESTABLISH A LODGMENT AREA

A lodgment area is a highly prepared position used as a base of operations in stability operations. Like an assembly area or defensive strongpoint, the lodgment provides a staging area for the occupying unit, affords a degree of force protection, and requires 360-degree security. At the same time, several important characteristics distinguish the lodgment area from less permanent positions. Most notable is the level of preparation and logistical support required for long-term occupation. The lodgment must have shelters and facilities that can support the occupying force and its attachments for an extended period. The area must be positioned and developed so the unit can effectively conduct its primary missions (such as peace enforcement or counterterrorism) throughout its area of responsibility. In establishing the lodgment, the company team may use existing facilities or request construction of new facilities. A key advantage in using existing structures is immediate availability; this also reduces or eliminates the need for construction support from engineers and members of the team. There are disadvantages as well. Existing facilities may be inadequate to meet the team's operational needs, and they may pose security problems because of their proximity to other structures.

The company team may establish and occupy a lodgment area as part of a task force or, with significant support from the controlling task force, as a separate element. Figure 14-1 depicts a company team lodgment area established using existing facilities.

a. **Planning the Lodgment.** Before preparation, construction, and occupation of the lodgment area, the commander must plan its general layout. He should evaluate the following factors:

- Location of the lodgment area.
- Effects of weather.
- Traffic patterns.
- OP sites and or patrol routes.
- Entry and exit procedures.
- Vehicle emplacement and orientation.
- Bunkers and fighting positions.
- Fire planning.
- Size and composition of the reserve.
- Location of possible LZs and PZs.
- CSS considerations, including locations of the following:
  —Mess areas, showers, and latrines (including drainage).
  —Storage bunkers for Class III, Class IV, and Class V supplies.
  —Maintenance and refueling areas.
  —Aid station.
- CP site security.
- Size, composition, and function of advance/reconnaissance parties.
- Nature and condition of existing facilities (quarters; water, sewer, and power utilities; reinforced *hard-stand* areas for maintenance).
- Proximity to structures and or roadways (including security factors).

b. **Priorities of Work.** The commander must designate priorities of work as the company team establishes the lodgment area. He should consider the following tasks:

- Establishment of security of the immediate area and the perimeter.
- Establishment of initial roadblocks to limit access to the area.
- Mine clearance.
- Construction of revetments to protect vehicles, generators, communications equipment, and other facilities.
- Construction of barriers or berms around the lodgment area to limit observation of the compound and provide protection for occupants.
- Construction of shelters for lodgment personnel.

- Construction of defensive positions.
- Construction of sanitation and personal hygiene facilities.
- Construction of hardened CP facilities.
- Continuing activities to improve the site (such as adding hard-wire electrical power or perimeter illumination).

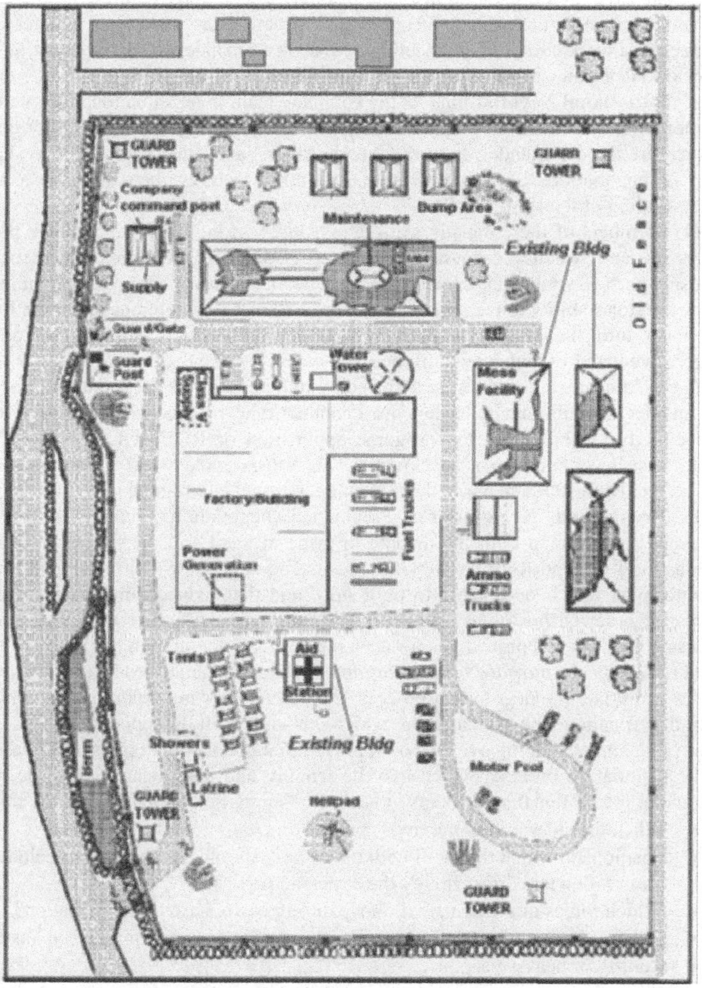

Figure 14-1. Example company lodgment using existing facilities.

14-4. CONDUCT NEGOTIATIONS

The company team may face a number of situations in which leaders will need to conduct negotiations. The two general types of negotiations are situational and preplanned. Situational negotiations are conducted in response to a requirement for on-the-spot discussion and resolution of a specific issue or problem. An example would be members of an advance guard negotiating the passage of a convoy through a checkpoint. Preplanned negotiations are conducted in such situations as a company team commander conducting a work coordination meeting between commanders of former warring factions (FWFs) to determine mine clearance responsibilities.

a. **Situational Negotiations.** At the company team level, situational negotiations are far more common than the preplanned type. In fact, employment in stability operations will require the commander, his subordinate leaders, and other soldiers to conduct some form of negotiations almost daily. This, in turn, requires them to have a thorough understanding of the ROE and ROI.

(1) Members of the company team apply this working knowledge to the process of discussing and, whenever possible, resolving issues and problems that arise between opposing parties, which may include the team itself. A critical aspect of this knowledge is the negotiator's ability to recognize that he has exhausted his options under the ROE/ROI and must turn the discussion over to a higher authority. Negotiations continue at progressive levels of authority until the issue is resolved.

(2) In preparing themselves and their soldiers for the negotiation process, the commander and subordinate leaders must conduct rehearsals covering the ROE and ROI. One effective technique is to war-game application of ROE/ROI in a given stability situation such as manning a checkpoint. This forces leaders and subordinates alike to analyze the ROE/ROI and apply them in an operational environment.

b. **Preplanned Negotiations.** Preplanned negotiations require negotiators to thoroughly understand both the dispute or issue at hand and the factors influencing it, such as the ROE and ROI, before talks begin. The negotiator's ultimate goal is to reach an agreement that is acceptable to both sides and that reduces antagonism (and or the chance of renewed hostilities) between the parties involved. The following paragraphs discuss guidelines and procedures for each phase of the negotiation process.

(1) ***Identify the purpose of negotiations.*** Before contacting leaders of the belligerent parties to initiate the negotiation process, the commander must familiarize himself with both the situation and the area in which his unit will be operating. This includes identifying and evaluating avenues of approach that connect the opposing forces. Results of the negotiation process, which may be lengthy and complicated, must be based on national or international agreements or accords. Negotiation topics include the following:

- When the sides will withdraw.
- Positions to which they will withdraw (these should be located to preclude observation and direct fire by the opposing parties).
- Which forces or elements will move during each phase of the operation?
- Pre-positioning of peace forces that can intervene in case of renewed hostilities.
- Control of heavy weapons.
- Mine clearance.
- Formal protest procedures for the belligerent parties.

(2) ***Establish the proper context.*** The next step in the process is to earn the trust and confidence of each opposing party. This includes establishing an atmosphere (and a physical setting) that participants will judge to be both fair and safe. These considerations apply:

- Always conduct joint negotiations on matters that affect both parties.
- When serving as a mediator, remain neutral at all times.
- Learn as much as possible about the belligerents, the details of the dispute or issue being negotiated, and other factors such as the geography of the area and specific limitations or restrictions (for example, the ROE and ROI).
- Gain and keep the trust of the opposing parties by being firm, fair, and polite.
- Use tact, remain patient, and be objective.
- Never deviate from applicable local and national laws and international agreements.

(3) ***Prepare for the negotiations.*** Thorough, exacting preparation is another important factor in ensuring the success of the negotiation process. Company team personnel should use the following guidelines:

- Negotiate sequentially, from subordinate level to senior level.
- Select and prepare a meeting place that is acceptable to all parties.
- Arrange for interpreters and adequate communications facilities as necessary.
- Ensure that all opposing parties, as well as the negotiating team, use a common map (edition and scale).
- Coordinate all necessary movement.
- Establish local security.
- Keep higher headquarters informed throughout preparation and during the negotiations.
- Make arrangements to record the negotiations (use audio or video recording equipment, if available).

(4) ***Conduct the negotiations.*** Negotiators must always strive to maintain control of the session. They must be firm, yet evenhanded, in leading the discussion. At the same time, they must be flexible, with a willingness to accept recommendations from the opposing parties and from their own assistants and advisors. The following procedures and guidelines apply:

- Exchange greetings.
- Introduce all participants by name, including negotiators and any advisors.
- Consider the use of small talk at the beginning of the session to put the participants at ease.
- Allow each side to state its case without interruption and without making premature judgments.
- Make a record of issues presented by both sides.
- If one side makes a statement that is incorrect, be prepared to produce evidence or proof to establish the facts.
- If the negotiating team or peacekeeping force has a preferred solution, present it and encourage both sides to accept it.

- Close the meeting by explaining to both sides what has been agreed upon and what actions they are expected to take. If necessary, be prepared to present this information in writing for their signatures.
- Do not negotiate or make deals in the presence of the media.
- Maintain the highest standards of conduct at all times.

14-5. MONITOR COMPLIANCE WITH AN AGREEMENT

Compliance monitoring involves observing FWFs and working with them to ensure they meet the conditions of one or more applicable agreements. Examples of the process include overseeing the separation of opposing combat elements, the withdrawal of heavy weapons from a sector, or the clearance of a minefield. Planning for compliance monitoring should cover, but is not limited to, the following considerations:

- Liaison teams, with suitable communications and transportation assets, are assigned to the headquarters of the opposing sides. Liaison personnel maintain communications with the leaders of their assigned element; they also talk directly to each other and to their mutual commander (the company team or task force commander).
- The commander positions himself at the point where it is most likely that violations could occur.
- He positions platoons and squads where they can observe the opposing parties, instructing them to assess compliance and report any violations.
- As directed, the commander keeps higher headquarters informed of all developments, including his assessment of compliance and or noncompliance.

14-6. ESTABLISH OBSERVATION POSTS

Construction and manning of OPs is a high-frequency task for company teams and subordinate elements when they must establish area security during stability operations. Each OP is established for a specified time and purpose. During most stability operations, OPs are both overt (conspicuously visible, unlike their tactical counterparts) and deliberately constructed. They are similar in construction to bunkers (see FM 5-103) and are supported by fighting positions, barriers, and patrols. If necessary, the company team can also employ hasty OPs, which are similar to individual fighting positions. Based on METT-TC factors, deliberate OPs may include specialized facilities such as the following:

- Observation tower.
- Ammunition and fuel storage area.
- Power sources.
- Supporting helipad.
- Kitchen, sleep area, shower, and or toilet.

Each OP must be integrated into supporting direct and indirect fire plans and into the overall observation plan. Figure 14-2 and Figure 14-3 (page 14-10), depict the general location and example layout of an overt, deliberately constructed OP. If OPs are established in buildings, they should be fortified and hardened IAW the information found in Chapter 3.

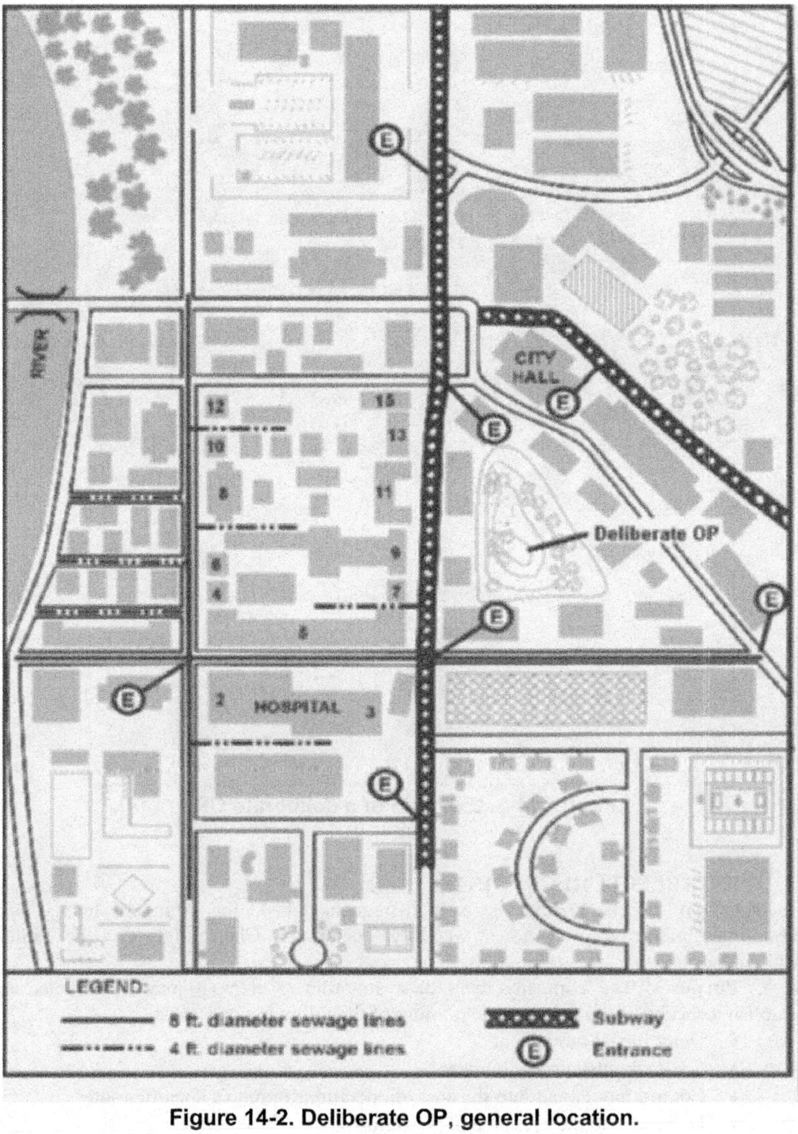

Figure 14-2. Deliberate OP, general location.

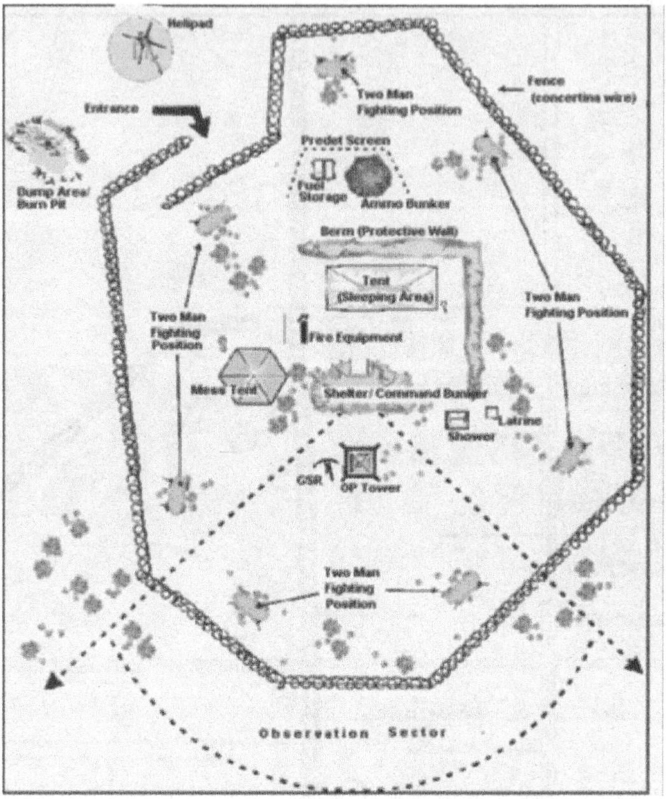

Figure 14-3. Example of a deliberate OP.

14-7. ESTABLISH CHECKPOINTS

Establishment of checkpoints is a high-frequency task for company teams and subordinate elements involved in stability operations. Checkpoints can be either deliberate or hasty.

a. **Purposes.** The company team or a subordinate element may be directed to establish a checkpoint to achieve one or more of the following purposes:

- Deter illegal movement.
- Create an instant roadblock.
- Control movement into the area of operations or onto a specific route.
- Demonstrate the presence of peace forces.
- Prevent smuggling of contraband.

- Enforce the terms of peace agreements.
- Serve as an OP and or patrol base.

b. **Checkpoint Procedures.** Checkpoint layout, construction, and manning should reflect METT-TC factors, including the amount of time available for emplacing it. The layout of a deliberate checkpoint is depicted in Figure 14-4, page 14-12. The following procedures and considerations may apply:

(1) Position the checkpoint in an area that is clear of hazards such as mines, booby traps, and unexploded ordnance (UXO).

(2) Position the checkpoint where it is visible and where traffic cannot turn back, get off the road, or bypass the checkpoint without being observed.

(3) Position a combat vehicle off the road, but within sight, to deter resistance to soldiers manning the checkpoint. The vehicle should be in a hull-down position and protected by local security. It must be able to engage vehicles attempting to break through or bypass the checkpoint.

(4) Place obstacles in the road to slow or canalize traffic into the search area.

(5) Establish a reserve.

(6) Establish a bypass lane for approved convoy traffic.

(7) Establish wire communications within the checkpoint area to connect the checkpoint bunker, the combat vehicle, the search area, security forces, the rest area, and any other elements involved in the operation.

(8) Designate the search area. If possible, it should be belowground to provide protection against such incidents as the explosion of a booby-trapped vehicle. Establish a parking area adjacent to the search area.

(9) If applicable, checkpoint personnel should include linguists.

(10) Properly construct and equip the checkpoint. Consider inclusion of the following items:

- Barrels filled with sand, concrete, or water (emplaced to slow and canalize vehicles).
- Concertina wire (emplaced to control movement around the checkpoint).
- Secure facilities for radio and wire communications with the controlling headquarters.
- First-aid kit.
- Sandbags for defensive positions.
- Wood or other materials for the checkpoint bunker.
- Binoculars, night vision devices, and or flashlights.
- Long-handled mirrors (these are used in inspections of vehicle undercarriages).

(10) Elements manning a deliberate CP may require access to specialized equipment such as the following:

- Floodlights.
- Duty log.
- Flag and unit sign.
- Barrier pole that can be raised and lowered.
- Generators with electric wire.

NOTE: Checkpoints that must be established inside an urban area must be overwatched with super surface (rooftop) security.

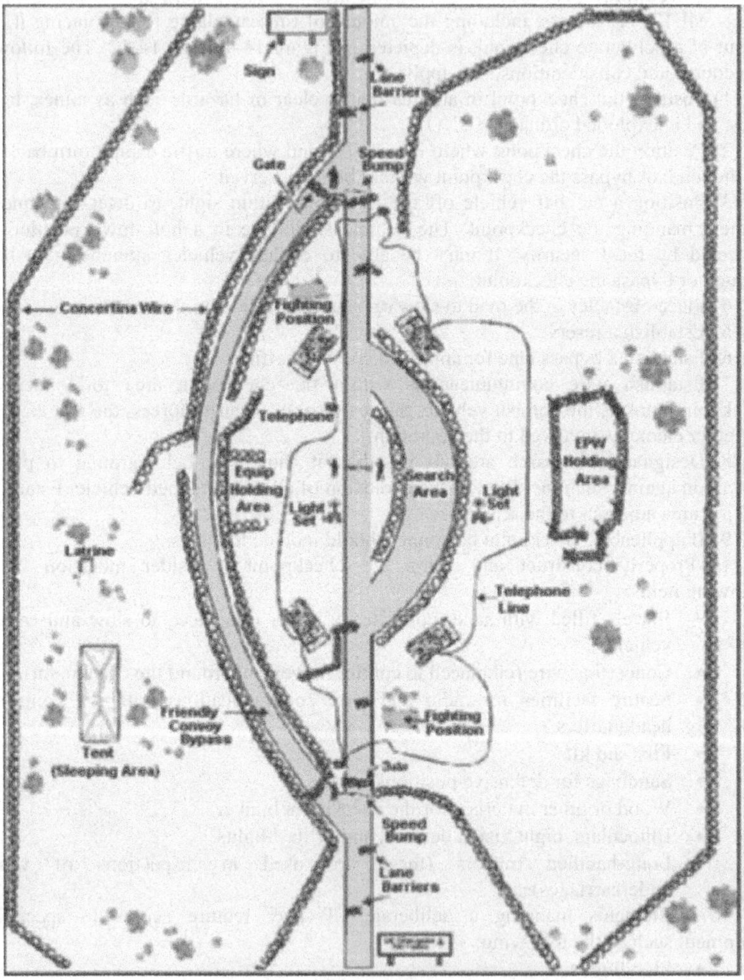

Figure 14-4. Deliberate checkpoint.

14-8. CONDUCT AREA SECURITY PATROLS

Patrolling is also a high-frequency task during stability operations. Planning and execution of an area security patrol are similar to procedures for other tactical patrols except that the patrol leaders must consider political implications and ROE. (Refer to FM 7-8 for a detailed discussion of patrol operations.) Figure 14-5 and Figure 14-6 (page 14-14), illustrate the use of area security patrols, in conjunction with checkpoints and OPs, in enforcing a zone of separation between belligerent forces.

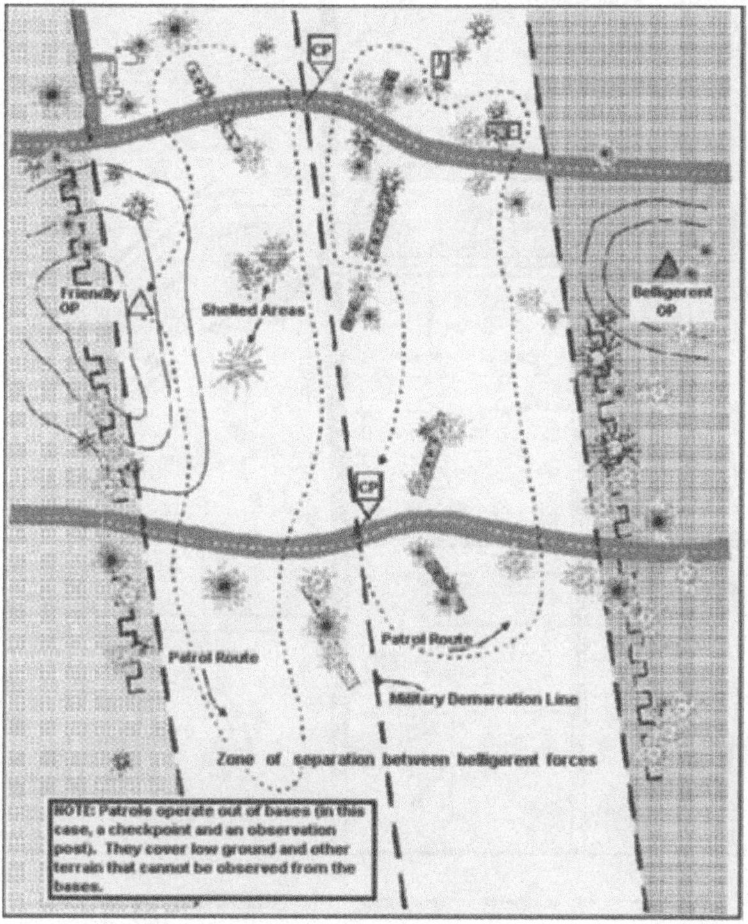

Figure 14-5. Example employment of checkpoints, OPs, and patrols to enforce a zone of separation.

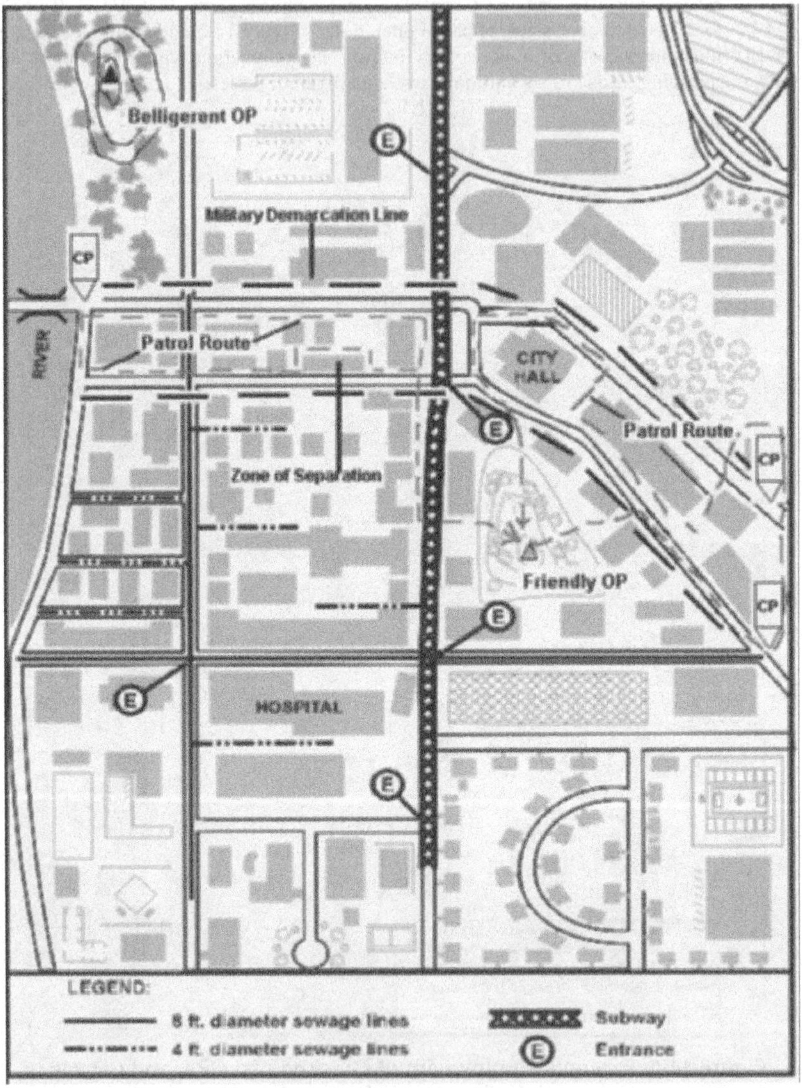

Figure 14-6. Example employment of checkpoints, OPs, and patrols to enforce a zone of separation.

14-9. CONDUCT CONVOY ESCORT

This mission requires the company team to provide a convoy with security and close-in protection from direct fire while on the move. The task force may choose this course of action, if enemy contact is imminent or when it anticipates a serious threat to the security of the convoy. Depending on METT-TC factors, the company team is capable of providing effective protection for a large convoy. (Smaller-scale convoy escort operations may be conducted by lighter security forces such as military police units.)

a. **Battle Command.** The task organization inherent in convoy escort missions makes battle command especially critical. The company team commander may serve either as the convoy security commander or as overall convoy commander. In the latter role, he is responsible for the employment not only of his own organic combat elements but also of CS and CSS attachments and drivers of the escorted vehicles. He must incorporate all of these elements into the various contingency plans developed for the operation. He must also maintain his link with the controlling tactical operations center (TOC).

(1) Effective SOPs and drills must supplement OPORD information for the convoy, and rehearsals should be conducted if time permits. Additionally, extensive precombat checks (PCCs) and inspections (PCIs) must be conducted, to include inspection of the escorted vehicles. The commander must also ensure that all required coordination is conducted with units and elements in areas through which the convoy will pass.

(2) Before the mission begins, the convoy commander should issue a complete OPORD to all vehicle commanders in the convoy. This is vital because the convoy may itself be task-organized from a variety of units and because some vehicles may not have tactical radios. The order should follow the standard five-paragraph OPORD format; it may place special emphasis on these subjects:

- Inspection of convoy vehicles.
- Route of march (including a strip map for each vehicle commander).
- Order of march.
- Actions at halts (scheduled and unscheduled).
- Actions in case of vehicle breakdown.
- Actions for a break in column.
- Actions in urban areas.
- Actions at danger areas (bridges, road intersections, defiles, and so forth).
- Actions on contact, covering such situations as snipers, enemy contact (including near or far ambush), indirect fire, and minefields.
- Riot drill.
- Refugee control drill.
- Evacuation drill.
- Actions at the delivery site.
- Chain of command.
- Guidelines and procedures for negotiating with local authorities.
- Communications and signal information.

b. **Tactical Disposition.** In any escort operation, the basic mission of the convoy commander (and, as applicable, the convoy security commander) is to establish and maintain security in all directions and throughout the length of the convoy. He must be prepared to adjust the disposition of the security force to fit the security requirements of each particular situation. Several factors affect this disposition, including METT-TC, convoy size, organization of the convoy, and types of vehicles involved. In some instances, the commander may position security elements, such as platoons, to the front, rear, and or flanks of the convoy. As an alternative, he may disperse the combat vehicles throughout the convoy body.

c. **Task Organization.** When sufficient escort assets are available, the convoy commander will usually organize convoy security into three distinct elements: advance guard, main body, and rear guard. He may also designate a reserve to handle contingency situations. Figure 14-7, on page 14-19, shows a company team escort force task-organized with an engineer platoon, an aerial scout section, a task force wheeled scout section, a BSFV air defense vehicle, a task force mortar section, and the team's normal maintenance and medical attachments.

NOTE: A convoy escort is normally provided with linguists as required.

(1) ***Advance Guard.*** The advance guard reconnoiters and proofs the convoy route. It searches for signs of enemy activity, such as ambushes and obstacles. Within its capabilities, it attempts to clear the route. The distance and time separation between the advance guard and the main body should be sufficient to provide the convoy commander with adequate early warning before the arrival of the vehicle column; however, the separation should be short enough that the route cannot be interdicted between the passage of the advance guard and the arrival of the main body. The advance guard should be task-organized with reconnaissance elements (wheeled scouts and aerial scouts, if available), combat elements (a tank or mechanized infantry platoon), and mobility assets (an engineer squad and a tank with plow or roller). It should also include linguists, as necessary.

(2) ***Main Body.*** The commander may choose to intersperse security elements with the vehicles of the convoy main body. These may include combat elements (including the rear guard), the convoy commander, additional linguists, mobility assets, and medical and maintenance support assets. Depending on METT-TC, the convoy commander may also consider the employment of flank security. The length of the convoy may dictate that he position the accompanying mortars with the main body.

(3) ***Reserve.*** In a company team escort mission, the reserve may consist of a tank or mechanized infantry platoon and the attached mortar section, if available. The reserve force will move with the convoy or locate at a staging area close enough to provide immediate interdiction against enemy forces. The supporting headquarters will normally designate an additional reserve, consisting of an additional company team or combat aviation assets, to support the convoy operation.

d. **Actions on Contact.** As the convoy moves to its new location, the enemy may attempt to harass or destroy it. This contact will usually occur in the form of an ambush, often executed in coordination with the use of a hasty obstacle. In such a situation, the safety of the convoy rests on the speed and effectiveness with which escort elements can execute appropriate actions on contact. Based on the factors of METT-TC, portions of the convoy security force, such as a tank platoon or tank section, may be designated as a reaction force. This element performs its normal escort duties, such as conducting tactical movement or occupying an assembly area, as required until enemy contact occurs; it then is given a reaction mission by the convoy commander.

(1) ***Actions at an Ambush.*** An ambush is one of the most effective ways to interdict a convoy. Conversely, reaction to an ambush must be immediate, overwhelming, and decisive. Actions on contact in response to an ambush must be planned for and rehearsed so they can be executed as a drill by all escort and convoy elements; particular attention should be given to fratricide prevention. In almost all situations, the security force will take several specific, instantaneous actions in reacting to an ambush. These steps include the following:

(a) As soon as they acquire an enemy force, the escort vehicles take action toward the enemy. They seek covered positions between the convoy and the enemy and suppress the enemy with the highest possible volume of fire permitted by the ROE. Contact reports are sent to higher headquarters as quickly as possible.

(b) The convoy commander retains control of the convoy vehicles and continues to move them along the route at the highest possible speed.

(c) Convoy vehicles, if armed, may return fire only if the security force has not positioned itself between the convoy and the enemy force.

(d) Subordinate leaders or the convoy commander may request that any damaged or disabled vehicles be abandoned and pushed off the route.

(e) The escort leader uses SPOTREPs to keep the convoy security commander informed. If necessary, the escort leader or the security commander can then request support from the reserve; he can also call for and adjust indirect fires.

(f) Once the convoy is clear of the kill zone, the escort element executes one of the following COAs based on the composition of the escort and reaction forces, the commander's intent, and the strength of the enemy force:

- Continue to suppress the enemy as the reserve moves to provide support.
- Assault the enemy.
- Break contact and move out of the kill zone.

(2) ***Actions at an Obstacle.*** Obstacles pose a major threat to convoy security. Obstacles can be used to harass the convoy by delaying it; if the terrain is favorable, the obstacle may stop the convoy altogether. In addition, obstacles can canalize or stop the convoy to set up an enemy ambush. The purpose of route reconnaissance ahead of a convoy is to identify obstacles and either breach them or find bypasses. In some cases, however, the enemy or its obstacles may avoid detection by the reconnaissance element. If this happens, the convoy must take actions to reduce or bypass the obstacle. When an obstacle is identified, the convoy escort faces two problems: reducing or bypassing the obstacle and maintaining protection for the convoy. Security becomes critical, and actions at the obstacle must be accomplished very quickly. The convoy commander must assume that the obstacle is overwatched and covered by enemy fires. To reduce the time the convoy is halted, and thus reduce its vulnerability, these actions should occur when the convoy escort encounters point-type obstacles:

(a) The lead element identifies the obstacle and directs the convoy to make a short halt and establish security. The escort overwatches the obstacle and requests that the breach force move forward.

(b) The escort maintains 360-degree security and provides overwatch as the breach force reconnoiters the obstacle in search of a bypass.

(c) Once all reconnaissance is complete, the convoy commander determines which of the following COAs he will take:

- Bypass the obstacle.
- Breach the obstacle with the assets on hand.
- Breach the obstacle with reinforcing assets.

(d) The commander relays a SPOTREP higher and, if necessary, requests support from combat reaction forces, engineer assets (if they are not part of the convoy), and aerial reconnaissance elements.

(e) The artillery units or the supporting mortar section is alerted to be prepared to provide fire support.

NOTE: Among the obstacles the convoy may encounter is an impromptu checkpoint established by civilians or noncombatants. If the checkpoint cannot be bypassed or breached, the commander must be prepared to negotiate passage for the convoy.

(3) ***Actions at a Halt.*** During a short halt, the convoy escort remains at REDCON-1 status regardless of what actions other convoy vehicles are taking. If the halt is for any reason other than an obstacle, the following actions should be taken:

(a) The convoy commander signals the short halt and transmits the order via tactical radio. Based on METT-TC factors, he directs all vehicles in the convoy to execute the designated formation or drill for the halt.

(b) Ideally, the convoy assumes a herringbone or coil formation. If the sides of the road are untrafficable or are mined; however, noncombat vehicles may simply pull over and establish 360-degree security as best they can. This procedure allows the movement of the escort vehicles as necessary through the convoy main body.

(c) If possible, escort vehicles are positioned up to 100 meters beyond other convoy vehicles, which are just clear of the route. Escort vehicles remain at REDCON-1 but establish local security based on the factors of METT-TC.

(d) When the order is given to move out, convoy vehicles reestablish the movement formation, leaving space for escort vehicles. Once the convoy is in column, local security elements (if used) return to their vehicles, and the escort vehicles rejoin the column.

(e) When all elements are in column, the convoy resumes movement.

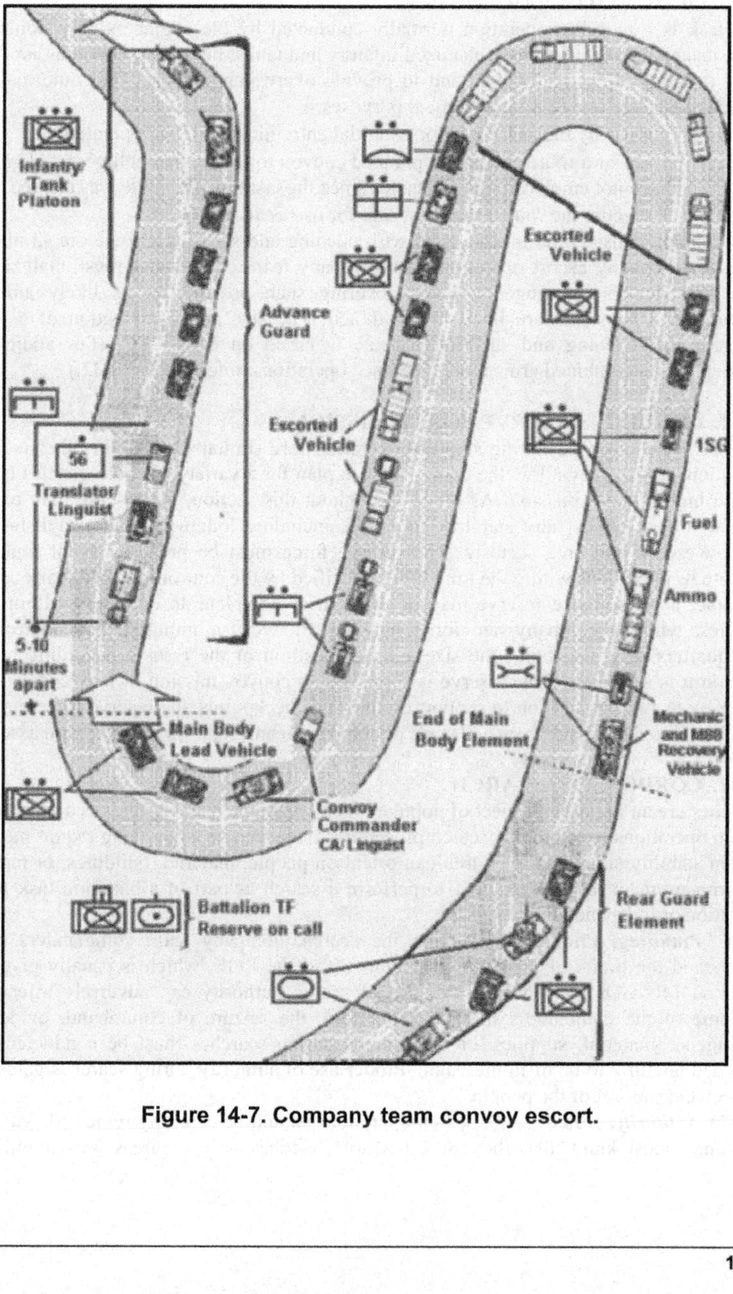

Figure 14-7. Company team convoy escort.

14-10. OPEN AND CLEAR ROUTES

This task is a mobility operation normally conducted by the engineers. The company team, tasked-organized with mechanized infantry and tanks, may be tasked to assist them using its mine plows and rollers and to provide overwatch support. The route may be cleared to achieve one of several tactical purposes:

- For use by the task force for its initial entry into an area of operations.
- To clear a route ahead of a planned convoy to ensure that belligerent elements have not emplaced new obstacles since the last time the route was cleared.
- To secure the route to make it safe for use as an MSR.

The planning considerations associated with opening and securing a route are similar to those for a convoy escort operation. The company team commander must analyze the route and develop contingency plans covering such possibilities as likely ambush locations and sites that are likely to be mined. The size and composition of a team charged with opening and securing a route is based on METT-TC. (For additional information on combined arms route clearance operations, refer to FM 20-32.)

14-11. CONDUCT RESERVE FORCE MISSION

Reserve force missions during stability operations are similar to those in other tactical operations in that they allow the commander to plan for a variety of contingencies based on the higher unit's mission. As noted throughout this section, the reserve may play a critical role in almost any stability operation, including lodgment area establishment, convoy escort, and area security. The reserve force must be prepared at all times to execute its operations within the time limits specified by the controlling headquarters. For example, a platoon-size reserve may be directed to complete an operation within five minutes, while a company-size force may be allotted ten minutes. The controlling headquarters may also tailor the size and composition of the reserve according to the mission it is assigned. If the reserve is supporting a convoy mission, it may consist of a company team; in a mission to support established checkpoints, the reserve force may be the dismounted elements from a platoon or company team, supported by aviation assets.

14-12. CORDON AND SEARCH

Searches are an important aspect of population and resource control. The need to conduct search operations or to employ search procedures may become an ongoing requirement in certain stability situations. A search can orient on people, materiel, buildings, or terrain. A company team may be required to perform a search as part of a battalion task force operation or independently.

a. **Planning.** Prior to conducting the search, company team commanders must understand the limits of their search authority and the ROE, which is usually given in battalion FRAGOs or OPORDs. Misuse of search authority can adversely affect the outcome of the command's mission. Therefore, the seizure of contraband, evidence, intelligence material, supplies, or other items during searches must be conducted and recorded lawfully to be of future value. Proper use of authority during searches gains the respect and support of the people.

(1) ***Authority.*** Authority for search operations should be carefully reviewed. Military personnel must know that they may perform searches only in areas within military

jurisdiction or where otherwise lawful. Searches may be conducted only to apprehend suspects or to secure evidence proving an offense has been committed.

(2) ***Instructions.*** Search teams should be given detailed instructions for handling controlled items. Lists of prohibited or controlled-distribution items should be widely disseminated and on hand during searches. The military or civil police who work with the populace and the resource control program are contacted before the search operations, or periodically if search operations are a continuing activity. This is normally coordinated by the battalion task force staff. Units must consider the effect of early warning on the effectiveness of their operation.

(3) ***Interpreters.*** Language difficulties can interfere when US forces interface with the local populace. Therefore, units given a search mission should be provided with interpreters as required.

(4) ***Tempo.*** Search operations are conducted slowly enough to allow for an effective search but rapidly enough to prevent the threat from reacting to the search.

(5) ***Use of Force.*** Under normal search conditions, minimum essential force is used to eliminate any active resistance encountered. Some situations may require the full shock effect of speed and surprise and limited violence of action. Company teams should be prepared to clear rooms under precision or high intensity conditions.

(6) ***Surprise.*** Searches may be conducted during limited visibility, early morning hours or multiple times to achieve surprise. Searchers should return to a searched area after the initial search to surprise and eliminate targeted individuals, groups, or their leaders who might have either returned or remained undetected during the search.

(7) ***Establishing a Cordon.*** Plans should be developed for securing the search area (establishing a cordon) and for handling detained personnel. Checkpoints can be employed to canalize traffic.

b. **Procedures.** The procedures for conducting a cordon and search are:

(1) ***Search of Individuals.*** The fact that anyone in an area to be searched could be an enemy or a sympathizer is stressed in all search operations. However, to avoid making an enemy out of a suspect, searchers must be tactful. The greatest caution is required during the initial handling of a person about to be searched. One member of the search team provides security while another member makes the actual search. Where appropriate, checkpoints are placed which allow controlling individuals with minimum force with maximum security.

(2) ***Search of Females.*** The threat may use females for all types of tasks when they think searches might be a threat. To counter this, female searchers should be used. (This should be coordinated by the battalion staff.) If male soldiers must search females, all possible measures must be taken to prevent any inference of sexual molestation or assault. Cultural differences may make this a particular problem, especially in Moslem communities.

(3) ***Search of Vehicles.*** Searching of vehicles may require that equipment such as detection devices, mirrors, and tools be made available. Occupants may need to be moved away from vehicles and individually searched, before the vehicle itself is searched. Specially trained dogs may be used to locate drugs or explosives. A thorough search of a vehicle is a time-consuming process. Effect on the population must be considered. A separate vehicle search area should be established to avoid unnecessary delays.

c. **Conduct of Cordon and Search.** When intelligence identifies and locates targeted individuals and groups, an operation is mounted to neutralize them. This should be done by police, acting on the warrant of a disinterested magistrate, and based on probable cause. Company teams will provide security and assist in this effort. In some cases, infantry units may have to conduct the actual search and apprehension. When the situation requires more aggressive action, emergency laws and regulations may dispense temporarily with some of these legal protections. The method used should be the least severe method that is adequate to accomplish the mission. Care should be taken to preserve evidence for future legal action. The area to be searched in a urban area should be divided into zones and a search party assigned to each. A search party consists of a security element to encircle the area, to prevent entrance and exit, and to secure open areas; a search element to conduct the search; and a reserve element to help as required (Figure 14-8).

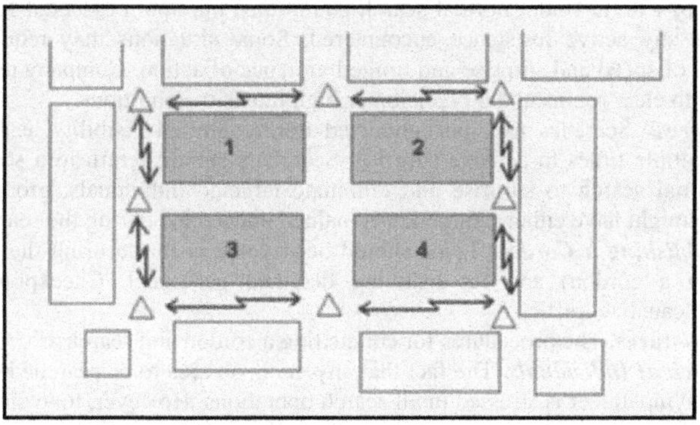

Figure 14-8. Urban cordon.

(1) ***Task Organization.*** The specific company team task organization will be determined by the factors of METT-TC. A sample company task organization for a cordon and search mission is shown below:

(a) *Security Element.* The security element consists of one platoon reinforced with the company AT; 60-mm mortar sections acting as infantry; and a tank platoon. This element establishes the checkpoints (shown in Figure 14-8) and conducts security patrols around the cordon preventing exit and entry. Depending on the enemy situation, hasty defensive positions can be assumed in buildings. In the example shown in Figure 14-8, the same security element can be employed to establish the OPs and roadblock.

(b) *Search Element.* The search element is one platoon reinforced with a squad. Each squad (search team) is given the mission to search buildings 1, 2, 3, and 4, respectively.

(c) *Reserve Element.* The reserve element is one platoon minus a squad. This platoon is given tasks as required by METT-TC. Part of the reserve may be located inside the cordon to be prepared to assist the search element; part may be located outside the cordon to assist the security element.

(2) ***Other Assets.*** A company team may be given other assets to assist them in this mission based on availability and METT-TC factors. These assets may be included as teams in the security element or they may remain outside the security element, on call. These assets may come from within the battalion task force, or attached or OPCON MP, engineer, CA, PSYOP, MI, or artillery units. Teams that can be formed from these assets include:

- Mine detection team.
- Demolition team.
- Interrogation team.
- Documentation team (utilizes a recorder with a camera).
- Scout dog team.
- PSYOP/CA augmentation team.
- FIST.
- Prisoner team.
- Tunnel reconnaissance team.
- Escort parties.
- Transportation teams.

Additionally, civilians or civilian authorities may assist or support the teams or elements mentioned above.

(3) ***Establishing a Cordon.*** An effective cordon is critical to the success of the search effort. Cordons are designed to prevent the escape of individuals to be searched, and to protect the forces conducting the operation. In remote areas, the cordon may be established without being detected. Limited visibility aids can be used in the establishment and security of the cordon.

(a) Plans should be developed to handle detained personnel. Infantrymen normally will provide security and accompany police and intelligence forces who will identify, question, and detain suspects. Infantry may also conduct searches and assist in detaining suspects, under police supervision; but, their principal role is to reduce any resistance that may develop and to provide security for the operation. Use of force is kept to a minimum.

(b) Deployment for the search should be rapid, especially if the threat is still in the area to be searched. Ideally, the entire area should be surrounded at once; observed fire covers any gaps.

(c) The security element surrounds the area while the search element moves in. Members of the security element orient mainly on people evading the search in the populated area; however, the security element can also cut off any belligerents trying to reinforce others within the area. Checkpoints and roadblocks may need to be established.

(d) Subsurface routes of escape in urban areas, such as subways and sewers, may also need to be cordoned and searched. The procedures below should be considered when preparing for the search of a urban area.

(4) ***Conducting the Search.*** A search of an urban area must be conducted with limited inconvenience to the populace. However, the populace should be inconvenienced enough to discourage targeted individuals and groups and their sympathizers from remaining in the locale, but not enough to drive the rest of the populace to collaborate with belligerents as a result of the search. A large-scale search of the urban area is a combined civil police and military operation. If this occurs, it is normally conducted at battalion task force level or higher. Such a search should be planned in detail and

rehearsed. Physical reconnaissance of the area just before a search is avoided. Information needed about the terrain can be obtained from aerial photographs. In larger towns or cities, the local police might have detailed maps showing relative sizes and locations of buildings. For success, the search plan must be simple and the search must be conducted swiftly. The search element conducts the mission assigned for the operation. The element is organized into special teams. These teams can include personnel and special equipment as previously discussed. Three basic methods are used to search the populated area.

(a) *Central Location.* Assemble inhabitants in a central location if they appear to be hostile. This method provides the most control; simplifies a thorough search; denies the belligerents an opportunity to conceal evidence; and allows for detailed interrogation. It has the disadvantage of taking the inhabitants away from their dwellings, thus encouraging looting, which, in turn, engenders ill feelings.

(b) *Home Restriction.* Restrict the inhabitants to their homes. This prohibits movement of civilians, allows them to stay in their dwellings, and discourages looting. The disadvantages of this method are that it makes control and interrogation difficult and gives inhabitants time to conceal evidence in their homes.

(c) *Control Heads of Households.* The head of each household is told to remain in the front of the house while everyone else in the house is brought to a central location. During the search, the head of the household can see that the search team will steal nothing. Often, this is the best method for controlling the populace during a search. This person can be used to open doors and containers to facilitate the search.

(5) ***Searching a House.*** Escort parties and transportation must be arranged before the search of a house. The object of a house search is to screen residents to determine if there are any targeted individuals and groups and their sympathizers, and to look for controlled items. A search party assigned to search an occupied building should consist of at least one local policeman, a protective escort (usually infantry), and a female searcher. Forced entry may be necessary if a house is vacant or if an occupant refuses to allow searchers to enter. If a house containing property is searched while its occupants are away, it should be secured to prevent looting. Before US forces depart, the commander should arrange for the community to protect such houses until the occupants return.

d. **Other Considerations.** Other considerations for conducting a cordon and search are:

(1) The reserve element is a mobile force positioned in a nearby area. Its mission is to help the other two elements if they meet resistance beyond their ability to handle. The reserve element can replace or reinforce either of the two elements if the need arises.

(2) Any objectionable material found, including propaganda signs and leaflets, should be treated as if it were booby-trapped until inspection proves it safe.

(3) Underground and underwater areas should be searched thoroughly. Any freshly excavated ground could be a hiding place. Mine detectors can be used to locate metal objects underground and underwater.

(4) Depending on the factors of METT-TC, a *graduated response* technique can be employed. This technique uses warnings and progressive amounts of force to obtain compliance. For example, warnings in the native language can be given announcing that some type of force, lethal or nonlethal, will be used in a given amount of time if the occupants do not exit the building.

(5) Before entering the area, psychological operations announcements can be made to encourage inhabitants to leave peacefully. While this technique minimizes collateral damage, it also gives the enemy time to react.

e. **Aerial Search Operations.** Search units mounted in helicopters can take full advantage of the mobility of these aircraft. Attack helicopters can also be used to help cordon off areas.

(1) Air assault combat patrols conducting an aerial search can reconnoiter an assigned area or route in search of targeted groups or individuals. When the patrols locate the threat, the patrol may engage it from the air or may land and engage it on the ground. This technique can be useful in urban areas that have open areas and wide streets. If an air defense threat is present, this technique will have limited value.

(2) Use of air assault patrols should be used only in operations when sufficient intelligence is available to justify their use. Even then, such patrols should be used along with ground operations.

f. **Cordon and Search of a Village.** The same techniques can be used to cordon and search a small urban area (village) surrounded by other terrain (Figure 14-9). In this situation, OPs are established on dominant terrain and a roadblock is established on the main road leading into the urban area.

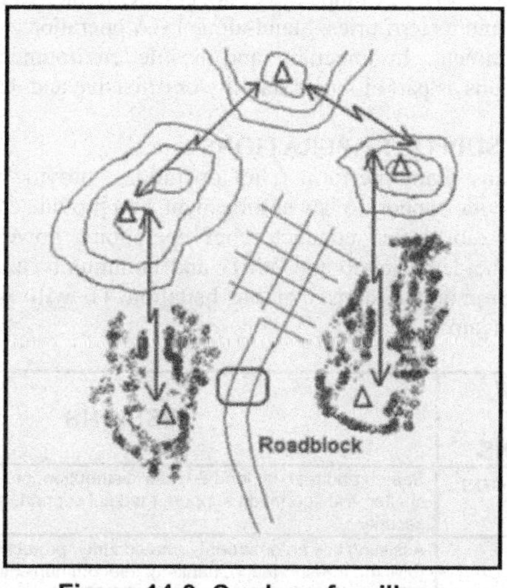

Figure 14-9. Cordon of a village.

NOTE: METT-TC factors may require a battalion TF to conduct the cordon and search. Inner and outer cordons may be required to properly secure the objective area. This will require the battalion TF to deploy its company teams as security, search, and reserve elements in accordance with the tactical situation. The techniques described in the preceding paragraph can be employed with modifications.

Section II. SUPPORT OPERATIONS

Support operations provide essential supplies and services to assist designated groups. They are conducted to help foreign and domestic civil authorities respond to crises. Company teams normally conduct support operations as part of a larger battalion operation to save or protect lives, reduce suffering, recover essential infrastructure, improve the quality of life, and restore situations to normal. Because of the nature of support operations, the company team can expect to interact with other units and agencies such as engineers, MPs, and nongovernment organizations (NGOs). Support actions rely on a partnership with other government and nongovernment agencies. Liaison with these agencies and between local governments is critical. Regardless of the positive relationships built, force protection always remains a top priority. Again, planning for support operations is fundamentally identical to planning for combat and stability operations. While each support operation is different, the military decision-making process (MDMP) and troop-leading procedures (TLPs) methodologies apply.

14-13. TYPES OF SUPPORT OPERATIONS

The two types of support operations are domestic support operations (DSO) and foreign humanitarian assistance (FHA). Companies conduct DSO in the US and its territories and FHA outside the US and its territories. Stand-alone FHA operations are conducted only in a permissive environment. In uncertain and hostile environments, company teams conduct FHA operations as part of larger stability or offensive and defensive operations.

14-14. FORMS OF SUPPORT OPERATIONS

During DSO, company teams perform relief operations, provide support to incidents involving WMD, provide support to law enforcement, and provide community assistance. In FHA, companies most often conduct relief operations; however, FHA may also involve support to incidents involving WMD and community assistance. Table 14-2 depicts the more common missions that the battalion TF will assign to subordinate companies/company teams.

FORMS OF SUPPORT OPERATIONS	MISSIONS
Relief Operations	Search and rescue, food & water distribution, providing temporary shelter, transportation support, medical support, sanitation, area security.
Support to Incidents Involving WMD	Assisting law enforcement, area security, protection of critical assets (utilities, transportation, banking, telecommunications), responding to WMD casualties, establishing roadblocks/checkpoints.
Support to Civil Law Enforcement	Civil disturbance missions; support to counterterrorism and counterdrug operations; providing resources, training, and augmentation; assisting with cordon and search; security patrols; establish roadblocks and checkpoints.
Community Assistance	Search and rescue, firefighting, assistance in safety and traffic control, emergency snow removal, providing temporary shelter.

Table 14-2. Forms of support operations, missions.

14-15. PHASES OF SUPPORT OPERATIONS

Although each operation is unique, support operations are generally conducted in three broad phases: response, recovery, and restoration. Army units can expect to be most heavily committed during the response phase. They will be progressively less involved during the recovery phase, with only very limited activity, if any, during the restoration phase.

a. **Response Phase.** In the response phase, commanders focus on the life-sustaining functions that are required by those in the disaster area. The following functions dominate these response operations:

- Search and rescue.
- Emergency flood control.
- Hazard identification.
- Food distribution.
- Water production, purification, and distribution.
- Temporary shelter construction and administration.
- Transportation support.
- Fire fighting.
- Medical support.
- Power generation.
- Communications support.

b. **Recovery Phase.** Recovery phase operations begin the process of returning the community infrastructure and related services to a status that meets the immediate needs of the population. Typical recovery operations include the following:

- Continuation of response operations as needed.
- Damage assessment.
- Power distribution.
- Water and sanitation services.
- Debris removal.

c. **Restoration Phase.** Restoration is a long-term process that returns the community to predisaster condition. Restoration activities do not generally involve large numbers of military forces. Army units generally would work with affected communities in the transfer of responsibility to other agencies, as military support forces redeploy.

Section III. TRANSITION TO COMBAT OPERATIONS

Stability, and to a lesser extent, support operations are missions that may transition to combat. The company team commander must always keep in mind that the pendulum can also shift from a stability or support operation to combat. An escalation to combat is a clear indicator that the stability or support operation failed. The company must always retain the ability to conduct offensive and defensive operations. Preserving the ability to transition allows the company to maintain initiative while providing force protection.

14-16. PLAN FOR CONTINGENCIES

The commander must plan for contingency operations that factor in what actions the company team will perform if combat cannot be averted; for example, reverting to a hasty defense in the event that a stability or support mission deteriorates.

14-17. BALANCED MINDSET

A balance must be achieved between the mindset of peace operations and the mindset of war fighting. Soldiers can not become too complacent in their warrior spirit, but also must not be too eager to rely on the use of force to resolve conflict. This balance is the essence of peace operations and the fundamental aspect that will enable the company team to perform its mission successfully and avoid an escalation to combat. Proactive leaders that are communicating and enforcing the ROE are instrumental to achieving this mindset.

14-18. COMBAT SKILLS TRAINING

If the stability or support operation extends over prolonged periods of time, training should be planned that focuses on the individual and collective combat tasks that would be performed during transition to offensive and or defensive missions.

APPENDIX A

URBAN OPERATIONS UNDER RESTRICTIVE CONDITIONS

During operations on urban terrain, commanders can expect to encounter restrictions on their use of firepower. Basic doctrinal principles remain the same, but the tactics, techniques, and procedures may have to be modified to stay within established rules of engagement (ROE) and to avoid unnecessary collateral damage.

A-1. HIGH-INTENSITY, PRECISION, AND SURGICAL CONDITIONS

At the company level and below, conditions of urban combat may change quickly. Soldiers conducting the operations often cannot distinguish what are high-intensity or precision conditions. The specific guidance that will assist them in performing their mission, while simultaneously avoiding civilian casualties and collateral damage, will be the ROE. Small-unit leaders will be responsible for enforcing the ROE because the urban battle is conducted at small-unit level. (Refer to Chapter 1 for detailed information concerning these terms.)

A-2. RULES OF ENGAGEMENT

The unified commander issues the ROE for tactical forces. The ROE are based on the commander's analysis of his guidance from the National Command Authority (NCA), the mission that he has been given, the existing threat, the laws of land warfare, and any host nation or third-world country constraints on US forces. Printed ROE will normally be distributed through command channels and delivered to units in sufficient quantities for each soldier to have a copy. Units will always be faced with adhering to ROE of some kind. ROE have a significant impact on how missions are executed during UO. They must provide clear guidance to soldiers about when and how to employ force to accomplish the mission and to defend themselves.

a. **Political Concerns.** The political concerns used while developing the ROE may appear to conflict with the physical security needs of units. Politically driven constraints must be weighed against the potential risks to mission accomplishment and to the force itself. ROE must be practical, realistic, understandable, and enforceable. If soldiers do not believe that they can survive under the ROE, the rules will be very difficult to enforce. Commanders can affect the ROE by suggesting changes or requesting clarification or modifications. Like the mission, ROE must be tailored to the day-to-day changes in the conditions and threats that face the US forces.

b. **Situation Dependent.** Whatever the situation that has called for restricted ROE, Infantry units will probably operate in a dangerous, yet highly constrained, environment. This demands a high degree of patience, training, and dedication on the part of the soldiers and leaders. An example of ROE used during Operation Just Cause is shown in Table A-1, page A-2. The example is not intended as a sole-source document for developing ROE. It is to be used as an example of how political considerations during a recent mission were translated by the commander into specific ROE.

(1) The ROE is much more restrictive under certain conditions of UO than under others. For example, a particular mission might require ROE that limit the use of indirect

fire weapons. On the other hand, a mission to clear a building may require ROE that authorize force to clear rooms, and include authoritative guidance concerning measures to protect noncombatants, breach obstacles, and react to snipers.

(2) One of the most significant issues of UO is collateral damage. Collateral damage is unintended and undesirable civilian personnel injuries or material damage adjacent to a target produced by the effects of friendly weapons. ROE will provide guidance concerning how to minimize collateral damage. For example, ROE may require use of nonlethal capabilities to the maximum extent possible before use of lethal weapons and munitions, or may restrict use of indirect fire weapons. The ROE will establish when certain types of weapons and munitions can be used (Table A-1).

(3) A mission can transition quickly from stability or support to offense or defense. This transition might be caused by threat actions or the actions of noncombatants. Commanders must be prepared to react to this situation and request changes in the ROE when necessary.

(4) ROE differentiate between the use of force for self-defense and for mission accomplishment. Commanders always retain the inherent authority and obligation to use necessary and proportional force for unit and individual self-defense in response to a hostile act or demonstrated hostile intent.

ALL ENEMY MILITARY PERSONNEL AND VEHICLES TRANSPORTING THE ENEMY OR THEIR SUPPLIES MAY BE ENGAGED SUBJECT TO THE FOLLOWING RESTRICTIONS:
a. Armed force is the last resort.
b. When possible, the enemy will be warned first and allowed to surrender.
c. Armed civilians will be engaged only in self-defense.
d. Civilian aircraft will not be engaged without approval from above division level unless it is in self-defense.
e. Avoid harming civilians unless necessary to save US lives. If possible, try to arrange for the evacuation of civilians prior to any US attack.
f. If civilians are in the area, do not use artillery, mortars, armed helicopters, AC-130s, tube- or rocket-launched weapons, or M551 main guns against known or suspected targets without the permission of a ground maneuver commander, LTC or higher (for any of these weapons).
g. If civilians are in the area, control all air attacks by FAC or FO.
h. If civilians are in the area, close air support (CAS), white phosphorus, and incendiary weapons are prohibited without approval from above division level.
i. If civilians are in the area, do not shoot except at known enemy locations.
j. If civilians are not in the area, you can shoot at suspected enemy locations.
k. Public works such as power stations, water treatment plants, dams, or other utilities may not be engaged without approval from above division level.
l. Hospitals, churches, shrines, schools, museums, and any other historical or cultural site will not be engaged except in self-defense.
m. All indirect fire and air attacks must be observed.
n. Pilots must be briefed for each mission on the location of civilians and friendly forces.
o. No booby traps. No mines except as approved by division commander. No riot control agents except with approval from above division level.
p. Avoid harming civilian property unless necessary to save US lives.

Table A-1. Example of Operation Just Cause ROE.

q. Treat all civilians and their property with respect and dignity. Before using privately owned property, check to see if any publicly owned property can substitute. No requisitioning of civilian property without permission of a company-level commander and without giving a receipt. If an ordering officer can contract for the property, then do not requisition it. No looting. Do not kick down doors unless necessary. Do not sleep in their houses. If you must sleep in privately owned buildings, have an ordering officer contract for it. r. Treat all prisoners humanely and with respect and dignity. s. Annex R to the OPLAN provides more detail. Conflicts between this card and the OPLAN should be resolved in favor of the OPLAN.	
DISTRIBUTION: 1 per trooper deployed to include all ranks.	

SUPPLEMENTAL RULES OF ENGAGEMENT FOR SELECTED RECURRING OPERATIONS:
1. CONTROL OF CIVILIANS ENGAGED IN LOOTING. a. Senior person in charge may order warning shots. b. Use minimum force but not deadly force to detain looters. c. Defend Panamanian (and other) lives with minimum force including deadly force when necessary.
2. ROADBLOCKS, CHECKPOINTS AND SECURE DEFENSIVE POSITIONS. a. Mark all perimeter barriers, wires, and limits. Erect warning signs. b. Establish second positions to hastily block those fleeing. c. Senior person in charge may order warning shots to deter breach. d. Control exfiltrating civilians with minimum force necessary. e. Use force necessary to disarm exfiltrating military and paramilitary. f. Attack to disable, not destroy, all vehicles attempting to breach or flee. g. Vehicle that returns or initiates fire is hostile. Fire to destroy hostile force. h. Vehicle that persists in breach attempt is presumed hostile. Fire to destroy hostile force. i. Vehicle that persists in flight after a blocking attempt IAW instruction 2b is presumed hostile. Fire to destroy hostile force.
3. CLEARING BUILDINGS NOT KNOWN TO CONTAIN HOSTILE FORCE. a. Warn all occupants to exit. b. Senior person in charge may order warning shots to induce occupants to exit c. Do not attack hospitals, churches, shrines, or schools, museums, and any historical or cultural sites except in self-defense. d. Respect and minimize damage to private property. e. Use minimum force necessary to control the situation and to ensure the area is free of hostile force.

Table A-1. Example of Operation Just Cause ROE (continued).

A-3. NONLETHAL CAPABILITIES

Nonlethal capabilities are weapon systems, munitions, and equipment required to improve the operational requirement for an enhanced capability to apply nonlethal force. The components are explicitly designed and primarily employed to incapacitate personnel or material, while minimizing fatalities or permanent injury to intended targets and collateral damage to property and the environment. Nonlethal capabilities are not guaranteed to be totally nonlethal; the use of certain capabilities under certain circumstances may, in fact, render some fatalities. The use of the term "nonlethal" is not intended to be misleading, but to covey the intention to be able to achieve military objectives while greatly reducing fatalities.

a. **Characteristics.** Unlike weapons that permanently destroy targets through blast fragmentation or penetration, nonlethal capabilities have one, or both, of the following characteristics:

(1) They use means other than physical destruction to prevent the target from functioning.

(2) They have relatively reversible effects. Even if they injure humans, the injured usually recover.

b. **Examples.** The Infantry has had some nonlethal capabilities for years. Other weapons represent new developing technology. Examples of nonlethal capabilities are:

- Riot control agents (RCA) such as CS (see Appendix F).
- Incapacitating sprays such as Mace and pepper spray.
- Kinetic stun projectiles such as rubber bullets, wooden baton rounds, and beanbag or ringfoil grenades.
- Rigid, sticky, or disorienting foams.
- Super-lubricants.
- Stun grenades.
- Acoustic distraction devices.

c. **Reasons for Using Nonlethal Capabilities.** Increased attention is being given to the use of nonlethal capabilities for the following reasons:

- Growing conviction about their potential military utility.
- Political sensitivity.
- New constraints imposed by arms control.
- Increasing interest by US allies and outside organizations concerned with international security.
- Recent advances in enabling technologies related to nonlethal capabilities.
- Emerging missions needing better nonlethal solutions such as crowd control, stability operations, and support operations in urban areas.

d. **Options.** Nonlethal capabilities provide commanders with multiple options ranging from not using military force at all through the use of lethal force. Nonlethal capabilities may be more appropriate for some missions than lethal weapons. Nonlethal capabilities can provide a more humane, discriminate, and reversible means of employing military force, with more precisely tailored and focused effects.

e. **Constraints.** Unless constrained by orders from higher headquarters, commanders are not obligated in any way to use only nonlethal capabilities, or to try nonlethal capabilities before resorting to more lethal means, in any military operation. The ROE and guidance from higher headquarters will determine appropriate use of nonlethal capabilities. Although US forces may wish to avoid casualties, many situations require overwhelming lethal force as the most effective and efficient means to accomplish the mission. In the final result, a swift victory by overwhelming force may actually involve fewer casualties on both sides.

f. **Troop Safety.** However the commander chooses to use nonlethal force, its use should be in a manner that minimizes additional risks to friendly forces. The right to use lethal force for self-defense against a deadly threat is unaffected by any earlier choice of nonlethal capabilities to achieve mission objectives.

g. **The Nonlethal Capability Set**. The nonlethal capability set (NLCS) is a systems approach containing weapon systems, munitions, and equipment required to improve the operational requirement for an enhanced capability to apply nonlethal force. Battalion NCLS were placed in contingency stocks in September 2000, and will be issued to units on an as needed basis. The NLCS contains:

- ***Individual Protection:*** Provides the individual ballistic and non-ballistic protective equipment that safeguards the soldier.
- ***Individual Enhancement:*** Provides the equipment that compliments individual protection and enhances the soldier's capability.
- ***Mission Enhancement:*** Provides the equipment that supports the commander in the accomplishment of the mission.
- ***Munitions:*** Provides the munitions that enable the accomplishment of the mission in compliance with current policy and guidance for the application of nonlethal force.
- ***Training:*** Provides the organization with the equipment that will assist in the training for operations involving the use of nonlethal force.

APPENDIX B

URBAN OPERATIONS UNDER CONDITIONS OF LIMITED VISIBILITY

With the rapid development of night vision devices throughout the world and doctrine that mandates continuous operations, US forces will continue to fight in built-up areas regardless of the weather or visibility conditions. To be successful, leaders must anticipate the effects of limited visibility conditions on operations and soldiers.

B-1. ADVANTAGES

When fighting in built-up areas during periods of limited visibility, attacking or defending forces have several advantages.

a. **Technological Advantages.** In most cases, US forces have a technological advantage in thermal imagery and light intensification over their opponents. This enables US forces to identify, engage, and destroy enemy targets before detection by the enemy.

b. **Continuous Operations.** Doctrine stresses continuous operations, day and night. This allows the attacking forces to conclude the battle decisively in a shorter period of time. It also allows the attacker to retain the initiative. Defending forces have the ability to see during conditions of limited visibility and also conduct continuous operations.

c. **Shorter Ranges.** Direct-fire target ranges in urban terrain are greatly reduced. During periods of limited visibility, effective target acquisition ranges are even further reduced. This enables attacking forces to close to shorter ranges, thus increasing the lethality and accuracy of weapons. Attacking forces can also take advantage of the enemy's reduced visibility and can engage before being detected with thermal imagery or light intensification devices. The defender also has the ability to take advantage of periods of limited visibility and effectively employ his weapons at shorter ranges and deliver lethal, accurate fire.

d. **Surprise.** Attacking during periods of limited visibility gives the attacker a greater chance of surprise.

e. **Speed.** Decreased visibility can facilitate speed for well-trained units accustomed to fighting under these conditions.

B-2. DISADVANTAGES

When fighting in built-up areas during limited visibility, attacking and defending forces also face some disadvantages.

a. **Command and Control.** Command and control is difficult in any operation in a built-up area, and periods of limited visibility increase this difficulty. This can be overcome by all leaders maintaining situational awareness.

b. **Dispersion.** Soldiers have an instinctive tendency to form groups during limited visibility. Leaders must ensure that soldiers do not "bunch-up."

c. **Confusion.** Due to the low visibility and the characteristics of built-up areas, soldiers become disoriented easily.

d. **Target Identification.** Target identification becomes difficult in limited visibility conditions. Depending on the individual, the soldier may fire at anything he sees, or he may hesitate too long before firing. This is one of the leading causes of fratricide, so leaders must pay close attention to target identification and engagement.

B-3. FRATRICIDE AVOIDANCE

The risk of fratricide is much greater during periods of limited visibility. The key to avoiding fratricide is situational awareness by leaders and individuals coupled with training. Other considerations include:

a. **Clear Graphic Control Measures.** Graphic control measures should be clearly defined and obvious. Examples include distinct buildings, large boulevards, streets, and so forth.

b. **Leader Control.** Leaders must exercise firm control when engaging targets. Movements should also be tightly controlled. Examples include code words for movement, use of pyrotechnic signals for shifting fire, and so forth.

c. **Marking Cleared Rooms.** Cleared rooms and buildings should be distinctly marked to identify cleared areas and the progress of clearing teams to any base of fire supporting the maneuver. Other examples include marking passage lanes, obstacles, booby traps, and so forth.

d. **Marking Soldiers.** Visible markers (for example, glint tape or thermal strips) should be attached to individual soldiers. Specific markings should be developed for distinguishing casualties.

e. **Recognition Symbols.** Far and near recognition symbols should be used properly. Examples include challenges and passwords, running passwords, building markings, other visual signals, and so forth.

f. **CAS.** Units using close air support (CAS) must exercise firm control and direct their firing. Failure to do so may lead to the pilot becoming disoriented and engaging friend and foe alike.

B-4. URBAN ENVIRONMENTAL EFFECTS ON NIGHT VISION DEVICES

The characteristics of urban areas affect standard US NVDs and sights differently than other areas. There are more lighted areas and not enough light inside of many structures, even in daytime. This may cause some confusion for soldiers operating during limited visibility, since the images they receive through their NVDs are unusual. They are subject to "washout" as well as insufficient ambient light.

a. **Aiming Lights.** Aiming lights include the AN/PEQ-2 family, AN/PAQ-4 family, AN/PAQ-4C, and GCP-1A.

(1) ***AN/PEQ-2 Family.*** The AN/PEQ-2 is a dual laser system developed to allow a combination of both pinpoint aiming and broad beam target illumination. It can be handheld or mounted to a weapon for operation. The AN/PEQ-2 is available in three models allowing for the selection of laser, infrared, or infrared/visible light illumination sources. Once mounted on a weapon, the lasers on the AN/PEQ-2 can be easily and individually boresighted using the independent azimuth and elevation adjustments. The unit is waterproof to 20 meters. Under ideal conditions, the range of the laser pointer exceeds ten miles (Figure B-1). Two AA batteries power the aiming light.

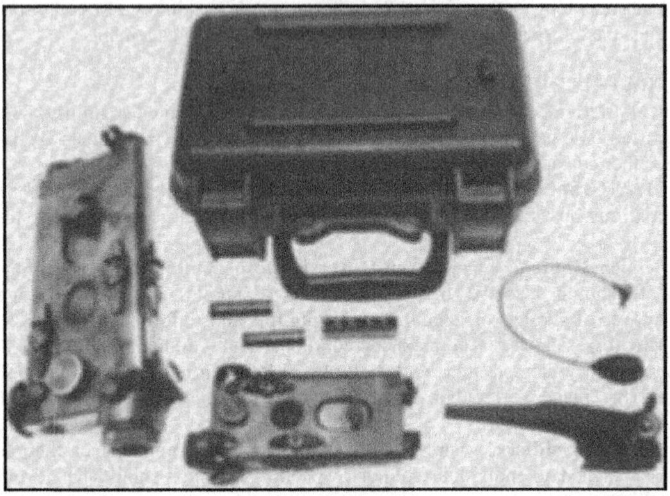

Figure B-1. AN/PEQ-2.

(2) ***AN/PAQ-4 Family.*** These devices project an invisible IR light along the weapon's line of sight, which can be seen with night vision devices, thus increasing the accuracy of night fire. These devices have a range of 600 meters. Depending on the model used, the IR beam can either be pulsating (AN/PAQ-4B) or steady beam (AN/PAQ-4C). Leaders and soldiers should ensure that the device is properly installed and zeroed to the weapon to enhance accuracy. These devices can be used in all night direct fire engagements or as a marking or signaling device during reduced visibility. Gunners with weapons equipped with the AN/PAQ-4 aiming light simply place the projected spot on the target and fire (Figure B-2). Two AA batteries power the aiming light.

Figure B-2. AN/PAQ-4B.

(3) ***AN/PAQ-4C.*** The AN/PAQ-4C (NSN 5855-01-361-1362) is an easy to mount, quick to zero aiming light with unmatched beam quality and range in an eye-safe device. It provides a rapid, accurate point for night engagements. This device, combined with the M16 mounting assembly, enables it to be easily zeroed when mounted on the M16A2 rifle. Windage and elevation adjustments enable fine zero adjustments to be made. Two AA batteries power the aiming light.

(4) ***GCP-1A.*** The ground commander's pointer (NSN 5855-01-420-0849) is an IR pointer and illuminator. Leaders use the GCP to designate targets, define sectors of fire, control fires, and illuminate targets. The light is invisible to the naked eye, but fully visible to NVGs and other NVDs. It is designed to provide clandestine target designation and illumination for night vision equipment users. The GCP IR light may be adjusted from a pencil beam, capable of pointing out targets up to 4,000 meters away, to a wide beam or flood light mode to illuminate large areas. The GCP has built in eye safe features. Two AA batteries power the GCP.

b. **Night Observation Devices.** Night observation devices (NODs) include:

(1) ***AN/PVS-4.*** The AN/PVS-4 night vision sight, for individual served weapons, is a self-contained night vision device that enables improved night vision using available ambient light. Leaders need to consider that the sight's effectiveness is impaired by rain, fog, sleet, snow, smoke, and other reflective matter. Leaders also must consider the effects that city lights, fires, and background illumination have on night vision devices. These elements could "white out" some NODs.

(2) ***AN/PVS-5 Night Vision Goggles.*** The AN/PVS-5 night vision goggles (NVG) are a lightweight, battery-powered binocular, passive night vision device worn on the head. The goggles have a 40-degree field of view. The system is normally operated in the passive mode but a built-in IR light source may be used to provide added illumination for close-up viewing. The AN/PVS-7 goggles are currently replacing this system. Leaders must consider the effects that city lights, fires, and background illumination have on night vision devices. These elements could "white out" some NVG.

(3) ***AN/PVS-7 Night Vision Goggles.*** The AN/PVS-7 NVG are lightweight, battery-powered passive devices worn on the head (Figure B-3). They have an IR-emitting light source for close-up illumination. The AN/PVS-7 has a much better night vision capability in lower light levels than the AN/PVS-5 goggles. Leaders must consider the effects that city lights, fires, and background illumination have on night vision devices. These elements could "white out" some NVG.

Figure B-3. AN/PVS-7 night vision goggles.

(4) ***AN/PVS-14 Monocular Night Vision Device.*** The AN/PVS-14 monocular night vision device is a hand-held, helmet-mounted, head-mounted, or weapon-mounted night vision device (Figure B-4) that enables walking, driving, weapon firing, short-range surveillance, map reading, vehicle maintenance, and administering first aid in moonlight and starlight. It has an IR light that provides illumination at close ranges (up to 3 meters in low ambient light conditions) and can also be used for signaling. The variable gain control is used to balance the illumination input to each eye. There is a "high light level" shut off if the device is exposed to damaging levels of bright light.

Figure B-4. AN/PVS-14 monocular night vision device.

(5) ***Three-Power Magnifier for the AN/PVS-7.*** This device is a three-power magnification lens (NSN 5855-01-391-7026) for the AN/PVS-7. The lens was developed to quickly convert the AN/PVS-7 NVG into a medium-range vision binocular surveillance system. The lightweight unit easily attaches to the NVG by screwing the threaded end of the lens assembly into the mating threads in the NVG's objective lens. The lens is also adaptable to the AN/PVS-14 monocular night vision device.

c. **White Lights.** These devices are small, lightweight, battery-powered white lights that can be attached to weapons. Either a pressure switch or an on/off switch activates the light. An IR filter can be attached to most white lights to provide IR and night vision device capability. These lights are sold commercially and are easily available. They can be attached to weapons using hose clamps or heavy tape. Another commonly used commercial light comes with rail adapters for the rail system and can be mounted with optional pressure switches for rapid ON/OFF capability. These lights must be checked periodically because they can loosen and shift.

d. **Red Dot Sights.** These devices are lightweight, battery-powered optical sights attached to the top of the weapon. A red dot in the sight aligns the weapon and the target. These sights are for use in low light levels, not in total darkness. They do not assist in identifying targets. The sights contain elevation and windage screws for zeroing the sight, and a rotary switch that contains several settings to increase the intensity of the red dot for use in various light conditions. These sights work well in precision clearing for quick target engagements.

e. **Active Laser Devices.** These devices are lightweight, battery-powered, visible light-emitting sights. These devices, when zeroed, project a red dot onto the target that corresponds to the point of bullet impact. These devices are not effective in sunlight.

f. **Thermal Weapons Sight**. The AN/PAS-13 thermal weapons sight (TWS) is used for detecting targets in total darkness. The TWS is also effective in detecting targets through smoke, obscurants, camouflage, and targets hidden in shadows, both day and night. There are three variants of the TWS—light, medium, and heavy. Leaders should consider the capabilities of thermal imaging to properly employ this system. The TWS's weight and bulk can be a disadvantage when performing reflexive firing techniques. With the sight in the "on" position, the TWS has a power saving feature that turns off the viewer after a period of inactivity. The soldier reactivates the sight by placing the eye against the rubber eye cup. When reactivated, it takes a few seconds for the sight to cool itself enough to gain an image. This delay is a disadvantage for soldiers trying to conduct precision clearing using the TWS. The TWS cannot detect targets through window glass.

g. **Tritium Sights.** These sights contain a light-emitting radioactive element that allows a firer to align the sights in total darkness. As long as the firer has a target in sight, he can effectively engage it as he would during daylight hours. These sights will not assist the firer in identifying the target.

h. **Hand-Held IR Flare/Smoke.** This is a hand-held tube with an IR flare on one end and smoke on the other end. This device can be used to illuminate LZs, friendly positions, or provide a smoke screen. It emits IR light and burns hot at the core. Burn time for the smoke is approximately 16 seconds, and burn time for the flare is approximately 20 seconds.

i. **Hand-Held IR Parachute Signal (M127A1).** This parachute signal is a rocket-propelled, fin-stabilized item that is hand fired from an expendable type launcher. It is used for ground-to-ground, as well as ground-to-air signaling. The M127A1 produces an average of 600 IR candlepower illumination with an average of 60 seconds burn time, and has a range of 300 meters. This is an excellent device for illuminating close-in target areas and can be used as a replacement for conventional visible light illuminators.

j. **BUDD Light.** The BUDD light is a compact near-IR source using a standard 9-volt battery (BA-3090) as its power source. Both the BUDD light and its power source fit in the palm of the hand. The average life span of the battery power for a BUDD light is eight hours of continuous use. The BUDD light comes in two configurations: a continuous beam of IR light and a pulsating light (every two seconds). It is invisible to the naked eye and thermal imagers. The light is clearly visible out to 4 kilometers under optimal conditions when pointing the beam directly at the viewer. The directional characteristics of the beam make it possible to limit observation by an enemy. It also limits the BUDD light's reliability for target identification unless multiple lights are visible to provide all-aspect coverage. This device is most effective for C2 purposes. The BUDD light is also very useful for operations at night (Figure B-5). It can be used to mark cleared areas, mark a path for other elements to follow, and so on.

Figure B-5. BUDD light.

k. **Phoenix IR Beacon**. The Phoenix IR beacon (NSN 5855-01-396-8732) is designed to be used with NVG and other NODs (Figure B-6, page B-8). A standard 9-volt battery (BA-3090) powers the light. The Phoenix light is ideal for use when positive identification at night must be made out to 4 kilometers under optimal conditions. It is a device that emits a codeable IR signal, which can be programmed. It flashes any code or sequence up to four seconds long. It is capable of instant code changes done by the individual soldier. The programming of a code can assist in distinguishing one individual, unit, and so on, from another. It can be used as an IR torch in the continuous mode. The light is weatherproof and sturdy in design. Other possible uses are to mark LZs, PZs, and main supply routes or to assist in marking passages of lines.

Figure B-6. Phoenix IR beacon.

D-5. CONSIDERATIONS

The urban environment presents special challenges and considerations during periods of limited visibility.

a. **Considerations for Use of Infrared Lights and NVG.** With its extensive night vision capability, the U. S. Army owns the night against most opposing forces. As with many other technical advances, when used in urban terrain, NODs may not give the same advantage as they do in open terrain. Leaders must consider all the factors and make a decision on how to use NODs.

(1) ***Advantages.*** The advantages of using IR and NVG in urban terrain are:

- Gives an assault team the ability to assault a structure in darkness, which could enhance the surprise (a fundamental of precision clearing) needed during the assault.
- No active light source is visible to the naked eye that could compromise the assault team's position. This could allow the assault element to move undetected up to and, depending on the situation, through the breach (entry) point.
- A cover could remain on the IR light until the operation is underway, limiting the chance of visually alerting the enemy to the assault teams' locations or intentions.

(2) ***Disadvantages.*** The disadvantages of using IR and NVG in urban terrain are:

- The use of an IR designator/illuminator requires a lot of familiarization and additional training.
- The use of IR and NVG may slow movement inside buildings due to the obstacles present and the lack of depth perception.
- IR illumination is an active light source that can be detected by an enemy with NODs. An active light source can compromise a clearing team's position inside a building or room, making them vulnerable to attack by an undetected assailant outside the building or room.

- NVG and mechanical IR sources are devices that can become inoperable due to mechanical failures, damage, or power failure. Firers must have both of these devices operational to be an effective fighter.
- Firers cannot use their iron sights while wearing NVG. Soldiers would be dependent upon both an aiming/illumination device and the NVG to be effective.
- NVG do not provide a wide field of view and there is a loss of depth perception associated with NVG.
- Early models of the NVG tend to "white out" from muzzle flash or flash bangs.
- If for some reason an assault team member has to remove his NVG, he is very vulnerable during the removal of the equipment and adjustment of his eyesight.
- In the absence of an IR source, NVG require some ambient light, which might not be available inside a building.
- Soldiers may move from a darkened area inside a structure into a white-lighted room, requiring the soldiers to remove their NVG, losing their night vision.
- Although target acquisition is possible, target identification is very difficult with IR and NVG inside a building/room.

b. **Considerations for Use of White Light or an Active Filtered Lens.** Use of white light can be effective during precision clearing and while searching an objective. As with all active light sources during limited visibility operations, the light must be used tactically to be effective. This means that individuals must control their light source and be aware of its effects at all times. Soldiers cannot be allowed to leave the light on constantly and wave it around. Tactical use of the light must be adhered to. Although not covert, white light has several advantages and disadvantages.

(1) ***Advantages.*** The advantages of using white light or an active filtered lens are:

- When attached to the weapon, the light can be used as a target designator. Where the light is oriented, the muzzle of the weapon is also.
- The equipment is readily available, inexpensive, reliable, and easily maintained.
- Little additional training is required to use the equipment.
- It offers the fastest means of identifying targets and searching a room, enhancing the speed and surprise of the assault element.
- It allows color vision.
- The firer uses his iron sights just as he does in daylight operations.
- After the structure has been dominated and if power is available, the building's light can be turned on so that the assault team(s) can make a thorough search if this does not compromise the unit or make the unit vulnerable to fire from outside the building.
- If white light is used, the assault team(s) will not have to remove their NVG and they won't be vulnerable to NVG whiteout.
- There is less possibility of equipment being inoperable with just one piece of equipment (the tactical light).

- A cover could remain on the tactical light until the operation is under way, limiting the chance of visually alerting the enemy to the assault teams locations or intentions.
- It may temporarily blind an opponent, causing disorientation.

(2) ***Disadvantages.*** The disadvantages of using white light or an active filtered lens are:

- An active light source can compromise an assault element's position inside a room or building making them vulnerable to attack by an undetected assailant from outside of the building or room.
- There will be a period of time for the eyes to adjust from the use of white light to NODs when the soldier turns off the light. The soldier will be ineffective for a short period of time due to this inability to see properly. This can be overcome with the soldier transitioning to NVG immediately after using his white light.
- A light may be activated too soon and alert the enemy to the clearing team's presence. If this occurs prior to the assault, the entire operation could be compromised.

c. **Techniques.** Techniques for using white light or an active filtered lens are:

(1) Soldiers should only activate the light when illumination of their line of sight/line of fire is desired. Care must be taken by all soldiers to understand the tactical situation. An example might be when a soldier is responsible for long security of a hallway. He turns his light on and off, as he deems necessary, to acquire movement at the far end of the hallway. A fire team is preparing to breach, enter, and clear a room off of the hallway. When the breach is initiated, the soldier on long security should extinguish his light so that he does not silhouette any member of the assault element moving through the breach into the room. Once the room is entered and the assault elements are in their points of domination, then the soldier can turn his light on for security of the hallway again.

(2) One technique to illuminate a room is to reflect the light from a source off of the ceiling. The reflected light provides enough illumination to allow a search of the room. The amount of light reflected depends upon the color of the ceiling and the amount of dust and smoke present in the room. This technique should not usually be considered in the conduct of precision clearing because it may silhouette soldiers in windows and doors and identify the location of the unit.

(3) Another technique to illuminate a room is to reflect the light from a source off of the center part of the floor. The reflected light will provide enough illumination to allow a search of the room. As in the ceiling technique, this technique can silhouette soldiers in the windows and doors and allow the enemy to acquire them.

B-6. COMBAT SUPPORT

Loss of synchronization is one of the major concerns to company commanders and leaders during limited visibility urban combat operations. The coordination of forces and fires at the point of decision can be facilitated by the technological advantage of US forces and by clear orders from leaders at all levels.

a. **Techniques.** Any degradation of artillery fire will be due to the limited target acquisition assets. While the field artillery FOs and combat observation and lasing teams (COLTS) have thermal sights and laser range finders, most soldiers on the battlefield do not have devices that will enable them to accurately call for fire. The following are some techniques to acquire targets for indirect fires.

(1) ***Preregistered TRPs.*** Preplanned and preregistered TRPs are effective only if the TRPs can be observed and the observer has clear communications to the firing unit.

(2) ***Reflective Surfaces.*** Reflective surfaces found in built-up areas may affect laser designators.

(3) ***Counterfire Radar***. Counterfire radar should be employed to cover likely areas of enemy mortar, cannon, and rocket use. Because of the masking effect of built-up areas, counterfire radars are not normally emplaced within the built-up area.

b. **Fixed-Wing Aviation Assets.** Fixed-wing aviation assets face a lower ADA threat during periods of limited visibility. However, the need for command and control is greater to prevent fratricide. The best fixed-wing aircraft available for fire support is the AC-130 because of its target acquisition capabilities, deadly and accurate fire, and long loiter time.

c. **Army Aviation.** Army aviation operates on similar limitations and considerations as fixed-wing aircraft. Most US Army attack helicopters have a forward-looking infrared (FLIR) night sight. Because of their slower speed and hover capability, helicopters can deliver highly accurate and responsive fire on enemy targets. However, helicopters are more susceptible to enemy ADA assets and, therefore, should only be employed where the enemy air defense threat is light. Commanders must identify clear landmarks for the pilots to navigate to and from the objective.

d. **Air Defense Artillery**. ADA is significantly degraded during periods of limited visibility. Visual detection, identification, and range estimation are all very difficult. Radar guidance systems have difficulty distinguishing the target from ground clutter.

e. **Engineer Units.** The lack of thermal imaging devices may hamper engineer units. Locating and clearing mines and booby traps also become more dangerous and difficult. The method of marking cleared lanes should be determined and coordinated in advance to avoid confusion with other limited visibility markers (glint tape, infrared strobe lights [BUDD lights], chemlites, and so forth).

f. **Military Intelligence.** Military intelligence relies primarily on human intelligence assets to gain information about the enemy in urban environments.

(1) GSR and remotely monitored battlefield sensor systems (REMBASS) have limited use in the center of built-up areas. They are best employed on the outskirts to monitor traffic into and out of the built-up area. If necessary, GSR can be used to cover large open areas such as parks and public squares. REMs can be used in subterranean areas such as sewers and utility tunnels.

(2) Based on the time available before the operation or the urgency of need, satellite photographs of the built-up area may be available.

B-7. SERVICE SUPPORT

Soldiers in maneuver platoons are not the only individuals that must adjust to combat under limited visibility conditions during urban combat. Company XOs, first sergeants, and supply sergeants must anticipate requirements for this unique environment.

a. **Resupply.** Units conducting resupply operations during periods of limited visibility should remember the following:

(1) Drivers and vehicle commanders should be issued NVDs so the vehicles going to and from logistic release points do not need any illumination. This also prevents the enemy from acquiring resupply locations by following vehicles with blackout lights on.

(2) Strict noise and light discipline should be maintained.

(3) Vehicles should follow a clearly marked route to avoid any obstacles and prevent the resupply vehicle(s) from becoming disoriented.

(4) Radios should be redistributed to resupply vehicles whenever possible, in order to maintain control.

(5) Each vehicle should have a map of the area of operations (preferably a city map with the street names).

(6) Prepackaged supplies going to forward units must be marked for quick ID of contents.

b. **Batteries.** Companies operating for extended periods during limited visibility should have enough batteries to keep the NVDs functioning at optimum power and sensitivity.

c. **Casualty Collection.** Casualty collection during periods of limited visibility is much more difficult. Clear methods for marking any casualties and casualty collection points must be established before the operation begins.

d. **Light Discipline.** Resupply operations in existing structures at night must not be visible from a long distance. This includes limiting vehicle traffic to an absolute minimum, sealing doors and windows to prevent light leakage, and dispersing assets as much as possible.

B-8. OFFENSIVE CONSIDERATIONS

Attacks are conducted during periods of limited visibility to achieve surprise and to gain or maintain the momentum of the attack. Before conducting a limited visibility attack, the commander must balance the risks and ensure that every soldier understands the intent and control measures. Rehearsals and strict command and control reduce casualties and greatly enhance the chances for mission accomplishment.

a. **Clearing Buildings and Rooms.** Soldiers should clear buildings and rooms using the same techniques as used during periods of unlimited visibility to reduce confusion. The only major difference is in equipment used. (See the paragraph on special equipment in this appendix.)

b. **Slower Movement.** Movement rates are slower. Each soldier must be alert for mines, booby traps, and enemy positions. Although thermal imaging devices can detect the difference in the temperature of the soil, light intensification devices are usually better for detecting recently disturbed dirt. Thermal imaging devices are better for identifying personnel; however, light intensifiers can identify friendly soldiers, noncombatants, and enemy troops better than the thermals in areas that have sufficient ambient light.

c. **Equipping Squads and Fire Teams.** Squads and fire teams should be equipped with a mixture of both thermal imaging and light intensifying devices whenever possible. This enables the squads and fire teams to obtain a better picture of the night environment and enables the soldiers to balance the strengths and weaknesses of each type of night vision device. Consideration must also be given to white lights and IR filters for them.

d. **Marking Cleared Rooms.** When moving through buildings, the assault teams must mark cleared rooms and buildings, and communicate with the support team(s). This communication is critical if more than one assault team is in the same building. (See Appendix I for more information on marking.)

e. **Communications with Supporting Units.** The assault team must have clear communications with all supporting elements, whether they are organic, in DS, under OPCON, or attached. Supporting units should not fire unless they have good communications with the assaulting elements and are sure the targets they are engaging or suppressing are the enemy.

f. **Situational Awareness.** Units must know where everyone is during offensive operations. Not only does this reduce the risk of fratricide, but it also increases the time of identifying, locating, and treating casualties. Also, it greatly reduces the chance of soldiers becoming disoriented and separated from the unit. This is even more critical during periods of limited visibility.

g. **Weapons Flashes and Flares.** Assault teams should be aware of adjacent fires that diminish the effectiveness of night vision devices. Weapons flashes within small rooms cause soldiers to lose their night vision and wash out light intensification devices. Also, enemy soldiers may use flares inside and outside of buildings to render some night vision devices ineffective.

B-9. DEFENSIVE CONSIDERATIONS

Threat forces can be expected to use periods of limited visibility for the same reasons US forces do (see paragraph B-1). Threat forces may have access to sophisticated NVDs manufactured in Europe, the United States, Japan, Korea, and the former Soviet Union. If this is the case, soldiers must be as disciplined with IR light as they are with white light. (See Chapter 5 for detailed information on defensive operations.)

APPENDIX C

LIGHT INFANTRY AND ARMORED VEHICLE TACTICS, TECHNIQUES, AND PROCEDURES

"...to breach these barricades and destroy (North Korean) defenders, the Marine and Army forces developed a highly effective combined arms team...Most UN forces quickly discovered that rifles or machine guns lacked the penetrating power and punch to overcome hardened (North Korean) barricade defenses."

Robert Tallent, "Street Fight in Seoul",
The Leathernecks: An Informal History of the US Marine Corps, as quoted in Armor Magazine, Sep-Oct 01

The first and most fundamental lesson learned from recent operations in urban areas is the value of the fully integrated combined arms team. There is no denying the value of light Infantry forces during urban combat. However, urban combat never should be considered a purely Infantry task. Urban combat by units composed entirely of Infantrymen is a historical anomaly. Across the spectrum of combat action in urban areas, powerful combined arms teams produce the best results. Infantry units operating alone suffer from critical shortcomings that can be compensated for only by appropriate task organization with mechanized Infantry, armor, and engineers. These teams must be supported by closely integrated aviation, fire support, communications, and logistical elements. This chapter discusses tactics, techniques, and procedures (TTP) that can be employed by light Infantry and armored vehicles during the execution of UO. With modifications, the TTP in this appendix can be used by IBCT companies and platoons.

C-1. LIMITATIONS AND STRENGTHS OF LIGHT INFANTRY AND ARMORED VEHICLES

Because of the decentralized nature of urban combat and the need for a high number of troops to conduct operations in dense, compact terrain, Infantrymen will always represent the bulk of forces. At the small-unit tactical level, light Infantry forces have disadvantages that can be compensated for by mechanized Infantry or armor units. Conversely, tanks and mechanized Infantry face problems in the confines of urban areas that place them at a severe disadvantage when operating alone. Only together can these forces accomplish their mission with minimal casualties, while avoiding unnecessary collateral damage.

a. **Light Infantry Limitations.** Light Infantry limitations include the following:

(1) Light Infantry forces lack heavy supporting firepower, protection, and long-range mobility.

(2) Exposed light Infantry forces are subject to taking a high number of casualties between buildings.

(3) Light Infantry forces are more subject to fratricide-related casualties from friendly direct and indirect fire.

b. **Armored Vehicle Limitations.** Armored vehicle limitations include:

(1) Crewmen in armored vehicles have poor all-round vision through their vision blocks; they are easily blinded by smoke or dust. Tanks cannot elevate or depress their main guns enough to engage targets very close to the vehicle or those high up in tall buildings.

(2) If isolated or unsupported by Infantry, armored vehicles are vulnerable to enemy hunter/killer teams firing light and medium antiarmor weapons. Because of the abundance of cover and concealment in urban terrain, armored vehicle gunners may not be able to easily identify enemy targets unless the commander exposes himself to fire by opening his hatch or Infantrymen directing the gunner to the target.

(3) Armored vehicles are noisy. Therefore, there is little chance of them arriving in an area undetected. Improvised barricades, narrow streets and alleyways, or large amounts of rubble can block armored vehicles.

(4) Due to the length of the tank main gun, the turret will not rotate if a solid object is encountered, for example, a wall, post, and so forth. Heavy fires from armored vehicles cause unwanted collateral damage or can destabilize basic structures.

(5) The main gun of an M1A2 can only elevate +20 degrees and depress –9 degrees. Examples of standoff distances for buildings where a HEAT round is used are:

- Ground floor—2.5 meters from the target.
- 3d story—23 meters from the target.
- 18th story—132 meters from the target.

NOTE: Figure C-1 shows the difference in the capabilities of the BFV and the M1 tank with regard to fields of fire on urban terrain. Note that the BFV can engage a target 9 to 10 stories high at 20 meters, whereas an M1 tank requires 90 meters. Although the tank main gun has these limitations, targets can be engaged by the M2HB and M240 machine guns that are part of the tank's weapon system.

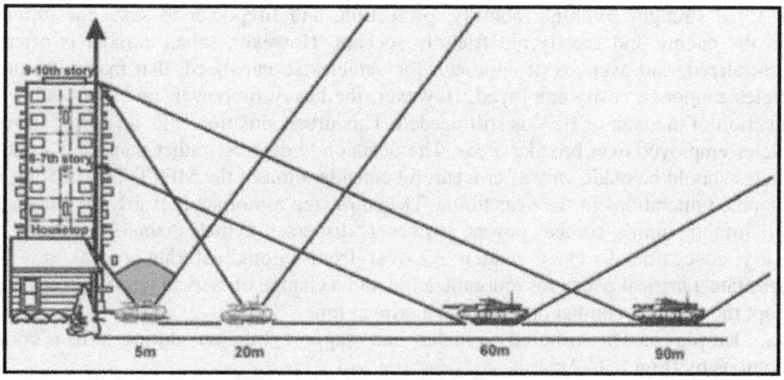

Figure C-1. Fields of fire on urban terrain.

c. **Light Infantry Strengths.** Light Infantry strengths include:

(1) Infantry small-arms fire within a building can eliminate resistance without seriously damaging the structure.

(2) Infantrymen can move stealthily into position without alerting the enemy. Infantrymen can move over or around most urban terrain, regardless of the amount of damage to buildings.

(3) Infantrymen have excellent all-round vision and can engage targets with small arms fire under almost all conditions.

d. **Armored Vehicle Strengths.** Armored vehicle strengths include:

(1) The thermal sights on armored vehicles can detect enemy activity through darkness and smoke, conditions that limit even the best-equipped Infantry.

(2) Armored forces, can deliver devastating fires, are fully protected against antipersonnel mines, fragments and small arms, and have excellent mobility along unblocked routes.

(3) Armored vehicles project a psychological presence, an aura of invulnerability that aids the friendly forces in deterring violence. Mounted patrols by armored vehicles can monitor large areas of a city while making their presence known to the entire populace, both friendly and unfriendly.

(4) Armored vehicles can move mounted Infantrymen rapidly to points where, together, they can dominate and isolate the cordoned area. With their long-range sights and weapons, armored vehicles can dominate large expanses of open area and thus free Infantry to isolate closer terrain and visual dead space.

(5) The mobile protected firepower of armored vehicles can be used to add security to resupply convoys and to extract wounded personnel under fire. The armored vehicle's smoke-generation capability can aid this and other small-unit actions.

C-2. ARMORED VEHICLE EMPLOYMENT CONSIDERATIONS

An effective use of armored combat vehicles in most tactical situations is en mass. Mechanized infantry/armored units operating in platoon, company team, and battalion task force strength combine mobility, protection, and firepower to seize the initiative from the enemy and greatly aid friendly success. However, urban combat is often so decentralized, and avenues of approach for vehicles so canalized, that massed armored vehicles cannot be easily employed. However, the heavy firepower, mobility, and armor protection of the tank or BFV is still needed. This urban situation calls for fewer armored vehicles employed over broader areas. The decision to disperse rather than mass armored vehicles should be made only after a careful consideration of the METT-TC situation and anticipated operations in the near future. Decentralized armor support greatly increases a small Infantry unit's combat power. However, dispersed vehicles cannot be easily and quickly concentrated. Their sudden removal from throughout the combat area will necessitate a tactical pause for reorganization and a change of tactical tempo, which could disrupt the ongoing combat operation at a critical time.

a. **Employment.** Armored vehicles can support Infantry during urban combat operations by (Figure C-2):

- Providing shock action and firepower.
- Isolating objectives with direct fire to prevent enemy withdrawal, reinforcement, or counterattack.
- Neutralizing or suppressing enemy positions with smoke, high-explosive (HE), and automatic weapons fire as Infantry closes with and destroys the enemy.
- Assisting opposed entry of Infantry into buildings when doorways are blocked by debris, obstacles, or enemy fire.
- Smashing through street barricades or reducing barricades by fire.
- Obscuring enemy observation using smoke grenade launchers.
- Holding cleared portions of the objective by covering avenues of approach.
- Attacking by fire any other targets designated by the Infantry.
- Establishing roadblocks or checkpoints.
- Suppressing identified sniper positions.

Figure C-2. Tank in direct fire, supported by Infantry.

> **CAUTION**
> When operating close to Infantry during combined arms urban operations, tanks should employ heat shields, normally used for towing, to deflect the intense heat caused by the exhaust.

c. **Vehicle Characteristics.** Fighting in urban areas is centered around prepared positions in houses and buildings. Such positions cover street approaches and are protected by mines, obstacles, and booby traps. Therefore, bridges, overpasses, and buildings must be inspected and cleared of mines before they are used. Reconnaissance parties must ascertain the weight-supporting capacity of roads, bridges, and floors to determine if they can support the weight of BFVs and tanks (Table C-1).

Vehicle	Weight (tons)	Height (feet)	Width (inches)
M1 Tank	68.7	10.14	143.75
BFV with reactive armor	33	11.3	142.2
BFV without reactive armor	28	11.3	130

Table C-1. Vehicle size and weight classification.

C-3. TASK ORGANIZATION WITH TANKS AT COMPANY TEAM LEVEL

The information in this paragraph refers to tank platoons. An attached or OPCON BFV platoon will have Infantry squads that can be employed in the scheme of maneuver. Therefore, platoon integrity with a BFV platoon should be maintained in urban combat and the BFV platoon should be used as a maneuver element. Normally, a tank platoon would be OPCON to a light, airborne, or air assault Infantry company during combined arms operations at the company team level. There are four basic techniques of task organizing the tank platoon into the light Infantry company for urban combat.

a. **Tank Platoon as a Maneuver Element.** In this technique, the tank platoon leader is responsible for maneuvering the tanks IAW the company team commander's intent. With this task organization, likely missions for the tanks would be to support by fire or to overwatch the movement of the Infantry. This task organization is the most difficult to maneuver tanks with the Infantry. However, the tank platoon leader can choose to maneuver the platoon by sections in order to execute the mission. This would provide greater flexibility in supporting the Infantry during the close fight.

b. **Tank Sections Under Infantry Platoon Control.** In this technique, tanks would be broken down into two sections and each section would be placed under the OPCON of an Infantry platoon, and maneuvered IAW the company team commander's intent. The company team commander relinquishes direct control of the tank maneuver to the Infantry platoon leaders. This technique is very effective in maintaining the same rate of progress between the tanks and the Infantry. However, Infantry platoon leaders are burdened with the additional responsibility of maneuvering tanks. The general lack of experience with tanks and the overall battlefield focus of the Infantry platoon leader can also affect this technique. This technique is best suited when contact with the enemy is expected and close continuous support is required for movement or clearing buildings.

c. **Tank Sections Under Company and Platoon Control.** The tank platoon can be broken down into two sections, one under company control, the other under platoon control. The selected maneuver Infantry platoon would have a tank section available to support the close fight. With this technique, the company team commander has a tank section to deploy at the critical place and time of his choosing. This task organization still allows support to the Infantry close fight while keeping additional support options in reserve for the commander to employ. The disadvantages to this technique are that an Infantry platoon leader is maneuvering tanks, instead of the tank platoon leader, and the tanks directly available to the company team commander are cut in half. This technique requires detailed planning, coordination, and rehearsals between the Infantry platoons and tank sections.

d. **Infantry Squads Under Tank Platoon Control.** In this technique, the company team commander has the option of placing one or more Infantry squads under the OPCON of the tank platoon leader. He may also retain all tanks under the control of the tank platoon leader or place a tank section under the OPCON of an Infantry platoon leader. This technique will give the company team commander a fourth maneuver platoon, and involves the tank platoon leader in the fight. It can work well in a situation where a mobile reserve that needs Infantry protection is required. This technique requires detailed planning, coordination, and rehearsals between the Infantry squads and tank platoon/sections.

e. **Guidelines.** None of the techniques described above are inherently better than the other one. The task organization must be tailored to accomplish the mission. Regardless of the technique selected, the guidelines below should be followed:

(1) Tanks should be used as sections. Single tanks may operate in support of Infantry, however it is preferable for tanks to operate as sections. If using tanks to shield squads and teams from building to building as part of the maneuver plan, the leader of the forward element needs to control the tanks.

(2) If the company commander is controlling the tanks, he needs to move forward to a position where he can effectively maneuver the tanks in support of the Infantry.

(3) The task organization should support the span of control. If the company commander is going to control the tanks, then there is no reason to task-organize the tanks by section under Infantry platoons.

(4) Tanks need Infantry support when the two elements are working together. Do not leave tanks alone because they are not prepared to provide local security during the operation. Tanks are extremely vulnerable to dismounted attack when operating on urban terrain. Tanks are most vulnerable and need local security when Infantry are in the process of clearing buildings. Tanks must remain relatively stationary for prolonged periods allowing threat AT teams to maneuver to a position of advantage.

f. **Mutual Support.** Infantry/tank teams work together to bring the maximum combat power to bear on the enemy. The Infantry provides the eyes and ears of the team. The Infantry locates and identifies targets for the tank to engage. It maneuvers along covered and concealed routes to assault enemy elements fixed and suppressed by tank fire. It provides protection for the tank against attack by enemy Infantry. Meanwhile, the tank provides heavy, continuous supporting fires against enemy strongpoints.

g. **Movement.** The Infantry normally leads movement through urban areas. The tanks follow and provide close overwatch. If the Infantry discovers an enemy position or encounters resistance, the tanks immediately respond with supporting fire to fix the enemy in place or suppress him and allow the Infantry to develop the situation. After sufficient time to develop the situation or conduct short-range reconnaissance, the Infantry squad leader directs the tank to move, if necessary, and identifies specific targets for the tank to engage.

h. **Coordination.** Coordination between tank and Infantry leaders must be close and continuous. The tank commander or driver may need to dismount and move, accompanied by the Infantry squad leader, to a position where the route or target can be seen better. Signals for initiating, shifting, or lifting fires must be understood by all. One of the greatest barriers to coordination and command and control in urban combat is the intense noise. Verbal commands should be backed up by simple, nonverbal signals.

i. **Communications.** The tank platoon leader and platoon sergeant maintain communications with the company team commander. Individual tanks and Infantrymen communicate with each other using one or more of these techniques.

(1) ***Visual Signals.*** Visual signals, either prescribed by SOP or coordinated during linkup, facilitate some simple communications.

(2) ***Wire.*** M1-series tank crewmen can route WD-1 wire from the AM-1780 through the loader's hatch or vision block and attach it to a field phone on the back of the tank. WD-1 wire can also be run on a more permanent basis starting through the engine compartment, run through the hull/subturret floor, and attached to the turret's intercom

system via the driver's communications box. These techniques work better in a defensive situation than in an attack, where it might hinder tank movement or reaction time.

(3) ***FM Radios.*** FM radios or other short-range hand-held radios can be distributed during the linkup to provide a reliable means of communications between Infantry and supporting TCs. These radios allow the Infantry to use terrain more effectively in providing close in protection for the tank; Infantrymen can watch for enemy elements while limiting exposure to enemy fires directed against the tank. SOI information can be used between the tank platoon/sections and the company team headquarters and or the Infantry platoons. This SOI information is a fast reliable method of communications that does not require additional assets.

j. **Smoke.** The tank's smoke grenade launchers may be used both to protect the tank from enemy fire and to provide concealment for the Infantry forces as they either move across open areas or recover wounded. The use of smoke must be carefully coordinated. Although the tanks' sights can see through most smoke, Infantrymen are at a significant disadvantage when enveloped in dense smoke clouds. The smoke grenade launchers on the tank provide excellent, rapidly developed local smoke clouds, but the grenades produce burning fragments that are hazardous to Infantrymen near the tank and that can ignite dangerous fires in urban areas.

k. **Heavy Direct Fire Support.** Tanks and BFVs are valuable tools for assisting the assaulting forces during isolation of the objective area and seizing a foothold. As the Infantry then moves to clear the position and expand the foothold, the tanks are left in their initial support by fire positions. When possible, tanks should move to subsequent positions where their fires can be used to prevent enemy reinforcement and engage enemy forces withdrawing from the objective. However, at this time the tank crew must be very alert. Because of the nonlinear nature of urban battles, enemy forces may move to the rear or flanks of the now-isolated tanks and destroy them. If a small element of Infantry cannot be spared to support the tanks, both vehicles in the section should move to positions of cover and mutual support. Loaders and vehicle commanders should be alert, especially for enemy Infantry approaching from above, the rear, or from the flanks.

l. **Other Considerations.** Other considerations for employing tanks at company team level are:

(1) In planning, pay close attention to available terrain that supports tank cross-country movement. While the pace may be slower, security may be significantly enhanced.

(2) Involve tank platoon leaders and sergeants in the Infantry company-level IPB process. Their expertise will hasten the understanding of what tanks can and cannot do and aid the Infantry company commander in making the best employment decision.

(3) Tanks and BFVs can be used to carry ammunition, water, and other supplies to support the urban fight.

(4) To keep tanks and BFVs mission capable requires planning for refueling and rearming. Also, there may be a requirement to recover disabled vehicles. The company XO must coordinate with the battalion S4 to ensure that the proper logistical support is provided for the tanks or BFVs.

(5) Infantry company commanders must specifically allocate time in the planning process for precombat inspections (PCIs) for the tanks or BFVs.

(6) Conduct a combined arms rehearsal at the level that the tanks are task-organized. Try to replicate conditions for mission execution during rehearsals for example, day, limited visibility, civilians on the battlefield, host nation support, and ROE. Include the following:

- Graphic and fire control measures.
- Communications.
- Direct fire plans.
- Breach drills.
- Procedures for Infantry riding on tanks. (Tanks can move a maximum of nine personnel.)
- Techniques for using tanks as Infantry shields.

(7) To minimize casualties when moving outside or between buildings, do the following:

(a) Cover all possible threat locations with either observation or fire.

(b) For those areas that cannot be covered with observation or fire, use smoke to set a screen to block enemy observation of friendly movement.

(c) Move tanks forward to support Infantry movement. Position the tanks before the Infantry begins moving, whether the tanks are supporting by fire, being used as shields, or both.

(d) Preplan positions, if possible, but devise a marking system and communication signals to designate situational dependent positions to help maintain momentum. (For example, *The VS-17 panel from Building 2 means move to SBF 3.*)

(e) When using tanks as a shield for Infantry, move the tanks as close as possible to the start point to allow the Infantry the freedom of movement when exiting the building. Tanks need to move at the Infantry's rate of movement.

(f) When the distance between buildings is short, tanks can position themselves to block the open area from enemy fire.

(8) Use simple, clearly understood graphic control measures. The following are particularly useful for light/heavy operations in urban combat (Figure C-3, page C-10):

- Phase lines.
- Number and lettering systems for buildings.
- Tentative support by fire positions.
- No fire areas.

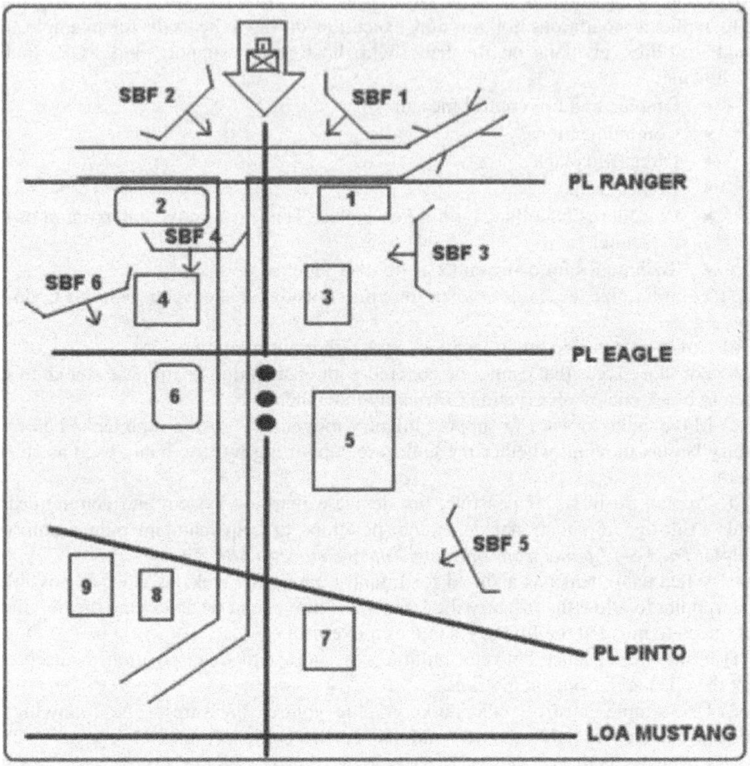

Figure C-3. Light/heavy graphic control measures.

(9) The company commander relies on the radio to help control the battle. It is essential that platoon leaders and RATELOs are well trained in sending reports. Constant reporting from the subordinate elements to the commander is critical for mission success.

C-4. ARMORED VEHICLE POSITIONS

Fighting positions for tanks and infantry fighting vehicles are essential to a complete and effective defensive plan in urban areas. Armored vehicle positions are selected and developed to obtain the best cover, concealment, observation, and fields of fire while retaining the vehicle's ability to move.

a. **Hull Down.** If fields of fire are restricted to streets, hull-down positions should be used to gain cover and fire directly down streets (Figure C-4). From those positions, tanks and BFVs are protected and can move to alternate positions rapidly. Buildings collapsing from enemy fires are a minimal hazard to the armored vehicle and crew.

Figure C-4. Hull-down position.

b. **Hide.** The hide position (Figure C-5) covers and conceals the vehicle until time to move into position for target engagement. Since the crew will not be able to see advancing enemy forces, an observer from the vehicle or a nearby infantry unit must be concealed in an adjacent building to alert the crew. The observer acquires the target and signals the armored vehicle to move to the firing position and to fire. After firing, the tank or BFV moves to an alternate position to avoid compromising one location.

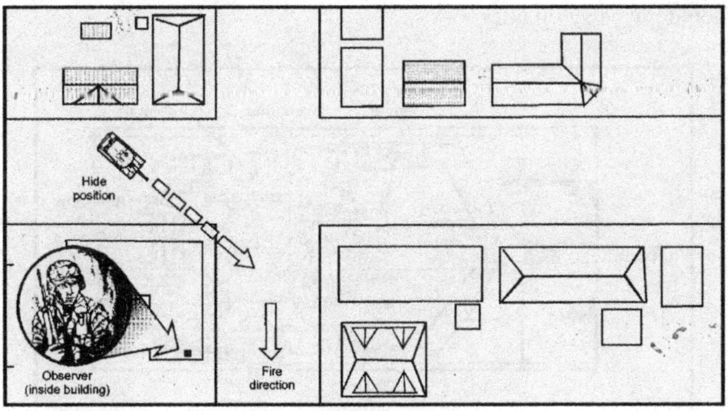

Figure C-5. Hide position.

c. **Building Hide.** The building hide position (Figure C-6, page C-12) conceals the vehicle inside a building. If basement hide positions are inaccessible, engineers must evaluate the building's floor strength and prepare for the vehicle. Once the position is detected, it should be evacuated to avoid enemy fires.

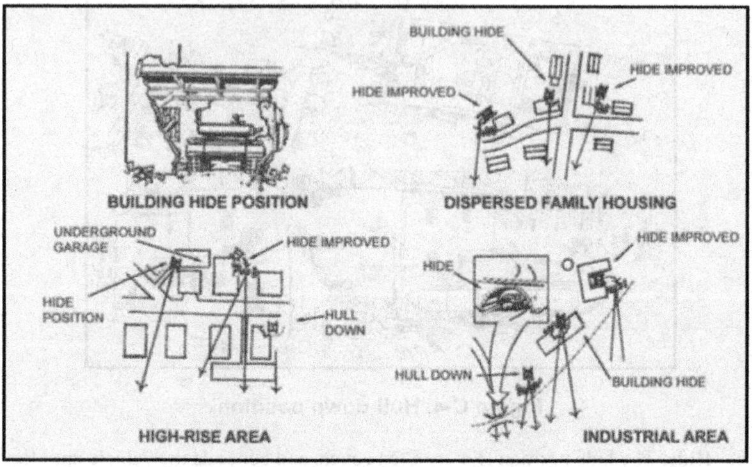

Figure C-6. Building hide position.

C-5. TRANSPORTING INFANTRY

At times, the tank platoon may be required to transport Infantrymen on its tanks (Figure C-7). This is done only when contact is not expected. If the tank platoon is moving as part of a larger force and is tasked to provide security for the move, the lead section or element should not carry infantry.

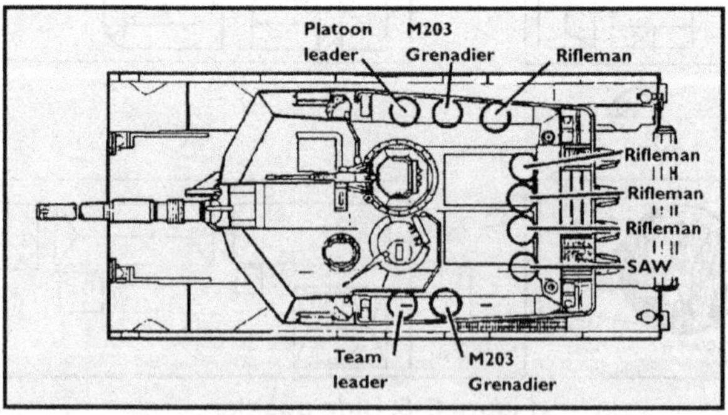

Figure C-7. Sample positions for Infantry riding on a tank.

a. **Procedures, Precautions, and Considerations.** Infantry and armor leaders must observe the following procedures, precautions, and considerations when Infantrymen ride on tanks:

(1) Infantrymen should thoroughly practice mounting and dismounting procedures and actions on contact.

(2) Infantrymen must always alert the TC before mounting or dismounting. They must follow the commands of the TC.

(3) Infantry platoons should be broken down by squads, similar to air assault chalks, with the infantry platoon leader on the armor platoon leader's vehicle and the infantry platoon sergeant on the armor platoon sergeant's vehicle.

(4) Platoon leaders, platoon sergeants, and team leaders should position themselves near the TC's hatch, using the external phone (if available) to talk to the TC and relay signals to the unit.

(5) If possible, the lead vehicle should not carry Infantrymen. Riders restrict turret movement and are more likely to be injured or killed on initial contact.

(6) Whenever possible, Infantrymen should mount and dismount over the left front slope of the vehicle. This ensures that the driver can see the infantrymen and that the infantrymen do not pass in front of the coax machine gun. Infantrymen must ensure that they remain behind the vehicle's smoke grenade launchers. This will automatically keep them clear of all weapon systems.

(7) Infantrymen must always have three points of contact with the vehicle, and they must watch for low-hanging objects such as tree branches.

(8) Infantrymen should wear hearing protection.

(9) Infantrymen should not ride with anything more than their battle gear. Rucksacks should be transported by other means.

(10) Infantrymen should scan in all directions while riding. They may be able to spot a target the vehicle crew does not see.

(11) Infantrymen should be prepared to take the following actions on contact:

- Wait for the vehicle to stop.
- At the TC's command, dismount **IMMEDIATELY** (one fire team on each side). **DO NOT** move forward of the turret. **DO NOT** dismount a vehicle unless ordered or given permission to do so.
- Move at least 5 meters to the either side of the vehicle. **DO NOT** move behind or forward of the vehicle.
- **DO NOT** move in front of vehicles unless ordered to do so. Main gun discharge overpressure can inflict sever injury or death to forward dismounted Infantrymen. (See Figure C-8 and the warning on page C-14.)
- **DO NOT** dangle arms or legs, equipment, or anything else off the side of a vehicle; they could get caught in the tracks, causing death, injury, or damage to the equipment or vehicle.
- **DO NOT** place too many riders on the vehicle.
- **DO NOT** fall asleep when riding. The warm engine may induce drowsiness; a fall could be fatal.
- **DO NOT** smoke when mounted on a vehicle.
- **DO NOT** stand near a moving or turning vehicle at any time. Tanks have a deceptively short turning radius.

DANGER

THE OVERPRESSURE FROM THE TANK'S 120-MM CANNON CAN KILL A DISMOUNTED INFANTRYMAN WITHIN A 90-DEGREE ARC EXTENDING FROM THE MUZZLE OF THE GUN TUBE OUT TO 200 METERS.

FROM 200 TO 1,000 METERS ALONG THE LINE OF FIRE, ON A FRONTAGE OF ABOUT 400 METERS, DISMOUNTED INFANTRY MUST BE AWARE OF THE DANGER FROM DISCARDING SABOT PETALS, WHICH CAN KILL OR SERIOUSLY INJURE PERSONNEL.

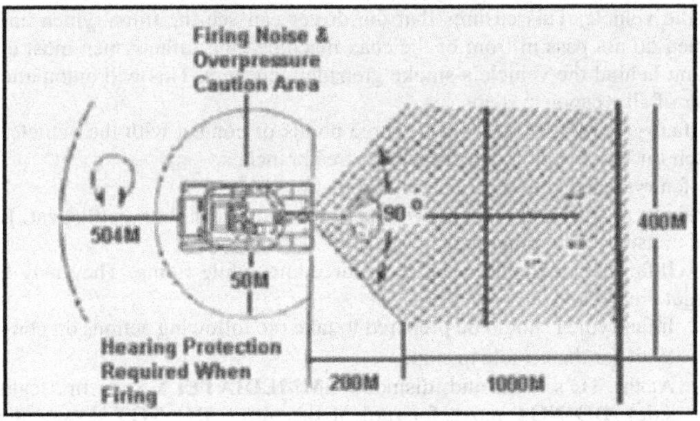

Figure C-8. Danger areas around a tank firing a 120-mm main gun.

b. **Additional Considerations and Precautions.** Additional considerations and preparations for transporting Infantrymen include the following:

(1) The armor—

- Uses main-gun fire to reduce obstacles or entrenched positions for the Infantry.
- Takes directions from the Infantry ground commander (platoon leader/platoon sergeant/squad leader) to support their fire and maneuver.
- Provides reconnaissance by fire for the Infantry.
- Should know and understand how the Infantry clears buildings, how they mark cleared buildings, the casualty evacuation plan, signal methods, engagement criteria for tank main gun, front line trace reporting, ground communication from the tank with the dismounted personnel.

- Uses its night vision capability to augment and supplement the Infantry's night vision capabilities.

(2) The Infantry—

- Provides real time information for the tank crewmen to help them overcome tank noise and the lack of ground situational understanding.
- Provides reconnaissance and fire direction of enemy positions for main gun attack.

(3) Considerations for dismounted tank security include the following:

- Tank crewmen should rehearse the mounting and dismounting of Infantrymen from their vehicle, briefing the Infantrymen on safety procedures for the vehicle and weapon systems.
- Tank commanders need to rehearse communicating with dismounted soldiers via TA-1 and DR-8 in the bustle rack.

(4) Vehicle preparation for combat in urban terrain should cover these procedures:

- Keep at least one ballistic shield to the "Dog House" closed (most engagements will be under boresight range and the battlesight technique will suffice).
- Place sandbags around antenna connections and electrical wiring on the turret top.
- Place extra coax ammunition inside the turret.
- Remove all highly flammable products from the outside of the vehicle and from the sponson boxes.

C-6. ARMOR VEHICULAR, WEAPONS, AND MUNITIONS CONSIDERATIONS

Numerous factors related to tanks and their organic weapons and munitions affect the tank platoon's urban operation planning and execution, including the following:

a. The preferred main gun rounds in the urban environment are HEAT, MPAT (ground mode), and MPAT-OR (XM908). These all perform much better than sabot rounds against bunkers and buildings.

b. HEAT ammunition will open a larger hole in reinforced concrete or masonry structures than MPAT or MPAT-OR (XM908). Both MPAT and MPAT-OR, however, offer greater incapacitation capability inside the structure.

c. HEAT ammunition arms approximately 60 feet from the gun muzzle. It loses most of its effectiveness against urban targets at ranges of less than 60 feet.

d. MPAT and MPAT-OR rounds arm approximately 100 feet from the muzzle of the gun. Because of the shape and metal components of the projectiles, however, this ammunition remains effective at ranges of less than 100 feet.

e. Sabot petals, including those on MPAT and MPAT-OR, endanger accompanying infantry elements. They create a hazard area extending 70 meters on either side of the gun-target line out to a range of 1 kilometer.

f. The tank's main gun can depress only to -10 degrees and can elevate only to +20 degrees, which creates considerable dead space for the crew at the close ranges that are typical in the urban environment.

g. The external M2 HB machine gun can deliver a heavy volume of suppressive fire and penetrate light construction, buildings, and most barricades. The M2 HB machine gun can elevate to +36 degrees; however, the TC must be unbuttoned to fire the M2 on the M1A2 or M1A2 SEP.

h. The M240 coax machine gun can effectively deliver suppressive fires against enemy personnel and against enemy positions that are behind light cover.

i. The loader's M240 machine gun can effectively deliver suppressive fire against enemy personnel and against enemy positions that are behind light cover; however, the loader must be unbuttoned to operate it. This weapon may be dismounted and used in a ground role if units are equipped with the M240 dismount kit.

j. When buttoned up, the tank crew has limited visibility to the sides and rear and no visibility to the top.

NOTE: FM 17-12-1-1 explains special uses for tank-mounted machine guns in the urban environment.

C-7. OFFENSIVE CONSIDERATIONS AND TACTICS, TECHNIQUES, AND PROCEDURES FOR THE BRADLEY FIGHTING VEHICLE

The mechanized Infantry platoon provides a very flexible heavy direct fire support asset to light Infantry companies conducting operations on urban terrain. The 25-mm cannon and 7.62-mm coax machine gun, combined with the additional Infantry, Javelin, and TOW ATGMs, provide the company team commander powerful combat multipliers during urban combat.

NOTE: While this paragraph specifically discusses BFVs, much of the information also applies to the use of tanks.

a. **Target Engagement.** Streets and alleys are natural firing lanes and killing zones. Because of this, all vehicular traffic is greatly restricted and canalized, and subject to ambush and short-range attack. Tanks are at a disadvantage because their main guns cannot be elevated enough to engage targets on the upper floors of tall buildings. The BFV, with +60 to -10 degrees elevation of the 25-mm gun and 7.62-mm coax machine gun, has a much greater ability to engage targets in urban terrain.

b. **General Considerations When Using BFVs.** Light Infantry companies may be task-organized with mechanized Infantry platoons when conducting operations in urban terrain. A BFV platoon is capable of providing its own Infantry support. Generally, BFVs should not be separated from their Infantry. Working as a team, Infantrymen (the rifle squads) provide security for the vehicles; the BFVs provide critical fire support for the Infantry company team.

(1) ***Movement.*** When moving, if the street is large enough, BFVs should stay close to a building on either side of the street. This allows each BFV to cover the opposite side of the street. BFVs can button up for protection, but the BFV crew must remain alert for signals from Infantry. Coordination between mounted and dismounted elements is critical in urban terrain.

(2) ***ATGMs.*** The BFV lacks adequate armor protection to withstand medium to heavy ATGM fire. It is normally employed after the area has been cleared of ATGM positions or on terrain dominating the city to provide long-range antiarmor support or fire suppression. LAWs, AT4s, Dragons, or Javelins provide a significant amount of the BFV

platoon's short-range antiarmor fires in urban areas; the TOWs provide long range antiarmor fires. The BFV's 25-mm gun and machine gun are employed while providing direct fire support.

c. **Organization and Tasks.** The BFV platoon is comprised of mounted and dismounted elements. Based on the company commander's guidance and the factors of METT-TC, the BFV platoon leader will normally determine how his elements will be deployed. (The organization of the BFV platoon, sections, and squads in the 3-by-9 configuration is shown in Figure C-9, page C-18.)

(1) ***Offensive Task Organization.*** During offensive operations, the BFV platoon is normally given the mission of providing support for the company team. The company team commander generally will not separate the dismounted element from the mounted element, since the BFVs must have Infantry support during urban combat. If the dismounted element is needed for other tasks, enough local security must be left with the BFVs in order to protect them against enemy counterattack or anti-armor ambushes.

(2) ***Assault Tasks.*** An Infantry company team commander may give the BFV platoon the mission of performing assault tasks. The BFV platoon's Infantry would perform these tasks operating in the same manner as light Infantry platoons and squads. (See Chapters 3 and 4.) If the Infantry is used in this role, enough local security must be left with the BFVs to protect them.

(3) ***Support Tasks.*** (See paragraph C-2a.). The most likely tasks that will be given to a BFV platoon supporting a light Infantry company in urban combat will be those assigned to the support element. Direct fire support and other assistance to facilitate the advance of the assault element is provided by the support element. The BFV platoon is well suited to act as the support element for the light Infantry company team during offensive operations. The BFV platoon leader, acting as the support element leader, can provide command and control over his platoon and other support element assets. Specific BFV platoon tasks include, but are not limited to, the following:

(a) Suppressing enemy gunners within the objective building(s) and adjacent structures. This is accomplished with the 25-mm gun and 7.62-mm coax machine gun, TOWs, Infantry antiarmor, and small arms weapons (Figure C-10, page C-19).

(b) Breaching walls en route to and in the objective buildings.

(c) Destroying enemy positions within a building with the direct fire of the 25-mm gun and the 7.62-mm coax machine gun (when the wall is constructed of light material).

(d) Providing replacements for the assault element.

(e) Providing a mobile reserve for the company team.

(f) Providing resupply of ammunition and explosives.

(g) Evacuating casualties, prisoners, and noncombatants.

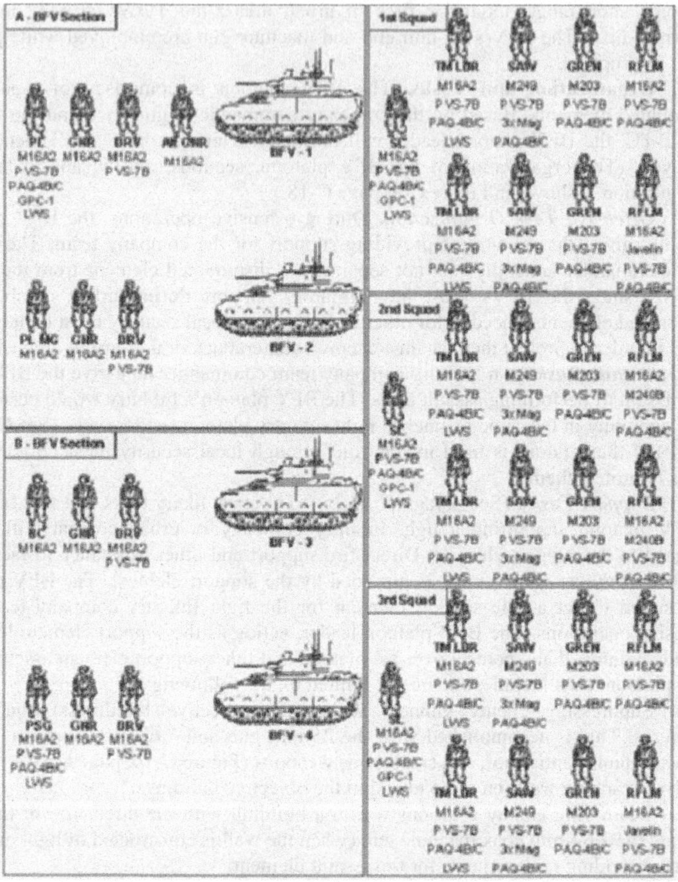

Figure C-9. Organization of BFV platoon.

Figure C-10. Suppression by 25-mm gun.

d. **Direct Fire Support.** The BFV is best used to provide direct fire support to Infantry. The BFV should move behind the Infantry, when required, to engage targets located by the rifle squads (Figure C-11). The dash speed (acceleration) of the BFV enables it to rapidly cross streets, open areas, or alleys.

Figure C-11. Moving with Infantry.

(1) ***Weapons***. The BFV mounted element provides fire with its 25-mm gun and 7.62-mm coax machine gun for Infantry on the opposite side of the street. The 25-mm gun is the most effective weapon on BFVs while fighting in urban terrain (Figure C-12).

Figure C-12. 25-mm gun support for Infantry.

(2) ***Safety Considerations.*** The use of the 25-mm gun in support of Infantry requires safety considerations.

- High-explosive 25-mm rounds arm 10 meters from the gun and explode on contact.
- APDS rounds discard their plastic sabots to the front of the gun when fired. This requires a 100-meter safety fan (17 degrees either side of the gun-target line for 100meters) to the front of 25-mm gun (Figure C-13). This means that exposed soldiers cannot go any farther forward than the end of the 25-mm's muzzle or must be a minimum of 100 meters from the muzzle blast.

Figure C-13. Safety fan for 25-mm gun.

(3) ***Use of Smoke.*** The BFVs' engine exhaust smoke system can be used in urban areas to cover the movement of Infantry. The BFV can also provide a smoke screen by using its smoke grenade launchers. This requires careful analysis of wind conditions to ensure that the smoke does not affect friendly units. This is a difficult task since wind currents tend to be erratic between buildings. The smoke can also screen the movements of the BFVs after the Infantry moves (Figure C-14).

NOTE: On-board smoke only works when the vehicle is using diesel fuel. It will not work with JP-8 fuel.

Figure C-14. Smoke screens movement of Infantry.

e. **Using the BFV to Isolate a Building.** To isolate a building, the BFVs take an overwatch position (Figure C-15). They fire the 25-mm gun and 7.62-mm coax machine gun, and adjust indirect fire to suppress enemy troops in the building and in nearby buildings who can fire at the assault element.

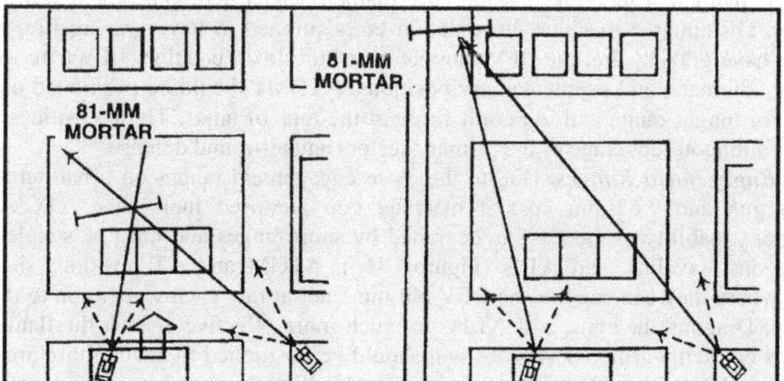

Figure C-15. Isolation of a building and shifting of fires.

C-8. DEFENSIVE CONSIDERATIONS AND TACTICS, TECHNIQUES, AND PROCEDURES FOR THE BFV

The BFV can provide a valuable combat multiplier in the defense. Infantry in the BFV platoon will defend in the same manner as light Infantry platoons. The following are typical defensive missions that may be given to a BFV platoon:

- Providing fire support for Infantry and mutual support to other BFV teams.
- Destroying enemy armored vehicles and direct fire artillery pieces.
- Destroying or making enemy footholds untenable by fire using the 25-mm gun.
- Providing rapid, protected transport for organic rifle teams or other Infantry elements.
- Reinforcing threatened areas by movement through covered and concealed routes to new firing positions.
- Providing mutual support to other antiarmor fires.
- Providing a mobile reserve and counterattack force.
- Providing resupply of ammunition and other supplies to the Infantry.
- Evacuating casualties, prisoners, and noncombatants.

NOTE: The overall value of the BFV to the defense must be weighed against the need to resupply or to evacuate casualties.

The BFVs are integrated into the company team defensive fire plan. The 25-mm gun and 7.62-mm coax machine gun fields of fire cover streets and open areas; TOWs are used to cover armor avenues of approach. Once placed in position, BFVs should not be moved for logistical or administrative functions. Other vehicles should accomplish these functions, when possible.

a. **Positioning of BFVs and Weapons.** Once the company team commander gives the BFV platoon leader his mission, the platoon leader will position his BFVs and Infantry. Dismounted machine guns should be positioned to have grazing fire. For the coax to have grazing fire, the BFV must be in a hull-down position. BFVs are assigned primary, alternate, and supplementary positions. ATGMs should be positioned on upper stories for longer range and to permit firing at the tops of tanks. These positions should permit continuous coverage of the primary sectors and all-round defense.

b. ***Engagement Ranges.*** Due to the close engagement ranges on urban terrain, the 25-mm gun and 7.62-mm coaxial machine gun are used more than ATGMs. The antiarmor capability of the BFV is degraded by short ranges and must be supplemented by Dragons, Javelins, and AT4s (Figure C-16). ATGM and AT positions should be placed where they can support the BFV but must not attract enemy attention to the BFV location. Dragons, Javelins, and AT4s are much more effective against the flanks, rear, and tops of enemy armored vehicles and should be positioned to attack those areas. The TOWs are also employed against enemy armored vehicles.

Figure C-16. AT4 position supporting BFV.

NOTE: TOWs cannot be dismounted. Company team commanders must determine which weapon will work most effectively against the threat he is anticipating. Ideally, the BFV should be positioned to take advantage of all the weapons on the vehicle.

c. **Integration of Fires.** All of the BFV's crew-served weapons are integrated with the rest of the company team's weapons and assets. The positions are recorded on a company sector sketch and forwarded to battalion.

C-9. STABILITY AND SUPPORT OPERATIONS CONSIDERATIONS AND TACTICS, TECHNIQUES, AND PROCEDURES

See Chapter 14 for information on stability operations and support operations.

APPENDIX D

INFORMATION OPERATIONS

"The side possessing better information and using that information more effectively to gain understanding has a major advantage over its opponent."

FM 3-0(FM 100-5), *Operations*

While many information operations (IO) will be planned and executed at levels above the brigade, brigades and below will often benefit from the results of properly executed IO, especially during urban operations. IO are primarily shaping operations that create and preserve opportunities for decisive operations. IO are both offensive and defensive. Related activities, such as public affairs and civil-military operations (CMO), support IO. This appendix will define IO, describe the elements of IO, and describe their effects on brigade urban operations.

D-1. DEFINITIONS

These IO terms are defined as follows:

a. **Information Operations.** IO are actions taken to affect the threat's, and influence others', decision-making processes, information, and information systems while protecting one's own information and information systems.

(1) The value of IO is not in their effect on how well an enemy transmits data. Their real value is measured only by their effect on the enemy's ability to execute military actions. Commanders use IO to attack enemy decision making processes, information, and information systems. Effective IO allow commanders to mass effects at decisive points more quickly than the enemy. IO are used to deny, destroy, degrade, disrupt, deceive, exploit, and influence the enemy's ability to exercise C2. To create this effect, friendly forces attempt to influence the enemy's perception of the situation. Similarly, IO and related activities affect the perceptions and attitudes of a host of others in the AO. These include the local population, displaced persons, and civilian leaders. IO are shaping operations that help commanders create favorable conditions for not only decisive operations but also sustaining operations. Commanders use IO and related activities to mitigate the effects of enemy IO, as well as adverse effects stemming from misinformation, rumors, confusion, and apprehension.

(2) Successful IO require a thorough and detailed IPB. IPB includes information about enemy capabilities, decision making style, and information systems. It also considers the effect of the media and the attitudes, culture, economy, demographics, politics, and personalities of people in the AO. Successful IO influences the perceptions, decisions, and will of enemies, adversaries, and others in the AO. Its primary goals are to produce a disparity in enemy commanders' minds between reality and their perception of reality and to disrupt their ability to exercise C2.

b. **Offensive IO.** Offensive IO are the integrated use of assigned and supporting capabilities and activities, mutually supported by intelligence, to affect enemy decision makers or to influence others to achieve or promote specific objectives. The desired effects of offensive IO are to destroy, degrade, disrupt, deny, deceive, exploit, and influence enemy functions. Concurrently, Army forces employ elements of offensive IO to affect the perceptions of adversaries and others within the AO. Using the elements of IO offensively, Army forces can either prevent the enemy from exercising effective C2 or leverage it to their advantage. Ultimately, IO targets are the human leaders and human decision making processes of adversaries, enemies, and others in the AO.

c. **Defensive IO.** Defensive IO are the integration and coordination of policies and procedures, operations, personnel, and technology to protect and defend friendly information and information systems. Defensive IO ensure timely, accurate, and relevant information access while denying adversaries the opportunity to exploit friendly information and information systems for their own purposes. Defensive IO protect friendly access to relevant information while denying adversaries and enemies the opportunity to affect friendly information and information systems. Defensive IO limit the vulnerability of C2 systems.

d. **Effects of Offensive and Defensive IO.** Offensive and defensive operations use complementary, reinforcing, and asymmetric effects to attack enemies, influence adversaries and others, and protect friendly forces. On a battlefield where concentrating forces is hazardous, IO can attack enemy C2 systems and undermine enemy capabilities and will to fight. It can reduce friendly vulnerabilities and exploit enemy weaknesses. Where the use of force is restricted or is not a viable option, IO can influence attitudes, reduce commitment to a hostile cause, and convey the willingness to use force without actually employing it. Information used in this manner allows friendly forces to accomplish missions faster, with fewer casualties.

D-2. ELEMENTS OF INFORMATION OPERATIONS

Integrating offensive and defensive IO is essential to success. Many activities or operations comprise IO. Each element may have offensive or defensive applications (Figure D-1).

ELEMENTS OF IO

- Military deception
- Counterdeception
- Operations security
- Physical security
- Electronic warfare
  - — *Electronic attack*
  - — *Electronic protection*
  - — *Electronic warfare support*
- Information assurance
- Physical destruction
- Psychological operations
- Counterpropaganda
- Counterintelligence
- Computer network attack
- Computer network defense

Figure D-1. Elements of information operations.

a. **Military Deception.** Military deception includes measures designed to mislead adversaries and enemies by manipulation, distortion, or falsification. Its aim is to influence the enemy's situational awareness and lead him to act in a manner that favors friendly forces.

b. **Counterdeception.** Counterdeception includes efforts to negate, neutralize, or diminish the effects of, or gain advantage from, a hostile deception operation. Counterdeception supports offensive IO by reducing harmful effects of enemy deception. Defensively, counterdeception can identify enemy attempts to mislead friendly forces.

c. **Operations Security.** Operations security (OPSEC) denies the enemy information critical to the success of friendly military operations. It contributes to the security of Army forces and their ability to surprise enemies and adversaries. OPSEC identifies routine activities that may telegraph friendly intentions, operations, capabilities, or military activities. It acts to suppress, conceal, control, or eliminate these indicators. OPSEC includes countersurveillance, signal security, and information security.

d. **Physical Security.** Physical security prevents unauthorized access to equipment, installations, and documents. It safeguards and protects information and information systems.

e. **Electronic Warfare.** Electronic warfare (EW) is military action involving the use of electromagnetic and directed energy to control the electromagnetic spectrum or to attack the enemy. EW can cause an enemy to misinterpret the information received by his electronic systems. EW includes:

(1) ***Electronic Attack.*** Electronic attack involves actions taken to degrade, neutralize, or destroy enemy electronic combat capabilities. Actions may include lethal attack, such as antiradiation missiles and directed energy weapons, and nonlethal electronic attack, such as jamming.

(2) ***Electronic Protection.*** Electronic protection involves actions taken to protect friendly use of the electronic spectrum by minimizing the effects of friendly or enemy EW. Actions may include radio silence and antijamming measures.

(3) ***Electronic Warfare Support.*** Electronic warfare support involves detecting, identifying, locating, and exploiting enemy signal emitters. It contributes to achieving situational awareness, target development and acquisition, damage assessment, and force protection.

f. **Information Assurance.** Information assurance protects and defends information systems. Threats to information systems include physical destruction, denial of service, capture, environmental damage, and malfunctions. Information assurance provides an enhanced degree of confidence that information and information systems possess the following characteristics: availability, integrity, authentication, confidentiality, and nonrepudiation. Computer network defense is part of this element.

g. **Physical Destruction.** Physical destruction applies combat power against IO-related targets. Targets include information systems, EW systems, and command posts. Physical destruction that supports IO is synchronized with other aspects of the operation. For example, when deciding whether to destroy an enemy command post, the friendly commander weighs the advantages gained from disrupting enemy C2 against those gained from collecting information from the command post's radio traffic.

h. **Psychological Operations.** Psychological operations (PSYOP) are planned operations that influence the behavior and actions of foreign audiences by conveying selected information and indicators to them (FM-05.30[FM 33-1]). The aim of PSYOP is to create behaviors that support US national interests and the mission of the force. PSYOP are closely integrated with OPSEC, military deception, physical destruction, and EW to create a perception of reality that supports friendly objectives.

i. **Counterpropaganda.** Counterpropaganda includes activities directed at an enemy or adversary conducting PSYOP against friendly forces. Counterpropaganda can contribute to situational awareness and expose enemy attempts to influence friendly populations and military forces. Preventive actions include propaganda awareness programs that inform US and friendly forces and friendly populations about hostile propaganda.

j. **Counterintelligence.** Counterintelligence consists of activities that identify and counteract threats to security posed by espionage, subversion, or terrorism. It detects, neutralizes, or prevents espionage or other intelligence activities. Counterintelligence supports the commander's requirements to preserve essential security and protect the force.

k. **Computer Network Attack.** Computer network attack consists of operations that disrupt, deny, degrade, or destroy information resident in computers and computer networks. It may also target computers and networks themselves. Although theater or national elements normally conduct computer network attack, the effects may be evident at corps and below.

D-3. RELATED ACTIVITIES

Public affairs and civil-military operations (CMO) are activities related to IO. Both communicate information to critical audiences to influence their understanding and perception of military operations. Related activities are distinct from IO because they do not manipulate or distort information; their effectiveness stems from their credibility with the local populace and news media. Public affairs and CMO—prime sources of information—link the force, the local populace, and the news media. They also provide assessments of the impact of military operations on civilians, neutrals, and others within the battlespace.

a. **Public Affairs.** Public affairs operations influence populations by transmitting information through the news media. They fulfill the Army's obligation to keep the American people and the Army informed. Public affairs help to establish conditions that lead to confidence in the Army and its readiness to conduct operations in peace, conflict, and war. Disseminating this information is desirable and consistent with security. Information disseminated through public affairs counters the effects of propaganda and misinformation.

b. **Civil-Military Operations**. CMO applies civil affairs to military operations. It encompasses activities that commanders take to establish, maintain, influence, or exploit relations between military forces and civil authorities—both governmental and nongovernmental—and the civilian populace. Commanders direct these activities in friendly, neutral, or hostile AOs to facilitate military operations and consolidate operational objectives. Civil affairs may include performance by military forces of activities and functions normally the responsibility of local government. These activities may occur before, during, or after other military actions. They may also occur as stand-alone operations. CMO is the decisive and timely application of planned activities that enhance the relationship between military forces and civilian authorities and population. They promote the development of favorable emotions, attitudes, or behavior in neutral, friendly, or hostile groups. CMO range from support to combat operations to assisting countries in establishing political, economic, and social stability.

D-4. INFORMATION OPERATIONS AND THEIR EFFECTS ON BRIGADE URBAN OPERATIONS

The IO, at brigade level, is planned and executed as a coordinated effort between the S2 and S3, and, when assigned, the S5. The IO affects all four elements of urban operations—assess, shape, dominate, and transition.

a. **Assess.** The brigade staff assesses the results of IO conducted by the division or JTF. For example, the results of military deception, PSYOP, electronic attack, counterpropaganda, and computer network attack conducted by higher headquarters may force the threat to act in a manner that facilitates brigade offensive operations and minimizes the need for close combat. During defensive operations, IO may force the threat to deploy into engagement areas that will facilitate the synchronization of fires and nonlethal capabilities.

b. **Shape.** Brigades can use IO to shape the urban battlespace. For example, continuous OPSEC and physical security can deny the threat valuable information concerning friendly forces. Likewise, brigades can use PSYOP teams, if available, to help shape desired threat behaviors prior to the execution of UO.

c. **Dominate.** Brigades may be given missions from higher headquarters to physically destroy or control IO related targets such as command posts or media centers. Moreover, the effects of IO during assessing and shaping may permit brigades to dominate without having to employ fires and close combat or by minimizing their use.

d. **Transition.** Brigades may use the results of IO to facilitate the transition to stability and support operations. For example, a civil-military operations center (CMOC) may be established within a brigade AO in order to permit restoration of utilities such as electricity, water, and sewage.

APPENDIX E

COALITION OPERATIONS

While formal alliances have developed effective methods for integrating forces, most coalitions are created on short notice. As a result, coalition members must quickly develop detailed plans. This coalition planning guide will assist commanders and their staffs during the decision-making process in organizing their efforts to integrate their forces into the overall coalition structure. This guide is not all-inclusive. Additional actions may be required, depending on the coalition's operational mission and or composition. Depending on the unique situation that exists in the urban area, many of the staff functions discussed below will be performed by the higher headquarters. If so, commanders at all levels must ensure that their staffs conduct adequate coordination to ensure smooth implementation of the higher headquarters' decisions.

E-1. MISSION ANALYSIS

Determine the desired political objective and end-state condition as defined by the legitimizing political authority.

- Establish the military role in attaining this political end state. Coordinate the military role with the roles of other federal agencies, nongovernment organizations (NGOs), and private voluntary organizations (PVOs) involved.
- Analyze the mission for clear and attainable military objectives. Ensure the commander's intent supports the desired political end state.
- Translate these objectives into tasks for subordinate commanders.

a. Select the nation with the greater national interest and or the best ability to plan for and direct the multinational forces (MNF) to lead the effort.

b. Establish the probable cost (in lives, money, resources) of the operation and determine if the cost is acceptable.

c. Determine which legitimizing political authority defined the desired end state and the relationship between the military force and this political entity.

d. Determine the national end-state criteria of each coalition partner and if it differs from the criteria of the MNF itself.

- Select alternate courses of action the MNF will follow when a national military element withdraws from the force (that is, actions following decomposition of the force).
- Determine courses of action to be executed if the sponsoring organization orders withdrawal of MNF in advance of end-state achievement.
- Determine redeployment and or withdrawal plans for MNF and how the departure of forces is accomplished under tactical conditions.
- Decide actions to be taken in the event one member of an MNF executes a unilateral withdrawal from the coalition (whether for political or other reasons).

E-2. CRITICAL OPERATING SYSTEMS

Select C2 structure and establish subordinate relationships (OPCON, TACON, national control retained).

- Establish a minimum communications suite for each of the coalition members to enable multinational operations.
- Identify the C2 systems support required for the diminishing MNF presence.

a. Determine how intelligence and information are shared.

b. Agree on a logistical support structure identifying capabilities and responsibilities.

E-3. CONTROL MECHANISMS

The following are control mechanisms:

a. **Language.**

- Agree on coalition language for force-wide use.
- Determine command level at which each force resorts to its national language and does not use interpreters.
- Select sufficient interpreters for planning and execution.

b. **Liaison.**

- Identify and interview key LNOs.
- Determine if LNOs will be sent to the various coalition headquarters. Identify their communication requirements. If coalition LNOs are sent to the US headquarters, determine communications support and augmentation.

c. **Foreign Disclosure.**

- Appoint foreign disclosure officers to oversee and control coordination of specific disclosures to coalition members.
- Disclose classified military information and controlled unclassified information to coalition members IAW AR 380-10.

E-4. RULES OF ENGAGEMENT

The following are rules of engagement:

a. Determine the ROE, who establishes them, and when they are to be reviewed.

b. Determine how national ROE affect the MNF organizations and operations.

c. Decide on procedures used by commanders to request a change to the current ROE.

E-5. STAFF TASKS

The following tasks are performed by the various staff agencies indicated.

a. **Personnel/S1.**

- Identify special skill requirements (unit and or individual). Establish requirement for RC units and or individuals. Devise a personnel replacement and rotation scheme.
- Request language-qualified personnel for augmentation. Request language survival kits Identify available training such as MLTs.
- Coordinate with G3 and request available augmentation, such as MLTs and joint movement control teams.

b. **Intelligence/S2.**

- Determine how US strategic intelligence is shared among other forces.

- Provide other national forces intelligence summaries, imagery, and threat assessments approved for dissemination.
- Assess overall MNF intelligence gathering and dissemination capabilities. Prescribe and disseminate procedures for intelligence sharing and reporting.
- Determine available linguistic support and or requirements. Coordinate with the G1.
- Determine any unique relationships (cultural, historical, other) between MNF partners and the adversary and or enemy.

c. **Operations/S3.**
- Prepare organization chart for the ARFOR and coalition forces.
- Prepare capabilities brief/description for ARFOR and coalition forces.
- Determine effect of national ROEs and objectives on force composition and mission assignment.
- Determine the current situation.
- Decide how mission, to include commander's intent, is disseminated and checked for understanding and consistency.
- Begin planning for transition to UN or other organizations that will take over from the MNF. Contact counterpart planning staffs.
- Determine logistical situation.
- Establish language and interpreter requirements.
- Determine special requirements.
- Review special customs and courtesies.
- Provide SOPs that include reporting requirements and procedures.
- Establish times when units are available.
- Coordinate ground and air reconnaissance and establish a common map data base.
- Coordinate staff visits.
- Coordinate unit commander's visit to higher headquarters.
- When redeploying or moving, determine what sector forces remain to support the coalition and how long they are required.
- If required, determine if foreign forces will interface with the joint movement control center.
- Determine training required before deployment and prepare for continued training, including multinational, after deployment.

d. **Logistics/S4.**
- Determine composition of multinational logistics command and or element.
- Divide responsibilities between multinational, national, and host nation (HN).
- Determine how each class of supply will be handled.
- Determine the combined transportation command and control structure.
- Determine the coalition reception, staging, onward movement, and integration process.
- Determine if coalition forces have tactical rotary- and fixed-wing assets for supply.

- Determine who supplies transportation supply throughput for coalition forces from the Joint Task Force(JTF) logistics center.
- Determine if coalition forces have transportation assets for the movement of troops.

e. **Supply/S4.**
- Identify the coalition forces' capabilities to receive, store, and issue dry cargo, fuel, and water to include water production and or purification capability.
- Determine if coalition forces have the means to communicate requirements to the JTF logistics center or coalition logistics management center.
- Identify availability of materiel handling equipment within the coalition.
- Identify coalition air and sea lines of communications (LOCs).
- Determine if coalition forces have a basic load of ammunition and their Class V procedures.
- Identify the coalition force's special requirements (tents, cots, reverse osmosis water purification units, laundry, latrines, batteries, etc.).
- Determine map requirements for coalition forces.

f. **Mortuary Affairs.**
- Determine capabilities of the coalition forces.
- Determine any specific religious requirements of the coalition forces.

g. **Maintenance.**
- Determine if coalition forces have maintenance support.
- Determine if coalition forces have the means to order and receive Class IX.
- Determine if coalition forces have wreckers, stake and platform trailers, or HETs.
- Determine if coalition forces have communications repair facilities.

h. **Civil Affairs/S5.**
- Establish and coordinate operations of the CMOC.
- Identify the NGOs and PVOs; determine their intentions and future plans; identify the key personnel in sector.

i. **Information Management.**
- Identify the command relationships, the location of the headquarters each unit reports to, and the type of service required (tactical satellite, telephone, facsimile, AM, FM).
- Articulate in the initiating directive that the multinational command channels are for the execution of military operations and national channels are for reporting status and requesting support.
- Establish a means and a plan to provide a common tactical picture to all forces.
- Provide coalition partners with a lesser C2 capability interpreters (if necessary), operators, and maintainers to enable interaction with the commander and other coalition members.
- Determine the frequency requirements and planning ranges for equipment.
- Consider the terrain while planning for the C2 network.
- Request frequencies from the JTF J6 or coalition communications coordinator.
- Provide for common data bases.

j. **Engineer.**
- Identify terrain visualization requirements.
- Determine types and capabilities of engineer units for the coalition and other services.
- Determine the facility support requirements from the MNF and its supported units (that is, latrines, base camp construction).
- Determine the condition of and requirements for infrastructure in the area of operations (AO) (roads, airfields, ports, power generation facilities).
- Identify the availability and type of engineer resources in the operating area.
- Determine real estate support requirements.
- Identify humanitarian and nation assistance engineering requirements.

k. **Aviation.**
- Determine the aviation assets, capabilities, and requirements of the coalition force.
- Identify the aviation logistics capabilities (fuel, spare parts) of the coalition force.
- Identify current and projected requirements for an air LOC.
- Determine aviation support required from US forces.
- Identify the intended base of operations.

l. **Public Affairs.**
- Determine PA coverage plan.
- Coordinate with other national PA officers or equivalents.
- Identify the procedures, if any, for a security review of media material.
- Establish a plan for handling publicity, news correspondents, and journalists.
- Determine the coalition senior leaders and their biographical backgrounds. Arrange face-to-face meetings.
- Identify the CMOC and the participants.
- Identify the senior operation spokesperson for the MNF.
- Establish a combined information center.

m. **Financial Operations.**
- Determine if a weapons bounty program is needed. Prepare procedures to support weapons bounties, claims.
- Determine limitations on the amounts of cash payments (including check cashing) that soldiers may receive in the AO. Determine who imposes limitations - the MNF commander or the HNs (SOFA).
- Establish requirement for US forces to provide other currencies for the operation and currency exchange.

n. **Legal.**
- Review and complete SOFA and ROE.
- Brief commander/S4 on legal aspects of supporting foreign forces.
- Determine if all soldiers are briefed on the ROE before the assumption of operational control by the coalition or deployment to the operational area.
- Establish a reporting procedure for all events that can be compensated, accidents, injuries to locals, deaths, or property damage for potential claims.

- Report all serious incidents with legal ramifications that involve HN and or coalition forces and the US force.
- Review the operations plan (OPLAN) to ensure compliance with US and international law.

o. **Surgeon.**

- Determine the capabilities for medical evacuation (MEDEVAC) and whether the US force will be supported by, or required to support, other coalition elements.
- Identify the MEDEVAC assets (air and ground).
- Identify the source of Class VIII supply and payment options.
- Identify procedures for tracking patients and coordination requirements for return-to-duty transportation.
- Determine if forces have organic Echelon I, II, or III combat health support. If not, what level will US forces provide.
- Determine the air evacuation procedures for US and coalition force personnel (evacuated to the continental US or their home country).
- Determine the policy and procedures for US medical personnel to use coalition forces Echelon II through V medical treatment facilities to provide medical treatment for US and coalition forces.

p. **Provost Marshal.**

- Determine if coalition forces require MP support.

APPENDIX F

WEAPONS OF MASS DESTRUCTION, TOXIC INDUSTRIAL MATERIALS, AND THE USE OF OBSCURATION

International and domestic law place severe limitations on the use of chemical and bacteriological weapons in armed conflict. The US condemns the use of all biological and bacteriological agents under any circumstances. Potential enemies may not operate under the same restrictions and may employ weapons of mass destruction (WMD) or release toxic industrial materials (TIM). Commanders and leaders must be prepared to assume an adequate NBC defensive posture when conducting urban operations. Leaders must be aware of how the urban environment affects the protection, detection, and decontamination process. Additionally, urban areas can contain many TIM. Release of these materials can pose a significant threat to friendly forces and noncombatants. This chapter will provide commanders and leaders with information and guidance concerning defense against WMD and TIM. (For additional information, see FM 3-9.)

Section I. NUCLEAR, BIOLOGICAL, AND CHEMICAL CONSIDERATIONS

Personnel who must move through a contaminated urban area should employ the procedures outlined in the appropriate NBC field manual.

F-1. PROTECTION FROM NBC WEAPONS EFFECTS

Buildings are usually not strong enough to provide shelter from a nuclear explosion but do provide some protection against fallout. They also have unique characteristics concerning the use of biological and chemical agents.

a. **Nuclear.** The lowest floor or basement of a reinforced concrete or steel-formed building offers good protection from nuclear hazards and liquid chemical contamination. Tunnels, storm drains, subway tubes, and sewers provide better protection than buildings. Tanks and BFVs provide limited protection.

b. **Biological.** Biological attacks are difficult to detect or recognize. Biological agents can be disseminated or dispersed using aerosols, vectors, and covert methods. (See FM 3-3 for more detailed information). Since biological agents can be sprayed or dropped in bomblets; personnel who observe such indicators should promptly report them. Prompt reporting and treatment of the sick speeds the employment of medical countermeasures. Although buildings and shelters provide some protection against spraying, they provide little protection against biological agents.

c. **Chemical.** Chemical agents cause casualties by being inhaled or by being absorbed through the skin. They may afford soldiers a few seconds to mask. Buildings have a channeling effect and tend to contain the effects of an agent, causing great variation in chemical concentration from room to room or from building to building. Chemical agents usually settle in low places, making basements, sewers and subways hazardous hiding places. A prepared defender should include some collective protective measures in the defensive network. Personnel using fans may be able to put enough overpressure into tunnels to keep some chemical agents from entering. The individual protective mask and battle dress overgarment provides the best protection against chemical agents.

d. **Personal Hygiene and Field Sanitation.** Personal hygiene is a critical defensive measure against infection and disease. Soldiers fighting in urban areas are exposed to many infectious diseases such as diarrhea, hepatitis, and even cholera. Lessons learned from recent Russian operations in Chechnya reinforce this point. An extremely high proportion of Russian troops suffered from diseases they had contracted from eating or drinking from unsafe sources. Modern urban areas are characterized by sophisticated sanitation systems. The populace relies on this human service. If those systems are destroyed or degraded, either by friendly or threat actions, the resulting sanitary conditions can become much worse than those in areas where sanitary facilities are more primitive. Commanders and leaders must ensure that personnel employ appropriate field sanitation measures and that their immunizations are current. Food and water sources must be kept sanitary and human waste disposed of correctly.

e. **Mission-Oriented Protective Posture.** Commanders should plan their mission-oriented protective posture (MOPP) realizing that urban area logistics also apply to NBC equipment. Protective clothing, detection and decontamination equipment, and sealed containers of food and water must be stockpiled the same as other supplies. When operating in protective clothing, commanders must make allowances for the strenuous activities normally associated with combat in urban areas.

(1) ***Detection.*** After an NBC attack, companies should dispatch their detection and survey teams. Detection in urban areas is complicated by the containing nature of buildings. Teams should conduct tests and surveys of major streets, intersections, and buildings in their area for inclusion in initial NBC reports. A systematic survey of all buildings, rooms, and underground facilities must be accomplished before occupation by unmasked personnel. All data should be forwarded using the appropriate NBC report.

(2) ***Decontamination.*** Personnel must begin decontamination operations as soon after an NBC attack as the mission allows. Personnel should conduct individual decontamination of themselves and their personal equipment. Unit commanders must determine the need for MOPP gear exchange and the requirements for a hasty or deliberate decontamination operation.

(a) *Radiological.* Personnel should wear wet-weather gear for certain decontamination operations (hosing down buildings) to prevent radioactive material from touching the skin.

(b) *Chemical and Biological.* Roads, sidewalks, and other hard surfaces are best decontaminated by weathering, if time permits. Agents can also be covered with several inches of dirt or sand to provide protection. Fragment testing should be conducted periodically to ensure that the agent has not seeped through the covering. For critical sections of terrain or roads, a mobile decontaminating system such as the M12A1 power driven decontaminating system can be used to spray a slurry of foam; this aids rapid decontamination. Buildings are difficult to decontaminate especially wooden ones. Some techniques for their decontamination are scrubbing with slurry; washing with hot, soapy water; washing or spraying with a soda solution; and airing.

F-2. SMOKE OPERATIONS

The use of smoke as an integral part of either offensive or defensive operations can complement missions in urban areas. Chemical support from smoke generator units, if available, can be employed for both offensive and defensive operations. Whenever smoke is employed, regardless of the source, combat patience must be practiced to insure that the effects of the smoke are maximized prior to moving, conducting assaults or counterattacks, or breaching operations.

a. In the offense, smoke can support the maneuver of combat elements and deception operations. Smoke employed in the defense obscures enemy air and ground observation, limiting the accuracy of enemy fires and target intelligence.

b. Smoke should not be used when it degrades the effectiveness of friendly forces. An extremely dense concentration of smoke in a closed room can displace the oxygen in the room, smothering soldiers even when they are wearing protective masks.

c. Smoke pots, generators, or artillery smoke munitions should be used to cover the withdrawal of defending forces or the movement of attacking forces. Artillery-delivered white phosphorus can also be effective on enemy forces by causing casualties and fires. The incendiary effects of both white phosphorus and base ejection munitions on the litter and debris of urban areas must be considered.

d. Smoke grenades can be used to provide a hasty screen for concealing personnel movement across streets and alleys. Smoke grenades can also be used for signaling; those launched by an M203 can be used to mark targets for attack helicopters or tactical air.

e. The use of smoke in urban areas is affected by complex wind patterns caused from buildings. Obscuration planning must include covering as much of the objective area as possible. Failure to obscure key structures provides enemy observers reference points for fire placement within the objective area.

F-3. RIOT CONTROL AGENTS

National Command Authority (NCA) approval is required prior to using riot control agents (RCAs).

a. **Executive Order 11850.** Executive Order (EO) 11850 prohibits the use of RCAs during offensive operations regardless of whether the enemy is using noncombatants as human shields. EO 11850 also renounces first use in armed conflict except in defensive military modes to save lives such as—

- Controlling riots.
- Dispersing civilians where the enemy uses them to mask or screen an attack.
- Rescue missions.

b. **Effects and Employment.** Riot control agents, such as CS, generally have no lasting effects. RCAs can be disseminated in hand grenades, ring airfoil projectiles, 40-mm and 66-mm cartridge grenades, or bulk agent aerial and ground dispersers. Units will normally employ RCAs, once permission is secured from the NCA through command channels, by using hand grenades or 40-mm rounds fired from the M203 grenade launcher. The protective mask and BDUs will protect soldiers from the effects of the RCAs. RCAs are not effective against a well trained threat that is equipped for chemical defense.

Section II. RELEASE OF TOXIC INDUSTRIAL MATERIALS (TIM) DURING URBAN OPERATIONS

In urban areas all over the world today, large amounts of very dangerous chemicals are routinely stored, handled and transported. There is a very real possibility that an accidental or deliberate release of toxic materials can occur during urban operations. Commanders must include an analysis of the threat of hazardous material release in the planning process and risk analysis prior to commencing operations. The inadvertent or planned release of a deadly chemical can present a threat to US soldiers and the local population and also create a negative effect on world opinion that may directly affect the urban operation.

F-4. TOXIC INDUSTRIAL MATERIALS IN URBAN AREAS

Toxic materials may be located throughout any large urban area (Figure F-1). The most common chemicals that pose a risk to friendly forces and noncombatants are irritant gases, especially chlorine, sulfur dioxide, ammonia and hydrogen chloride. These substances have relatively high toxicity when inhaled and they are produced, stored and transported in large volumes. Production sources and stockpiles are frequently located near inhabited areas. Most toxic industrial chemicals are released as vapors. These vapors tend to remain concentrated downwind from the release point and in natural low-lying areas such as valleys, ravines, and man-made underground structures. Irritants are not the only toxic materials likely to be in urban areas. The list in paragraph 9-5 shows common sources of large amounts of chemicals in urban areas. In addition to any irritants that may be released, many substances generated during fires in chemical stockpiles create a special problem not only to troops and non-combatants on the ground, but also to the pilots and crew of low flying aircraft.

Figure F-1. Large urban chemical manufacturing plant.

F-5. MOST COMMON TYPES AND LOCATIONS OF TOXIC INDUSTRIAL MATERIALS

The most common types of TIM are listed below. The most common locations of TIM are shown in Table F-1. Barges, pipelines, rail cars, and tank trucks deliver large quantities of chemicals directly to cities located great distances from stationary sources of chemicals. Transportation assets frequently traverse routes near highly populated areas (Figures F-2 and F-3, page F-6). Pipelines remain the most vulnerable chemical transportation assets, because they pass through areas beyond the influence of security assets located at fixed facilities and in populated areas.

- Irritants (acids, ammonia, acrylates, aldehydes, and isocyanates)
- Choking agents (chlorine, hydrogen sulfide, and phosgene)
- Flammable industrial gases (acetone, alkenes, alkyl halides, amines)
- Water supplies contaminants (aromatic hydrocarbons, benzene, etc.)
- Oxidizers that increase explosion dangers (oxygen, butadiene, and peroxides)
- Chemical asphyxiants (Aniline, nitrile, and cyanide compounds)
- Incendiary gases (compressed isobutane, liquefied natural gas, propane,)
- Incendiary liquids (liquid hydrocarbons, gasoline, diesel, jet fuel)
- Industrial compounds that act much like blister agents (dimethyl sulfate)
- Organophosphate pesticides that act as low-grade nerve agents.

LOCATION	TYPE OF TIM
Airports	Aviation gasoline, jet fuel.
Farm and garden supply warehouses	Pesticides
Shipping terminals	Bulk petroleum and chemicals
College laboratories	Organic chemicals, radioactive materials
Electronics manufacturers	Arsine, arsenic trichloride
Food processing and storage areas	Ammonia
Glass and mirror plants	Fluorine, hydrofluoric acid
Pipelines and propane storage tanks	Ammonia, methane, and propane
Plastic manufacturers	Isocyanates, cyanide compunds
Landscaping businesses	Ricin (a food and water poison)
Medical facilities	Radioactive isotopes, mercury
Inorganic chemical plants	Chlorine
Hard rock ore mines	Potassium and sodium cyanide
Pesticide plants	Organophosphate pesticides
Petroleum storage tanks	Gasoline, diesel fuel
Photographic supply distributors	Cyanides, heavy metals
Rail and trucking lines	Anhydrous ammonia; sulfuric phosphoric ans hydrochloric acids, and flammable liquids
Chemical manufacturing plants	Chlorine. Peroxides, and other industrial gases
Power stations and transformers	Polychlorinated biphenyls (PCBs)
Large refrigeration units (grocery stores, dairy processing plants)	Anhydrous ammonia

Table F-1. Location and types of TIM.

Figure F-2. TIM transported on a rail line.

Figure F-3. Barge carrying fuel oil.

F-6. CHEMICALS MOST OFTEN CAUSING DEATH OR SERIOUS INJURY

A study by the US Environmental Protection Agency, analyzing over 7,000 large scale releases of industrial chemicals, indicated four chemicals accounted for almost 30 percent of all the deaths and serious injuries. These were:

- Chlorine.
- Hydrochloric acid.
- Sulfuric acid.
- Anhydrous ammonia.

F-7. CHLORINE GAS

Chlorine hazards should not be underestimated. Large chlorine releases from rail cars, large storage tanks, or tank trucks in enclosed areas such as narrow streets pose substantial hazards (Figure F-4). In these situations, chlorine can be a very dangerous choking agent.

a. Such a situation was illustrated at the Pliva Pharmaceutical Factory in the capital of Croatia, during the war between Croats and Serbs. This factory used acids, ammonia, bases, chlorine, and other toxic substances to produce pharmaceutical products. Croatian modeling indicated that, in the event of a major attack, concentrations of chlorine and stannic acid lethal for 50 percent of the population would extend up to 4 km away from the facility.

b. Croatian industrial engineers also studied the potential effects of an instantaneous release of chlorine from a single rail tank car damaged by conventional weapons such as bombs or artillery fire. The models indicated that with a normal load of 16 cubic meters per railcar, a lethal concentration of chlorine could extend up to 5 kilometers downwind and that serious adverse health effects could occur as far as 12 kilometers downwind.

c. These distances provide a good idea of the ranges that TIM can travel and affect friendly personnel or noncombatants. Commanders should request this type of specific information, whenever conducting UO near facilities that may contain TIM.

Figure F-4. Standard rail car filled with chlorine gas.

F-8. THE EFFECTS OF A LARGE RELEASE OF TOXIC INDUSTRIAL MATERIALS

The damage caused by the release of TIM during urban operations will depend on the type and size of the discharge, the period during which personnel are exposed, and the length of time between exposure and treatment. The most effective action is immediate evacuation outside the hazard's path. Most military protective masks, respirators, and protective clothing provide only limited protection against industrial chemicals, although they are much better than no protection at all.

a. A massive discharge of a hazardous chemical in an urban area took place at Union Carbide's factory at Bhopal, India, in 1984. A tank containing methylisocyanate, an extremely reactive chemical used in the production of insecticide, leaked for about one hour. This developed into one of the largest industrial disaster ever to occur.

b. Methylisocyanate (CH3NCO) is used in several chemical manufacturing processes. It has an extremely irritating effect on mucous membrane and is highly toxic when inhaled. Symptoms accompanying exposure are coughing, increased saliva production, tear flow and difficulty in keeping the eyes open, much like the effects of riot control agents.

c. About one-third of the town's total population of 800,000 were affected. About 100,000 of these required some kind of medical treatment, about 50,000 were hospitalized, and about 2,500 immediately inhaled lethal doses. Approximately 16,000 people eventually died.

F-9. INTERNATIONAL HAZARDOUS MATERIAL SYMBOLS

There is a standard, internationally recognized, marking scheme for TIM in use worldwide. It uses a combination of colors, shapes, and symbols. These symbols may not be found on all TIM or cargo, but most modern manufacturing and transportation facilities will have at least some markings similar to these. Figure F-5 contains examples.

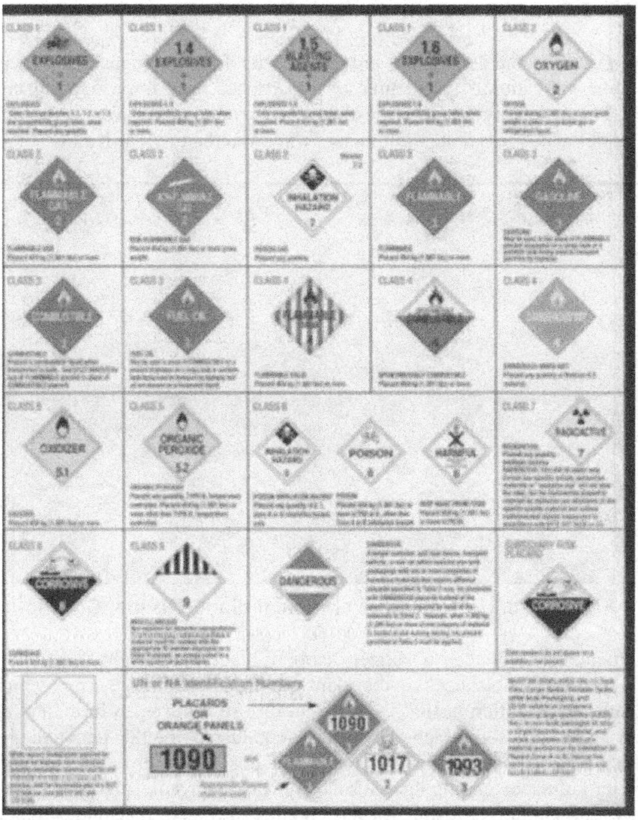

Figure F-5. Example of standard hazardous material markings.

F-10. THE RELEASE OF HAZARDOUS CHEMICALS DURING URBAN COMBAT

The fighting in Croatia during the breakup of the former Yugoslavian Republic has provided some indication of the degree of threat faced by a force fighting in a modern urban area. This is best typified by the situation surrounding attacks on the Petrochemia chemical plant located less than 1 kilometer away from the town of Kutina (Figure F-6).

Figure F-6. Petrochemia chemical plant, Kutina, Croatia.

a. Petrochemia produced fertilizer, carbon black, and light fraction petroleum products. Hazardous substances at the plant included ammonia; sulfur, which poses a hydrogen sulfide inhalation hazard in the event of a fire; nitric, sulfuric, and phosphoric acids; heavy oil; and formaldehyde. Studies conducted by the Croatian government indicated that a massive fire at Petrochemia would pose a danger to public health across a 100-kilometer radius, extending into Bosnia, Hungary, Slovenia, and Italy. The plant was attacked by Serbian forces using rockets, bombs, artillery, machine gun tracers, and mortars on six occasions during 1993 to 1995. During a missile attack in 1995, Petrochemia was hit 32 times with rockets in the area where anhydrous ammonia, sulfur, heavy oil, and other chemicals were stored (Figure F-7).

b. The Croatian Ministry of Defense took extensive actions to mitigate and prevent public health hazards from attacks on the plant. These included:

- Installing area alarm signals.
- Organizing special fire brigades and hazardous materials response units.
- Conducting mass casualty training exercises.
- Stationing a Croatian army field decontamination unit near the plant.
- Creating special helicopter fire suppression and casualty evacuation units.
- Preparing local emergency rooms to treat contaminated patients.
- Conducting special training for police to control rail and road traffic.

In spite of these preparations, the manager of the Petrochemia plant admitted that, in the event of repeated attacks, "...uncontrolled negative consequences may develop."

c. Other chemical plants were also attacked during the war in Croatia.

(1) Serbs using rockets containing cluster bombs attacked a natural gas refinery in eastern Slavonia, located only 1 kilometer from the center of a city, and which produced ethane, propane, and butane. Fortunately, the government had reduced the amounts of dangerous chemicals on hand and had curtailed plant operations as the Serb forces advanced.

Figure F-7. Petrochemia chemical plant under attack by Serbian forces.

(2) An attack on a chemical plant near the town of Jovan resulted in an instantaneous release of 72 tons of anhydrous ammonia. Fortunately, that plant was located 30 kilometers from the town and local public safety officials had time to evacuate its 32,000 residents.

(3) Mortar attacks were launched on the Herbos pesticide plant, located in Croatia's industrial center at Sisak. Fortunately, these attacks did not hit vital process control and chemical storage areas.

(4) During the war, Serbian forces attacked large fuel storage tanks along the highway from Belgrade to the outskirts of Zagreb and started large fires at Osijek, Sisak, and Karlovak. The refinery, at Sisak, which produced liquefied petroleum gases (LPG), fuels, petroleum coke, and solvents, was hit hard by 400 to 500 Serbian artillery rounds; 38 storage tanks were destroyed. Refineries are usually designed so that two fires can be controlled and suppressed at one time, but at this refinery firefighters had to fight as many as five major fires simultaneously (Figure F-8, page F-12).

d. Croatian defense and public safety officials modeled potential releases that might result from military attacks to determine the potential health effects. They determined that if isopropylamine were released from a standard 150-cubic-meter tank, it would result in concentrations immediately dangerous to life and health up to 145 meters downwind. Similar analyses for ethylamine reflected dangerous concentrations 1,760 meters downwind.

Figure F-8. Specialized equipment used in an oil well fire in Croatia.

e. These historical examples provide some lessons for leaders and commanders.

(1) Infantry units conducting UO near potentially hazardous sites must be aware of the potential TIM and be prepared to initiate appropriate MOPP levels or evacuate the area based on the tactical situation.

(2) TIM sites are vulnerable targets and may be used as an asymmetric threat against friendly forces.

(3) Medical personnel need to be briefed on the potential hazards and carry medical supplies appropriate to the situation.

(4) If stability and support operations will be conducted near a possible TIM site, contingency plans need to be made for the evacuation of noncombatants in the event of a TIM release.

(5) Units should avoid putting logistical and medical assets near a TIM site, if possible.

(6) UO can be seriously affected by the release of TIM and may result in mass CASEVAC.

(7) Unit chemical officers can provide assistance in analyzing the possible effects of TIM and provide advice on hazards and methods to mitigate effects.

F-11. MINIMUM DOWNWIND HAZARD DISTANCES

For planning, Table F-2 shows minimum downwind hazard distances (day and night) from chemical production or storage sites for selected toxic industrial chemicals. These are the distances a lethal exposure level could reach if a massive release occurred. Note that toxic industrial chemicals normally have a greater downwind hazard distance at night.

CHEMICAL	QUANTITY	DISTANCE	
		DAY	NIGHT
Chlorine	Up to 100 Tons	2 km	5 km
Phosgene	Up to 50 Tons	2 km	5 km
Ammonia	Up to 500 Tons	2 km	5 km
Hydrogen Cyanide	Up to 50 Tons		
(Hot Weather)		2 km	5 km
(Cold Weather)		1 km	2.5 km
Hydrogen Sulfide	Up to 50 Tons	2 km	5 km
Methyl Isocyanate	Up to 50 Tons	2 km	5 km
Hydrogen Flouride	Up to 100 Tons	1 km	2.5 km
Sulfur Trioxide	Up to 50 Tons	1 km	2.5 km
Nitrogen Tetroxide	Up to 50 Tons	1 km	2.5 km
Hydrogen Chloride	Up to 100 Tons	1 km	2.5 km
Ammonia	Up to 50 Tons	1 km	2.5 km
Bromide	Up to 50 Tons	1 km	2.5 km
Sulfur Dioxide	Up to 50 Tons	1 km	2.5 km

Table F-2. Minimum downwind distances.

9-12. OTHER THREATS FROM TOXIC INDUSTRIAL MATERIALS

TIM present many other threats to include the following.

a. Water contamination hazards were present in Croatia during the war. Transformers were destroyed at power stations located along the Dalmatian coast. Ground water is highly susceptible to contamination. In certain areas, contaminants can move several hundred feet or even miles per day, posing public health hazards at great distances from the source of contamination. Croatian hydrogeologists found concentrations of polychlorinated biphenyls (PCBs), flame-retardants, and explosives in several areas.

b. Lightning rods in Croatia contain sources of radioactive cobalt that are much stronger than those used by other countries. Attacks on industries at Osijek in Eastern Slavonia released radiation from these rods. Many of the lightning rods were found because snow had melted around them.

c. Radioactive materials used in medical treatment or research present a significant threat of accidental release during urban combat operations. Cleanup after release is an extensive task. For example, releases of radioactive cesium from an abandoned piece of medical equipment caused an environmental disaster in the Brazilian town of Goiania.

d. Liquid propane (LP) and other fuel storage tanks present a serious threat to forces engaged in UO. Some urban industrial areas have large concentrations of LP gas and other fuel storage tanks. Fires in these tanks are almost impossible to extinguish without special equipment and must be allowed to burn out. If the fire superheats the tanks, boiling liquid expanding vapor explosions (BLEVE) can occur. BLEVE can throw huge pieces of the tank well over 100 meters (Figure F-9, page F-14).

Figure F-9. 18,000-gallon liquid propane tank in industrial area.

APPENDIX G

INTELLIGENCE REQUIREMENTS CHECKLISTS FOR URBAN OPERATIONS

Commanders use priority intelligence requirements (PIR) and essential elements of information (EEI) to facilitate rapid crisis planning. Intelligence staffs at all levels must be prepared to react to crisis situations and provide commanders with accurate, timely, and detailed intelligence and information in support of urban operations. (See FM 34-130 for more detailed information.)

Section I. CULTURAL INTELLIGENCE REQUIREMENTS

Accommodating the social fabric of a city is potentially the most influential factor in the conduct of urban operations (UO). Unfortunately, this accommodation is often the most neglected factor. Social factors have greater impact in UO than in any other environment. The density of civilians and the constant interaction between them and US forces greatly increase the importance of social considerations. The fastest way to damage the legitimacy of an operation is to ignore or violate the social mores or precepts of a particular population.

G-1. CULTURAL NORMS

The interaction of different cultures during UO demands much greater recognition than in other environments. This greater need for understanding comes from the increased interaction with the civilian populace. Every culture has a set of norms and values, which could involve such diverse areas as food, sleep patterns, casual and close relationships, manners, and cleanliness. Understanding these differences is only the start in preparing for cultural differences.

G-2. RELIGIOUS BELIEFS

Religious beliefs and practices are among the most important and least understood aspects of the cultures of other peoples. In many parts of the world, religious norms are a matter of life and death. In many religious wars, it is not uncommon to find suicidal acts in the name of a particular god. In those situations, religious beliefs may be considered more important than life itself. Failure to recognize and respect religious beliefs will quickly erode the legitimacy of the US mission.

G-3. LOCAL GOVERNMENT

In many US military missions, forces respond in support of a given political entity. Consequently, promoting the supported government aids in legitimizing the military mission. While it is important to articulate US contributions, it is also important to advocate the accomplishments of the native government for long-term success. If legitimacy is not established for the native government, stability may be only temporary. US military planners must identify key governmental officials and integrate them as appropriate early in the operation. There are two benefits to this early integration. First, they can provide valuable information needed for successful completion of the operations to include city infrastructure, locations of enemy concentration, and a common picture of cultural norms. Second, close cooperation with government officials provides the host government the catalyst to attain legitimacy with the populace for involvement of US forces.

a. In some developing countries, governments are characterized by nepotism, favor-trading, subtle sabotage, and indifference. Corruption is sometimes pervasive and institutionalized as a practical way to manage excess demand for city services. The power of officials is primarily based on family connections, personal power base, and age, and only after that on education, training, and competence.

b. A local government's breakdown from its previous level of effectiveness quickly exacerbates problems of public health and mobility. Attempts to get the local-level bureaucracy to function along US lines produce further breakdown or passive indifference. Any unintentional or intentional threat to the privileges of ranking local officials or to members of their families will be stubbornly resisted. Avoiding such threats and assessing the importance of particular officials requires knowledge of family ties.

G-4. LOCAL POPULATION

US military planners must also recognize that the urban populace will behave as they see their own self-interest. They are keenly aware of four sets of interests at work: those of the US forces; those of hostile elements; those of local *opportunists*; and those of the general population. They size up these interests constantly in order to ascertain their own stakes, risks, and payoffs.

G-5. REFUGEES

Another significant cultural problem is the presence of refugees within an urban area. Rural immigrants, combined with city residents displaced by urban conflict, can create a significant strategic problem. Noncombatants and refugees without hostile intent can overwhelm the force of an advancing platoon. Additionally, there may be enemy troops, criminal gangs, vigilantes, paramilitary factions, and factions within those factions hiding in the waves of the displaced. The threat knows that it is impossible to tell friend from foe from disinterested. Local combat situations can change with bewildering speed, as the supposed innocent becomes the opposition within close quarters and indefensible positions. Chechen rebels and the Hezbollah effectively used the cover of refugees to attack occupying forces and counted on heavy civilian casualties in the counterattack to gain support with the native population. The goal is to place incalculable stresses on the individual soldier in order to break down discipline and operational integrity. From Belfast to Lebanon, the constant pressure of differentiating friend from foe taxed and sometimes undermined rules of engagement, and in some cases, entire missions.

G-5. INTELLIGENCE REQUIREMENTS

Tables G-1 and G-2 display intelligence requirement checklists for the population and urban social structure.

What is the urban population? • Total. • City core. • Trend (increasing/decreasing for each area).
What is the age structure? • Males and females under age 15 or over age 65. • Males age 15-49 (potential military age). • Youth bulge (percent under 15).
Is there a refugee/displaced people situation? • Where are the refugees originally from? • What is the size of the refugee population? • What is the size of the original population? • Why did they come here? • What is the relationship between the refugees and the city populace? • Do they support each other? • Are they hostile towards each other? • Is there a segment of the population that fled the city? • Why did they leave? • Where did they go? • Under what circumstances will they return?
What is the population of the rural areas surrounding the urban area? • Size. • Location and distance from urban area. • Do they travel to the urban area? • Frequency of travel. • What is their relationship with the urban population? • Role in conflict.

Table G-1. Population intelligence requirements checklist.

What is the ethnic structure? • Percent of population by ethnic group. • Is one ethnic group dominant? If so, why? • Bias of one group towards another. • Physical boundaries of influence. • Boundary overlaps. • Role in conflict. • Key personnel and location.
What is the racial structure? • Percent of population. • Is one racial group dominant? If so, why? • Bias of one race towards another. • Physical boundaries of influence. • Boundary overlaps. • Role in conflict. • Key personnel and location.

Table G-2. Urban social structure intelligence requirements checklist.

What is the religious structure? • Beliefs. • Percent of population. • Importance in society. • Leaders. • Practices. • Bias of one religion towards another. • Is one religious group dominant? If so, why? • Physical boundaries of influence. • Boundary overlaps. • Role in conflict. • Key personnel and location.
What is the structure of national origin? • Percent of population. • Bias of one group towards another. • Physical boundaries of influence. • Boundary overlaps. • Role in conflict. • Key personnel and location.
What is the tribal/clan structure? • Basis of affiliation. • Percent of population. • Is one tribe/clan dominant? If so, why? • Bias of one tribe/clan towards another. • Physical boundaries of influence. • Boundary overlaps. • Role in conflict. • Key personnel and location.
Is there a terrorist structure? • Motivation. • Activities confined to urban area. • National connections. • International connections, external support. • State sponsored. • Popular support. • Connections to political parties, militias, terrorist groups. • Strength: active fighters, supporters. • Membership profile: age, sex, education, political ideology, economic background. • Type of support: arms, explosives, technical expertise, money, intelligence, safe haven, cache. • Modus operandi. • Physical boundaries/areas of operation. • Weapons and equipment. • Role in conflict. • Key personnel and location.

Table G-2. Urban social structure intelligence requirements checklist (continued).

Is there a gang structure?
• Basis of affiliation. • Criminal behavior. • Are activities confined to urban areas? • Is there more then one gang? If so, do they compete or cooperate? • Is competition violent? • Support of local populace. • Physical boundaries of influence. • Boundary overlaps. • Strength. • Weapons and equipment. • Role in conflict. • Key personnel and location.
Is there an organized crime structure? • Type of crimes. • Are activities confined to urban areas? • National connections. • International connections. • Is there more than one crime organization? If so, do they compete or cooperate? • Is the competition violent? • Do city, military, terrorist or other officials/members aid or otherwise support an organized crime organization. • Support of local populace. • Physical boundaries of influence. • Boundary overlaps. • Strength. • Weapons and equipment. • Role in conflict. • Key personnel and location.
What is the structure of the economic classes? • Distribution of wealth. • Per capita income. • Number per family employed. • Percentage of population by economic sector (industry, service, and so forth). • Percentage of population living in poverty. • Percentage of population dependent on economic aid. • Type of aid. • Unemployment rate. • Under employment rate. • Are their trade unions? • Is membership in union's compulsory? • How do unions influence the urban area (politically, economically, and so forth)? • Union key personnel and location.

Table G-2. Urban social structure intelligence requirements checklist (continued).

What is the structure of the economic classes? (Continued)
• Is there a management class? • How does the management class influence the urban area? • Management class key personnel and location. • Is there an urban economic elite class? • How does the economic elite influence the urban area? • Does the economic elite control the urban area? • Economic elite key personnel and location. • Physical boundaries based on economics. • Boundary overlaps. • Economic status role in conflict.
What is the political structure?
• Factions within urban area based on political beliefs. • Structured political parties. • Percent of population belonging to political faction or party. • Is one faction or party dominant? If so, how? • Are any political factions, parties or beliefs banned or illegal? • Suffrage. • Election turnout. • Election fraud. • Other election irregularities. • Physical boundaries based on political beliefs. • Boundary overlaps. • Key personnel and location. • Role in conflict.
What is the structure of the city government?
• Elected or appointed? If appointed, by whom? • Executive branch. • Judicial branch. • Key personnel and location. • Administrative divisions. • Legislative branch. • Physical boundaries of administrative divisions.
What social actions are taboo or insulting to the urban population?
• Verbal. • Nonverbal (body gestures/manner of dress).
What languages are spoken in the urban area?
• Dialects. • Common languages for business transactions.
Are there any overlaps and or splits among the social divisions?
• Do union members belong to one or a few religious or racial groups? • Are there ideological divisions within a profession?
Are the composite groups based on political behavior?
• Those who actively or passively support the government, an economic or social urban elite organization, an insurgency, a criminal organization, a terrorist organization, or are they neutral? • Component. • Composite strength.

Table G-2. Urban social structure intelligence requirements checklist (continued).

What are the active or potential issues motivating the political behavior of each group/subgroup? • Economic benefits. • Social prestige. • Political participation. • Perception of relative deprivation. • Population growth or decline. • Average distribution. • Changes in location of each group.
What is the history of conflict in the state? • Internal and external. • Recent conflicts.
Does the city, other local, or national government control the city? If not, who does? • Key personnel and location. • Reason for control (economic elite, military, criminal, other).
What is the popular support of the city, other local and national governments? • Total. • By social structure group. • By overlap in social structure groups.
What is the popular support of US/coalition government? • Total. • By social structure group. • By overlap in social structure groups. • If coalition, posture by each country.
What is the local posture toward US/coalition government? • Active or passive. • Violent or non-violent. • If coalition, posture by country. • What would change this posture? • By total, group, and overlap in social structures.
What posture should US/coalition forces adopt in order to gain maximum local support? • Total. • By social structure group. • Identify contradictions: particular postures favorable to one group, unfavorable to another. • What forms of posture ought to be avoided?
What is the influence of the media? • By form: print, radio, television. • By origin: local, national, international. • What media is controlled or influenced by a social structure group? • Describe any media bias.
What propaganda is currently being disseminated? • Type. • How is it disseminated? • How effective is it? • Why is it effective or not effective?

Table G-2. Urban social structure intelligence requirements checklist (continued).

What is the relationship between the urban and rural areas? • Social structure ties, bias, or conflict. • Dependency of one to another for critical goods or services.
Describe the urban social make-up, to include: • City core. • Commercial ribbon. • Core periphery. • Residential sprawl. • Outlying industrial areas. • Outlying high-rise areas.
What is the urban area's importance? • Internationally. • Nationally. • Religiously. • Militarily. • Economically. • Socially. • Historically.
Are there any historic structures? • Religious. • Historic. • Cultural. • Other. • Are any significant to one portion of populace but offensive to another? • Location.
What types of identification are required or in use in the urban area? • National. • Local. • Government service. • Military. • Driver's license. • Professional/trade/union. • Voters. • Other. • Photo IDs. • Who issues/What are the requirements? • Where are they issued? • Are forgeries in use?
What crisis management procedures exist within the urban area? • Who are the decision-makers? • Is there a civil alert system? • How is an alert communicated? • Has it been tested? • Is it effective? • Is there a civil evacuation plan? • Has it been tested? • Is it effective?

Table G-2. Urban social structure intelligence requirements checklist (continued).

Section II. CITY INFRASTRUCTURE AND SERVICES

A city is more than a change in terrain on which to apply conventional tactics. A city is a system of systems that supports the total functioning of an urban area—no part functions independent of the others. The systems within the city include its physical composition, supporting utilities, and social factors. Each component impacts on the population, the normal operation of the city, and potentially the long-term success of military operations conducted there. Military planners must understand the functions and interrelationships of these systems in order to achieve success. (Tables G-3 through G-24, pages G-12 through G-34, provide intelligence requirements checklists for urban operations.)

G-6. CITY INFRASTRUCTURE SYSTEMS

Urban infrastructure is the city's foundation. This infrastructure includes buildings, bridges, roads, airfields, ports, subways, and similar physical structures. These structures provide the base on which the rest of the city is developed.

a. **Transportation**. The transportation network of a city is an integral part of its operation. This network includes roads, railways, subways, and ports (air and sea). Transportation facilitates the inter- and intra-movement of material and personnel that form the lifeblood of the city.

(1) Control of these transportation nodes may be important for both a given military operation and the normal functioning of the city. Supplies, which travel through this transportation system, could include food, medicine, heating oil and gas, and military supplies such as ammunition and spare parts. Personnel who are moved along this transportation network could include people with various skills and intent such as doctors, government officials, repairmen or military reinforcements. US military forces may have to limit the transit of enemy supplies and reinforcements while facilitating the transport of critical civilian supplies and personnel.

(2) Operationally, securing air and seaports may be imperative for follow-on forces and supplies, but there are many possible implications of securing all the transportation nodes and stopping all inter- and intra-city movement. While the US mission may be immediately facilitated, there are critical needs of the noncombatant populace that would go unmet. Although it might be attractive to isolate the city, military planners must be aware of the noncombatants' needs for medical personnel and supplies, heating supplies, food, transit to work and school, and all the other items that minimize hardship and promote normalcy within the city. Reducing this hardship contributes to gaining the mission's legitimacy.

(3) Most developing country urban areas have two transportation systems: formal and paratransit. Formal systems are characterized by large organizations, bureaucracy, imported technology, scheduled services, fixed fares/rates, and limited employment opportunities for the urban populace. Paratransit is characterized by great decentralization; low barriers to entry, family and individual entrepreneur organization, adapted technology, negotiated prices, and flexible routes, destinations, and times of service. Paratransit tends to be labor absorbing and covers a much greater area of the city than the formal system. Therefore, paratransit is more likely to function through turbulence and conflict. Paratransit often includes a waterborne element. Together with elements of the formal system, paratransit plays a key role in the movement of goods and people into, out of, and within the urban area. This key role also includes the city's food

supply zone that may extend up to 100 kilometers from the urban center. Understanding both systems will help US Forces monitor or control movement in a developing country city.

b. **Physical Composition**. The physical composition of the city provides the fundamental structure in which the city community conducts normal activities. Physical features of the city have more than military significance. In addition to housing an enemy, the buildings of the city also accommodate the businesses, government, noncombatants, schools, and similar functions critical to the normal conditions of the city. Military planners and personnel must restrain the urge to rubble structures, even when they identify enemy within. There are both legal and moral reasons for this restraint. The Geneva Convention states "any destruction by the occupying power of real or personal property belonging individually or collectively to private persons, or to the state, or to other public authorities, or to social or cooperative organizations, is prohibited, except where such destruction is rendered absolutely necessary by military operations." This restriction also has a moral consideration. While the enemy may be inside the building, so too may be innocent civilians. Therefore, the tactical commander must carefully consider a full range of implication before leveling a building housing the enemy. Military success may well be measured by how a mission is accomplished while minimizing the destruction. Minimizing collateral damage reduces the hardship within the city and leads to a faster return to normalcy during the post-hostilities phase.

c. **Utilities**. Urban utilities include communications, natural gas, electricity, water, and health services.

(1) ***Communications***. Communications is a utility that impacts the military mission and the civilian populace. Besides face-to-face conversation, the communications system controls the information flow within the city. Telephones (wire and cellular), radio, television, newspapers, and the Internet provide a community information and thus influences individual perspectives. Information can relieve much of the populace's tension and at the same time provide essential intelligence for friendly forces. The management of information to enhance legitimacy may occur in three distinct ways.

(a) First, communication with the local populace serves to enhance the legitimacy of the mission to the population, which includes factors such as the intent of the mission, locations of services available, or the manner in which the population can assist the mission. If the populace does not understand the mission, false expectations may be created that US forces may not be able to meet. The use of civil affairs personnel is the most effective way to communicate with the civilian populace on these matters.

(b) A second function of information management is to enhance the legitimacy of the US involvement in the eyes of the international community. This function is done through the media and may include a candid assessment of the current operation, changes in the mission, or any other newsworthy story. The public affairs office (PAO) can use the communication system to link with the media and subsequently the international community.

(c) Third, information management affords the US commander the ability to carry out his psychological operations (PSYOP) plan. Used as a force multiplier, PSYOP has the potential to convince enemy combatants that further resistance would be futile. Controlling the city's communication facilitates this and, at the same time, limits the enemy's use for their operations.

(2) ***Natural Gas***. Natural gas provides the basic heating for the population in many parts of the world. The natural gas industry consists of three components: production, transmission, and local distribution. The gas companies must transport the gas to central areas and then store it in numerous facilities before pumping it to homes and businesses for use. From a tactical and operational perspective, control of this system provides minimal advantage to friendly forces, but protecting its destruction or damage would prevent unnecessary hardship to the civilian population.

(3) ***Electricity***. Electricity is critical to the normal state within a city. Power companies in a community provide a basic service which allows the population to cook, communicate, heat water, and see at night.

(a) There are three stages to this process: generation, transmission, and distribution. Generation is the process of producing electricity. Transmission connects power systems to the market areas. Distribution is the process of delivering the electricity to the consumer. A key factor is that electricity cannot be stored in any sizable amount and that damage to any portion of this utility causes an immediate impact on the population.

(b) While electricity facilitates many functions of normality, there are also military considerations. For example, the combination of equipment and training affords US forces a marked advantage over most adversaries during night operations. Consequently, US forces may want to control the electric system so that they may maintain this advantage during certain time periods. Likewise, the commander may want to deny enemy access to services provided through electricity. Rather than destroying a power generation capability, forces may gain an advantage by selectively turning off power for a specified time, retaining the ability to return to normal operation at a moment's notice.

(4) ***Water***. Water is essential to many basic human needs. Water companies provide the population clean water to drink, cook, bathe, and wash. Water production and distribution are also basic processes. The water companies refine the water, pump it to storage facilities, and finally pump it to the consumer. The tactical implications of controlling this system are similar to that of natural gas. US forces may gain no marked tactical advantage by controlling this system, but its protection minimizes the population's hardship and thus contributes to overall mission success.

(5) ***Health Services***. Health services are significantly lacking in many countries. Compounding this problem is the presence of deadly parasites and diseases that are abundant in many areas. Disasters (natural or manmade) can significantly worsen the already poor health services condition. Contaminated water, the lack of fuel for heating and sterilization, and increased injuries could easily overwhelm the medical infrastructure of a city. Support to an existing medical system may enhance the US mission, as well as foster its legitimacy.

Police or military units with police authority/mission. • Station locations. • Substation locations. • Headquarters location. • Operating borders. • Jurisdiction borders. • Number of police—active and reserve. • Number of administrative personnel. • Types of units—uniformed, plainclothes, special. • Weapons and equipment. • Communications equipment, procedures, and frequencies/phone numbers. • Training. • Proficiency. • Capabilities. • Limitations.
Fire. • Station locations. • Substation locations. • Operating borders. • Number of firefighters—active and reserve. • Full-time or volunteer. • Number and types of vehicles (pump, ladder, and so forth). • Other fire fighting equipment. • Location and condition of fire hydrants. • Communications equipment, procedures, and frequencies/phone numbers. • Training. • Proficiency. • Capabilities • Limitations. • Airport/airfield fire fighting capabilities. • Port fire-fighting capabilities. • Fire fighting units outside urban area that assist in city emergencies.
Medical. • Hospital/Medical clinic locations. • Number of trained medical personnel. • Doctors per population. • Number of emergency rooms and equipment capabilities. • Number of operating rooms and equipment capabilities. • Number of hospital beds. • Capabilities. • Limitations.

Table G-3. Critical urban services checklist.

Water. • Types and locations of water sources (intake crib on lake, piped or trucked in). • Location and condition of filtration plants. • Location and condition of pumping stations. • Location and condition of distribution shafts and tunnels. • Water quality. • Other water distribution methods (trucks, bulls, bottles), location of distribution points.
Graves/casualty disposal. • Morgues. • Cemeteries. • Capacity.
Trash disposal. • Collection. • Transport. • Incinerators. • Dumps/landfills. • Dumps/landfills outside urban area that are used by the city. • Toxic waste. • Medical waste.
Food supply. • Location of markets. • Location of storage facilities.
Schools and churches. • Location. • Types.
Other structures of political or social significance. • Locations. • Types. • Significance.

Table G-3. Critical urban services checklist (continued).

What are the capabilities concerning fuel resources? • Storage. • Transportation. • Emergency supply. • Access. • Distribution. • Production. • Types.
Where are the material-producing factories in the urban area? • Location. • Material produced. • Size of factory. • Number of employees. • Sector of production.

Table G-4. Resources and material production checklist.

Are any materials produced used for a military purpose? • Type. • Purpose. • Location of military production. • Local use or export.
What makes the production significant to the urban area?
How can materials be used in conflict?
Where are the machine shops? • Location. • Owners. • Capacity. • Normal activity.
Where are the foundries? • Location. • Owner. • Capacity. • Normal activity. • Condition.
What is the fire protection? • Fire boats (type, power, location, number, pumping capacity, condition). • Shore fire-fighting equipment (location, type, number, and condition). • Water supply (source, adequacy, and distribution system). • Equipment available. • Dredging requirements. • Rehabilitation requirements.
Where is the water supply? • Source. • Potability. • Distribution. • Capacity. • Adequacy. • Storage (location, type, capacity).
What is the source of electricity? • Current characteristics. • Substations. • Labor resources.

Table G-4. Resources and material production checklist (continued).

What is the location of the airfield?
Description. • Construction. • Security. • Area. • Access.
What is the status of the airfield?
Who is in control of the airfield?

Table G-5. Airfields checklist.

What type of airfield is it? • Civilian. • Military. • Joint.
What is the principle use?
Where are the storage areas? • Jet fuel. • Aviation gas. • Jet oil. • Aviation oil. • Lubricants, manifolds, and filters. • Pipelines. • Above and below ground storage and capacity. • Pumps. • Fuel trucks.
Where are the maintenance facilities? • Size. • Capacity.
Where is the administration building located?
Where is the electrical power source? • Location of master switch. • Current characteristics. • Emergency power available (type, location, generating capacity and delay). • Location of transformers.
Where is the natural gas source?
Where are the medical facilities? • Type and location. • Capability and characteristics. • Prevailing weather conditions.
Describe any terrain key to both the airfield and urban area.
Are there any vertical obstructions, not contained in published data, that pose a threat to helicopters or other aircraft at/under 50 feet above ground level (AGL)?
Are there any obstacles to flight within 5 to 10 kilometers of the airfield?
What radar equipment and type are associated with the airfield?
Do we have access to blueprints of the airfield and its facilities?
Where are the emergency facilities located? • Type. • Equipment. • Capabilities. • Limitations.
What kind of and how many ground security forces are routinely present? • Civilian. • Military.

Table G-5. Airfields checklist (continued).

What is/are the locations of LZs?
What are the characteristics of the LZ? • Site. • Dimensions. • Landing points • Capacity by type. • Surface material. • Soil trafficability. • Obstacles (existing, reinforcing). • Slope (direction, degree). • Lighting conditions.
Identify any buildings with rooftop HLZs. • Location. • Dimensions. • Landing points. • Capacity by type. • Obstacles.
What are the distance and direction from the designated objectives?
What/where is the anti-air threat? • Small arms. • Shoulder-fired antitank grenade launchers (RPGs). • MANPADs. • SAMs (mobile/fixed). • Anti-helicopter mines. • Anti-aircraft artillery (AAA) (mobile/fixed).

Table G-6. Helicopter landing zones checklist.

What are the classifications of routes in the urban area? • Highways. • Streets. • Alleys. • Trails. • Bike paths. • Pedestrian paths.
What is the location?
Is the route trafficable?
What is the length of the route?
What are the roadway characteristics? • Width. • Surface material. • Maximum wheel load.
Where are the side hill cuts?
Where are the through cuts?

Table G-7. Roadways checklist.

Where are the bridges? • Total number of segments. • Number greater than 18 meters. • Load bearing capacity. • Width. • Location. • Vehicle capacity. • Facilities. • Surface material.
Where are the fords?
Where are the ferries?
Where are the tunnels, galleries and snowsheds? • Dimensions. • Capacity.
Where are the underpasses/vulnerable points?
Where are the areas subject to blockade?
Where are the checkpoints?
Where are the obstacles? • Type. • Effort required to remove.
What is the civilian use? • Importance to economy. • Possible rerouting.
What are the street patterns? • Radial. • Rectangular. • Concentric. • Contour conforming. • Medieval irregular. • Planned irregular. • Numbering and mailing system. • Traffic control system.
Do maps correspond with the routes?
Are routes compartmentalized?

Table G-7. Roadways checklist (continued).

Where are the freight handling facilities?
Where are the repair shops/locomotive terminals?
Where are the fuel facilities? • Location. • Type of fuel. • Storage and capacity. • Quantity of fuel on-hand. • Method of loading.
Where is the rolling stock? • Types of trains. • Types of railcars and carrying capacity.
Describe railheads. • Supply transfer points. • Characteristics (spurs, sidings, piles of materials, tract trucks, wagons, tents, huts, guards and supply handlers).

Table G-8. Railways checklist.

Location?
What obstacle is crossed?
What is the route designation?
What are the military load classifications?
What is the maximum load capacity?
What is the condition of the bridge?
What extent of effort would be required to repair it?
Where and what condition are the bypasses?
What is the condition of the approaches?
What is the condition of the banks?
What is the overall bridge length?
What is the structure type?
What is the military nomenclature?
What is the number of spans? • Type of control (mechanical, electronic). • Time required to move.
What is the condition of the abutments/piers?
What is the width of the roadway?
What is the under-bridge clearance?
Where are the walkways?
What is the condition of the spans?
What is the condition of the suspension system?
What are the characteristics of any moveable spans?
What is the condition of the intermediate supports?
What are the characteristics of a floating bridge?
What are the safety and security features?
Where are the traffic control markings?
What is the effect of weather and climate?

Table G-9. Bridges checklist.

What type of bridge is it?
Is the bridge significant in linking the urban area together?
If the use of the bridge is denied, what is the result?
Are there any alternate routes?
What is the horizontal clearance?
What is the vertical clearance?

Table G-9. Bridges checklist (continued).

Are maps available of the subway system? • Are the maps of the actual tracks or an altered rendition of lines and stations? • Are blueprints available of stations, maintenance areas, and so forth?
Where are the entrances and exits of the system?
What is the schedule of trains?
How can the system be accessed? • Pedestrian traffic routes. • Stairs, escalators, elevators. • Security obstructions (gates, turnstiles). • Maintenance and utility tunnels. • Drains. • Transition areas between above/below ground portions of the lines.
Describe the tunnels.
Describe the rails.
Describe the trains. • Type. • Speed. • Size and capacity of cars.
What is the soil, rock or compound that the tunnels are cut through?
How would demolition in the subway affect the urban area? (surface structure, subterranean foundation).
What is the composition of the subway system?
What is the electrical source?
How is the subway operated and managed?
What is the emergency response?
What is the condition of the subway system?
What is the decibel level around the trains? At what distance is noise a factor to ground operations?
Type of construction. • Typical subway near surface with flat roof and I-beams for roof and sides, supported between tracks with steel bulb-angle columns. • Flat roof typical subway of reinforced concrete construction supported between tracks by steel bulb-angle columns, used for short distances. • Concrete lined tunnel of open cutwork, and rock tunnel work. • Elevated road on steel viaduct. • Cast iron tubes used under water (river, lakes).

Table G-10. Subways checklist.

What is the location of crossover sections?
Where are the controls for the crossover sections?
What reinforcements are part of the construction to add support and waterproofing? (Example: hard burned brick laid in hot asphalt.)
Describe the layers of the inner subway construction.
Locate manholes.
What is the evacuation plan?
What is the construction of beams and shafts?
Where are any turns or corners?
Where are terminals located?
Where are the maintenance facilities located?
Describe the railbed (grade, subgrade, rail embedding).
Describe any subterranean structures.
What is the security of the subway?
Describe the ventilation system.
Is there a vagrant/squatter population residing in the subway system? • Approximate numbers. • Primary locations/concentrations.

Table G-10. Subways checklist (continued).

What is the location of the tunnel/passage?
What is the location of the portals on roof?
What does the tunnel/passage connect in the urban area?
If the use of the tunnel were denied, what would result? Where would redirection occur?
What is the length?
What is the type? • Maintenance access tunnel. • Electric grid/utility line. • Pedestrian passageway. • Sewer drainage systems and waterways. • If sewer, is it a sanitary, storm or combination? • Natural underground passage.
What are the features? • Shape. • Width at narrowest point. • Width at widest point. • Maximum height. • Minimum height. • Rise of arch.
Where are the horizontal and vertical constructions? (Type, least clearance, location from portal).

Table G-11. Other subterranean features checklist.

What are the features of the sewer/drainage systems? • Cross-section, dimensional. • Sides. • Bottom. • Normal depth. • Normal current velocity.
What is the alignment? • Horizontal (position, curve radius, curve location). • Vertical (grade percent, length, location).
How many man-ways are there? • Dimensions. • Spacing.
What is the obstacle tunneled?
What are the features of the portals? • Design. • Materials. • Dimensions.
What are the features of the lining materials? • Type. • Thickness. • Condition. • Point of change.
Where are the shoring and bracing? • Location of spacing. • Design. • Materials. • Dimensions. • Arrangements.
What is the geological data?
What is the overburden? (Material, depth).
Where are the demolition chambers? (Location, dimension).
Where is the ventilation? (Description, adequacy).
What is the drainage system? (Description, location, power source).
Where are the lighting facilities? (Type, location, power source).
What is the year of construction?
Where are the bypasses? • Location. • Condition. • Effort required to establish.
Where are the alternate routes?
What are the traffic control markings?
What are the surface features over tunnels?
What are the effects of climate and weather?
What are the special geophysical phenomena?
What is the susceptibilities to above ground demolitions?
Describe adjacent key terrain.

Table G-11. Other subterranean features checklist (continued).

Do buildings in the urban area have basements/subbasements/sub-subbasements? • Residential. • Government. • Military. • Commercial. • Industrial. • How deep/how many layers? • Access points/methods. • Stairs, ladders, ramps, elevators, and dumb waiters. • Hidden access points. • Above ground windows. • Wall, floor and ceiling construction. • Ceiling support. • Typical use (storage, living quarters, office, and so forth).

Table G-11. Other subterranean features checklist (continued).

What type of plant is it? • Conventional. • Nuclear.
What is the economic significance of the power plant?
What is the social significance of the power plant?
What is the fuel source? • Type. • Location. • Available reserve. • Usage rate.
Where are the generators? • Quantity. • Power output.
Where are the exciters? (Quantity).
Where are the turbines? (Quantity).
Where are the boilers? (Steam power plant).
Where are the compressors? (Gas power plant).
Where are the combustion chambers? (Gas power plant).
Where are the diesel engines? (Diesel power plant).
Where are the reactor containment structures? (Nuclear power plant).
Where are the water intakes and outlets? (Nuclear power plant).
Where are the power transformers?
Where are the switchyards?
What is the threat to the surrounding area should this facility be damaged? (Nuclear power plant). • Inhabitants. • Animal life. • Surrounding ecosystem • No effect on surrounding area.

Table G-12. Power plants checklist.

Is there a power generator backup?
What is the power grid that the power plant services and its location? • Power generation system. • Power transmission system. • Power distribution system. • Streetlights, time on/off.
What is the best component to disable the facility, and for how long will it be disabled?
Where are the power plant step-down transformers for plant power distribution located? • Control rooms. • Condensers. • Water pumps.
Who controls the power plant?
Who controls the maintenance?
Where are the substations located?
How are the substations powered?
What (sub) station is the primary power plant for the incident urban area?
Does the power plant control interconnected energy?
Where is the water storage center?
Where are the fuel storage centers? • Size/capacity. • Fuel type. • Usage rate.
Where are log booms located?
What kind of physical security measures are present? • Military/civilian personnel. • Fencing and or other physical security obstruction.
Describe any other components of the facility's layout.

Table G-12. Power plants (continued).

Where is the water control center located? • Size/output. • Method of purification. • Grid/pipe layout.
Are there any substations?
Does the incident urban area have water supply towers?
Does the incident urban area have wells available for use?
How available are local streams, rivers, lakes and ocean waters?
How effective is the incident country at leak detection?
Is the incident urban area currently experiencing any water shortage?
How is irrigation in the area? (Poor irrigation practices consume 90 percent of all water used in poor countries.)
Is incident country or urban area receiving aid to be spent on projects to increase potable water output?

Table G-13. Water systems checklist.

How reliable are the engineering and environmental testing?
Is water treatment privatized?
Is there any trenchless construction?
What type of pipes have been installed?
What are the health risks from raw water?
What type of security is present at the facility?
Do any rivers/canals run through the city? • Location. • Depth. • Width. • Tide. • Current. • Seasonal changes (for example, time frozen, flooded, and or dried out). • Dams.

Table G-13. Water systems checklist (continued).

How adequate is the sewage and waste disposal system?
What action (for example, combat) will lead to the breakdown of the waste disposal system?
Who/where are the points of contact to ensure the sewage and waste disposal system is maintained?
Where is the control center of the sewage system?
How is it operated?
What is the schedule of operation?
What is the security of the facility?

Table G-14. Sewage and waste disposal checklist.

Where are the fields located?
Who owns the fields?
What type of field is it?
What is the size of the field?
What is the status of the field?
What is the level of production? (Barrels, tons, cubic feet per time period).
What are the number and location of the producing wells?
What percent of national production does this field produce?
What is the product? • Type. • Characteristic.
Where are the reserves? • Proven. • Unproven.

Table G-15. Petroleum and natural gas facilities checklist.

What is the planned expansion? • Expected increase. • Date. • Method.
What is the transportation method? • Method. • Identification. • Destination.
What is the easiest method for rendering the field inoperable? For what period of time?
What is the percentage of infield processing?
What is the percentage of refinery processing?
What is the distance to the nearest refinery?
What are the transportation methods for product movement to the refinery?
What is the location of the processing plants?
What type of plant is it?
What percent of national refining/cracking does this plant represent?
When was it completed?
What is the general condition?
What is the rated production capacity?
Where is the refinery processing area?
Where are the atmospheric distillation towers?
Where are the crude oil feed furnaces? • Type. • Number feeding towers.
Where are the receiving facilities and crude oil storage?
Where are the catalyst vessels? • Type. • Function.
Where are the vapor vessels? • Type. • Function. • Location.
Where is the ancillary equipment? • Type. • Function. • Location.
What is the output? • Product. • Quantity. • Quality.
What is the power source of the plant?
What is the water source of the plant?
What type of transportation is used? • Raw materials in. (Identification, method, origin.) • Finished products out. (Identification, method, origin.)

Table G-15. Petroleum and natural gas facilities checklist (continued).

What is the planned expansion?
Where are the administration and maintenance buildings?
Where are the finished product storage areas? • Type. • Location. • Function.
Are refinery flow charts/diagrams available to include distribution facilities?
Where are the critical damage points?
What type of storage is there?
Who owns the storage facility?
What is the total storage capacity?
What percent of the national total is stored?
What is the general condition?
Where are the storage tanks? • Location. • Displacement (above/below ground). • Shape. • Top.
Where are the storage drums? • Manufacturing (location and capacity). • Cleaning and reclamation (location and capacity). • Filling facilities (location, equipment and capacity).
What is the transportation method? • Number. • Type. • Characteristics.
Where are the receiving and distribution facilities?
Where are the support facilities?
What is the easiest method for rendering the storage facility inoperable? Time?

Table G-15. Petroleum and natural gas facilities checklist (continued).

What are the locations of telecommunication and broadcasting assets (such as radio and television) and their primary use?
What are their economic and strategic/tactical importances?
What is the controlling government administration? • Agency involved. • Function. • Location. • Relationship to military.
Where is the transmitting/receiving equipment? • Location. • Type. • Frequency range.

Table G-16. Communications checklist.

Where are the control buildings?
What is the power source? • Type. • Voltage. • Transformer requirement.
What are the auxiliary power sources? • Type. • Voltage. • Duration of usage.
Where are the antenna fields? • Location. • Dimensions. • Feed system. • Use. • Polarization. • Mounting position. • Band width capacity. • Operating frequency. • Mounting structure. • Radiation pattern.
Where is the support area? • Location. • Housing area. • Equipment storage.
Where are the radio relay stations?
What is the military communications establishment? • Background. • Staff control. • Type. • Other facilities under military control. • Source. • Key personnel. • Training.
Where is the research and development? • Installation. • Projects (type, location and purpose). • Performance data. • Progress. • Foreign contributions.
Where are the intercept and direction-finding stations? • Location. • Line of communication. • Type shelter. • Antenna layout. • Rotating antenna (type, number, description and pattern).

Table G-16. Communications checklist (continued).

Where are the telephone and telegraph facilities? • Line routes. • Construction type. • Exchanges and offices. • Repair facilities. • Interconnection of system.
Where is the construction line? • Open-wire line (material, spacing, treatment and conductor). • Poles (materials, treatment, spacing and number of cross-arms). • Underground and underwater cables (method, type, and location).
Where are the submarine cable facilities? • Cable (type, location, length and description). • Terminal and operation equipment (type, dimension, power requirements, technical characteristics, and land line connections).
What are the technical details? • Equipment identification. • Frequency range. • Function. • Operator requirements. • Operator status. • Maximum reliable range. • Power source requirements. • Transmitter specifications. • Receiver specifications. • Antennas.
Is there any electromagnetic (EM) activity site areas that would hamper communications?
Is there an in-city trucking capability to include cross-country and cross-border trucking?
Who provides or produces the military and commercial communication equipment? • Security at the installation. • Power sources.
Where are the high voltage/em radiation hazard areas?
Are there schematics or diagrams of the facility? Where can they be obtained?
What is the connectivity to local/national communication systems?
What is the data transmission capability through satellite communications (SATCOM)?
What are the locations and capabilities of technicians to repair the facility?
Where are central locations for the control facilities?
How many television channels are there?
How many satellite earth stations are there? • Number. • Locations.
What computer networks are available? • Types of computers and locations. • Types of networks and control centers.

Table G-16. Communications checklist (continued).

Where are the medical facilities?
How many patients can the facility accommodate? • Number of beds. • Number of surgical operating rooms.
Who manages the facility?
Is the facility accessible to the public?
Are private medical facilities available?
How advanced is the medicine being practiced?
Is medicine practiced different from that in the US?
How adequate are the facilities supplies? Resupply?
How well staffed is the facility?
At what level will medical care become over whelmed should combat occur?
What is the predicted infectious disease risk and occurrence?
Where are the blood banks?
How much blood is stored?
Is blood adequately screened for infectious disease?
Where are the medical manufacturers?
What medical capable transport is available from the host nation?
What is the environmental health risk?
What is the acquired immune deficiency syndrome (aids) risk?

Table G-17. Medical facilities checklist.

Where are all the entrances to the building? (Include fire escapes.)
What size is the building? • Dimensions. • Stories.
What is the building used for?
Describe the occupants.
What type of security system or other protective devices are in use?
Are there security personnel? • Quantity. • Civilian/military.
What is the composition of the outer walls?
What type of insulation is used in the walls? Is it flammable?
Can small arms fire penetrate the external walls?
Can the utilities be regulated from outside the building?
What types of utilities are in the building?
Are there flammables or fuel stored in or around the building?
What are the dimensions of the inside rooms?
Where are the inside doorways?
Can small arms fire penetrate the inner walls? Are they reinforced?
Is there a basement or cellar?

Table G-18. Building construction checklist.

Is there an attic or roof crawl space?
Are there any telephones?
Are there other communications means? • Radios. • Internet. • Cellular phones.
Where are the stairwells?
Are there blueprints, engineer plans or wiring diagrams of the building?
Are there photographs of the building and surrounding area?
Identify personnel who are able to describe the building and it's interior.
Describe the pattern of construction.
Describe the roof.
Describe any subterranean construction. • Sewers. • Subways. • Basements/cellars. • Other utility tunnels (water, maintenance, electric, gas or telephone).
Describe any stadiums. • Location. • Area. • Capacity. • Routes in/out. • Height. • Protection from observation. • Observation posts. • Levels. • Type of construction.
What type of support structure is used in the buildings?
Are there any mouseholes?
NOTE: The design and construction of buildings within a certain urban area are influenced by numerous factors to include climate, materials available, function, and cultural development of the region. Critical factors to be considered in evaluating the construction of a building for attack, defense or destruction include: 1. The protective value offered by walls, roof, ceilings and doors. 2. The ease with which it may be demolished. 3. The availability of internal lines of communications and the effort required to breach exterior walls. 4. The time, effort, and material required to use the building. 5. Potential fire hazard.

Table G-18. Building construction checklist (continued).

Are civil defense plans in place and accessible?
What is the organization of the civil defense agency?
Who are the key personnel in the civil defense?
What facilities are available for civil defense? • Key industry sites. • Underground shelters. • Food storage sites. • Potable water sites. • Medical supplies. • Hospitals.
What are the warning and alert procedures?
What are the evacuation routes and capacities?

Table G-19. Civil defense checklist.

Has the enemy fought in an urban area?
What is the enemy experience in an urban environment? • Urban training. • Tactics, techniques and procedures.
What is the enemy center of gravity in urban warfare?
What is the enemy's critical vulnerability?
What is the enemy sniper capability? • Training. • Weapons. • Tactics, techniques, and procedures of employment.
Does the enemy possess mechanized assets? How do they intend to employ mech.?
Will they attack or defend?
How will they fight? (attrit, destroy, clear, and so forth).
Will the enemy use the local populace? Will the local populace cooperate?
Where will they employ crew-served weapons from?
Does the enemy have indirect fire support?
Are there mines and booby traps in the urban area?
How large of a force can the enemy deploy?
How will they reinforce?

Table G-20. Conventional military threat checklist.

Why does the organization exist?
What is their political ideology?
How many members are in the organization?
Are they a threat to US forces? To local government forces? To police?
Identification features. • Clothing. • Propaganda.

Table G-21. Paramilitary threat checklist.

What external groups are they involved with?
Who do they fight for?
What do they fight for?
Who are their leaders?
How can we recognize them?
Is the group demanding anything?
How do they operate?
What is their command and control?
Are they fighting against any certain group?
Where is the organization's headquarters?
Do they have allies? (who)
Do security forces support them?
What types of weapons do they employ?
What do they use for transportation?
Where do they fight?
What tactics do they employ?
How are the forces organized?
What type of weapons technology do they have access to?
How are they trained?
How well are they trained?
Describe their weapons. • Firearms. • Explosives. • Incendiary. • Indirect fire weapons. • Mechanized.

Table G-21. Paramilitary threat checklist (continued).

What group/subgroup does the insurgent identify with?
What political ideology is the insurgency supporting?
What issues is/are driving the insurgency?
What is the desired result of the insurgency?
What allows the insurgents to sustain the conflict? (support?)
Who are the leaders?
Describe the command and control.
How do they recruit supporters?
What is the reaction of the genuine opposition?
Is the opposition a one-party state?
What type of governmental rule is established?
What fighting strategy and tactics do they employ?
What kind of military capabilities do they have?

Table G-22. Insurgent threat checklist.

What is the crime rate within the urban area? By section?
What criminal organizations are known to be in the urban area? • Local. • Regional. • International.
What activities are criminal organizations involved in? • Finance. • Narcotics. • Technology transfer. • Arms dealing. • Illegal immigration. • Counterfeiting operations. • Armed robbery. • Prostitution. • Gambling (if illegal). • Other.
What are the structure and organization of criminal organizations?
What are the territorial divisions?
What weapons and equipment do the criminal organizations possess?
What are the management and organization of law enforcement agencies? • Personnel. • Weapons and equipment. • Key leaders. • Organizational structure. • Training of police. • Tactics, techniques and procedures. • Locations of police stations. • Where are their facilities? • What are their shipment routes? • What front companies are known to be involved?
What are their essential chemicals and where/how are they acquired/ produced? • Cocaine hydrogen chloride. • Heroin hydrogen chloride. • Ether. • Methyl ethyl ketone. • Acetic anhydride.
Describe any prior civilian riot activity.
Does the population own/carry any weapons on a large scale?

Table G-23. Crime threat checklist.

What are the units in the urban area? • Committed. • Reserves. • Reinforcements. • Composition, location, commander and headquarters.
What are the weapons and equipment? • Organic weapons. • Attached weapons. • Locations. • Communications equipment. • Frequencies used. • Special equipment.
What units can reinforce within 3, 6, 12, 24 and 72 hours? • Composition. • Location. • Commander. • Time to reinforce. • Most likely routes.
What are their logistics capabilities?
What tactics, techniques and procedures does the enemy employ in an urban environment?
Will the enemy use the locals to blend in?
How does the enemy use snipers?
How does the enemy use mech.?
How does the enemy use indirect fires?
What area of the city does the enemy control? Occupy? Influence?
Where are the hardened defensive sites within the urban area?
What weapons systems and artillery can deliver fire within the urban area?
Where are the obstacles other then those of natural structure and part of the construction?
Where are the mined areas?
Are there any forces in the urban area that might fight with the enemy? • Police forces. • Paramilitary.

Table G-24. Ground order of battle checklist.

APPENDIX H

LESSONS LEARNED FROM MODERN URBAN COMBAT

Although the US Army has a long history of combat operations in urban areas, it is not alone in conducting UO. Other armies have also conducted extensive urban combat, some very recently. Just as each war is unique, each nation's army is a unique reflection of its national strategy, government, economy, demographics, and culture. For that reason, no one lesson learned can be valid for all cases of urban combat. It is important, however, to study and to learn from the experiences of others. This appendix presents abstracts from various sources of information on the lessons learned by non-US armies in recent urban combat. Because of the subjective nature of such abstracts, no attempt is made to validate these lessons against US experience.

H-1. RUSSIA AND THE WAR IN CHECHNYA

Following the collapse of the Soviet Union, the people of Chechnya began to seek full independence. By 1994, Chechnya had fallen into a civil war between pro-independence and pro-Russian factions. In December 1994, Russia sent 40,000 troops into Chechnya to restore Russian primacy over the breakaway republic. An attack was launched by 6,000 mechanized troops against the Chechen capital of Grozny. Instead of the anticipated light resistance, Russian forces encountered heavy resistance from the Chechens, armed with "massive amounts" of antitank weapons. The Russians were repulsed with shockingly high loses. It took them another two months of heavy fighting, and changing their tactics, before they were able to capture Grozny. Between January and May 1995, Russian losses in Chechnya were approximately 2,800 killed, 10,000 wounded, and over 500 missing or captured. Chechen casualties were also high, especially among noncombatants.

a. **General Analysis.**[i] A reversal of fortune so astonishing and unprecedented as the Chechen victory of the Russians should make this war a major and cautionary episode in military history. The large-scale lessons of Chechnya lie in three areas:[ii]

- It showed, again, the limited effectiveness of heavy weaponry in urban terrain and, by extension, the crucial importance of well-trained, well-led, well-equipped, and highly motivated infantry.
- It validated, once more, the continued relevance of Clausewitz's dictum to seek decisive battle.
- It proved again that a society judged "primitive" or "chaotic" by Western standards can still generate a tremendous fighting spirit and very effective military discipline. This is not a new lesson. The "primitive" Cheyenne, Apache, Nez Pierce, Seminole, and Moro tribesmen; the uneducated North Korean, North Vietnamese and Chinese peasants; and the ragtag Somali clansmen have all taught it to the Americans they faced, just as the Chechens streetfighters taught the conscript soldiers of the Russian Army.

The following lessons learned are from The World Turned Upside Down: Military Lessons of the Chechen War, by Mr. Anatol Lieven[iii].

> *"It cannot be emphasized too strongly, therefore, that the key to success in urban warfare is good infantry. And the key to good infantry, rather than good weaponry, is a traditional mixture of training, leadership qualities in NCOs and junior officers, and morale – implying a readiness to take casualties."*
>
> *"The Russians faced an opponent who was singularly determined not to make peace and retained the means to go on fighting. The lesson to be learned by armies everywhere is that, especially against such an opponent, there is no valid strategic alternative to seeking decisive battle..."*
>
> *"The US will not always have the ability to pick and choose its wars, and the key lesson Chechnya is that there will always be military actions in which a determined infantryman will remain the greatest asset."*

b. **US Marine Corps Analysis.**[iv]

(1) ***Strategic Lessons.*** Military operations alone cannot solve deep-seated political problems.

(a) Military commanders need clear policy guidance from which they could work steadily and logically.

(b) Confusion generated by missing or conflicting policy guidance is made worse by poorly defined lines of command and control.

(c) Russian senior command lacked continuity and was plagued by too much senior leadership involvement at the lower operational level.

(d) Contrary to initial expectation, operations were neither of short duration or low cost.

(e) When Russian security operations began achieving results, the Chechens started attacking targets within Russia.

(f) It was difficult to unite police and military units into a single, cohesive force.

(g) Distinct tactical advantages accrue to the side with less concern for the safety of the civilian population.

(h) Concern about civilian casualties and property destruction declined as casualties among the Russian forces rose.

(i) Chechen forces received extensive outside assistance despite rigorous attempts to stop it.

(2) ***Operational Lessons.***

(a) Having well-developed military doctrine for urban warfare is not enough in and of itself.

(b) Situation-oriented training would have improved Russian military effectiveness.

(c) Inadequate training in the most basic maneuver and combat skills inhibited Russian operations.

(d) Urban combat is extremely manpower intensive and produces significant attrition of men and materiel among the attackers.

(e) Overwhelming firepower can make up for organizational and tactical deficiencies the short-run if one is willing to disregard collateral damage.

(f) The sudden requirement to deploy to Chechnya, coupled with the unique supply problems posed by the weather and the urban environment, overwhelmed the already fragile Russian-military logistics system.

(g) A lack of high-quality intelligence made operations more difficult and dangerous the Russian forces.

(h) The geometry and perspectives of urban combat are very different from combat in the open area. Urban combat is much more vertically oriented.

(i) Composite units were generally unsatisfactory.

(j) Fratricide was a serious and continuing problem throughout the campaign in Chechnya because it was difficult to tell friend from foe.

(k) Standard The Russian-military unit configurations were inappropriate for urban combat.

(l) Foregoing peacetime maintenance is a false economy.

(m)The potential of special forces for urban operations was never realized in Chechnya.

(n) The nature of cities tends to channel combat operations along narrow lanes of activity.

(o) Strategic bombing can be used in urban operations to shape the battlefield, especially during the early phases.

(3) ***Tactical Lessons.***

(a) Rigorous communications security is essential, even against relatively primitive enemies.

(b) Night fighting was the single most difficult operation in Chechnya for Infantry forces.

(c) Tanks and APCs cannot operate in cities without extensive dismounted Infantry support.

(d) Forces, operating in cities, need special equipment not usually found in Russian TO&Es. Lightweight ladders were invaluable for assaulting Infantry.

(e) Firing tracer ammunition in cities makes the user a target for snipers.

(f) Trained snipers were essential, but in short supply.

(g) Obscurants are especially useful when fighting in cities.

(h) Armored combat engineer vehicles can perform important, specialized urban combat tasks.

(i) Recovering damaged armored vehicles is especially difficult in cities.

(j) Hit-and-run ambush attacks by small groups were the most favored and effective of the Chechen tactics.

(k) Direct-fire artillery can be a valuable tool in urban combat, provided collateral damage is not a major concern.

(l) A failure of small unit leadership, especially at the NCO level, was a primary cause of Russian tactical failures in Grozny.

(m)Tracked armored vehicles are preferable to wheeled armored vehicles in urban combat.

(n) When operating in urban areas, armored vehicles require more protection and that protection needs to be distributed differently than for combat in the open.

(o) RPGs can be used against helicopters.

(p) Air defense guns are valuable for suppressing ground targets.

(q) Heavy machine guns still offer good defense against close air attack, especially from helicopters.

(r) Both sides employed commercial off-the-shelf technologies for military purposes.

(s) Non-lethal technologies were seldom used.

(t) Tactical communications proved very difficult in Grozny.

(u) Indigenous forces can improvise crude chemical weapons using hazardous materials from the urban area.

(v) The cabs of supply trucks must be armored.

(w) Bunker busting weapons are invaluable for urban combat.

(x) Some Russian equipment needed to be modified in the field to counter enemy tactics and equipment.

(y) Helicopters need standoff weapons.

(z) Helicopters are not well suited for urban combat.

(aa) Precision guided weapons were used extensively by the Russian Air Force.

(bb) Inadequate on-board navigation systems and poor radar limited the use of helicopters in adverse weather and at night.

(cc) Precision-guided artillery projectiles were considered too expensive to "waste". Direct fire artillery was often substituted for precision-guided indirect fire.

(dd) UAVs were used extensively and were very effective.

c. **US Army Infantry School Analysis.**[v] Russian Army Lessons Learned from the Battle of Grozny.

(1) You need to culturally orient your forces so you don't end up being your own worst enemy simply out of cultural ignorance. Many times, Russian soldiers made serious cultural errors in dealing with the Chechen civilians. Once insulted or mistreated, the Chechens became active fighters or, at least, supported the active fighters. Russians admit they underestimated the affect of religion on the conflict.

(2) You need some way of sorting out combatants from non-combatants. The Russians were forced to resort to searching the pockets of civilians for military equipment and to sniffing then for the smell of gunpowder and gun oil. This was crude, and not very reliable. Trained dogs were used to detect the smell of gunpowder or explosives, but were not always effective. Nevertheless, specially trained dogs probably are the best way to determine if a person has been using explosives or firing a weapon recently.

(3) The psychological impact of high intensity urban combat is so intense units should maintain a large reserve that will allow them to rotate units in and out of combat. If a commander does this, he can preserve a unit for a long time. If he doesn't, once it gets used up, it can't be rebuilt.

(4) Training and discipline are paramount. You can accomplish nothing without them. You may need to do the training in the combat zone. Discipline must be demanded. Once it begins to slip, the results are disastrous.

(5) The Russians were surprised and embarrassed at the degree to which the Chechens exploited the use of cell phones, Motorola® radios, improvised TV stations,

lightweight video cameras and the internet to win the information war. The Russians admitted that they lost control of the information coming out of Grozny early in the operation and never regained it.

(6) The proliferation of rocket propelled grenade launchers surprised them, as well as the diversity of uses to which they were put. RPGs were shot at everything that moved. They were fired at high angle over low buildings and from around buildings with little or no attempt made to aim. They were sometimes fired in very disciplined volleys and were the weapon of choice for the Chechens, along with the sniper rifle. Not only were the Russians faced with well-trained, well equipped Chechen military snipers, there were also large numbers of designated marksmen who were very good shots using standard military rifles. These were very hard to deal with and usually required massive firepower to overcome.

(7) As expected, the Russians reiterated the need for large numbers of trained Infantrymen. They said that some tasks, such as conducting logpac operations, could only be conducted by infantrymen. The logistical unit soldiers were hopelessly inept at basic military skills, such as perimeter defense, establishing security overwatch, and so forth, and thereby fell easy prey to the Chechens.

(8) They found that boundaries between units were still tactical-weak points, but that it wasn't just horizontal boundaries they had to worry about. In some cases, the Chechens held the third floor and above, while the Russians held the first two floors and sometimes the roof. If a unit holding the second floor evacuated parts of it without telling the unit on the ground floor, the Chechens would move troops in and attack the ground floor unit through the ceiling. Often this resulted in fratricide as the ground floor unit responded with uncontrolled fire through all of the ceilings, including the ones below that section of the building still occupied by Russians. Entire battles were fought through floors, ceilings, and walls without visual contact.

(9) Ambushes were common. Sometimes they actually had three tiers. Chechens would be underground, on the ground floor, and on the roof. Each group had a different task in the ambush.

(10) The most common response by the Chechens to the increasingly powerful Russian indirect and aerial firepower was hugging the Russian unit. If the hugging tactics caused the Russians to cease artillery and air fires, it became a man-to-man fight and the Chechens were well equipped to win it. If they didn't cease the supporting fires, the Russian units suffered just as much as the Chechen fighters did, sometimes even more, and the morale effect was much worse on the Russians.

(11) Both the physical and mental health of the Russian units began to decline almost immediately upon initiation of high intensity combat. In less than a month, almost 20% of the Russian soldiers were suffering from viral hepatitis (very serious, very debilitating, slow recovery). Most had chronic diarrhea and upper respiratory infections that turned to pneumonia easily. This was blamed on the breakdown of logistical support that meant units had to drink contaminated water. Unit sanitary discipline broke down almost completely.

(12) According to a survey of over 1300 troops made immediately after the fighting, about 72 percent had some sort of psychological disorder. Almost 75 percent had an exaggerated startle response. About 28 percent had what was described as neuro-emotional and almost 10 percent had acute emotional reactions. The Russians

recommended two psycho-physiologists, one psycho-pharmacologist, one psychiatrist, and one medical psychologist at each (US) Corps-sized unit. Although their experience in Afghanistan prepared them somewhat for the physical health problems, they were not prepared for this level of mental health treatment. Many permanent-combat stressed casualties resulted from the soldiers not being provided proper immediate treatment.

(13) Chechens weren't afraid of tanks or BMPs. They assigned groups of RPG gunners to fire volleys at the lead and trail vehicles. Once these were destroyed, the others were picked off, one-by-one. The Russian forces lost 20 of 26 tanks, 102 of 120 BMPs, and 6 of 6 ZSU-23s in the first three days of fighting. Chechens chose firing positions high enough or low enough to stay out of the fields of fire of the tank and BMP weapons.

(14) Russian conscript infantry sometimes refused to dismount and often died in their BMP without ever firing a shot. Russian elite infantry did much better, but didn't coordinate well with armored vehicles initially.

(15) Chechens were brutish, especially with prisoners. (Some reports say the Russians were no better, but most say the Chechens were the worse of the two sides.) Whoever was at fault, the battle degenerated quickly to one of "no quarter asked, none given." Russian wounded and dead were hung upside down in windows of defended Chechen positions. Russians had to shoot at the bodies to engage the Chechens. Russian prisoners were decapitated and at night their heads were placed on stakes beside roads leading into the city, over which Russian replacements and reinforcements had to travel. Both Russian and Chechen dead were routinely booby-trapped.

(16) The Russians were not surprised by the ferocity and brutality of the Chechens; they expected them to be "criminals and animal brutes". But they were surprised by the sophistication of the Chechen use of booby-traps and mines. Chechens mined and booby-trapped everything, showing excellent insight into the actions and reactions of the average Russian soldier. Mine and booby-trap awareness was hard to maintain.

(17) The Russians were satisfied with the combat performance of most of their infantry weapons. The T-72 tank was unsatisfactory, often called, "dead meat". It was too vulnerable, too awkward, not agile, had poor visibility, and poor weapons coverage at close ranges. The Russians removed them from the battle and replaced them with smaller numbers of older tanks and more self-propelled artillery, more ADA weapons, and more BMPs.

(18) Precision guided weapons and UAVs were very useful. There was some need for non-lethal weapons, but mostly riot gas and tranquilizer gas, not stuff like sticky foam. The Russian equivalent of the M202 Flash flame projector and the MK 19 grenade launcher were very useful weapons.

(19) Ultimately, a strong combined arms team and flexible command and control meant more that the individual weapons used by each side.

H-2. BEIRUT (1982), MOGADISHU (1993), AND GROZNY (1995)[vi]

This paragraph draws together lessons learned on the organization, equipment, and training of the forces involved and draws conclusions about what types of military systems, munitions, and force structure were effective and why.

a. **Beirut.**

(1) ***Armor***. Tanks were under the command of the Israeli Defense Force (IDF) Infantry companies. The armor of the Merkava, with an internal troop compartment, proved excellent protection against RPGs, mines, and small arms fire. The IDF felt that tanks were the most useful weapon in Beirut, both in terms of delivering firepower on specific targets and protecting the Infantry. The IDF concluded that the M113 family of armored vehicles was too unreliable, unmaneuverable, and vulnerable.

(2) ***Infantry.*** Clearing buildings presented a major problem for the IDF. In the words of one IDF brigade commander, "Every room is a new battle." Once troops are inside a building, it is impossible for a commander to understand what his troops are confronting, "The battlefield is invisible." In his judgment, the dangers of clearing buildings are so great that they should never be entered unless absolutely necessary: "Avoid cities if you can. If you can't, avoid enemy areas. If you can't do that, avoid entering buildings."

(3) ***Artillery.*** The IDF found the American-made M109 155-mm self-propelled howitzer extremely effective when using high explosives in a direct-fire role. Both sides employed Katyusha multiple rocket launchers, which had tremendous psychological effect on noncombatants. PLO antiaircraft artillery (mostly heavy automatic weapons) was not effective against the Israeli Air Force.

(4) ***Munitions.*** Air-delivered cluster bombs, smart 1000 pound bombs, rockets, missiles and other munitions were precise and effective. Cluster bomb units (CBUs) were highly effective in destroying antiaircraft artillery. However, the CBUs' sub-munitions were responsible for many civilian casualties. Large proportions of the shells used were white phosphorous (WP). The IDF appeared to use WP primarily for the psychological effect it generated, including fear.

b. **Mogadishu.**

(1) ***Rules of Engagement***. Despite strict rules of engagement that severely limited the use of mortars and artillery, the US forces inflicted significant collateral damage in Mogadishu. During the 17 June attack, for example, helicopter gunships pounded an Aideed stronghold with TOW missiles and aerial rockets, killing at least sixty Somali noncombatants. Although the Cobra gunships and the AC 130s were removed in August, the lavish use of firepower during the first few months of UNOSOM II significantly alienated the civilian population. One analyst described the larger consequences this way:

> *"By the time the American resorted to the use of anti-tank guided missiles to root our snipers, it had become apparent that the firepower which had demolished the Iraqi Republican Guards was ill-suited to the streets of Mogadishu.....The Gulf War's promise of a style of fighting that minimized noncombatant casualties was a long way from fulfillment."*

(2) ***Helicopter Close Air Support***. Given the constraints on indirect fire, the only significant fire support element available to the 10th Mountain Division was an attack helicopter company, equipped with AH-1 Cobras. One participant noted, "Air strikes are only suppressive fire…and did not completely destroy enemy positions or buildings. Many building that were struck were reoccupied by Somali guerrillas within minutes." The Somali fighters' skill with RPGs made all rotary wing aircraft vulnerable. Regardless of this, however, the bravery, skill and combat effectiveness of the pilots flying the AH-6

helicopters were a major factor in the successful defense of several buildings by elements of Task Force Ranger during the intense battles of 3-4 October 1993.

c. **Grozny.**

(1) ***Russian Readiness and Training***. Russian performance was severely hampered by the fact that its poorly trained troops were forced to serve together in hastily assembled units. One observer described them as, "Untrained kids" and a "shapeless and disorganized groups of men which does not know itself where it is going." Additionally, troops received little or no specialized urban warfare training. Both sides employed snipers effectively, but both sides also experienced shortages of these trained personnel.

(2) ***Munitions***. The RPG, brilliantly employed by the Chechens, was perhaps the single most useful weapon in the conflict. Smoke, white phosphorus rounds and tear gas were used extensively by the Russians, and they proved useful. According to one analyst, every fourth or fifth artillery round was white phosphorus, which burns upon explosion. The resulting smoke provided cover for the movement of Russian forces. However, these munitions, like much else, were in short supply due to logistical breakdowns.

(3) ***Aircraft***. The Russians had large numbers of fixed-wing aircraft, but they proved of limited tactical value. For most of the conflict, poor weather kept Russian fixed-wing aircraft grounded. Helicopter gunships proved more useful, particularly against snipers on the upper floors of buildings.

(4) ***Psychological Operations (PSYOP).*** Disinformation, deception, and other forms of information warfare were used extensively by both sides during the battle for Grozny. PSYOP ranged from the tactical; for example, the Russian use of loudspeakers and leaflets to encourage the Chechens to lay down their weapons, to the strategic; for example, Russians claim to the world press that its military activities in Chechnya were *peacekeeping* operations. The Chechen commanders organized civil disobedience actions; claimed falsely that the Russians were employing chemical weapons; and claimed, apparently falsely, that Chechnya possessed nuclear weapons. The utility of all this information warfare techniques is difficult to assess. It is clear however, that both the Russians and the Chechens believed that they are important military instruments.

H-3. GENERAL OBSERVATIONS ON URBAN WARFARE.[vii]

This paragraph is extracted from an article in which the author draws together lessons learned from urban combat around the world and generates a set of conclusions about the nature of the urban battlefield. The conclusions are solely the author's but they provide interesting professional reading.

a. **Urban Warfare - Different and Demanding.**

(1) Conventional warfare on open terrain—the preferred form of combat for all modern military forces, is a complex and challenging undertaking, requiring vast resources, training, and excellent organization to perform well. Whatever challenges are inherent in this mode of warfare are magnified significantly in the city environment. From Stalingrad to Inchon to Panama to Grozny, urban combat has been characterized by:

- Poor communications
- Difficult command and control
- Difficult target acquisition
- Short engagement ranges

- Reduced effectiveness for transportation and fire support assets
- Difficulties in providing logistical support to the front line.

(2) Tall buildings, sewers, and tunnels give the battlefield dimensions of height and depth that are absent on open terrain. In essence, city fighting is primordial combat. It is clearly distinct from the elegant maneuver warfare that characterized the US conduct of the Gulf War.

- The enemy is at close range.
- Snipers are almost always present.
- Stress is extremely high.
- The opposing force is frequently indistinguishable from the civilian population.

(3) Demographic and urbanization trends in the developing world are likely to lead to city environments that are even more stressful and more difficult to operate in.

b. **No US Comparative Advantage.**

(1) The historical data suggest that it is extremely difficult for modern forces to leverage their technological advantages against a determined adversary in an urban environment. To be sure, the US military is highly motivated, well trained and well equipped, but not for urban warfare per se.

(2) The city environment, with its high population density and multistory buildings, tends to negate the technological advantages, for example, close air support, mobility, communications, enjoyed by modern military forces. Some US military technology, designed for large scale war in the open areas of central Europe or the desert, is not well suited for urban combat. The US technological advantage, typically associated with long range, high-technology weapons platforms that use mass and mobility, is significantly reduced in urban environments.

(3) It is precisely for this reason that less sophisticated forces are drawn to cities. Urban battles in the recent past, such as Grozny and Mogadishu, have been characterized by conflict between modern combined arms forces and informally organized irregulars. The battle of Inchon was the last significant urban engagement in which US forces fought a remotely comparable force in an urban environment. Aware of our increasing unwillingness to take casualties or cause major collateral damage, and understanding our lack of comparative advantage in the urban environment, US adversaries are increasingly likely to engage our forces in cities.

c. **Small-Unit Operations.**

(1) The degeneration of urban warfare into a series of small-group—or even of individual-battles was evident in operations as different as Stalingrad, Hue and Beirut. The nature of cities themselves is responsible for this fragmentation process. As battles wear on, the streets and building blocks of the urban physical morphology fragment urban warfare into conflict between units usually of squad or platoon size, with generally insufficient space for the deployment and maneuvering of larger units. The battle rapidly disintegrates into a series of more or less separate and isolated conflicts around such "fortresses".

(2) Given that much of an urban battle is likely to take place inside buildings or underground, it is likely to be invisible to all except the men actually fighting it on the ground. As a result, it is difficult for higher-level commanders to maintain an accurate picture of the battlefield at any given time.

(3) Given this inevitable fragmentation, operational effectiveness will depend greatly on the quality of leadership at lower echelons, for example, at battalion and below. Leadership shortfalls were apparent among US Marines in Hue and among the Russian army in Grozny. In both cases, the generous use of firepower served as a substitute. The relatively successful Israeli operation in Beirut can be attributed in part to the attention the IDF high command paid to developing small-unit leadership, which long stressed the importance of initiative and independence among junior commanders.

d. **Importance of Armor.**

(1) Tanks, as one Operation Just Cause participant has written, "are an infantryman's friend in city fighting." They can go anywhere. They can deliver steel on target and they scare the enemy. Their firepower can be used precisely, thus minimizing collateral damage; they can serve as troop carriers, as the IDF discovered in Beirut; and they can be useful for shocking opposition forces and less-than-friendly noncombatants.

(2) To be effective, however, they must be supported by dismounted infantry. In the absence of such support, tanks are vulnerable to rocket-propelled grenades, Molotov cocktails, and other systems and munitions. Once a tank is destroyed, it loses its psychological shock value among combatants and noncombatants alike.

(3) As mentioned above, small units are the norm in urban warfare. Given this reality, it may make sense to assign tanks to smaller units than is the norm:

(4) Tanks habitually operate in section or platoon formations. Tank communications procedures are designed for this. Support of a dismounted unit in a city, however, often involves only single or paired armored vehicles. Tanks might be assigned to units as small as a squad.

e. **The Primacy of Infantry.**

(1) The historical record suggests that urban warfare is manpower intensive. Large numbers of ground troops are needed to attack, clear, and hold cities. No attacking force has ever succeeded in the city environment without using large amounts of infantry.

(2) No firm rule of thumb exists, but according to one source, "a commander is left with the prospect of needing between 9 and 27 attackers per defender in an urban environment-Significantly more than is required for open terrain."

(3) Placing significant numbers of men on the around is equally important in stability or support operations. In such situations, human intelligence is a critical capability, requiring large numbers of collectors and analysts. As Army planners begin to create a task force for involvement, they will need to increase the intelligence assets for that task force. Infantrymen, provided they avoid a garrison mentality and patrol often, could be an essential component of this intelligence collection process.

f. **Aircraft - A Mixed Blessing.**

(1) Since the battle of Stalingrad, aircraft have been employed in city battles. Their use has been completely lopsided. Defenders have never been able to employ either fixed-wing or rotary-wing aircraft. During high intensity urban combat, with liberal rules of engagement, for example, Stalingrad, aerial bombardment has been very useful.

(2) In all other cases, however, airpower has not been terribly effective. The need to limit collateral damage has been the most significant factor in this regard. To date, air-delivered munitions, rockets, and other systems and munitions have lacked the low circular error probable (CEP) that is needed to minimize such damage.

(3) Even in the case of Beirut, where the Israeli Air Force reportedly employed smart munitions, collateral damage, while relatively low, was still significant. Similarly, fixed-wing aircraft, such as the AC-130 gunship, caused significant collateral damage in Panama City, Panama. In the case of Mogadishu, rotary-wing aircraft were vulnerable to RPG rounds, which reduced their utility.

(4) Finally, it should be mentioned that poor weather kept aircraft on the ground at least part of the time during the battles of both Hue and Grozny. As all-weather capability becomes a reality, this should be less of a concern. Similarly, new generations of precision-guided munitions could conceivably turn helicopters into "flying artillery" capable of great precision.

(5) Such munitions are likely to be very expensive, however, and this may limit their utility. During the battle of Grozny, for example, Russian commanders reportedly were unwilling to "waste" precious PGMs on the Chechen rebel forces.

g. **Population Control Is Critical.**

(1) In every major urban battle in modern times, the presence of noncombatants has affected the course of the operation. At Stalingrad, they served as a force multiplier for the Red Army. In Seoul, friendly noncombatants, exultant at the arrival of their US liberators, slowed the course of the American advance. In Hue, the Viet Cong and North Vietnamese Army forced civilians to construct defensive positions. In Mogadishu, Aideed employed "rent-a-mobs" to hold anti-UN demonstrations and to serve as human shields.

(2) Civil affairs (CA) as we know it today is a relatively recent concept—there is no evidence, for example, that civil affairs units were used widely in urban battles during World War II or the Korean War. The evidence suggests that a robust CA (and with it, PSYOPS) capability will be necessary during future urban conflicts.

(4) Cities are unlikely to be empty. In recent conflicts, such as the battle of Mogadishu, and in Gorazde, Bosnia, in 1994, urban populations have sometimes even increased as the battle has progressed. What is more, Western morality (if not the Law of Land Warfare) will require Western military forces to ensure that noncombatants are protected and properly cared for.

(4) During the battle of Manila, for example, US forces spent two days battling fires set by fleeing Japanese forces. More recently, in the aftermath of the battle of Panama, a breakdown in public order forced US troops to conduct emergency law enforcement operations.

(5) Ensuring public safety, although an unpalatable task for US armed forces is inevitable, given the fact that it is unlikely that civilians will be able to provide these services during and immediately after an urban battle. Inevitably, military resources will have to be diverted to perform these public-safety tasks.

h. **Key Munitions and Systems.**

(1) As suggested earlier, urban warfare has been primordial, characterized by the use of such relatively simple systems as tanks, artillery, rockets, heavy machine guns, rocket-propelled grenades and mortars. However, in terms of utility these have above all been small arms. It is not surprising that in a manpower-intensive environment, the soldier's most basic weapon should prove to have been of such importance.

(2) When used in a direct-fire mode, artillery has proved useful in destroying fortified targets, although their relative lack of accuracy has limited their use in recent battles, at

least among Western forces. Mortars, with their high trajectory, have also proved useful in the urban environment, but as with artillery, concerns about collateral damage have often limited their use.

(3) Flame, napalm, shotguns, recoilless rifles, and other low-technology systems and munitions that have proven highly effective in the urban environment are no longer readily available in the US military inventory. If the United States and other nations decide to become serious about improving their ability to fight on urban terrain, they have to look closely at their existing inventories and explore the possibility of reinstating such "quaint" systems as flame-throwers.

(4) The US may also need to examine the size of the explosives it uses in its artillery rounds, rockets, and other munitions. HEAT (high-explosive antitank) rounds used in M-1 tanks, for example, were designed to defeat enemy armored vehicles. As such, their lethal energy is directed forward, and the resultant projectile penetration power is such that they may go through several rooms or buildings before stopping.

(5) Similarly, Hellfire and missiles, launched from attack helicopters are antiarmor systems whose utility is limited in an environment where collateral damage needs to be minimized.

(6) Defeating the enemy while reducing collateral damage and friendly casualties requires a new set of munitions. This technique includes highly accurate mortars, lightweight charges for creating *breach holes*, and low yield, low-collateral damage munitions. These perhaps may be delivered by a system such as the Fiber-Optic Guided Missile (FOG-M) that would give US forces the ability to target much more precisely.

i. **Rules of Engagement.**

(1) In almost every modern urban battle, the attacking force-which is always the more modern force-has entered the battle with a set of strict rules of engagement designed to minimize collateral damage. Even in the case of total war, for example, the United States in Manila, rules of engagement, at least initially, have prohibited unobserved artillery fire, wholesale aerial bombardment, and other techniques of modern war.

(2) However, in each case, these rules of engagement have eased as the battle wore on. The explanation is straightforward: strong resistance and mounting friendly casualties lead inexorably to a relaxation of the earlier prohibitions.

(3) This suggests a tension between the desire to reduce civilian deaths and the destruction of infrastructure and the requirement to reduce friendly casualties. The days of using troops as cannon fodder (as at the Somme, for example) have long passed. In Western democracies, relatively low birth rates have made large numbers of casualties among one's own forces completely unacceptable politically.

(4) Massive destruction of civilian populations and the vast destruction of city infrastructure are equally unacceptable. If the recent past is any guide, it seems fair to assume that the urban battlespace of the future will be characterized by even greater media transparency. Given that cities are increasingly the world's centers of commerce, politics, and media, it is likely that warfare will be conducted there under even greater international scrutiny.

(5) In short, the battlefield will no longer be invisible to outside observers. Limiting the use of violence will be even more important in stability and support operations, where the goal of developing and maintaining political legitimacy could be undermined by the excessive use of force.

(6) The challenge, then, for military commanders will be to square the circle. Minimizing friendly casualties and reducing collateral damage have been mutually exclusive in. the past. Commanders have resolved this tension in favor of the former.

(7) In the future, however, such a resolution is unlikely to be acceptable. Two possible answers suggest themselves. The first is technological. Advances in nonlethal technology, or in the ability to scan an urban structure's interior may make it possible to keep one's own casualties down while reducing collateral damage.

(8) The second possibility is operational. The ancient technique of laying siege to a city, although by definition time-consuming and thus difficult to sell politically, should be reexamined. A *humane* siege, bolstered by a robust strategic PSYOPs campaign designed to de-legitimize the defending force could minimize both friendly casualties and collateral damage.

j. **Difficult for Attackers tó Prevail but They Almost Always Do.**

(1) As in any mode of warfare, defenders in urban battles enjoy distinct advantages. Intimate knowledge of the buildings, alleyways, tunnels, and rooftops that are a feature of most cities-perhaps gained over the course of a lifetime is one obvious advantage. In many cases, such as the shantytowns surrounding cities in the developing world, maps are likely to be outdated or even nonexistent.

(2) During Operation Urgent Fury, for example, the lack of official maps of Grenada forced troops to rely on tourist maps. Similar shortages reportedly plagued US forces in Somalia and Russian troops in Chechnya.

(3) Cities, particularly capital cities, are the locus of economic, political, and social power, and are becoming more so. It is not surprising that cities serve as critical arenas for those fighting to preserve national, ethnic, or religious identity. Put another way, urban areas are the key battlegrounds in any significant defense of the homeland. Forces claiming to defend that homeland from invasion, as in the cases of Stalingrad, Mogadishu, and Grozny, enjoy a tremendous advantage over attacking forces.

(4) All other things being equal, defending forces are much more likely to be able to gain the allegiance of the local population and use it as a source of food, munitions, shelter, and information.

(5) These observations lead to several conclusions. The first is that it is very difficult for attacking forces in an urban environment to prevail. However, if they are willing to accept high casualties, and can either focus their firepower or simply mass it regardless of collateral damage, they will normally prevail eventually. The second is that in the future, attackers will have to employ effective strategic-level PSYOPS and other techniques of political warfare if they hope to win in cities.

(6) As the Israelis discovered in Beirut, simply crushing an adversary no longer guarantees victory—the attacker must also win the international propaganda battle. A well crafted, effective political warfare campaign, being essentially nonviolent in nature, could also contribute to resolving the friendly casualties—collateral damage tension described above.

[i]Extracted from The World Turned Upside Down: Military Lessons of the Chechen War, by Anatol Lieven, Armed Forces Journal International, August 1998, pp 40-43

[ii]Ibid.

[iii]Armed Forces Journal International, August 1998

[iv]Extracted from Russia's War in Chechnya, Urban Warfare Lessons Learned 1994 – 1996, prepared by the USMC Intelligence Activity, Nov 1998

[v]Extracted from a briefing by Mr. Timothy Thomas of FMSO, Fort Leavenworth, KS to the *RAND-DBBL Conference on Military Operations on Urbanized Terrain,* 24-25 Feb 1998, Washington, DC, and from personal interviews by USAIS representative. Included as part of USAIS trip report.

[vi]Extracted in part from *"Every Room is a New Battle": The Lessons of Modern Urban Warfare,* by William G. Rosenau, SAIC, McLean, VA, included in a collection of professional military reading originally published by the USMC Intelligence Activity, Quantico, VA.

[vii]Ibid.

APPENDIX I

PLATOON URBAN OPERATIONS KIT AND TACTICS, TECHNIQUES, AND PROCEDURES FOR MARKING BUILDINGS AND ROOMS

Urban operations present many unique challenges for the infantry platoon and squad. They face obstacles and hazards not found in any other environment. The infantry platoon/squads are required to breach obstacles, enter and clear buildings, and cross streets and open areas when conducting urban operations. To increase effectiveness of the unit and decrease risk to individuals, the infantry platoon/squad uses special equipment in the urban environment.

I-1. EXAMPLE PLATOON URBAN OPERATIONS KIT

When developing an urban operations kit SOP, brigades and battalions must consider their METL and the unit tactical SOP. Input from platoon and squad leadership is vital when identifying the proper equipment for the urban operations kit. These leaders know best what equipment is needed for closing with and destroying the enemy and facilitating consolidation and reorganization actions. Once this task is accomplished, the commander ensures that the contents of the platoon kit are standardized. This standardization allows the command to develop an effective training program keying on the proper uses, storage, and accountability of all items listed in the urban operations kit.

Breaching Items:

Axes.. 1
Bolt cutters.. 2
Crowbars.. 1
Sledgehammers.. 1
Grappling hooks.. 2
120' nylon ropes.. 2
12' sling ropes.. 4
Snap links.. 1 per man
Lineman's pliers w/cutter.. 6
Wire-handling gloves.. 3 pair
Fireman's tool (hand pick-ax).. 3
Ladders (folding, collapsible, lightweight).. 1

Signaling and Marking Items:

Chalk (large, sidewalk).. 2 boxes
Spray Paint (assorted colors).. 5 cans
Chemlites (assorted colors).. 3 boxes
IR Chemlites.. 3 boxes
Signaling mirrors (can be used for observing around corners).. 6

Signaling and Marking Items (continued):

Cone flashlights w/extra batteries	12
Flashlights (Magnum and Mini-Mag) w/extra batteries	1 per man
NATO marking squares	200
100mph tape	5 rolls
Masking tape	2 rolls
2 sided tape	3 rolls
Engineer tape	2 rolls

Other Items:

Urban-specific sand table kit	
Hammers	2
Saws	2
Nails (various sizes)	4,000
Steel wire (16 gauge)	1,000ft
Hose clamps (6 inch)	300
Extra batteries (various sizes)	3 additional for every one in use

NOTE: The preceding list of items and quantities is not all inclusive or exclusive to urban operations. It is merely a guide to aid commanders in developing a standard platoon UO kit for their unit.

I-2. INDIVIDUAL PROTECTIVE EQUIPMENT

When planning for urban operations leaders must ensure that each soldier receives the following protective equipment in addition to standard TA-50 or unit issue items.

- Knee pads.
- Elbow pads.
- Eye protection.
- Hearing protection.

I-3. TACTICS, TECHNIQUES, AND PROCEDURES FOR MARKING BUILDINGS AND ROOMS

Units have long identified a need to mark specific buildings and rooms during UO. Sometimes rooms need to be marked as having been cleared, or buildings need to be marked as containing friendly forces. The US Army Infantry School is currently testing a remote marking device that can be used to mark doors from as far away as across a wide street. In the past, units have tried several different field-expedient marking devices; some with more success than others. Chalk has been the most common. It is light and easily obtained but not as visible as other markings. Some of the other techniques have been to use spray paint, and paintball guns.

a. **Spray Paint.** Canned spray paint is easily obtained and comes in a wide assortment of colors including florescent shades that are highly visible in daylight. It cannot be removed once used. Cans of spray paint are bulky and hard to carry with other combat equipment. Paint is not visible during darkness nor does it show up well through thermal sights.

b. **Paintball Guns.** Commercial paintball guns have been purchased by some units and issued to small unit leaders. Some models can be carried in standard military holsters. They can mark a building or door from about 30 meters. The ammunition and propellant gas is not easily obtainable. The ammunition is fragile and often jams the gun if it gets wet. The available colors are not very bright, and just like spray paint, cannot be seen at night or through thermal sights.

c. **Wolf Tail.** A simple, effective, easy-to-make, lightweight device called a "wolf tail" can be fabricated to mark buildings, doorways, and windows (Figure I-1, page I-4). A unit has changed its tactical TSOP to require that each Infantryman carry one of these devices in his BDU cargo pocket. Wolf tails, when used IAW a simple signaling plan understood by all members of the unit, can aid in command and control, reduce the chances of fratricide, and speed up casualty collection during urban combat.

(1) The wolf tail marking device is simple to make and versatile. It can be used together with the NATO marking scheme. Rolled up, it makes a small, easily accessible package that can be carried in the cargo pocket of the BDUs. It can be recovered easily and used again if the situation changes. All its components can be easily obtained through unit supply. It combines a variety of visual signals (colored strapping and one or more chemlites of varying colors) with a distinctive heat signature that is easily identified through a thermal weapon sight. An infrared chemlite can be used either as a substitute for the colored chemlite(s) or in addition to them.

(2) Constructing the wolf tail marking device requires the following material:

- A 2-foot length of nylon strap (the type used for cargo tie-downs) (engineer tape can be substituted).
- About 5 feet of 550 cord.
- A small weight such as a bolt or similar object.
- Duct tape.
- Chemlites (colored and or IR).
- Two 9-volt batteries

(3) Assemble the items by tying or taping the cord to the small weight. Attach the other end of the cord to the nylon strapping, securing it with duct tape. Attach the 9-volt batteries in pairs to the lower end of the strapping with several wraps of duct tape, making sure that the negative terminals are opposite the positive, but not actually touching. Use more duct tape to attach the chemlites, approximately 2 inches above the batteries, to the strapping.

(4) When you want to mark your position, push the batteries together firmly until the male and female plugs lock. This shorts out the battery, causing it to heat up rapidly. The hot battery is easily identified through the thermal sights of tanks or BFVs. The batteries will remain visible for about 45 minutes. Activating the chemlites provides an easily identified light source visible to the naked eye. You can use infrared chemlites if you want them to be seen through night vision devices but not with the naked eye.

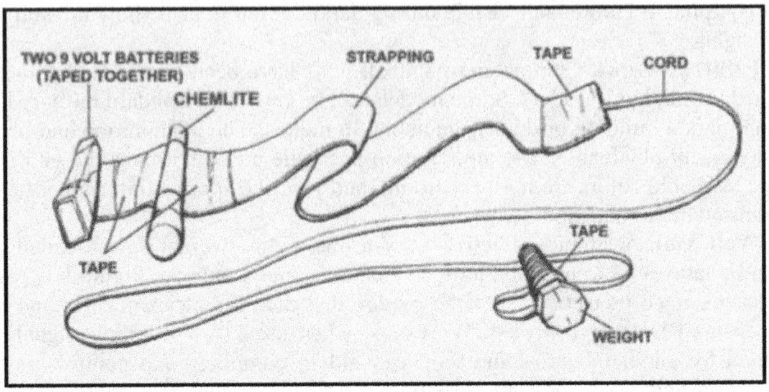

Figure I-1. Example of a wolf tail marking device.

NOTE: An option is to place chemlites and batteries at both ends of the wolf tail to mark the inside and outside of a building or room.

(5) Use the cord and the small weight to hold the wolf tail in position by tying or draping it out a window or hanging it on a door, wherever it is best seen by other friendly troops. Squads or platoons can vary the numbers and colors of chemlites, or use multiple battery sets to identify precisely what unit is in which building.

(6) Medics and combat lifesavers can carry a standardized variation that can be used to clearly identify a building as containing wounded personnel needing evacuation. This could be a white strap with multiple red chemlites, or any other easily identified combination.

I-4. NATO STANDARD MARKING SOP

The North Atlantic Treaty Organization (NATO) has developed a standard marking SOP for use during urban combat. It uses a combination of colors, shapes, and symbols. These markings can be fabricated from any material available. (Figure I-2 shows examples.)

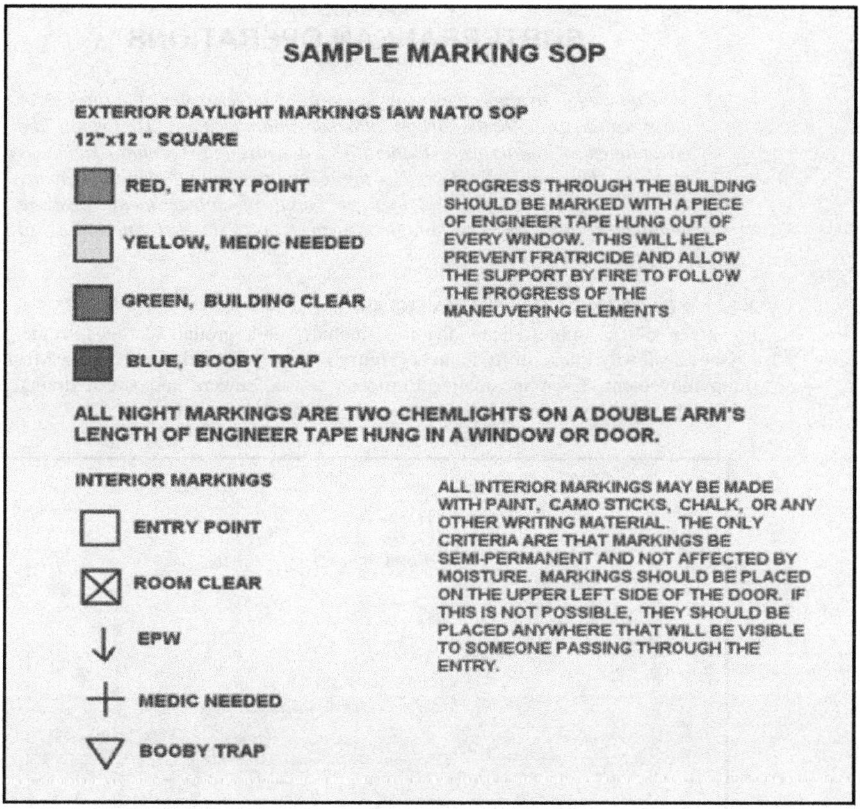

Figure I-2. Sample marking SOP.

APPENDIX J

SUBTERRANEAN OPERATIONS

Knowledge of the nature and location of underground facilities is of great value to both the urban attacker and defender. To exploit the advantages of underground facilities, a thorough reconnaissance is required. This appendix describes the techniques used to deny the enemy use of these features, the tactical value of subterranean passage techniques, and the psychological aspects of extended operations in subterranean passages.

J-1. PLANNING CONSIDERATIONS

In larger cities, subterranean features include underground garages, underground passages, subway lines, utility tunnels (Figure J-1), sewers, and storm drains. Most allow troop movement. Even in smaller European towns, sewers and storm drains permit soldiers to move beneath the fighting to surface behind the enemy.

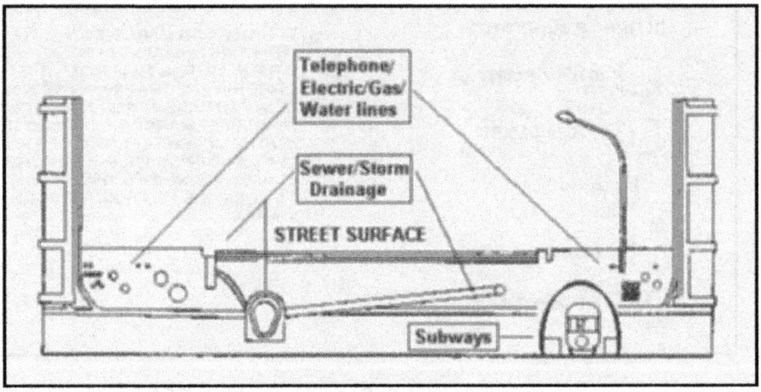

Figure J-1. Tunnels.

a. **General.** Subterranean passages provide covered and concealed routes of movement throughout the urban areas. A detailed knowledge of the nature and location of underground facilities is potentially of great importance to both the attacker and the defender. Maximizing the use of these facilities could prove to be a decisive factor during UO. Units planning to conduct subterranean operations should:

- Conduct a thorough analysis of the reasons to place troops in a subterranean environment. Subterranean combat is physically and psychologically demanding and must be carefully thought out prior to committing troops.
- Determine whether the need to use subterranean avenues of approach or occupy subterranean areas will facilitate mission accomplishment.
- Plan for redundant communications (messengers, wire, radios).

- Plan for additional weapons and ammunition that may be required for subterranean operations (shotguns, pistols, distraction devices, early warning, and so forth).
- Plan for and provide support above ground for those elements that are deployed in subterranean areas. Insure that situational awareness is maintained both above, as well as below ground.
- Consider that it may be easier to seal off access routes to underground passages and use smoke to flush out anyone hiding in them. Smoke displaces oxygen in enclosed areas like tunnels; even protective masks are useless in these situations.

DANGER

LARGE AMOUNTS OF ANY TYPE OF GAS CAN DISPLACE THE OXYGEN IN AN ENCLOSED SPACE. THIS CONDITION RENDERS PROTECTIVE MASKS USELESS AND ENDANGERS THE LIVES OF ANYONE OPERATING IN THIS TYPE OF ENVIRONMENT. RESPIRATORS THAT HAVE THEIR OWN OXYGEN SUPPLY ARE THE ONLY ACCEPTABLE SOLUTION WHEN OPERATING IN THIS TYPE OF ENVIRONMENT. THE PRESENCE OF RODENTS AND OTHER PESTS IN A SUBTERRANEAN ENVIRONMENT INDICATE THAT THERE IS AN ADEQUATE AMOUNT OF OXYGEN. SMOKE GRENADES MAY DISPLACE OXYGEN IN CONFINED SPACES.

THE PRESENCE OF FLAMMABLE GASES CAN CAUSE A MAJOR EXPLOSION WITH THE SLIGHTEST SPARK. THE FIRING OF A WEAPON COULD CAUSE AN EXPLOSION.

SOME GASES CANNOT BE DETECTED BY SMELL. THE ONLY SURE WAY TO PROTECT SOLDIERS FROM HARMFUL GASES IS TO VENTILATE THE PASSAGEWAY BY FORCING FRESH AIR INTO THE SITE. REMOVING A MANHOLE COVER DOES NOT ADEQUATELY VENTILATE A SUBTERRANEAN PASSAGEWAY.

b. **Sewers.** Sewers are separated into sanitary, storm, or combined systems. Sanitary sewers carry wastes and are normally too small for troop movement or protection. Storm sewers, however, provide rainfall removal and are often large enough to permit troop and occasional vehicle movement and protection. Except for groundwater, these sewers are dry during periods of no precipitation. During rain-storms, however, sewers fill rapidly and, though normally drained by electrical pumps, may overflow. During winter, melting snow may preclude their use. Another hazard is poor ventilation and the resultant toxic fume build-up that occurs in sewer tunnels and subways. The conditions, in sewers, provide an excellent breeding ground for disease, which demands proper troop hygiene and immunization.

c. **Subways.** Subways tend to run under main roadways and have the potential hazard of having electrified rails and power leads. Passageways often extend outward from underground malls or storage areas, and catacombs are sometimes encountered in older sections of cities.

d. **Subterranean Characteristics.** Underground passageways provide tight fields of fire. They amplify the effect of munitions such as grenades. If the tunnels are to be blocked, early warning devices and obstacles should be employed and the entry points should be secured. If tunnels are not to be used, entry points should be sealed. Manhole covers can be blocked with heavy weights or can be tack-welded if the capability is available. The insides of tunnels provide little or no cover and concealment except for the darkness and any man-made barriers. Obstacles at intersections in the tunnels set up excellent ambush sites and turn the subterranean passages into a deadly maze. These obstacles can be quickly created using fencing, barbed or concertina wire, along with rubble, furniture, and parts of abandoned vehicles interspersed with command-detonated explosives or mines. A thorough reconnaissance of the subterranean or sewer system must be made first. As opposed to storm systems, sewers contain various types of contamination. Careful consideration should be given before entering such systems. To be effective, obstacles must be located at critical intersections in the passage network so that they trap attackers in a kill zone but allow defenders freedom of movement.

e. **Tactical Value.** The tactical values of underground facilities are:

(1) ***Offensive Operations.*** During offensive operations the use of subterranean routes could enable the attacker to use both surface and subterranean avenues of approach, thus being able to infiltrate a smaller force behind the enemy's defenses. The purpose of this maneuver is to disrupt his defense and obstruct avenues of withdrawal for his forward defense. Depending upon the strength and depth of the above ground defense, the attack along the subterranean avenue of approach could become the main attack. Even if the subterranean effort is not immediately successful, it forces the defender to fight on two levels and to extend his resources too more than just street-level fighting.

(2) ***Defensive Operations.*** The existence of subterranean passages forces the defender to cover the urban area above and below ground with observation and fire. Subterranean passages are more a disadvantage to the defender than the attacker. They do offer some advantages when thoroughly reconnoitered and controlled by the defender. Subterranean passages provide covered and concealed routes to move reinforcements or to launch counterattacks. They can be used as lines of communication, for the movement of supplies and evacuation of casualties. Tunnels can also be used to cache supplies for

forward companies. Subterranean passages also offer the defender a ready-made conduit for communications wire, protecting it from tracked vehicles and indirect fires.

f. **Use of Weapons.** The confined space amplifies the sound of weapons firing to a dangerous level. Hearing protection is critical to allow soldiers to continue to function. The overpressure from grenades, mines, and booby traps exploding in a sewer or tunnel can have an adverse effect on troops. Also, gases found in sewers can be flammable, making this a double-edged weapon for both attackers and defenders. For these reasons, small-arms weapons should be employed as the main weapon system in tunnels and sewers. Friendly personnel should be outside tunnels or out of range of the effects when mines or demolitions are detonated.

g. **Local Knowledge/Navigation.** Prior to conducting an urban operation and especially a tunnel patrol, it is imperative that up-to-date local town plans and sewer maps are acquired if at all possible. Any locals with knowledge of underground routes must be questioned in detail. Without this kind of information, you must map as you go. Once below ground, pacing must be used as a guide to location.

h. **Threat.** The following threats must be considered for both offensive and defensive operations in subterranean passageways:

(1) ***Enemy Presence in Tunnels.*** It is likely that the enemy will want to use tunnels and they may have the advantage of marked routes and detailed reconnaissance. They may have the element of surprise, being able to select ambush positions and withdrawal routes. A defended position in an underground facility could be very effective in countering enemy subterranean operations. It should be well protected, canalizing the enemy into a killing zone to inflict maximum casualties.

(2) ***Booby Traps.*** When moving through tunnels, great care must be taken to avoid booby traps. These will normally be deployed near junctions and will often be operated by trip wires. Standing water in tunnels provides excellent camouflage for AP mines and booby traps scattered on likely routes. If moving without light, the lead man should feel for wires using a tripwire feeler (Figure J-2). He should also avoid walking in water if possible.

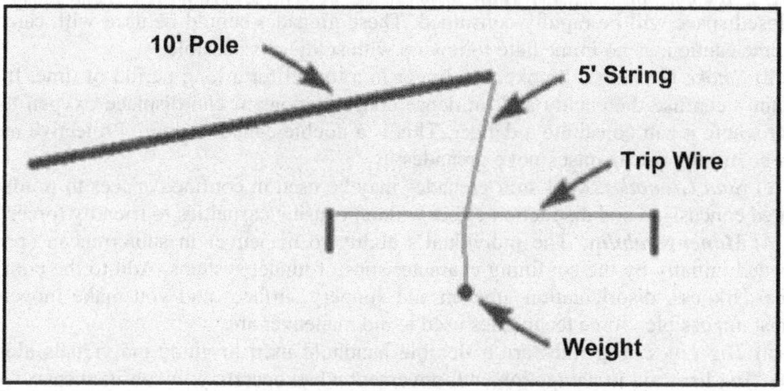

Figure J-2. Tripwire feeler.

(3) ***Residual Dangers.*** With the battle above continuing, and the possibility of artillery barrages and the use of demolitions, there is a strong possibility of flooding and a cave in. It is essential to identify escape routes.

(4) ***Individual Protective Equipment.*** To aid in the protection against small arms fire and fragmentation, it is advisable to wear all combat body armors available, as cover is limited inside tunnels. To avoid being stunned by blast and noise, all personnel that enter subterranean facilities must wear at least one set of hearing protection.

i. **Problems Involved with Tunnel Fighting.** There are several factors peculiar to tunnel fighting that restricts the soldier in his efforts to accomplish the mission. Some of those factors and solutions are:

(1) ***Target Detection.*** In the close confines of a tunnel, passive vision equipment such as NVGs, which require ambient light, are of little use. In order to quickly identify enemy personnel or other threats, an IR or white light should be used. The following equipment could be useful:

- NVG with IR source.
- IR filtered lights.
- White light flashlights.

(2) ***Visual Aids When Defending.*** When preparing defensive positions in the subterranean environment soldiers must be able to see what they are constructing. Battery-powered floodlights or bright flashlights can be used to do this. Because of the angles and lengths of tunnels, the ability to view around a corner without exposing oneself is extremely difficult. Mirrors can be used to look around corners and monitor major junctions for a defended position.

(3) ***Use of Grenades.*** Concussion and fragmentation grenades will produce a large shock wave and could, if used excessively, collapse the tunnel. The following alternatives should be considered:

(a) *WP Grenades.* Careful consideration must be given prior to the use of the white phosphorus grenade inside an enclosed space as the white phosphorus spreads, ignites, burns in the air, and causes extreme burns to the body. Any available oxygen in such an enclosed space will be rapidly consumed. These grenades should be used with care and extreme caution, as no immediate follow up with soldiers is possible.

(b) *Smoke Grenades.* Smoke will linger in a tunnel for a long period of time. It will certainly confuse the enemy and, in dense concentrations, it can displace oxygen to the point where it can constitute a danger. This is a double-edged weapon. Protective masks may be of little use against smoke grenades.

(c) *Stun Grenades.* M84 stun grenades may be used in confined spaces to produce a limited concussion and distraction effect without causing casualties to friendly forces.

(4) ***Maneuverability.*** The individual's ability to maneuver in subterranean areas is impeded initially by the confining characteristics of tunnel systems. Add to the confined space darkness, disorientation, uneven and slippery surfaces and you make movement almost impossible. Some techniques used to aid maneuver are:

(a) *Tag Lines.* Tag lines are a flexible handhold used to guide individuals along a route. Tag lines aid in navigation and movement when operating in confined spaces such as buildings, tunnel systems and caverns where visibility is limited and sense of direction can be lost. These lines can be made of rope, string, cable, wire and so forth. The most effective item to be used as a tag line is WD-1A communications wire. Along with serving as a tag line it can be used as a primary means of communicating to elements above ground.

(b) *Safety Lines.* Ropes can be used not only as tag lines to ease movement through tunnels, but also to attach team members together to avoid breaks in contact. This safety line should be tied to team members leaving 5-meter intervals between them.

(c) *Wire Mesh.* Attaching wire mesh to the bottom of combat boots will enhance traction in some situations.

(5) ***Degradation of Radio Communications.*** Communications inside the tunnels will be severely degraded. Several methods to overcome this are:

- Land line and field telephone.
- Use of messengers.
- Increased power setting on radios.

(6) ***Limitation of Firepower.*** The limited use of available firepower will favor the defender. Prepared positions should be able to hold off a much larger force. Weapons best suited to the defender are:

- Automatic weapons.
- Shotguns or other scatter type munitions.
- Flame producing weapons.

NOTE: Any type of flame or incendiary weapon needs to be used in a well-vented area because of the smoke created and the oxygen used.

(7) ***Chemical Hazard.*** A constant concern for troops conducting subterranean operations is chemical defense. Enemy chemical warfare (CW) agents used in tunnels may be encountered in dense concentrations, with little chance of it dispersion (no wind). A chemical agent alarm system, carried by the point man, will provide instantaneous warning of the presence of CW agents. M8 and M9 detection papers should also be used to test for the presence of chemical agents. Noxious gases from decomposing sewage can also pose a threat because they displace oxygen and because they are not filtered by the soldier's protective mask. Unit leaders must be constantly alert for the physical presence and symptoms of gases on soldiers.

(8) ***Psychological Considerations.*** Combat operations in subterranean passages are much like night combat operations. The psychological factors of night operations reduce confidence, cause fear, and increase a sense of isolation. This is further magnified in the confines of the tunnels. As was discovered during tunnel operations in the Vietnam War, many personnel are unsuited for operations below ground. The layout of tunnels could require greater dispersion between positions, further enhancing the feeling of isolation.

(a) *Overcoming Fear and Anxiety.* Leaders must conduct reinforcement training to help suppress the fear and anxiety experienced when isolated in tunnels. These measures include leadership training, physical and mental fitness, sleep discipline, and stress management.

(b) *Communications.* Leaders maintain communication with soldiers manning positions in the tunnels either by personal visits or by field telephone. Communications inform leaders of the tactical situation as well as the mental state of their soldiers. Training during combat operations is limited; however, soldiers manning positions below ground should be given as much information as possible on the organization of the tunnels and the importance of the mission. They should be briefed on contingency plans and alternate positions should their primary positions become untenable. All members, both above and below ground (Figure J-3), must know recognition signals.

Figure J-3. Recognition signals.

(c) *Stress.* Physical and mental fitness can be maintained by periodically rotating soldiers out of tunnels so they can stand and walk in fresh air and sunlight. Stress management is also a factor of operations in tunnels. Historically, combat in urban areas has been one of the most stressful forms of combat. Continuous darkness and restricted maneuver space cause more stress to soldiers than street fighting.

J-2. SUBTERRANEAN RECONNAISSANCE TECHNIQUES

Local reconnaissance should be given to a squad-size element (seven to nine personnel). Enough soldiers are in a squad to gather the required data without getting in each other's way in the confines of the tunnel. Only in extremely large subterranean facilities should the size of a patrol be increased.

a. **Task Organization.** The squad leader organizes his squad for movement. One fire team leads with the rifleman as the point man, tasked with security to the front. The A-team leader moves directly behind the point man, controlling movement. Behind the team leader is the M249 gunner, ready to move forward and provide suppressive automatic fire if required. The squad leader moves directly behind the M249 gunner and records data. The team grenadier follows the squad leader and is the pace man during movement. Following the grenadier is the rifleman of the other team; he is responsible for rear security and rolling out the "tag line." The team leader, M249 gunner, and grenadier remain at the point of entry as a security post. They are responsible for detecting enemy and serving as the communications link between the squad leader and his higher headquarters.

b. **Special Equipment.** The squad leader should carry a map, compass, street plan, and notebook in which he has written the information he must gather for the platoon leader. The grenadier should carry tools needed to open manhole covers. If the patrol is to move more than 200 to 300 meters or if the platoon leader directs, the rear team rifleman should also carry a sound-powered telephone (TA-1) and wire dispenser for communications because radios are unreliable in this environment. The point man should be equipped with night vision goggles (NVGs) and an IR source to maintain surveillance within the sewer.

c. **Specific Individual Equipment.** All soldiers entering the sewer should carry a sketch of the sewer system to include magnetic north, azimuths, distances, and manholes. They should also carry protective masks, flashlights, gloves, and chalk for marking features along the route. If available, all soldiers entering subterranean facilities should be equipped with night vision goggles.

d. **Initial Entry.** Once the squad is organized and equipped, it moves to the entrance of the tunnel, which is usually a manhole. With the manhole cover removed, the patrol waits 15 minutes before entry to allow any gases to dissipate. The point man descends into the tunnel (safety line attached), to determine if the air is safe to breathe and if movement is restricted. The point man should remain in the tunnel 10 minutes before the rest of the squad follows. If the point man becomes ill or is exposed to danger, he can be retrieved by use of the safety rope. (See Figure J-4.)

Figure J-4. Squad subterranean reconnaissance patrol.

e. **Movement Techniques.** When the squad is moving through the tunnel, the point man moves about 10 meters in front of the team leader. Other squad members maintain 5-meter intervals. If water in the tunnel is flowing faster than 2.5 meters per second or if the sewer contains slippery obstacles, those intervals should be increased to allow for squad members to react if one man slips. If using a safety rope, all squad members should remain tied in so that they can be retrieved from danger. The rear security man marks the route with the "tag line" so other troops can find the squad.

f. **Mapping.** The squad leader should note the azimuth and pace count of each turn he takes in the tunnel. When he encounters a manhole to the surface, the point man should open it and determine the location, which the squad leader then records. The use of recognition signals prevents friendly troops from accidentally shooting the point man as he appears at a manhole.

g. **Follow-on Tasks.** Once the squad has returned and submitted its report, the platoon leader must decide how to use the tunnel. In the offense, the tunnel could provide a covered route to move behind the enemy's defenses. In the defense, the tunnel could provide a covered passage between positions. In either mission, the squad members can serve as guides along the route.

J-3. DENIAL TO THE ENEMY

Subterranean passages are useful to the defender only to the extent the attacker can be denied their use. The defender has an advantage, given the confining, dark environment of these passages; a small group of determined soldiers in a prepared defensive position can defeat a numerically superior force. Tunnels afford the attacker little cover and concealment except for the darkness and any man-made barriers. The passageways provide tight fields of fire and amplify the effect of grenades. Obstacles at intersections in the tunnels set up excellent ambush sites turning the subterranean passages into a deadly maze. These obstacles can be quickly created using fencing, barbed or concertina wire, along with rubble, furniture, and parts of abandoned vehicles interspersed with Claymore mines. A thorough reconnaissance of the subterranean passages or sewer system must be made first. Once the reconnaissance has been conducted, obstacles must be located at critical intersections in the tunnel network so they trap attackers in a kill zone but allow defenders freedom of movement (Figure J-5).

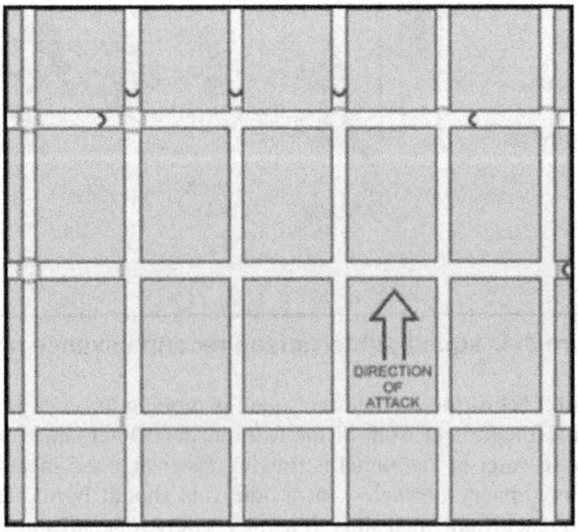

Figure J-5. Defense of a sewer system.

APPENDIX K

TACTICS, TECHNIQUES, AND PROCEDURES FOR THE EMPLOYMENT OF MORTARS ON URBAN TERRAIN

The following information was extracted from the Berlin Brigade, USCOB/USAB Pamphlet 350-1, Combat in Cities Procedures, dated 13 July 1984. The information contained herein was updated and reviewed by subject matter experts (SMEs). This appendix is provided as additional TTP for the employment of mortars during urban combat.

K-1. TARGET ENGAGEMENT

This paragraph describes the high angle fire, range and deflection probable errors, shapes of targets, limitations on the use of mortars, round adjustment, and effective use of mortars.

a. **High Angle.** The masking caused by the buildings necessitates the use of high angle fires to attack targets in the streets or behind buildings. Firing, at high angles, decreases the range, but this is often offset by the fact that proximity to the enemy generally will be close. In urban terrain, all mortar sections should be firing at the highest elevation possible. Use of high angle fires increases the probability of acquisition of the firing unit by enemy counterfire radar and therefore increases the frequency and accuracy of counter-mortar fires. This increased threat will be reduced by careful positioning of the mortars, avoiding fire missions which would not be effective, and moving mortar systems into covered hide positions or to alternate positions.

b. **Range and Deflection Probable Error.** The mortar is an area weapon. The single-mortar, firing multiple rounds, lands dispersed even if the gun is laid identical each time it is fired. One range probable error is the distance which, when added to and subtracted from range to the planned point of impact, will include 50 percent of the rounds fired at the planned point of impact. One deflection probable error is the distance which when added left and right of the gun target line will include 50 percent of the rounds fired at the planned point of impact. When the target lies between buildings, the variance might cause the rounds to land on top of a building or on the other side of a building even if the mortar was adjusted correctly. This may even cause the round to be unobserved. For example, a 60-mm mortar firing at a range of 3,000 meters at a high elevation of fire has a range probable error of 20 meters. This means 25 percent of the rounds fired could land more than 20 meters beyond the target and 25 percent could land less than 20 meters short of the target. The expected result is that only 50 percent of the rounds fired could land in an oval area that is 40 meters long. Human error will decrease the number of rounds that land close to the target. Deflection probable errors are much smaller and not as much concern for the 60-mm and the 81-mm mortar.

c. **Shapes of Targets.**

(1) ***Point Targets.*** Point targets are the most common type of target that mortars will engage due to restricted sight lines, numerous street intersections used as adjustment points, and kill zones as wide as a street. Point targets can be engaged by a single gun, a section, or platoon with a converged sheaf. Using a high rate of fire from a single tube

will put a higher percentage of rounds in a small target area as opposed to using the same number of rounds from multiple tubes.

(2) ***Linear Targets***. Linear targets occur along streets. The streets may be perpendicular to the gun-target lines such as an FPF directly in front of a battle position or parallel or at some other angle to the gun-target line, such as a street which approaches the objective. For linear targets, the FO includes the attitude of the target with the call for fire. The FDC may have to issue separate gun data to each mortar to orient the sheaf correctly to bring effective fire on the targets.

(3) ***Area Targets***. Area targets are not as common in urban combat as point or linear targets. However, area targets may occur in parks which may be used as staging areas, or behind friendly lines during enemy air assault operations, or during an attempt to mass forces such as in an assembly area or an assault position.

d. **Limitations on the Use of Mortars.**

(1) ***Dead Space***. Enemy targets close to the base of buildings on the side away from the firing mortars cannot be effectively engaged. When firing at the maximum elevation possible, the dead space behind the building is greatly reduced. However, while firing at high angles decreases the amount of dead space, it cannot be eliminated.

(2) ***Observation***. Observation is severely restricted. Enemy targets are often only visible when they are within one block of the observer or on the same street as the observer. Positions in tall buildings can provide long range observation, but normally only from the tops of buildings. To engage the enemy, the FO should be positioned forward. Many fire missions will either be on streets that lead up to friendly positions or will be called in on targets that are within one block of the observer or friendly positions. Probable errors associated with mortars necessitate friendly positions be chosen and constructed to withstand an occasional mortar round striking the top or rear of the building in which they are constructed

(3) ***Penetration***. Mortar rounds are not effective in penetrating buildings. Rounds with delay fuzes sometimes penetrate the top floor, but damage is limited by interior walls, and the rounds do not normally penetrate to lower floors. Mortar rounds that do not impact directly against mass construction walls with the fuze may not cause much damage to personnel in the interior and in many cases will fail to explode. Mortar rounds will penetrate light walls on frame construction buildings but the damage is usually confined to one or two floors because of the heavy floor construction normally found in high rise frame buildings. For deeper penetration, HE delay can be used. HE quick is effective in lightly built structures found in some suburban areas. HE quick would also be effective in the flimsy construction usually found in shantytowns.

(4) ***Proximity Fuzes***. Proximity fuzes function erratically when used in the vicinity of buildings. They are still effective if used in large open areas (for example, parks, parking lots, and so forth).

(5) ***White Phosphorous***. White phosphorous (WP) smoke rounds will often start fires. This may interfere with the tactical plan and may cause civilian casualties. Note that all mortar rounds currently in the inventory are WP.

e. **Round Adjustment.** Adjustment of rounds will be very difficult because of the large percentage of the urban area that probably cannot be observed. If an adjustment round is 50 meters off, it may land a full block away. The FO would hear echoes, but have no indication where the round landed and what corrections are necessary. Using time fuzes to cause adjusting rounds to burst just above the height of the buildings will ensure the FO can observe the rounds. Time fuzes will be used until the rounds are adjusted above the target.

f. **Effective Use of Mortars**. Mortar rounds must be carefully used and not wasted. Resupply of mortar ammunition may not be immediately available. Unnecessary firing will increase the likelihood of being engaged with accurate counter-mortar fire. Ineffective mortar fires may cause collateral damage and civilian casualties.

(1) ***Armored Vehicles***. Mortar fire normally will not disable armored vehicles, but it can be a combat multiplier when used with direct fire weapons. Indirect fire is effective in forcing the enemy to button up during movement, will slow his advance, make it hard for him to determine his exact location, decrease the probability he will see our mines, and assist friendly light antiarmor weapons fire before the enemy can return accurate fire. The enemy can take casualties from mortar fire during Infantry assaults against our positions or if he exposes himself.

(2) ***Smoke Missions***. Smoke missions are vital in urban combat in order to provide obscuration for assaulting or withdrawing forces. When planning fire support, careful analysis needs to be given to the length of required missions and the amount of rounds needed to support the coverage requested. Planners must also account for the duration the smoke lasts when planning for WP smoke, which is much longer in an urban area than in open terrain. In some cases, smoke mission requests may have to be modified based on the amount of rounds available for the mission. It is important to remember that when requesting smoke missions from mortars, WP rounds are used. These rounds burn until all the oxygen in the immediate area or the white phosphorous is exhausted.

(3) ***Obstacles and Minefields***. Covering obstacles and minefields with mortar fire will decrease the enemy's ability to view the extent of the barrier, slow his breaching efforts, and cause casualties if he tries to use Infantry to bypass or clear the obstacle. consistent with the ROE. Airbursts should be employed if the obstacle includes mines or concertina wire. Ground burst rounds would destroy much of the barbed wire and detonate many of the mines. Massed fires of mixed point detonating fuzes and time fuzes are effective if the enemy is breaching existing obstacles.

(4) ***Tops of Buildings***. Enemy soldiers can be forced off building tops by using HE rounds with time or proximity fuzes. When firing at the top or upper stories of buildings, it is most important the FO provide the vertical interval.

(5) ***Attics***. Targets located in an attic or on the floor immediately below the location of the enemy in mass construction buildings can be engaged with delay fuzes. Mass construction buildings have weak roofs and attic floors. Because these are point targets, only one gun should be used.

(6) ***Fronts of Buildings***. Enemy hasty positions or observers in the front side of buildings, or in a large open area in front of the building, can be engaged using proximity fuzes. Effectiveness will depend on the amount of window surface. Shrapnel normally will not penetrate walls and most casualties will be caused by the secondary shrapnel hazard of flying glass. For this use, the trajectory of the rounds should be the lowest point possible which clears the buildings along the gun target line, and which enables the rounds to reach down the building far enough to hit the target. If the goal is to blow glass into the street to cause casualties, delay fuzes should be used. (The 60-mm mortar in the handheld mode can be very effective against this type of target.)

(7) ***Enemy Air Assaults***. Enemy air assaults or raids behind friendly lines can be broken up and affected if immediately engaged with massed indirect fires using HE with PD and proximity fuzes

(8) ***Final Protective Fires***. Final protective fires planned immediately in front of friendly positions will cause casualties if the enemy attempts to move dismounted across the final gap before reaching friendly positions. Proximity fuses should be used to avoid destroying obstacle belts. The same fires are necessary if the enemy gains a foothold. Mortar fires will aid in sealing the breach to prevent reinforcement.

(9) ***Counterattacks***. If the enemy counterattacks, the commander may call for mortar fire on his own position. Friendly soldiers will be exposed to this fire and protection should be sought inside buildings. Smoke rounds should not be used because of the fire hazard

(10) ***Use of WP or Illumination***. WP or illumination rounds ignite fires that may burn or smoke enemy out of buildings. Because of heat, the building may be unusable to the enemy for days. Effects to be considered before using this technique are the possibility of friendly casualties from stray rounds and large fires, the possibility that burning buildings or heavy smoke will interfere with planned operations, the possibility of collateral damage and civilian casualties, and whether use is permitted by the ROE. Illumination is greatly influenced by the presence of buildings. Deep canyons formed by buildings severely limit the effect and duration of illumination on the target even if properly placed. Illumination rounds should be planned to place friendly soldiers in the shadows and place enemy troops in the light (Figure K-1). Because of the shadows produced by the buildings and the drift of the illumination round, effective illumination may be for a short duration. The FDC needs to calculate where the illumination shell casings will impact and inform friendly units in their path.

Figure K-1. Mortar illumination.

K-2. CONTROL

Mortars are responsive to calls for fire from platoon FOs and are capable of massing fires to have a decisive effect at key points in the battle. Control of the 60-mm section is normally retained at company level. It may be consolidated at battalion if the commander determines the primary use of mortars in his sector is to support the effort of an assaulting company, or to defeat breaching efforts by the enemy at a major obstacle. Most of time the massing of fire can be achieved without having to consolidate the individual mortar sections by having the mortars fire under the battalion FSO's control from company positions. In these special cases, the loss of responsiveness to the company commander may be outweighed by the need for the battalion commander to be able to immediately mass the battalion's mortar fires. Frontages covered by companies are small enough that there is no advantage to be gained by attaching individual 60-mm mortar tubes to individual rifle platoons.

a. **Communications and Fire Missions.** The mortar section maintains communications with its company commander and with the FIST chief on the company command net. When a wire line has been run from the company to the mortar section, the radio can be placed on an alternate frequency which can provide a dedicated fire support net or it can be used as redundant communications. This provides a radio net for the FO or units to use in order to make calls for fire if the wire line to the company becomes overloaded. Each company commander has a dedicated indirect fire system. The company FIST chief will plan and coordinate the company's calls for fire. Most needs are met by the company mortar section. If additional support is needed, it will be requested through the battalion FSCOORD. The battalion commander, through the FSCOORD, can take direct control of indirect fire assets to meet an urgent tactical situation.

b. **Displacements.** The company mortars can displace with authorization from the company commander, but coordination with the battalion FSO should also be made. The battalion FSO will ensure mortars from more than one company are not displacing at the same time and will coordinate availability of additional fire support, if needed, while some mortars are out of action during movement. This support may be provided by the battalion mortar platoon or an adjacent company's mortars.

c. **Preplanned Targets.** Targets should be preplanned prior to any operation. Target lists and overlays of the entire battalion sector will be prepared for each mortar FDC, to aid sections providing temporary support for adjacent companies and responding to situations that require immediate massed fires from all the battalion's mortars. This is especially helpful if a company has been given the mission to support by fire and isolate an urban objective. Having accurate information at each FDC will decrease the likelihood of causing friendly mortar casualties and enhance situational awareness.

K-3. MORTAR POSITIONS

The distance that mortars are kept behind friendly positions depends on the observation available to the FO and how far forward the FO is located in the offense, or how far forward the LP/OP and or FO are located in the defense. The maximum range of the platoon is reduced if it needs to fire at high angles to engage targets. Mortars should not be so close to friendly positions that they will be overrun or forced to displace if the enemy achieves a minor breakthrough in the defense or counterattacks in the offense. The position should enable fire to be brought on friendly positions to counter enemy breaches. Logistical support of the 60-mm mortar section and wire communications is less difficult if the section is located relatively close to the company trains and CP.

a. **Use of Buildings for Protection.**

(1) ***Choosing Positions.*** By properly choosing mortar positions, protection can be obtained from enemy counter-mortar fires without restricting the ability to engage targets

with indirect fire. Mass construction buildings to the front and buildings to the sides and rear of positions can limit the mortars' exposure. If friendly mortars are close to a tall mass construction building and firing at near maximum elevation, they will be fairly well protected from counterbattery fires. Friendly mortars should move as close to the buildings as possible while maintaining clearance to fire over the building. The laid gun should be checked to ensure the rounds will clear the frontal mask. This positioning puts the weapon systems well within the enemy's dead space, if he is using flatter firing trajectories. Mortars should not be positioned close to buildings with a large percentage of surface area made of glass, because of the secondary shrapnel hazard of glass.

(2) ***Low Buildings.*** Low buildings close to the mortars on any side do not provide much dead space in which to position mortars, however, they will offer some protection from shrapnel from that side. If the area from which the mortar is firing is relatively small and the adjacent walls are fortified or strong enough to stop shrapnel, then the incoming rounds have to be almost a direct hit in order to damage the mortars or crew.

(3) ***Hide Positions.*** Hide positions (for example, large garages, store fronts, underpasses) can be used to protect the mortars before firing.

(4) ***FDC Location.*** The FDC can be placed in a parking garage, ground floor garage in a larger building, or covered vehicle passageway to protect it.

(5) ***Buildings Near Mortar Positions.*** Buildings near mortar positions obstruct the view of the mortars, making the position less likely to be discovered by the enemy.

b. **Ground Mounting Mortars.** There are only limited patches of soil available to ground mount mortars. These patches are rarely in the optimum location to ensure the mortar is protected by surrounding buildings and not masked. If mortars must be fired on a paved surface, baseplates should be supported with sandbags or any other non-skid surface that may be available.

c. **Positions.** Possible positions for 60-mm section and 81-mm mortar platoons are shown in Figures K-2 and K-3 on pages K-7 and K-8 respectively.

(1) ***Courtyard Positions.*** Positions in courtyards in the center of city blocks provide excellent protection from indirect fire. They aid local security by concealing the mortars (unless an observer is on the same block). By shielding the sound of firing, the surrounding buildings make it difficult to locate the guns by listening to fire. The primary concern with this position is ensuring adequate access routes. A single small passage through a building that could be blocked by minor rubbling is unacceptable. There should be at least two access routes not susceptible to blockage from minor rubbling.

(2) ***Gun Positions.*** Because many targets will be on streets, linear fire missions will be routinely planned. Consideration should be given to emplacing mortars at street intersections, so linear targets can be engaged quickly without the FDC having to adjust the sheaf. The shapes shown in Figure K-2 on page K-7, and Figure K-3 on page K-8, can assist in decreasing ammunition waste, provide adequate access routes, and simplify local security. Positions should be chosen away from likely enemy avenues of approach.

(3) ***Local Security.*** Local security is necessary to guard against destruction by the enemy and to protect supplies from pilferage and vandalism by local personnel. If an LP/OP is positioned to observe approaches to the section's positions, it must be close enough to return to the guns rapidly and conduct fire missions. Section members will have assigned sectors of fire and a position defense plan will be prepared. Claymores are set out to cover possible Infantry approaches. It is preferable for the mortar section to

displace rather then stay in position and fight a decisive battle. Underground approaches are identified and either blocked or kept under observation.

(4) ***Vehicles.*** If vehicles are used, they should be parked under cover and camouflaged against ground and aerial observation. Debris, cars, trucks, and other items found in the vicinity can be used to break up the outline of friendly vehicles without causing the position to attract attention by not matching the surroundings.

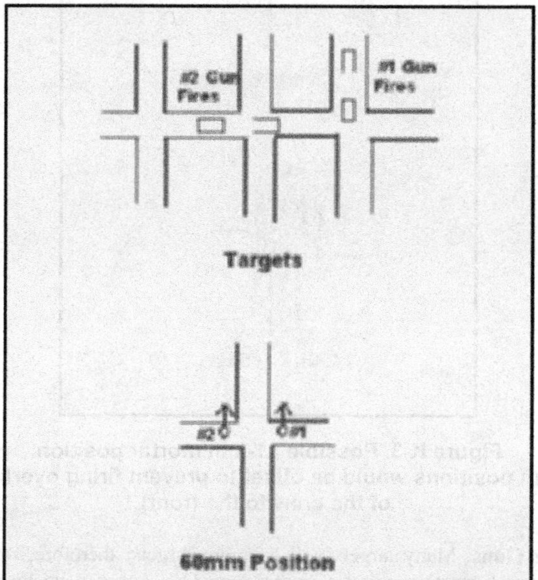

Figure K-2. Possible 60-mm mortar position.

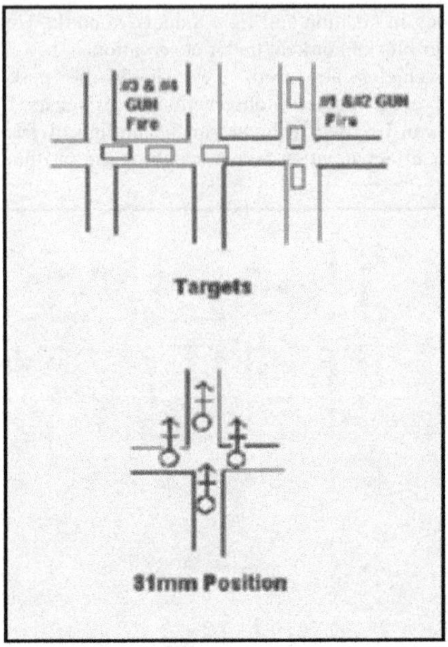

Figure K-3. Possible 81-mm mortar position.
(Gun positions would be offset to prevent firing overhead of the crew to the front)

d. **Roving Guns.** Many targets will be point targets; therefore, it is possible for 60-mm mortar sections to employ a roving gun. One gun can be located in a firing position with the FDC and the other gun can be located either forward or rearward with wire communications to the FDC. (For an 81-mm platoon, two guns would he located in a firing position with the FDC and the remaining gun(s) may be 400 meters away, with wire communications to the FDC.) The other gun(s) would engage point targets without drawing counter-mortar fire on the rest of the section or platoon. If a linear fire mission or higher rate of fire were required, the two-gun section would fire that mission. Because it would fire more missions, the single gun would be expending the most ammunition and would be the most probable target for counter-mortar fire. When it ran low on ammunition or had to displace due to incoming indirect fire, it would displace back to the position near the FDC and the other gun would assume the role of roving gun. This method provides the section the most survivability from counter-mortar fire and ensures continuous fire support will be available, but it increases the difficulty of maintaining local security because the section or platoon is in two locations.

e. **Laying the Section.** Magnetic instruments, for example, compasses, are affected by the presence of massive amounts of structural steel and electrical cables usually found in urban terrain. During one test in Berlin conducted in the 1980's, the variation in deflection on an aiming circle was more than 70 mils.

(1) ***Orienting Line Method.*** Use of the orienting line method with the aiming circle avoids reliance on magnetic azimuths. The steps are:

- Set up the aiming circle on a point that can be located to 8-digit accuracy on a map.
- Locate a point several hundred meters away (a minimum of 200 meters away, but the farther the better) that can be seen from the aiming circle and located to 8-digit accuracy on a map.
- Determine the grid azimuth on the map from the aiming circle location to the distant point.
- Index the grid azimuth on the aiming circle.
- Use the non-recording motion to place the vertical crosshair on the distant point.
- Index zero on the aiming circle and place an aiming stake to serve as a north reference point.
- Subtract the mounting azimuth from 6400 and place the remainder on the aiming circle. Use the non-recording motion to place the vertical crosshair on the left edge of the north reference stake.
- Lay the section using the aiming circle.

(2) ***Laying the Mortar Without a Compass.*** If an aiming circle is not available, the mortar can be laid without using a compass by a similar procedure.

- Set up the mortar on a point that can be located to 8-digit accuracy on a map.
- Locate a point several hundred meters away (a minimum of 200 meters, the farther the better) that can be seen from the mortar sight and located to 8-digit accuracy on a map.
- Determine the grid azimuth on the map from the mortar location to the distant aiming point. The direction of a street may be used instead of a distant aiming point, if necessary.
- Subtract the back azimuth of the direction of fire from the grid azimuth.
- Index the difference on the red scale and the mortar sight and manipulate the gun until the vertical crosshair of the sight is on the aiming point. The gun is now laid.
- Use the sight of this mortar to lay other mortars in the same position.

(3) ***Aiming Stakes.*** It is usually impossible to drive in aiming stakes because of the limited amount of open soil. Two ammunition cans filled with sand can be used on each gun. The stakes stand in the open can supported by the sand. If soil is used instead of sand, the stake will fall when it rains. Distant aiming points can be used instead of aiming stakes, but the distant point may not be consistently visible during the battle because of the large amount of smoke.

(4) ***Identification of Key Locations.*** During planning and identification of mortar positions, the exact location of the aiming circle or mortar and the distant aiming point must be identified. These identified locations and the grid azimuth can be calculated and included in prepared orders, this should assist in position occupation.

f. **Continuing Preparation.** After the position has been occupied, communications established, and the initial fire coordination conducted with the company FIST and the battalion FSO, the section leader can conduct reconnaissance of alternate and supplementary positions. Alternate positions are used to avoid destruction by counter-mortar fires. Supplementary positions are used if the section is forced to withdraw from its battle positions or sector. This reconnaissance is to review the sites and confirm there have been no major changes since the section's plans were last updated.

APPENDIX L

COMMUNICATIONS DURING URBAN OPERATIONS

The complexities of the urban environment such as line-of-sight restrictions, inherent fortifications provided by structures, limited intelligence, densely constructed areas, and the presence of noncombatants can restrict military technology. Communications in an urban environment are extremely important. Prior planning is essential to ensure continuous communications. Units should anticipate possible communications failures during UO. Nonelectronic communication signals must be planned for and practiced as alternative methods. This appendix addresses communications planning, communications methods, alternative communications, problems and possible solutions encountered with planning communications during UO.

L-1. INTRODUCTION

Each urban area and operation is unique and prescribing specific doctrinal *solutions* for communications is virtually impossible. The information in this appendix provides a basis for approaching UO, which, combined with other appropriate Army doctrine, will assist units in achieving mission success during UO.

L-2. URBAN INFRASTRUCTURE: COMMUNICATIONS AND INFORMATION ELEMENT

This is the element of the urban infrastructure that is comprised of the facilities and means to transmit information from place to place. It includes telecommunications such as telephones (to include wireless), telegraphs, radios, televisions, and computers; newspapers, magazines, and other forms of written communications; as well as human interaction that conveys information. Perhaps more than any other element of the infrastructure, communications and information link all the other elements together in an interdependent "system of systems."

a. **Computers.** In many urban areas, computers are an essential link between other elements of the urban infrastructure. They link not only functions and systems within the urban area, but also connect the urban area to other parts of the world. This latter aspect creates important implications for commanders. Operations involving this cybernetic function may produce undesirable effects on a greater scale than initially intended. For example, commanders may be able to close or obstruct an urban area's banking system; however, this system may impact the international monetary exchange with unwanted or even unknown effects. Decisions to conduct this type of information operations may often be retained at the strategic level. Conversely, shaping the urban battlespace through the electronic isolation of selected decisive points may often be a key factor in permitting units to dominate their objectives.

b. **Alternate Communications Systems.** Urban communications and information systems can serve as an alternate or backup means of communications for both friendly and threat forces and can be easily secured with civilian, off-the-shelf technologies. Adversaries may make use of commercial systems intertwined with legitimate civilian users, making it undesirable to prevent use of these assets. Army forces can also use these systems to influence public opinion, gain intelligence information, support deception efforts, or otherwise support information operations.

L-3. COMMAND AND CONTROL

The urban environment challenges C2 systems that support the commander. Perhaps one of the biggest challenges will be horizontal and vertical communications between and within units.

a. **FM Limitations.** The physical characteristics of manmade construction may significantly degrade FM communications. The frequency spectrum becomes more confined based on structures that would degrade all communications, especially line of sight (LOS) within the urban environment. This causes problems at brigade-level and below where commanders rely heavily on constant FM radio contact with subordinates. Tactical communications problems can result in many other issues including an inability to maintain a common operational picture (COP), give orders and guidance, request support, or coordinate and synchronize elements of the combined arms team. Communications problems within urban areas can contribute directly to mission failure.

b. **Measures to Mitigate Urban FM Limitations.** Units and staffs that properly prepare can mitigate the communications problems within urban areas. During offensive operations, use of retransmission and relay sites and equipment, careful positioning of commanders and command posts, careful placement of antennas, and proper communications procedures will permit adequate FM communications for mission accomplishment. In defensive, stability, or support operations, where positions do not change frequently without preparation, units should consider increasing their reliance on wire, commercial communications, and messengers. Even under combat conditions, some or all of the urban area's organic communications structure may be intact and available for use. C2 systems must use these alternatives to FM communications utilizing proper security procedures.

c. **Information Overload.** Urban areas have the potential of overloading C2 systems with information. Full spectrum urban operations can generate large volumes of information when crises threaten. The sheer volume of tasks that require attention in UO can overwhelm commanders and command posts. Training must prepare command posts to handle this volume of information and to filter the critical from the merely informative.

L-4. URBAN COMMUNICATIONS PLANNING

Communications planning for UO will assist in minimizing the effects of the urban environment. VHF radios are screened and communication ranges reduced in urban areas. Radios must be carefully located to maximize their effectiveness. Retransmission stations and remoting of antennas on taller structures maximize the communications range. Ground units attempting to communicate with aircraft or other ground units should use the upper end of the VHF band and high power switches on radios. Commanders must set limited objectives covering a small area and plan for the frequent relocation of retransmission stations. If time and METT-TC factors permit, maximum use should be made of the civilian telephone system.

L-5. COMUNICATIONS METHODS

During Operation Just Cause, the two most significant elements of the Army communications network were the airborne command and control center (ABCCC) and satellite communications (SATCOM) radios utilized by the ground and air elements. In addition, multi-channel unit radios provided essential command and control links. Several units augmented their administrative communications with commercial "walkie-talkies" (brick radios). Brick radios were used for perimeter security and communications within logistical areas.

a. **Retransmitters.** Currently, radio retransmitters are used to communicate with elements on perpendicular and other streets. If the urban terrain has tall buildings with iron support beams, or if the buildings have metal roofs, position UHF transmitters some 2 to 5 times the height of the intervening building away from that building.

b. **Directional Antennas.** A directional antenna can use stone or brick walls as retransmitters to bounce signals down a street. When trying to bounce radio signals off adjacent buildings, avoid aiming antennas at windows. The reflective properties of glass are different from brick or stone. Communications using a directional antenna with a clear LOS to the other station is best. When a tall object such as a church is located between the stations, the stations should aim their directional antennas at a common point. Avoid positioning radios near power and telephone lines. If a radio is located inside a building, position the antenna on an upper-story or window facing the receiving station or on the rooftop. A directional antenna has better range than a whip antenna. Use a 10- to 15-meter-long cable to connect the radio to the antenna, but avoid longer cables as they weaken the transmission signal.

L-6. ALTERNATIVES TO FM RADIO

Units should consider alternatives to FM radio communications.

a. **Wire.** Wire is a more secure and effective means of communications in urban areas. Wire should be laid overhead on existing poles, underground, or through buildings to prevent vehicles from cutting it.

b. **Messengers and Visual Signals.** Messengers and visual signals can also be used in urban areas. Messengers must plan routes that avoid the enemy. Routes and time schedules should be varied to avoid establishing a pattern. Visual signals must be planned so they can be seen from the buildings.

c. **Sound.** Sound signals are normally not effective in urban areas due to the amount of surrounding noise.

d. **Satellite Communications.** SATCOM would alleviate many communications problems, keeping in mind the space needed for proper take off angles for the satellite orientation. A retransmission or radio relay site is another alternative communications method.

e. **Existing Systems.** If existing civil or military communications facilities can be captured intact, they can also be used by units. An operable civilian phone system, for instance, can provide a reliable means of communication. Telephones should not be considered secure. Other civilian media can also be used to broadcast messages to the public.

(1) Evacuation notices, evacuation routes, and other emergency notices designed to warn or advise the civilian population must be coordinated at the battalion and brigade

levels through the S1 or civil affairs officer. Such notices should be issued by the local civil government through printed or electronic news media.

(2) Use of news media channels in the immediate area of combat operations for other than emergency communications must also be coordinated through the S1 or civil affairs officer.

f. **Civilian Telephone Network.** Units can use the civilian telephone networks when METT-TC factors permit. Apartment buildings, stores, and factories are wired into the civilian telephone system. Every building has a telephone distribution box, which controls many (up to 200) individual telephone lines. Setting up wire communications using these points is relatively simple. Computers hooked to the Internet through existing telephone lines can relay a commander's order to his subordinates and can include audio, video, and graphics.

g. **Cellular Telephones.** Cellular telephones work well in urban areas, but are disabled easily by taking out the repeater stations or destroying the central cellular telephone system. In addition, the signal from an omnidirectional antenna can be used to locate and target the site with indirect fire. Cellular telephones as alternative communications method should be considered as a less desirable alternative.

h. **Message Pagers.** Message pagers can be used by leaders. They provide flexibility for short message communications. Disadvantages to using the pager are the beeping sound and the visual light indicator. Leaders should ensure that the pager is in the vibrate mode and, if necessary, that the visual light indicator is covered.

L-7. POTENTIAL PROBLEMS AND SOLUTIONS

The problems encountered in an urban environment may be different from other tactical environments. Units must contend with buildings, electrical structures, and noise.

a. **Potential Problems.** Large cities have power lines, electric trains, trolley lines, and industrial power lines, which can interfere with communications. Potential problems that might be encountered when communicating in an urban environment include:

(1) Dead spots in communications caused by structures that block or absorb signals.

(2) Fading of communications caused by the reflection of the signal off urban terrain. The signal and one or more of its reflected versions meet at the antenna causing constructive or destructive interference. The severity of fading increases with frequency. For most VHF radios, fading is usually less than 10 decibels. Higher frequency radios may experience fades as severe as 20 to 30 decibels.

(3) Manmade electrical noise concentrated at low frequencies. This will cause communications to decline rapidly as frequency increases, but still will be 15 to 20 decibels higher than open terrain.

(4) Mechanical noise problems caused by heavy concentration of manmade noise associated with heavy industry or manufacturing. The more audible the noise level, the higher the volume must be adjusted on communications equipment.

b. **Solutions.** Urban communications problems vary depending on the area. The following solutions are provided for most of the problems units are likely to encounter.

(1) Increasing the power or antenna gain may increase the communications capability. Trade-off is decreased battery life and increased susceptibility to EW.

(2) HF is susceptible to noise but best in reducing dead spots and multi-path effects. UHF/SHF (above 500 megahertz) is very susceptible to dead spots and multipath but best

in reducing noise interference. VHF to low UHF range offsets various problems associated with operating in an urban environment.

(3) Diversity can eliminate multipath effects by using multiple receive antennas or multiple frequency transmissions. Frequency hopping or direct sequence spread spectrum can also provide diversity.

(4) Retransmission stations shorten distances and can overcome noise problems. Retransmission stations used on higher urban terrain/structures unmanned aerial vehicles, or air platforms reduce dead spots.

(5) Commercial systems developed for robotic use in manufacturing may have military application. These systems use spread spectrum modulation (direct sequence or frequency hopping) to overcome fading and interference.

GLOSSARY

AA	avenue of approach; assembly area
AASLT	air assault
ABF	attack by fire
ACE	armored combat earth mover
ACF	aviation close fires
ACM	airspace control measures
ADA	air defense artillery
ADO	air defense officer
AI	area of interest
ALO	air liaison officer
AO	area of operations
AP	antipersonnel
APDS	armor piercing discarding sabot
APDS-T	armor piercing discarding sabot with tracer
APC	armored personnel carrier
APFSDS	armor piercing, fin stabilized, discarding sabot
ARFOR	Army forces
ASAP	as soon as possible
AT	antitank
ATGM	antitank guided missile
ATM	advanced trauma management
AVLB	armored vehicle launched bridge
AXP	air exchange point
BAS	battalion aid station
BDA	battle damage assessment
BDAR	battle damage assessment and repair
bde	brigade
BDU	battle dress uniform
BFV	Bradley Fighting Vehicle
BSFV	Bradley Stinger Fighting Vehicle
BMNT	beginning of morning nautical twilight
bn	battalion
BOS	battlefield operation system
BP	battle position
BSA	brigade support area
C2	command and control
C3	command, control, and communications
C4	command, control, communications, and computers
C4ISR	command, control, communications, computers, intelligence, surveillance, and reconnaissance
CA	Civil Affairs
cal	caliber

CAS	close air support
CASEVAC	casualty evacuation
CCIR	commander's critical information requirements
CCP	casualty collection point
CFL	coordinated fire line
CHS	combat health support
COA	course of action
COLTS	combat observation and lasing teams
CME	counter mine equipment
CMO	civil military operation
CMOC	civil military operations center
COMSEC	communications security
co	company
COP	common operational picture
CP	command post; concrete piercing
CPA	controlled penetration ammunition
CS	combat support
CSS	combat service support
CST	company support team
CTC	combat training center
DFSCOORD	deputy fire support coordinator
DNBI	disease and nonbattle injury
DPRE	displaced persons, refugees, and evacuees
DS	direct support
DZ	drop zone
EA	engagement area
ECC	effects coordination cell
ECM	electronic countermeasures
EEFI	essential elements of friendly information
EEI	essential elements of information
EENT	end of evening nautical twilight
EFAT	essential field artillery task
EFST	essential fire support task
EMT	emergency medical treatment
EOD	explosive ordnance disposal
EPLRS	enhanced position location system
EPW	enemy prisoner of war
EW	electronic warfare
FA	field artillery
FARP	forward arming and refueling point
FASCAM	family of scatterable mines
FASP	field artillery support plan
FBCB2	Force XXI Battalion Command Brigade and Below

FCL	final coordination line
FDC	fire direction center
FEBA	forward edge of the battle area
FFIR	friendly forces information requirements
FIST	fire support team
1SG	first sergeant
FLOT	forward line of own troops
FLSC	flexible linear shaped charge
FM	frequency modulation
FO	forward observer
FPF	final protection fires
FPL	final protective line
FRAGO	fragmentary order
FS	fire support
FSCL	fire support coordination line
FSCM	fire support control measures
FSCOORD	fire support coordinator
FSE	fire support element
FSEM	fire support execution matrix
FSMC	forward support medical company
FSO	fire support officer
FW	fixed wing
G/VLLD	ground/vehicular laser locator designator
GPS	global positioning system
GS	general support
GSR	ground surveillance radar
HE	high-explosive
HEAT	high-explosive antitank
HEDP	high-explosive, dual-purpose
HEI-T	high-explosive incendiary with tracer
HMMWV	high-mobility, multipurpose wheeled vehicle
HPT	high payoff target
HUMINT	human intelligence
IAW	in accordance with
ICM	improved conventional munition
ID	identification
IFV	infantry fighting vehicle
ILLUM	illumination
IO	information operations
IPB	intelligence preparation of the battlefield
IR	infrared
ISR	intelligence, surveillance, reconnaissance
IV	intravenous

JA	judge advocate
JTF	joint task force
LAW	light antitank weapon
LCE	load-carrying equipment
LD	line of departure
LOA	limit of advance
LOC	lines of communications
LOGPAC	logistics package
LOS	line of sight
LNO	liaison officer
LP	listening post
LRP	logistics release point
LZ	landing zone
MANPADS	man-portable air defense system
MBA	main battle area
MDI	modernized demolition initiator
MDMP	military decision-making process
M/E	main effort
MEDEVAC	medical evacuation
METT-TC	mission, enemy, terrain, troops and time available, civil considerations
MG	machine gun
MI	military intelligence
MICLIC	mine-clearing line charge
MLRS	multiple-launch rocket system
mm	millimeter
MNF	multinational forces
MP	military police
MRE	meal, ready to eat
MSD	minimum safe distance
msn	mission
MSR	main supply route
MTF	medical treatment facility
NAI	named area of interest
NATO	North Atlantic Treaty Organization
NBC	nuclear, biological, chemical
NCA	National Command Authority
NEO	noncombatant evacuation operation
NGF	naval gunfire
NGO	nongovernmental organization
NLOS	non-line-of-sight
NOD	night observation device
NVD	night vision device
NVG	night vision goggles

OKOCA	observation and field of fire, key terrain, obstacles, cover and concealment, avenues of approach
obj	objective
ODSS	offense, defensive, stability, and support
O/O	on order
OP	observation post
OPCON	operational control
OPLAN	operation plan
OPORD	operation order
OPSEC	operational security
PA	public affairs; physician's assistant
PAA	position area for artillery
PAC	personnel administration center
PCC	precombat checks
PCI	precombat inspection
PD	point detonating
PE	probable error
PIR	priority intelligence requirements
PL	phase line
plt	platoon
POL	petroleum, oils, and lubricants
PSYOP	psychological operations
PTL	primary target line
PVNTMED	preventive medicine
PVO	private voluntary organization
PZ	pickup zone
QRF	quick reaction force
RAAWS	ranger antiarmor weapon system (Carl Gustaf)
RATELO	radio-telephone operator
RCA	riot control agent
REM	remote sensor
REMBASS	remotely monitored battlefield sensor system
RFL	restrictive fire line
RLEM	rifle-launched entry munitions
ROE	rules of engagement
ROI	rules of interaction
RP	red phosphorous
RPG	rocket-propelled grenade
R&S	reconnaissance and surveillance
S1	adjutant (US Army)
S2	intelligence officer (US Army)

S3	operations and training officer (US Army)
S4	supply officer (US Army)
S5	civil affairs officer (US Army)
SATCOM	satellite communications
SAW	squad automatic weapon
SBF	support by fire (position)
SCATMINE	scatterable mines
S/E	supporting effort
SEAD	suppression of enemy air defenses
SHORAD	short-range air defense artillery
SINCGARS	single-channel ground/airborne radio system
SITREP	situation report
SITEMP	situation template
SMAW	shoulder-launched, multipurpose assault weapon
SF	special forces
SJA	staff judge advocate
SOF	special operations forces
SOP	standing operating procedure
SOSR	suppress, obscure, secure, reduce
sqd	squad
SWAT	Special Weapons and Tactics
TACON	tactical control
TACP	tactical air control party
TACSOP	tactical standard operating procedures
TCF	tactical combat force
TCP	traffic control post
TI	time
TIM	toxic industrial materials
TF	task force
tm	team
TO&E	table of organization and equipment
TOW	tube-launched, optically tracked, wire-guided (heavy antitank missile system)
TPT	target practice with tracer
TRP	target reference point
TTP	tactics, techniques, and procedures
UAV	unmanned aerial vehicle
UCMJ	Uniform Code of Military Justice
UGV	unmanned ground vehicle
UHF	ultra-high frequency
UO	urban operations
UXO	unexploded ordnance

VHF	very high frequency
VT	variable time
WMD	weapons of mass destruction
WP	white phosphorous
XO	executive officer

REFERENCES

SOURCES USED
These are the sources quoted or paraphrased in this publication.

FM 3-0(FM 100-5) Operations. 14 June 2001.

FM 3-06.1 Aviation Urban Operations – Multiservice Procedures for Aviation Urban Operations. 15 April 2001.

FM 3-90 Tactics. 04 July 2001.

FM 7-7J The Mechanized Infantry Platoon and Squad (Bradley). 07 May 1993.

FM 7-8 Infantry Rifle Platoon and Squad. 22 April 1992.

FM 7-10 The Infantry Rifle Company. 14 December 1990.

FM 7-20 The Infantry Battalion. 06 April 1992.

FM 7-30 The Infantry Brigade. 03 October 1995.

FM 34-130 Intelligence Preparation of the Battlefield. 08 July 1994.

FM 71-1 Tank and Mechanized Infantry Company Team. 26 January 1998.

FM 71-2 The Tank and Mechanized Infantry Battalion Task Force. 27 September 1988.

FM 71-3 The Armored and Mechanized Infantry Brigade. 08 January 1996.

FM 101-5 Staff Organization and Operations. 31 May 1997.

FM 101-5-1 Operational Terms and Graphics. 30 September 1997.

READINGS RECOMMENDED
These readings contain relevant supplemental information.

AR 385-62 Regulations for Firing Guided Missiles and Heavy Rockets for Training, Target Practice and Combat. 05 January 1977.

AR 385-63	Policies and Procedures for Firing Ammunition for Training, Target Practice and Combat. 15 October 1983.
AR 600-8-1	Army Casualty Operations/Assistance/Insurance. 20 October 1994.
FM 1-100	Army Aviation Operations. 21 February 1997.
FM 1-112	Attack Helicopter Operations. 02 April 1997.
FM 3-05.30(FM 33-1)	Psychological Operations. 19 June 2000.
FM 3-3	Chemical and Biological Contamination Avoidance. 16 November 1992.
FM 3-4	NBC Protection. 29 May 1992.
FM 3-5	NBC Decontamination. 28 July 2000.
FM 3-6	Field Behavior of NBC Agents (Including Smoke and Incendiaries). 3 November 1986.
FM 3-9	Potential Military Chemical/Biological Agents and Compounds. 12 December 1990.
FM 3-11	Flame, Riot Control Agents and Herbicide Operations. 19 August 1996.
FM 3-19.1(FM 19-1)	Military Police Operations. 22 Mar 2001.
FM 3-19.30(FM 19-30)	Physical Security. 08 January 2001.
FM 3-19.40(FM 19-40)	Military Police Internment/Resettlement Operations. 21 August 2001.
FM 3-23.25(FM 23-25)	Light Antiarmor Weapons. 30 August 2001.
FM 3-23.30(FM 23-30)	Grenades and Pyrotechnic Signals. 01 September 2000.
FM 3-34.2	Combined-Arms Breaching Operations. 31 August 2000.
FM 3-34.230(FM 5-105)	Topographic Operations. 3 August 2000.
FM 3-100	Chemical Operations Principles and Fundamentals. 08 May 1996.

FM 4-30.16	Explosive Ordnance Disposal of Multiservice Procedures for EOD in a Joint Environment. 15 February 2001.
FM 5-25	Explosives and Demolitions. 10 March 1986.
FM 5-33	Terrain Analysis. 11 July 1990.
FM 5-34	Engineer Field Data. 30 August 1999.
FM 5-102	Countermobility. 14 March 1985.
FM 5-103	Survivability. 10 June 1985.
FM 5-250	Explosives and Demolitions. 30 July 1998.
FM 6-20-40	Tactics, Techniques, and Procedures for Fire Support for Brigade Operations (Heavy). 05 January 1990.
FM 6-20-50	Tactics, Techniques, and Procedures for Fire Support for Brigade Operations (Light). 05 January 1990.
FM 6-22.5(FM 22-5)	Combat Stress. 23 June 2000.
FM 6-50	Tactics, Techniques, and Procedures for the Field Artillery Cannon Battery. 23 December 1996.
FM 7-90	Tactical Employment of Mortars. 09 October 1992.
FM 7-91	Tactical Employment of Antiarmor Platoons, Companies, and Battalions. 30 September 1987.
FM 7-92	The Infantry Reconnaissance Platoon and Squad (Airborne, Air Assault, Light Infantry). 23 December 1992.
FM 7-98	Operations in Low-Intensity Conflict. 19 October 1992.
FM 8-10	Health Service Support in a Theater of Operations. 01 March 1991.
FM 8-10-1	The Medical Company Tactics, Techniques, and Procedures. 29 December 1994.
FM 8-10-6	Medical Evacuation in a Theater of Operations Tactics, Techniques, and Procedures. 14 April 2000.

FM 10-64	Mortuary Affairs Operations, 16 February 1999.
FM 10-286	Identification of Deceased Personnel, 30 June 1976.
FM 11-50	Combat Communications Within the Division (Heavy and Light). 04 April 1991.
FM 17-12-1-1	Tank Gunnery (Abrams), Volume 1. 05 May 1998.
FM 17-95	Cavalry Operations. 24 December 1996.
FM 17-98	Scout Platoon. 10 April 1999.
FM 19-15	Civil Disturbances. 25 November 1985.
FM 20-32	Mine/Countermine Operations. 29 May 1998.
FM 21-10	Field Hygiene and Sanitation. 21 June 2000.
FM 21-11	First Aid for Soldiers. 27 October 1988.
FM 21-60	Visual Signals. 30 September 1987.
FM 21-75	Combat Skills of the Soldier. 03 August 1984.
FM 22-51	Leaders' Manual for Combat Stress Control. 29 September 1994
FM 23-1	Bradley Gunnery. 18 March 1996.
FM 23-9	M16A1 Rifle and M16A2 Rifle Marksmanship. 03 July 1989.
FM 23-90	Mortars. 01 March 2000.
FM 27-10	The Law of Land Warfare. 18 July 1956.
FM 27-100	Legal Support to Operations. 01 March 2000.
FM 34-1	Intelligence and Electronic Warfare Operations. 27 September 1994.
FM 41-10	Civil Affairs Operations. 14 February 2000.
FM 44-8	Combined Arms for Air Defense. 01 June 1999.
FM 44-18-1	Stinger Team Operations. 31 December 1984

FM 44-43	Bradley Stinger Fighting Vehicle for Platoon and Squad Operations. 03 October 1995.
FM 44-44	Avenger Platoon, Section, and Squad Operations. 03 October 1995.
FM 63-1	Support Battalions and Squadrons, Separate Brigades and Armored Cavalry Regiment. 30 September 1993.
FM 63-2	Division Support Command, Armored, Infantry, and Mechanized Infantry Divisions. 20 May 1991.
FM 71-100	Division Operations. 28 August 1996.
FM 71-100-2	Infantry Division Operations, Tactics, Techniques and Procedures. 31 August 1993.
FM 71-100-3	Air Assault Operations for Tactics, Techniques, and Procedures. 29 October 1996.
FM 71-123	Tactics and Techniques and Combined Arms Heavy Forces: Armored Brigade, Battalion/Task Force, and Company/Team. 30 Sep 1992.
FM 90-8	Counterguerilla Operations. 29 August 1986.
FM 90-13-1	Combined Arms Breaching Operations. 28 February 1991.
FM 90-29	Noncombatant Evacuation Operations. 17 October 1994.
FM 100-10	Combat Service Support. 03 October 1995.
FM 100-20	Military Operations in Low Intensity Conflict. 05 December 1990.
TC 7-98-1	Stability and Support Operations Training Support Package. 05 June 1997.
TC 21-24	Rappelling. 10 September 1997.
TC 23-14	Sniper Training and Employment. 14 June 1989.
TC 24-20	Tactical Wire and Cable Techniques. 03 October 1988.

INDEX